Annals of Mathematics Studies

Number 138

Prospects in Topology

PROCEEDINGS OF A CONFERENCE
IN HONOR OF
WILLIAM BROWDER

edited by

Frank Quinn

PRINCETON UNIVERSITY PRESS

PRINCETON, NEW JERSEY
1995

The Annals of Mathematics Studies are edited by
Luis A. Caffarelli, John N. Mather, and Elias M. Stein

Princeton University Press books are printed on acid-free paper and meet
the guidelines for permanence and durability of the Committee on Production
Guidelines for Book Longevity of the Council on Library Resources

Printed in the United States of America by Princeton Academic Press

10 9 8 7 6 5 4 3 2 1

Library of Congress Cataloging-in-Publication Data
Prospects in topology : proceedings of a conference in honor of William
Browder / Edited by Frank Quinn.
p. cm. — (Annals of mathematics studies ; no. 138)
Conference held Mar. 1994, at Princeton University.
Includes bibliographical references.
ISBN 0-691-02729-3 (alk. paper). — ISBN 0-691-02728-5 (pbk. : alk. paper)
1. Topology—Congresses. I. Browder, William. II. Quinn, F. (Frank), 1946–
. III. Series.
QA611.A1P76   1996
514—dc20   95-25751

The publisher would like to acknowledge the editor of this volume for
providing the camera-ready copy from which this book was printed

PROSPECTS IN TOPOLOGY

Frank Quinn, Editor

Proceedings of a conference in honor of William Browder
Princeton, March 1994

Contents

Foreword

To the great benefit of Mathematics, William Browder was born in New York City on January 6, 1934. Sixty years later a conference was convened in Princeton to celebrate this event. This volume is the proceedings of that conference.

In these sixty years Bill has made remarkable contributions to Mathematics. He has written over n books and research papers. He has made fundamental contributions to three areas: homotopy theory, transformation groups, and the topology of manifolds. He has served as President of the American Mathematical Society, chair of the department at Princeton, editor of the Annals of Mathematics, and on many influential committees for the Mathematical Society and the National Academy of Sciences. But as remarkable as these achievements are, there is is another which may be even more profound: his influence on younger mathematicians. He has had 28 PhD students, who have in their turn contributed greatly to mathematics. And beyond that he has served as mentor and role model for hundreds of students and young faculty at Princeton and institutions he has visited, and at the Institute for Advanced Study. His penetrating insight, good humor, and evident love of mathematics have inspired an entire generation of mathematicians.

Program of the Conference

SCIENTIFIC PROGRAM COMMITEE. Sylvain Cappell, Wu-Chung Hsiang, Frank Quinn (chair), Dennis Sullivan

LOCAL ARRANGEMENTS COMMITTEE. Tony Bahri (chair), Natasha Brunswick, Joe Kohn, Scott Kenney

PROCEEDINGS. Frank Quinn, editor; Sarah Jaffe, file and copy editor.

LECTURES.

Frank Quinn: *Welcome, and low-dimensional quantum invariants*

John Morgan: *Applications of Gauge Theory to four-manifold topology*

Sylvain Cappell: *Singularities of analytic functions and counting of lattice points*

William Pardon: *Hodge structure on the L^2-cohomology of an algebraic surface*

Ib Madsen: *On the K-theory of complete local rings*

Shmuel Weinberger: *Large Scale Topology and Geometry*

Soren Illman: *Every proper smooth action of a Lie group is equivalent to a real analytic action: A contribution to Hilbert's fifth problem*

Andrew Casson: *Speculations on the geometrization conjecture*

Edward Witten: *Quantum Gauge Theories In Two And Four Dimensions*

Amir Assadi: *Invariants for Finite Group Actions*

Julius Shaneson: *An Euler-MacLaurin expansion for lattice sums in dimensions above one*

Dennis Sullivan: *A local chart formula for characteristic classes which is algebraic in the $*$-operator*

Serguei Novikov: *Some applications of hamiltonian foliations of surfaces*

Andrew Ranicki: *Overview of Browder's work on surgery*

Alejandro Adem: *Overview of Browder's work on group actions*

Martin Bendersky: *Overview of Browder's work on homotopy theory*

Alejandro Adem: *Topological K-theory of Arithmetic Groups*

Louis Kauffman: *Knot Theory — What Next?*

Michael Freedman: *Applied Topology*

The organizers would like to express their gratitude to the National Science Foundation and the Departments of Mathematics at Princeton University, the University of Wisconsin, and Virginia Tech for partial financial support of the conference.

Mathematical descendants of William Browder

Alejandro Adem
>> Daniel Juan-Pineda

Amir Assadi
>> Mingli Chen,
>> Joseph Dolinak II
>> Hamid Egbalnia
>> Semra Ozturk
>> Ergun Yalcin

Sylvan Cappell
>> Oliver Attie
>> Ricardo Cruz
>> Amy Davidow
>> William Homer
>> David Miller
>> Washington Mio
>> Stanley Ocken
>> Faiz al-Rubaii
>> Justin Smith
>>> Yixin Zhang
>> Rachel Sturm
>> Shmuel Weinberger
>>> Steven Curran
>>> Min Yan

Daniel Chess

George Cooke

Ken Dahlberg

Roy DeMeo

Stefano DeMichelis

Lanh Dhang

Michael Freedman
>> Thomas Kwok Keung Au
>> Stephen H. Brindle
>> Michael Patrick Casey
>> Slava Krushkal
>> Zheng-Xu He
>> Hickling, Fred
>> Huang, Wei
>> Xiao-Song Lin

(Michael Freedman, cont.)
> Feng Luo
> Ping-Zen Ong
> Zhenghan Wang
> Shu Yan

Paul Green

Soren Illman
> Osmo-Jukka Kanerva
> Marja Kankaanrinta
> Vesa Pollanen
> Jussi Talsi

Louis Kaufmann
> Steven Winker
> Randall Weiss

Norman Levitt
> Regina Mladineo

Santiago Lopez de Medrano

George Lusztig
> R. Bedard
> C. De Concini
> I. Grojnowski
> J. Kelly
> O. K. Kwon.
> G. Lawton
> P. Lees
> N. O'Brien
> J. Smelt
> N. Spaltenstein,

James Maiorana

Stavros Papastavridis

William Pardon
> Fernando Fernandez-Carmena
> David Massey
> Les Reid

Gerald J. Porter
> Curtis Murley
> Albert Shar

Frank Quinn
> Ivelina Bobtcheva
> Masayuki Yamasaki

Neal Stoltzfus

David Stone
Dennis Sullivan
       Elmar Winkelnkemper
       Edward Hendricks
       R. Bruce Williams
             George Kennedy
             Thomas Zukowski
             Nancy Cardim
             Damjan Kobal
       Curt McMullen
             Kevin Pilgrim
             Jeremy Kahn
       Yunpin Jiang
       Waldek Paluba
       Edson deFaria
       Adam Epstein
       Andre Carvalho
       Jun Hu
       Jim Gendron
John Wagoner
       Gene Raymond Hall
       Laurence Robert Taylor
             Constance Elko
             Emanouil Magiroupoulos
             Žarko Bižaca
       Sandor Howard Strauss
       Henry Churchill King
             L. Christine Kinsey
       Ross Evon Staffeldt
       Karen Lee Vogtmann
             Martin Bridson
             Thomas Brady
       Randolph Stacey Tuler
       Robert William Dussault
       Deirdre Wynne Dobbs Paul
       Frank Zizza
       Jeffey Lawrence McIver
       Lars Kadison
       William Geller
       Leslie Badoian
Elmar Winkelnkemper
Yuen Fat Wong
Alexander Zabrodsky

Prospects in Topology

The Mod 2 Cohomology Rings
of Rank 3 Simple Groups
are Cohen–Macaulay

Alejandro Adem
R. James Milgram

Introduction

In recent years we have been studying the cohomology rings of the sporadic simple groups and their relations with problems in algebra and homotopy theory. This work was motivated in part by results of D. Quillen [Q3] and J. Carlson [C], connecting $H^*(G; \mathbb{K})$ ($\mathbb{K}$ a field of characteristic p) with the structure of modular representations of G, and also by the connection of finite group theory with stable homotopy theory through the identification $B_{S_\infty}^+ \simeq Q(S^0)$.

The rank (or 2-rank) of a finite group G is the dimension of the largest 2-elementary subgroup $(\mathbb{Z}/2)^m$ contained in G, and this provides, at least in low rank, a convenient method for organizing the simple groups. The only simple group of rank 1 is $\mathbb{Z}/2$. For the simple groups of rank two the classification is given by the following theorem due to Alperin, Brauer, Gorenstein, Lyons, and Walter.

Theorem. *The simple groups of rank 2 are* $PSL_2(\mathbb{F}_q)$, *q odd and* $q \geq 5$, $PSL_3(\mathbb{F}_q)$, $PSU_3(\mathbb{F}_q)$ *for q odd*, $PSU_3(\mathbb{F}_4)$, *the alternating group* $\mathcal{A}_7$, *and the Mathieu group*, M_{11}.

For the simple groups of rank 3 the classification is given by O'Nan and Stroth.

Theorem. *The simple groups of rank 3 are* $G_2(q)$, $^3D_4(q)$, *for q odd*, $^2G_2(3^n)$, *n odd*, $n \geq 3$, $PSL_2(\mathbb{F}_8)$, $Sz(8)$, $PSU_3(\mathbb{F}_8)$, J_1, M_{12}, $O'N$.

(We review these classification results in §1.)

In this list, the mod(2) cohomology of the groups of Lie type *when the characteristic q is odd* are closely tied to the cohomology of the classifying space B_G, where G is the associated complex group, at least when the resulting graph automorphism allows a twisting over $\mathbb{C}$; this happens to be the case for PSU_3 but not for the other twisted groups in the list above (see [Q1], [Q2]). However, the cohomology of the triality twisted groups $^3D_4(q)$ is determined

Both authors were partially supported by grants from the National Science Foundation and by the ETH–Zürich; the first author was also supported by an NSF Young Investigator Award.

in [FM], and, since $Syl_2(^2G_2(3^n))$ is elementary abelian $H^*(^2G_2(3^n); \mathbb{F}_2) \cong H^*(Syl_2(^2G_2(3^n)); \mathbb{F}_2)^N$ where $N = \mathbb{Z}/7 \times_T \mathbb{Z}/3$ is the Weyl group of $(\mathbb{Z}/2)^3$ in $^2G_2(3^n)$. Note that this is also isomorphic to $H^*(J_1; \mathbb{F}_2)$ (see [W]).

The mod 2 cohomology of $\mathcal{A}_7$ is quite easy since $Syl_2(\mathcal{A}_7) = D_8$, the dihedral group; the result is given in several sources (see [AM1]). $H^*(M_{11}; \mathbb{F}_2)$ was originally determined by Webb [W] and Benson and Carlson [BC], though a much simpler derivation is given in [AM1]. Similarly, $H^*(M_{12}; \mathbb{F}_2)$ was determined in [AMM], and recently we determined $H^*(O'N; \mathbb{F}_2)$ ([AM4] and [M]).

Thus, the complete determination of the cohomology rings $H^*(G; \mathbb{F}_2)$ for all simple groups of 2-rank ≤ 3 reduces to the study of the groups $PSU_3(\mathbb{F}_{2^n})$ and $Sz(2^n)$. Since we are working here at the characteristic we cannot take advantage of the étale techniques which were so successful previously. However, it turns out that the Sylow 2-subgroups in these cases are sufficiently simple that we can study their cohomology rings anyway. Although we are specifically concerned with $PSU_3(\mathbb{F}_4)$, $PSU_3(\mathbb{F}_8)$ and $Sz(8)$, we provide an analysis for the general situation.

The cohomology ring of a finite group is often quite complicated, which makes it exceedingly hard to obtain general structural results. Perhaps the nicest property one could hope for is that $H^*(G, \mathbb{F}_p)$ satisfy the *Cohen–Macaulay condition*: that is to say, it is free and finitely generated over a polynomial subalgebra of finite rank. In fact we will prove the following result.

Theorem. *If G is a finite simple group and its 2-rank is at most three, then $H^*(G, \mathbb{F}_2)$ is Cohen–Macaulay.*

The theorem above is a corollary of the general result that is the culmination of the arguments in sections 2 and 3:

Theorem. *If G is one of the simple groups of Lie type, $Sz(2^{2n+1})$ or $PSU_3(\mathbb{F}_{2^n})$, then $H^*(G; \mathbb{F}_2)$ is Cohen-Macaulay.*

In turn, this follows from the determination, in §2, of the structure of $Syl_2(G)$ for G as above as a central extension, with center an elementary 2-group, and showing that for these $Syl_2(G)$, any element of order two is contained in the center. In §3 we study the cohomology rings of finite 2-groups with the property above and prove that $H^*(G; \mathbb{F}_2)$ is always Cohen-Macaulay, thus proving our result. We would like to point out that after this paper was written we learned that J. Clark [Cl] in his Oxford Ph. D. thesis has explicitly computed the mod 2 cohomology of $PSU_3(\mathbb{F}_4)$ and $Sz(8)$. Hence the mod 2 cohomology of practically all low rank simple groups has been explicitly calculated.

In the final section we describe how this result breaks down for groups of rank 4, as illustrated for example by the Mathieu group M_{22} or $PSL_3(\mathbb{F}_4)$. Work in progress indicates that a better grasp of the rank 4 situation will soon be available, but it will be of far greater complexity than what we present here.

This paper is dedicated to Bill Browder in celebration of his sixtieth birthday. In addition to his well–known geometric insight, he has had a long-standing interest in the cohomology of finite groups, which was a source of motivation for both authors. We owe him a debt of gratitude for his advice and encouragement.

1   Low rank simple groups

In this section we will recall a few basic facts about simple groups. The main reference for them are Gorenstein's book [G1] and the Atlas [Co].

Let p be a prime dividing the order of a finite group G; the p–rank of G is by definition the maximal rank of a p–elementary abelian subgroup in G. As we will be dealing exclusively with the case $p = 2$, we will simply call this the rank of G, and abbreviate it by $r(G)$. Note that as a consequence of the Feit–Thompson Theorem, all simple groups are of *even order*, and hence $r(G) > 0$ if G is simple.

To begin, we first observe that there are no simple groups of rank one except $\mathbb{Z}/2$. In this situation a 2–Sylow subgroup of G is either cyclic or generalized quaternion. In the first case, Burnside's transfer theorem implies that G has a normal 2–complement and hence is solvable, whereas in the other case, the Brauer–Suzuki theorem shows that G has a unique, hence central, involution.

For the rank two case, a theorem of J. Alperin implies that $Syl_2(G)$ is either dihedral, quasi–dihedral, wreathed, or isomorphic to $Syl_2(PSU_3(\mathbb{F}_4))$. This was used as a basis for the classification of simple groups of rank 2, which we now describe.

Theorem 1.1 (Gorenstein–Walter). *If G is a finite simple group with a di-hedral 2–Sylow subgroup, then it is isomorphic to one of the following groups: $PSL_2(\mathbb{F}_q)$, $q > 3$ odd or A_7.*

Next we have

Theorem 1.2 (Alperin–Brauer–Gorenstein). *If G is a finite simple group with a quasi–dihedral or wreathed 2–Sylow subgroup, then it is isomorphic to one of the following groups:*
$$PSL_3(\mathbb{F}_q), PSU_3(\mathbb{F}_q), q \text{ odd or } M_{11}, \text{ the first Mathieu group.}$$

Finally, we have the result from Lyons' thesis

Theorem 1.3 (Lyons). *If G is a simple group with $Syl_2(G) \cong Syl_2(PSU_3(\mathbb{F}_4))$, then $G \cong PSU_3(\mathbb{F}_4)$.*

Combining these results, we obtain the following

(1.4) Classification of rank 2 simple groups. If G is a simple group of rank 2, then $G \cong PSL_2(\mathbb{F}_q)$, ($q$ odd, $q \geq 5$), $PSL_3(\mathbb{F}_q)$ or $PSU_3(\mathbb{F}_q)$ (q odd), $PSU_3(\mathbb{F}_4)$, A_7 or M_{11}.

For the rank 3 case there is a similar result, proved by O'Nan and Stroth

(1.5) Classification of rank 3 simple groups. If G is a simple group of rank 3, then $G \cong G_2(q)$ or $^3D_4(q)$ (triality twisted) for q odd, $^2G_2(3^n)$, n odd, $n > 1$, $PSL_2(\mathbb{F}_8)$, $Sz(8)$, $PSU_3(\mathbb{F}_8)$, J_1, M_{12} or $O'N$.

Note that in the list above, the last three are sporadic groups, the rest are finite Chevalley groups, or twisted versions of them.

Now we briefly review the preceding classification results from the point of view of group cohomology. First we recall that the cohomology ring $H^*(G, \mathbb{F}_2)$ is said to be *Cohen–Macaulay* (abbreviated from now on as *CM*) if there exists a polynomial subalgebra $\mathcal{R} \subset H^*(G, \mathbb{F}_2)$ over which the ring is *free and finitely generated*. *It follows from the finite generation in group cohomology that if $H^*(Syl_2(G), \mathbb{F}_2)$ is CM, then so is $H^*(G, \mathbb{F}_2)$.* Also note that if $\mathbb{K}$ is an extension of $\mathbb{F}_2$, then $H^*(G, \mathbb{K}) \cong H^*(G, \mathbb{F}_2) \otimes \mathbb{K}$, so that the *CM* property over $\mathbb{K}$ is equivalent to having it over the ground field.

We begin our analysis by first pointing out that the cohomology of the dihedral groups and of the "wreathed groups" $(\mathbb{Z}/2^n) \wr \mathbb{Z}/2$ is well-known (see [AM1], Chapters III and IV) and easily verified to be *CM*. Next we observe that the groups $PSL_3(\mathbb{F}_q)$ and $PSU_3(\mathbb{F}_q)$ (q odd) are quotients of $SL_3(\mathbb{F}_q)$ and $SU_3(\mathbb{F}_q)$ (respectively) by subgroups of *odd* order, and hence have the same mod 2 cohomology. They are, consequently, *CM*. (We can also check this from the explicit calculations in [Q1], [FP].) Since $Syl_2(\mathcal{A}_7) = D_8$, it too is *CM*.

The ring $H^*(Syl_2(M_{11}); \mathbb{F}_2)$ is *not CM*. However, $H^*(M_{11}; \mathbb{F}_2)$ is explicitly determined in [AM1] and is *CM*. We conclude that in the rank two case, the only possible group not having a *CM* cohomology ring is $PSU_3(\mathbb{F}_4)$.

We defer this case for the moment and turn to the rank 3 case: first note that the groups $^2G_2(3^n)$, $n > 1$ odd, $PSL_2(\mathbb{F}_8)$ and J_1 have $(\mathbb{Z}/2)^3$ as their 2–Sylow subgroup. Hence they have *CM* cohomology rings. The Chevalley groups of finite type have been extensively studied using étale are well as other methods; in particular the explicit calculations in [Q2], [K] and [FM] show that $G_2(q)$ and $^3D_4(q)$ (q odd) have *CM* cohomology rings. In [AMM] it was shown that $H^*(M_{12}, \mathbb{F}_2)$ is *CM*, and very recently we verified by an explicit calculation [AM4] that the same is true for the O'Nan group O'N. Hence we conclude that the only two possible rank 3 simple groups with cohomology not satisfying *CM* are $PSU_3(\mathbb{F}_8)$ and $Sz(8)$.

2 The 2–Sylow subgroups of $PSU_3(\mathbb{F}_{2^n})$ and $Sz(2^{2n+1})$

In this section we study the structure of the groups $Syl_2(G)$ when $G = PSU_3(\mathbb{F}_{2^n})$ or $Sz(2^{2n+1})$. Our main objective will be to prove that in each case every element of order two in the Sylow subgroup is contained in its center.

The groups $PSU_3(\mathbb{F}_{2^n})$

Consider the non-singular Hermitian form over $\mathbb{F}_{2^{2n}}$

$$H = \begin{pmatrix} 0 & 1 & 0 \\ 1 & 0 & 0 \\ 0 & 0 & 1 \end{pmatrix}.$$

$U_3(\mathbb{F}_{2^n})$ is the subgroup of $GL_3(\mathbb{F}_{2^{2n}})$ consisting of matrices A so that $AH\bar{A}^t = H$, where $(\bar{A}^t)_{i,j} = (A_{j,i})^{2^n}$. Then, for A diagonal, $A = \begin{pmatrix} a & 0 & 0 \\ 0 & b & 0 \\ 0 & 0 & c \end{pmatrix}$, we see that $A \in U_3(\mathbb{F}_{2^n})$ if and only if $c\bar{c} = 1$, and $\bar{b} = a^{-1}$. Now, suppose that A is central so $a = b = c$. Then $a^{2^n} = a^{-1}$ so $a^{2^n+1} = 1$, and conversely, if $a^{2^n+1} = 1$ then the diagonal matrix with a's along the diagonal is contained in $U_3(\mathbb{F}_{2^n})$. On the other hand, if we concentrate on $SU_3(\mathbb{F}_{2^n})$, we must also have $a^3 = 1$, so a is a third root of unity, and, since $a^{2^n} = a$ for $a \in \mathbb{F}_{2^n}$ we must have n odd in order that $SU_3(\mathbb{F}_{2^n})$ contain a non-trivial center. Hence the simple group

$$PSU_3(\mathbb{F}_{2^n}) = \begin{cases} SU_3(\mathbb{F}_{2^n}) & \text{if } n \text{ is even,} \\ SU_3(\mathbb{F}_{2^n})/(\mathbb{Z}/3) & \text{if } n \text{ is odd.} \end{cases}$$

Now we construct the Sylow 2-subgroup of $PSU_3(\mathbb{F}_{2^n})$ and its normalizer explicitly, as this is needed to understand its cohomology.

For $\theta, \gamma \in \mathbb{F}_{2^{2n}}$, define the element $\mathcal{G}_{\theta,\gamma} \in SL_3(\mathbb{F}_{2^{2n}})$ as the matrix

$$\mathcal{G}_{\theta,\gamma} = \begin{pmatrix} 1 & \gamma & \bar{\theta} \\ 0 & 1 & 0 \\ 0 & \theta & 1 \end{pmatrix}.$$

Note that

$$\mathcal{G}_{\theta,\gamma} H \bar{\mathcal{G}}_{\theta,\gamma}^t = \begin{pmatrix} \gamma & 1 & \bar{\theta} \\ 1 & 0 & 0 \\ \theta & 0 & 1 \end{pmatrix} \begin{pmatrix} 1 & 0 & 0 \\ \bar{\gamma} & 1 & \bar{\theta} \\ \theta & 0 & 1 \end{pmatrix}$$

$$= \begin{pmatrix} \gamma + \bar{\gamma} + \theta\bar{\theta} & 1 & 0 \\ 1 & 0 & 0 \\ 0 & 0 & 1 \end{pmatrix},$$

so $\theta_{\theta,\gamma} \in SU_3(\mathbb{F}_{2^{2n}})$ if and only if $\gamma + \bar{\gamma} + \theta\bar{\theta} = 0$. However, given any $\theta \in \mathbb{F}_{2^{2n}}$ there is at least one γ so that $tr(\gamma) = N(\theta) \in \mathbb{F}_{2^n}$. Any two such γ's differ by an element in $\mathbb{F}_{2^n}$, so, in fact, there are exactly 2^n suitable choices for γ. We also have

$$\begin{pmatrix} 1 & \gamma & \bar{\theta} \\ 0 & 1 & 0 \\ 0 & \theta & 1 \end{pmatrix} \begin{pmatrix} 1 & \tau & \bar{\mu} \\ 0 & 1 & 0 \\ 0 & \mu & 1 \end{pmatrix} = \begin{pmatrix} 1 & \gamma + \tau + \bar{\theta}\mu & \bar{\mu} + \bar{\theta} \\ 0 & 1 & 0 \\ 0 & \mu + \theta & 1 \end{pmatrix}$$

so $\mathcal{G}_{\theta,\gamma}\mathcal{G}_{\mu,\tau} = \mathcal{G}_{\theta+\mu,\gamma+\tau+\bar{\theta}\mu}$. It follows that $K = \{\mathcal{G}_{\theta,\gamma} \mid \gamma + \bar{\gamma} + \theta\bar{\theta} = 0\}$ is a subgroup of $SU_3(\mathbb{F}_{2^n})$.

Define $Z \subset K$ to be the subgroup of $Z = \{\mathcal{G}_{0,\gamma} \mid \gamma \in \mathbb{F}_{2^n}\}$. Then Z is central in K with quotient isomorphic to $\mathbb{F}_{2^{2n}}^+$. Moreover, $Z \cong \mathbb{F}_{2^n}^+$, and we have that $|K| = 2^{3n}$, and K is given as a central extension of the form

$$1 \longrightarrow \mathbb{F}_{2^n}^+ \overset{\vartriangleleft}{\longrightarrow} K \overset{\pi}{\longrightarrow} \mathbb{F}_{2^{2n}}^+ \longrightarrow 1. \tag{2.1}$$

The extension data for this extension is given by

$$\begin{aligned}
\mathcal{G}_{\theta,\gamma}^2 &= \mathcal{G}_{0,\theta\bar{\theta}} \\
[\mathcal{G}_{\theta,\gamma}, \mathcal{G}_{\mu,\tau}] &= \mathcal{G}_{0,\bar{\mu}\theta+\mu\bar{\theta}}
\end{aligned}$$

and gives K as an explicit central extension. Moreover, from the expression for the order of $PSU_n(\mathbb{F}_q)$ given in [Co], page x, we see that $K = Syl_2(PSU_3(\mathbb{F}_{2^n}))$.

We can make the extension data more explicit. Let $\rho \in \mathbb{F}_{2^{2n}}$ be a primitive $(2^n+1)^{st}$ root of unity. Then $\rho\bar{\rho} = 1$ and, if $\lambda = \rho + \bar{\rho}$ we have that $x^2 + \lambda x + 1$ is irreducible over $\mathbb{F}_{2^n}$ with roots ρ and $\bar{\rho}$. Moreover, λ cannot lie in any proper subfield of $\mathbb{F}_{2^l} \subset \mathbb{F}_{2^n}$ since, if it did, then $(2^n + 1)$ would have to divide $2^{2l} - 1$, and since l divides n this is impossible.

It follows that a basis for $\mathbb{F}_{2^n}$ over $\mathbb{F}_2$ is

$$\{1, \lambda, \lambda^2, \ldots, \lambda^{n-1}\},$$

and this completes to a basis for $\mathbb{F}_{2^{2n}}$ over $\mathbb{F}_2$ of the form

$$\{1, \lambda, \lambda^2, \ldots, \lambda^{n-1}, \rho, \lambda\rho, \lambda^2\rho, \ldots, \lambda^{n-1}\rho\}.$$

Specializing the relations above to this basis we have

$$\begin{aligned}
\left[\mathcal{G}_{\lambda^i\rho,-}, \mathcal{G}_{\lambda^j,-}\right] &= \mathcal{G}_{0,\lambda^{i+j+1}} \\
\left[\mathcal{G}_{\lambda^i\rho,-}, \mathcal{G}_{\lambda^j\rho,-}\right] &= 1 \\
\left[\mathcal{G}_{\lambda^i,-}, \mathcal{G}_{\lambda^j,-}\right] &= 1
\end{aligned}$$

while

$$\begin{aligned}
(\mathcal{G}_{r\lambda^i,-})^2 &= \mathcal{G}_{0,\lambda^{2i}} \\
(\mathcal{G}_{\lambda^i,-})^2 &= \mathcal{G}_{0,\lambda^{2i}},
\end{aligned}$$

which gives an explicit presentation of K.

As a direct consequence we have

Lemma 2.2. $Z \subset K$ *is the unique maximal elementary 2-subgroup of K. In particular, every element of order two in K is contained in Z.*

We now turn our attention to the remaining case, the Suzuki groups.

The group $Syl_2(Sz(2^{2n+1}))$

From [G2], p. 153, we see that $Syl_2(Sz(2^{2n+1}))$ is the subgroup of $GL_3(\mathbb{F}_{2^{2n+1}})$ with generators

$$\begin{pmatrix} 1 & 0 & 0 \\ a^\theta & 1 & 0 \\ b & a & 1 \end{pmatrix}$$

where $a, b \in \mathbb{F}_{2^{2n+1}}$, and $\theta = 2^s$ with $2^{2s} \cong 2 \mod (2^{2n+1} - 1)$. Thus $\theta = 4$ for $2^{2n+1} = 8$, the case of most interest to us, but more generally, θ can be taken to be 2^{n+1}. Consequently $|Syl_2(Sz(2^{2n+1}))| = 2^{4n+2}$, and it is given as a central extension

$$1 \longrightarrow \mathbb{F}_{2^{2n+1}}^+ \overset{\vartriangleleft}{\longrightarrow} Syl_2(Sz(2^{2n+1})) \overset{\pi}{\longrightarrow} \mathbb{F}_{2^{2n+1}}^+ \longrightarrow 1.$$

Here, the extension data is

$$\{a\}^2 = a^{1+\theta}$$
$$[\{a\}, \{b\}] = a^\theta b + a b^\theta$$

where $\{a\} = \begin{pmatrix} 1 & 0 & 0 \\ a^\theta & 1 & 0 \\ 0 & a & 1 \end{pmatrix}$, while $b = \begin{pmatrix} 1 & 0 & 0 \\ 0 & 1 & 0 \\ b & 0 & 0 \end{pmatrix}$ give explicit representatives

for the two parts of the central extension. In particular, the squaring relation above shows that once more we have only one maximal elementary 2-subgroup in $Syl_2(Sz(2^{2n+1}))$, since, if $m = (2^{2n+1} - 1, 1 + 2^{n+1})$, then m divides both $2^{2n+1} - 1$ and $2^{2n+2} - 1$, so m divides $2^{2n+2} - 2^{2n-1} = 2^{2n-1}$ and m is 1. We have proved

Lemma 2.3. *The center of $Syl_2(Sz(2^{2n+1}))$ is an elementary 2-group, 2^{2n+1}, and every element of order two in $Syl_2(Sz(2^{2n+1}))$ is contained in the center.*

3   The cohomology calculations

We consider a finite 2-group G with center Z, and suppose that Z is 2-elementary for simplicity in what follows. Specifically, assume $Z = 2^n$. Multiplication induces a *homomorphism*

$$Z \times G \overset{\mu}{\longrightarrow} G$$

and there is a finite $l \geq 1$ so that, for each $x \in H^*(Z; \mathbb{F}_2)$, $x^{2^l} \in res_G^Z(H^*(G; \mathbb{F}_2))$. In particular, the polynomial algebra $\tilde{\mathcal{P}} = \mathbb{F}_2[x_1^{2^l}, x_2^{2^l}, \ldots, x_n^{2^l}]$ is in the image of restriction from $H^*(G; \mathbb{F}_2)$, and it follows that there is a polynomial subalgebra $\mathcal{P} \subset H^*(G; \mathbb{F}_2)$ which injects isomorphically to $\tilde{\mathcal{P}}$ under restriction.

Lemma 3.1. *Under the assumptions above, $H^*(G; \mathbb{F}_2)$ is free as a module over $\mathcal{P}$.*

Proof. The composition of homomorphisms

$$G \xrightarrow{i_2} Z \times G \xrightarrow{\mu} G$$

is the identity, where i_2 is injection to the second factor. Thus $H^*(G; \mathbb{F}_2)$ is a direct summand of $H^*(Z \times G; \mathbb{F}_2)$ as a $\mathcal{P}$-module. Moreover, $H^*(Z \times G; \mathbb{F}_2)$ is free over $\mathcal{P}$ on generators

$$\left(H^*(Z; \mathbb{F}_2) / I(\tilde{\mathcal{P}}) \right) \otimes H^*(G; \mathbb{F}_2).$$

Thus $H^*(G; \mathbb{F}_2)$ is projective over $\mathcal{P}$, but an easy induction on dimension then shows it is free as well. □

Corollary 3.2. *For each $n \geq 1$, the rings*

$$H^*(Syl_2(PSU_3(\mathbb{F}_{2^n})); \mathbb{F}_2), \qquad H^*(Syl_2(Sz(2^{2n+1})); \mathbb{F}_2)$$

are CM. Consequently, the rings $H^*(PSU_3(\mathbb{F}_{2^n}); \mathbb{F}_2)$ and $H^*(Sz(2^{2n+1}); \mathbb{F}_2)$ are CM as well.

Proof. Let G be one of the Sylow subgroups above. It suffices to show that $H^*(G; \mathbb{F}_2)$ is CM. We know that $H^*(G; \mathbb{F}_2)$ is finitely generated, and from the facts that an element is nilpotent if it does not restrict non-trivially to $H^*(Z; \mathbb{F}_2)$, and the restriction of x^{2^l} is in the polynomial algebra $\tilde{\mathcal{P}}$, we see that at most a finite number of the generators and their products are needed to describe $H^*(G; \mathbb{F}_2)$ as a module over $\mathcal{P}$. □

Remark 3.3. In the three cases of most interest to us here, $PSU_3(\mathbb{F}_4)$, $PSU_3(\mathbb{F}_8)$, and $Sz(8)$, it is possible to make the calculations explicit and from this obtain, at least in the case of $Sz(8)$, the actual cohomology rings (see [Cl]).

Remark 3.4. The results in this section are actually a special case of a general theorem proved by J. Duflot [D]: *if $E \subset G$ is a maximal p–elementary abelian subgroup of G, then $H^*(C_G(E), \mathbb{F}_p)$ is Cohen–Macaulay.*

4   Conclusions and final remarks

Combining the results in the previous three sections, we deduce the following

Theorem 4.1. *The mod 2 cohomology ring of every simple group of rank ≤ 3 is Cohen–Macaulay.*

This result would seem to indicate to the casual reader that perhaps simple groups have cohomology rings which are particularly nice. Unfortunately, this naive suggestion collapses at the next rank. Note that for $H^*(G, \mathbb{F}_2)$ to be CM it is necessary that its maximal elementary abelian subgroups (at p=2) be of the same rank. This fails to be true for M_{22}, the next Mathieu group; it has maximal elementaries of ranks three and four. The cohomology of this group is quite complicated, but nevertheless computable (see [AM2]). On the other hand, $PSL_3(\mathbb{F}_4)$ is a simple group with maximal elementaries all of rank four, but its cohomology is not CM, as can be verified from the explicit calculation in [AM3].

The main result in this paper shows that low rank simple groups are particularly accessible, and from a topological point of view their classifying spaces can often be modelled (2–locally) using fibrations

$$F \longrightarrow BG \longrightarrow X$$

where the cohomology of X embeds as a polynomial subalgebra realizing the CM property and F has the homotopy type of a finite complex. Examples of this can be found in [AM], page 279–80 and in [BW].

In contrast, the situation at rank four and beyond is a lot more complex, and we are only starting to obtain some idea (based on very recent calculations or work in progress) as to how involved the cohomology of simple groups can truly be.

References

[AM1] A. Adem and R. J. Milgram, *Cohomology of Finite Groups*, Springer–Verlag *Grundlehren 309*, in press.

[AM2] A. Adem and R. J. Milgram, "The Mod 2 Cohomology of the Mathieu Group M_{22}," *Topology*, to appear.

[AM3] A. Adem and R. J. Milgram, "A_5–invariants, the Cohomology of $L_3(4)$ and Related Extensions," Proc. London Math. Soc. **66** (1993), 187–224.

[AM4] A. Adem and R. J. Milgram, "The Subgroup Structure and Mod 2 Cohomology of O'Nan's Sporadic Simple Group," *Journal of Algebra*, to appear.

[AMM] A. Adem, J. Maginnis and R. J. Milgram, "The Geometry and Cohomology of the Mathieu Group M_{12}," *Journal of Algebra* **139** (1991), 90–133.

[BC] D. Benson, J. Carlson, "Diagrammatic Methods for Modular Representations and Cohomology," *Comm. Algebra*, **15** (1987), 53-121.

[BW] D. Benson, C. Wilkerson, "Finite Simple Groups and Dickson Invariants," *Contemporary Mathematics* (Gitler Volume), to appear.

[C] J. Carlson, "The Varieties and Cohomology Ring of a Module," *Journal of Algebra*, **85** (1983), 104-143.

[Cl]   J. Clark, Oxford Ph.D. Thesis 1992.

[Co]   J. Conway *et al*, *Atlas of Finite Groups*, Oxford University Press (1985).

[D]   J. Duflot, "Depth and Equivariant Cohomology," *Comment. Math. Helv.*, **56** (1981), 627–637.

[G1]   D. Gorenstein, *The Classification of Finite Simple Groups*, Univ. Series in Mathematics, Plenum Press 1983.

[G2]   D. Gorenstein, *Finite Simple Groups*, Univ. Series in Mathematics, Plenum Press 1982.

[FM]   P. Fong and R. J. Milgram, "On the Geometry and Cohomology of the Simple Groups $G_2(q)$ and $^3D_4(q)$," preprint, Stanford University (1990).

[FP]   S. Fiedorowicz and S. Priddy, *Homology of Classical Groups over Finite Fields and their Associated Infinite Loop Spaces*, Lecture Notes in Mathematics **674**, Springer–Verlag, 1978.

[K]   S. N. Kleinerman, *The Cohomology of Chevalley Groups of Exceptional Lie Type*, AMS Memoir 1982.

[M]   R. J. Milgram, "The Cohomology of $Syl_2(O'N)$ and Some Associated Groups," *Preprint: U. of New Mexico*, (1994).

[Q1]   D. Quillen, "On the Cohomology and K–theory of the General Linear Groups over a Finite Field," *Annals of Mathematics* **96** (1972), 552–586.

[Q2]   D. Quillen, "Cohomology of Groups," ICM Proceedings, Nice 1970, Gauthier Villars (1971), Vol. II, 47–51.

[Q3]   D.Quillen, "The Spectrum of an Equivariant Cohomology Ring I & II," *Annals of Mathematics* **94** (1971), 549-602.

[W]   P. Webb, "A Local Method in Group Cohomology," *Comm. Math. Helv.* **62** (1987), 135-167.

Algebraic Geometric Invariants for Homotopy Actions

Amir H. Assadi

1 Introduction

Let X be a paracompact topological space, and let $\mathcal{H}(X)$ denote the monoid of self-homotopy equivalences of X. The homotopy equivalences which are homotopy equivalent to the identity map of X form a submonoid $\mathcal{H}_1(X)$. The quotient $\mathcal{E}(X) \equiv \mathcal{H}(X)/\mathcal{H}_1(X)$ is a group, called *the group of self-equivalences of X*. Subgroups of $\mathcal{E}(X)$ represent "homotopy symmetries" of X, or symmetries of the homotopy type of X in the homotopy category. More generally, any group homomorphism $\rho : G \to \mathcal{E}(X)$ leads to a *homotopy action of G on X* by choosing representatives $\tilde{g} \in \mathcal{H}(X)$ for $\rho(g) \in \mathcal{E}(X)$ for each $g \in G$, and considering $\tilde{g} : X \to X$ acting on X.

If $\varphi : G \times X \to X$ is a topological action, then φ induces a homotopy action. Conversely, given a homotopy action $\alpha : G \to \mathcal{E}(X)$, one is interested in finding a topological action $\psi : G \times X' \to X'$ and a homotopy equivalence $f : X' \to X$ which is equivariant up to homotopy, i.e. $f \circ g$ and $\tilde{g} \circ f$ are homotopic (using the above-mentioned notation). If such an (X', ψ) exists, then ψ is called a *topological replacement* (following George Cooke [10]) or a *topological realization* for α. According to [10], the homotopy action (X, α) is equivalent to a topological G-action if and only if the map $B_\alpha : B_G \to B_{\mathcal{E}(X)}$ lifts to $B_{\mathcal{H}(X)}$ in the fibration $B_{\mathcal{H}(X)} \to B_{\mathcal{E}(X)}$. In practice, lifting problems of [10] are difficult to study via obstruction theory. In general, the intermediate obstructions to a lift may not have any useful geometric or algebraic interpretation beyond the first stage. An alternative approach is to study the existence of topological replacements via constructing global invariants that encapsulate geometric characteristics of the potential G-actions replacing (X, α). In this paper, such invariants are constructed from certain coherent algebraic sheaves associated to $H^*(X)$, and their geometric interpretations are in terms of fixed point sets and stability subgroups when (X, α) is equivalent to a finite dimensional topological G-action.

While homotopy actions and topological actions of a group G on a space X are geometrically very different, they share some common algebraic features.

The author is grateful to Professor Abdus Salam and Professor Friedrich Hirzebruch for their kind invitations to visit ICTP (Trieste, Italy) and Max-Planck-Institute für Mathematik (Bonn, Germany) during this research.

This research is partially funded by a grant from The National Science Foundation whose support is gratefully acknowledged.

Namely, if $\mathcal{F}$ is a homotopy functor from spaces to abelian groups, then $\mathcal{F}(X)$ affords a G-representation in both cases. When $\mathcal{F}(X)$ admits a richer algebraic structure, the G-representation $\mathcal{F}(X)$ carries more sophisticated information. In the case of topological G-actions, the RG-modules $H^*(X;R)$ and $H_*(X;R)$ are called *homology representations* of X. The problems of characterizing the homology representations of a group G acting on a Moore space (e.g. as in a problem of Steenrod [26]) and topological replacements for homotopy actions are closely related. (Observe that any R-free RG-module gives rise to a homotopy G-action on a suitable bouquet of n-spheres and vice versa.) Homology representations arise naturally in group theory, algebraic geometry, number theory and transformation groups. Cf. [1] [5][15][26][27] for a sample of examples in this context.

The concept of a homotopy-action and its first applications to algebraic topology appear in the 1978 paper of George Cooke [10]. Further generalization of Cooke's work (via an obstruction theoretic approach to finding a topological replacement for homotopy actions) and its applications to homotopy theory are due to Alex Zabrodsky [29] [30]. Bill Browder and the author were led to homotopy actions as a tool to study and construct finite group actions on simply-connected manifolds [3] [6]. These efforts motivated [4][5] and the author's present paper. There are related research articles by Frank Quinn [20] [21] , Justin Smith [25], Peter Kahn [15] and Schwänzl-Vogt [22].

Background. The earliest applications of commutative algebra and algebraic geometry to fixed point theory in topological transformation groups are due to Borel, Wu Yi Hsiang and Quillen through localization in equivariant cohomolgy [14][19]. In the geometric direction, Quillen [19] proved that the closed points of the spectrum of the (Borel) equivariant cohomology ring of a finite dimensional G-space X (G compact Lie) is stratified by similar spaces corresponding to the induced actions of elementary abelian p-subgroups of G on X. As a result, the dimension of the spectrum is the same as the largest rank of the collection of elementary abelian p-subgroups of G with a fixed point. When X is one point, Quillen's theorem shows that the Krull dimension of the cohomolgy ring of G (mod p) and the p-rank of G (the rank of a maximal elementary abelian p-subgroup) coincide, proving a conjecture of Atiyah and Swan. On the other hand, the Krull dimension of the cohomology ring of G is the same as its (polynomial) growth rate, which is known to coincide with growth rate of any finitely generated resolution that computes the cohomology. Quillen's dimension theorem inspired a number of "local-to-global" results in topology and algebra, in particular modular representation theory. Among such developments in topology, Jean Duflot, Peter Landweber and Bob Stong [16] investigated further properties of the equivariant cohomolgy ring of a G-space as a commutative ring. Among homological applications of Quillen's stratification theorem, Chouinard's theorem (to the effect that the G-projectivity of G-modules can be detected by restriction to the family of elementary abelian subgroups of G) influenced a far-reaching sequence of applications of cohomology in modular representation theory of finite groups, algebraic groups and resricted Lie algebras, cf. [1][7] and *Proc. Sympos. Pure Math.* **47** (1986) for survey articles.

On the algebraic side, Alperin [1][2] introduced the notion of complexity of modules over a finite group as the growth rate of their minimal projective resolution. Alperin and Evens [2] proved that the complexity of a G-module is the same as the maximum complexity of the module restricted to the elementary abelian p-subgroups of G, thus generalizing Quillen's dimension theorem to the cohomology of G with coefficients in a G-module. Since projective modules are exactly those with complexity zero, the Alperin-Evens theorem generalizes Chouinard's theorem as well. The Alperin-Evens complexity theorem was further strengthened by Avrunin-Scott [7], who described the stratification of the support of the cohomolgy of G with coefficients in a G-module (as a Zariski subspace of $MaxH^*(BG)$) in terms of similar spaces arising from the cohomolgy of elementary abelian p-subgroups of G. Avrunin-Scott's stratification theorem provides a purely algebraic approach to Quillen's stratification theorem (for twisted coefficients) via Green's theory of vertices and sources and Carlson's rank varieties [8]. A generalization of Quillen's argument to the case of cohomology with twisted coefficients is also obtained by Stefan Jackowsky, giving a topological proof of Avrunin-Scott's theorem. Note that the Alperin complexity of G-modules and the dimension of the support of their cohomology coincide, giving different interpretations and generalizations of Quillen's theorem. The rank variety is defined for a module over an elementary abelian p-group in terms of the action of a family of well-behaved cyclic subgroups of the group algebra that were discovered earlier by E. Dade [7]. It is an affine algebraic variety with dimension equal to the Alperin complexity of the G-module [8]. These numerical invariants and their (isomorphic) geometric counterparts, namely rank and support varieties do not distinguish any properties of G-modules when their complexity is maximal, i.e. equal to the rank of the group. Equivalently, the support variety of all G-modules with cohomology containing a non-torsion element is $MaxH^*(BG)$.

The author's papers [4][5] adapt and extend the above-mentioned results to transformation groups. Invariants from modular representation theory in conjunction with affine algebraic geometry were introduced for homotopy actions [4]. The purpose of the invariants in [4][5] was to propose an extension of the notion of inertia groups and isotropy groups to the category of spaces (or homotopy types) with homotopy actions and possibly infinite-dimensional G-spaces. As in the case of G-modules, our invariants coincide for all G-actions with G-fixed points or their analogues in the extended categories, including homotopy actions

Outline. The present paper proposes a new approach to both the algebraic and geometric theories mentioned above. In Section Two, we outline the construction of new invariants for some categories of (algebraic or geometric) objects with symmetries. The new invariants are defined via coherent algebraic sheaves on certain projective schemes. They encode all the above-mentioned geometric and numerical invariants via (essentially forgetful) functors to the category of affine algebraic varieties. The case of homotopy actions relies on the stabilization methods introduced in [4], and it is treated in Section Four. In Section Three, we prove that the sheaves arising from homolgy representations satisfy strong restrictions (e.g. a splitting theorem) that could be used to impose new necessary conditions for topological realizability of homotopy actions. Such applications are discussed in Section Four. Besides applications to homotopy actions and topological transformation groups, our approach produces new invariants in algebra. In the case of modular representations, these invariants differentiate representations with maximal complexity, while the support and rank varieties do not. For example, all Heller shifts of the trivial kG-module k have the same support and rank varieties, hence the same complexity. In contrast, our invariants for different Heller shifts correspond to non-isomorphic line bundles that can be distinguished by their class in the appropriate Picard group (or their degrees in the sense of algebraic geometry).

Similar refinements of the previous invariants follow for other categories. Based on the rich theory developed for coherent algebraic sheaves, we are able to associate other useful invariants to modular representations, e.g. characteristic classes and moduli schemes. More details and applications of these ideas will appear elsewhere.

2   Algebraic geometric invariants

Let $\mathcal{C}$ be any of the following categories: (i) $\mathbf{T}$ = category of paracompact topological spaces and continuous maps; (ii) $\mathbf{T}^h$ = the homotopy category corresponding to $\mathbf{T}$; (iii) $\mathbf{M}$ = the category of R-modules (finitely generated) and R-homomorphisms; (iv) $\mathbf{C}$ = the category of R-chain complexes and chain homomorphism. Let G be a finite group. The G-equivariant category corresponding to $\mathcal{C}$ is denoted by $\mathcal{C}_G$. Thus objects of $\mathcal{C}_G$ are endowed with a G-action and the morphisms are G-equivariant (homotopy-actions and homotopy-equivariant maps for $\mathbf{T}^h$). Given objects X and Y in $\mathcal{C}_G$ and a G-morphism $f : X \to Y$, we wish to attach invariants for the G-actions on X and Y and establish a relationship between them via f. First, consider $\mathbf{T}_G$, where the fixed point set X^G, $X^H (H \subseteq G)$, the isotropy subgroups G_x for $x \in X$, the orbit space X/G, the Borel-construction $X_G \equiv E_G \times_G X$ (the fibre bundle associated to the universal principal $E_G \to BG$) and their many algebraic-topological invariants are defined as usual. When $\dim X < \infty$, X^G and its topological invariants are useful invariants of the G-action on X. The isotropy groups are rather weak invariants when one deals with homological properties of the G-action as opposed to more subtle geometric circumstances (such as G-manifolds). A better substitute is the set $\mathcal{A}(X) = \{ A \subseteq G \mid H^*(X^A, X^{A'}; \mathbb{Z}_p) \neq 0$, where A and

A' are elementary abelian p-subgroups and $A' \not\supseteq A\}$ which we call *the set of essential stabilizers*. When only one prime $p \mid |G|$ must be considered, we use the notation $\mathcal{A}_p(X)$. Among the algebraic topological invariants in $\mathbf{T}_G$, the Borel equivariant cohomology $H^*_G(X; R) \equiv H^*(X_G; R)$ is one of the most useful and common invariants. Many of the invariants of G-actions on X and Y are obtained from the functor $E_G \times_G (-) : \mathbf{T}_G \to \mathbf{T}$ and the usual algebraic-topological invariants of $\mathbf{T}$. Note that $H^*_G(-; R)$ yields a graded module over $H^*(BG; R)$ and $f^*_G \equiv H^*_G(f; R)$ is $H^*(BG; R)$-linear.

In the following, we wish to construct analogues of the above-mentioned invariants of $\mathbf{T}_G$ for the remaining categories. The concepts corresponding to $E_G \times_G (-)$ and $H^*_G(-)$ for $\mathbf{M}$ and $\mathbf{C}$ were introduced and studied in homological algebra in the same period. The set-theoretic notion "G-fixed points" when applied to an RG-module M, with the same definition as X^G, leads to group cohomology with twisted coefficients $H^*(G; M)$. Let Q^* be an RG-free acyclic complex. $Q^* \otimes_G M$ in $\mathbf{M}_G$ is analogous to the Borel construction in $\mathbf{T}_G$. Thus, $H(Q^* \otimes_G M) \cong H^*(G; M)$ is a good candidate to imitate $H^*_G(X)$. This construction carries over to $\mathbf{C}_G$, and leads to the notion of group-cohomology with coefficients in a chain complex, originally called hypercohomology. For an RG-chain complex C^*, the cohomology of the total complex of the double complex $Q^* \otimes C^*$ is denoted by $\mathbb{H}^*(G; C^*)$. If C^* is the singular cochain complex of a paracompact G-space X, then $\mathbb{H}^*(G; C^*)$ agrees with $H^*_G(X; R)$, confirming the expected requirements. Moreover, if $C^0 = M$ and $C^i = 0$ for all $i > 0$, then $\mathbb{H}^*(G; C^*) = H^*(G; M)$. Thus, constructions in $\mathbf{C}_G$ yield generalizations for both $\mathbf{T}_G$ and $\mathbf{M}_G$. The case of homotopy actions $\mathbf{T}^h_G$ is not clear as readily, even if we ask for the analogue of the Borel equivariant cohomology $H^*_G(-)$. It is worth mentioning that the direct generalization of fixed point set X^G to the algebraic set-up yields an algebraic object only. The question of finding a *geometric generalization* still remains to be explored. For example, consider a finite dimensional G-space X and its singular cochain complex $C^*(X; R)$. The derived functors of $(-)^G$ in $\mathbf{C}_G$ give back $H^*_G(X; R)$ and not $H^*(X^G; R)$. Thus, analogues of fixed point sets and essential stabilizer subgroups from a geometric point of view remain to be formulated for $\mathbf{M}_G$ and $\mathbf{C}_G$ besides $\mathbf{T}^h_G$.

In the sequel, we will assume that all RG-modules, including total homologies of G-spaces and G-chain complexes, are finitely generated. This ensures that the necessary finiteness conditions in algebraic geometry are satisfied as in the following:

2.1. Proposition[2][19]. *Let R be a commutative ring , let G be a finite group and M a finitely generated RG-module. Then $H^*(G; R)$ is a (graded-commutative) noetherian R-algebra and $H^*(G; M)$ is a finitely generated $H^*(G; R)$-module. Similarly, assume that $H^*(X; R)$ and $H^*(C^*; R)$ are finitely generated R-modules where X is a G-space and C^* is a G-complex. Then $H^*_G(X; R)$ and $\mathbb{H}^*(G; C^*)$ are finitely generated as $H^*(G; R)$-modules.*

We shall restrict attention to the graded R-algebra H_G defined to be

$\oplus_i H^{2i}(G; R)$ whenever R is not a ring of characteristic 2. There are two main reasons for this restriction. First, the ring $H^*(G; R)$ is not strictly commutative in general. One may consider the $\mathbb{Z}_2$-graded R-algebra corresponding to H_G and $\oplus_i H^{2i+1}(G; R)$ as a super-ring, and apply constructions from super-algebraic geometry. There are many important technical differences between the commutative and the $\mathbb{Z}_2$-graded theories. We shall study the super-algebraic geometric invariants analogous to the invariants of H_G elsewhere. The second reason is the more satisfactory geometric situation that arises for the case of elementary abelian p-groups when we consider the polynomial subalgebra of $H^*(G; R)$.

Consider the projective scheme $(S, \mathcal{O}_S)$ over $\mathrm{Spec}(R)$, where $S = \mathrm{Proj}(H_G) =$ the projective scheme consisting of all proper homogeneous prime ideals of H_G. A finitely generated graded H_G-module F^* gives rise to a coherent algebraic sheaf $\mathcal{F}$ of modules over $(S, \mathcal{O}_S)$, also called an $\mathcal{O}_S$-module for short. Let us recall briefly this construction due to Serre and Grothendieck [12] [23]. A sub-basis for the Zariski topology on $\mathrm{Proj}(H_G)$ consists of open sets of the form $\mathrm{Spec}((H_G)_{(f)})$ which are denoted by S_f. Here, $(H_G)_{(f)}$ is the homogeneous localization of H_G with respect to the homogeneous element $f \in H_G$ and it consists of all (equivalence classes of) elements $\frac{a}{f^n}$ with $a \in H_G$ homogeneous, $\deg(a) = \deg(f^n)$ and $n \geq 0$. Given F^*, let $\mathcal{F}(S_f) = \left\{ \frac{b}{f^n} \mid b \in F^* \text{ is homogeneous}, \deg b = \deg f^n \text{ and } n \geq 0 \right\}$ define the presheaf associated to $\mathcal{F}$. Given a coherent $\mathcal{O}_S$-module $\mathcal{F}$, define the following exact sequence: $0 \to \mathcal{F}_\tau \to \mathcal{F} \to \mathcal{F}_\varphi \to 0$. The subsheaf $\mathcal{F}_\tau$ is defined via $\mathcal{F}_\tau(S_f) = $ torsion $(H_G)_{(f)}$-submodule of $\mathcal{F}(S_f)$, and $\mathcal{F}_\varphi$ is the quotient sheaf $\mathcal{F}/\mathcal{F}_\tau$. $\mathcal{F}_\tau$ is called the torsion subsheaf of $\mathcal{F}$ and $\mathcal{F}_\varphi$ is the largest torsion-free quotient of $\mathcal{F}$. The support of $\mathcal{F}_\tau$ is defined to be $\mathrm{supp}(\mathcal{F}_\tau) = \{s \in S \mid \mathcal{F}_s \neq 0\}$. It defines a closed subscheme of $(S, \mathcal{O}_S)$ whose underlying topological space is $\mathrm{supp}(\mathcal{F}_\tau)$.

Definition. Let W be any of the following: (i) a topological space with an action $G \times X \to X$; (ii) an R-free RG-chain complex C^*; (iii) an RG-module M. Let F^* be the H_G-module in each case, respectively: (i) $H_G^*(X; R)$; (ii) $\mathbb{H}^*(G; C^*)$; (iii) $H^*(G; M)$. Let $\mathcal{F}$ be the $\mathcal{O}_S$-module corresponding to F^*. $\mathcal{F}$ is called *the characteristic sheaf* of W. $\mathcal{F}_\tau$ and $\mathcal{F}_\varphi$ are called respectively *the characteristic torsion* and *the characteristic torsion-free sheaves* of W.

The characteristic sheaves of a finite dimensional G-space are related to its more familiar geometric invariants. Let k be a field of characteristic p and $A \cong (\mathbb{Z}_p)^{n+1}$, $n \geq 0$. It turns out that the set of essential stabilizers $\mathcal{A}_p(X)$ and the corresponding $\dim_k H^*(X^A; k)$ for each $A \in \mathcal{A}_p(X)$ can be recovered from the characteristic sheaves associated to the G-action on X. To see this, we need to look at the case $G \cong (\mathbb{Z}_p)^{n+1}$. For simplicity of notation, we shall consider the case $p = 2$. For $p = $ odd prime, one has the same statements by replacing H_G and $\mathrm{Proj}(H_G)$ in the following with the reduced k-algebra

H_G/radical and the corresponding reduced scheme. By considering the odd and even degrees separately, one can obtain more detailed information, such as Euler characteristics of the fixed point sets. Cf. [5] for related results.

2.2. Theorem. *Let $G = (\mathbb{Z}_2)^{n+1}$ and k be a field of characteristic 2. Let X be a finite-dimensional G-space with characteristic sheaf $\mathcal{F}$. Then:*

(a) $\mathcal{F}_\tau$ *determine* $\mathcal{A}_2(X) = $ *set of essential stabilizers of X, and conversely* $\mathcal{A}_2(X)$ *determines* $\mathrm{supp}(\mathcal{F}_\tau)$.

(b) $\dim_k H^*(X^G; k) = \mathrm{rank}\ \mathcal{F}_\varphi = \mathrm{rank}\ \mathcal{F}$.

Proof. First consider the case $X^G = \emptyset$. Then $H_G^*(X)$ (k-coefficients understood) is a torsion H_G-module by Borel-Quillen-Hsiang localization theorem (cf. [19] [14], for example). Hence $\mathcal{F}_\varphi = 0$ and $\mathcal{F}_\tau = \mathcal{F}$. Since $H_G \cong k[t_0, \ldots, t_n]$ is a polynomial ring, $\mathrm{Proj}(H_G) = \mathbb{P}^n(k) = \mathbb{P}^n(\mathbb{F}_2) \times_{\mathrm{Spec}\ \mathbb{F}_2} \mathrm{Spec}\ k$. The linear subspaces of $\mathbb{P}^n(\mathbb{F}_2)$ are seen to be in one-to-one correspondence with subgroups of G, while points of $\mathbb{P}^n(\mathbb{F}_2)$ correspond to cyclic subgroups of G. The defining ideal for the linear subspace corresponding to the subgroup $A \subsetneq G$ is given by the kernel of the induced homomorphism $H^*(G; \mathbb{F}_2) \to H^*(A; \mathbb{F}_2)$. The formulation of the localization theorem in equivariant cohomology by Hsiang [14] implies that $A \subsetneq G$ is an essential stabilizer if and only if there exists $u \in H_G(X; \mathbb{F}_2)$ such that $\mathrm{ann}(u) = \mathrm{Ker}(H^*(G; \mathbb{F}_2) \to H^*(A; \mathbb{F}_2))$. Hence, corresponding to each $A \in \mathcal{A}_2(X)$, there exists a subsheaf of $\mathcal{F}_\tau$ whose support is the $\mathbb{F}_2$-rational linear subspace of $\mathbb{P}^n(k)$ defined by the homogeneous prime ideal $I_A = \mathrm{Ker}(H^*(G; k) \to H^*(A; k))$. Further, I_A occurs as the annihilating ideal for a global section $u \in H^0(\mathbb{P}^n(k), \mathcal{F}_\tau(d))$ for a sufficiently large d. Therefore, $\mathcal{A}_2(X)$ determines $\mathrm{supp}(\mathcal{F}_\tau)$. Conversely, for any homogeneous element $u \in H_G^r(X; k)$, the annihilating ideal $\mathrm{ann}(u) \subseteq H_G$ is seen to be invariant under the action of the Steenrod algebra, i.e. invariant under the substitution $t_i \mapsto t_i + t_i^2$ for all polynomials generators of H_G. According to Serre [24], the variety defined by such an ideal is the union of the $\mathbb{F}_2$-rational linear subspaces of $\mathbb{P}^n(k)$. Consider an element $v \in H^0(\mathbb{P}^n(k); \mathcal{F}_\tau(d))$ (for a sufficiently high d) whose annihilating ideal is a homogeneous prime $I \subseteq H_G$. Accordingly, I determines a subgroup $A \subseteq G$ via, $I \cong I_A = \mathrm{Ker}(H^*(G; k) \to H^*(A; k))$. Therefore, there is a one-one correspondence between $\mathcal{A}_2(X)$ and the homogeneous prime ideals of H_G which occur as annihilating ideals for some $u \in H^0(\mathbb{P}^n(k), \mathcal{F}(d))$ for a sufficiently large d.

(b) $\mathrm{rank}\ \mathcal{F}_\varphi$ is equal to the dimension of the $(\mathcal{F}_\varphi)_x$ at the generic point x of $\mathbb{P}^n(k)$ as a vector space over K (the quotient field of H_G). By the localization theorem, $H_G^*(X) \otimes_{H_G} K \cong H_G^*(X^G) \otimes_{H_G} K \cong H^*(X^G) \otimes_k K$. Hence $\dim_k H^*(X^G) = \mathrm{rank}\ \mathcal{F}_\varphi = \mathrm{rank}\ \mathcal{F}$. $\square$

3　Invariants for homotopy actions

In this section, we extend the construction of the previous section to homotopy actions. This uses an appropriate stabilization procedure from [4] which we briefly recall for reader's convenience.

Preliminaries. Let M_i be RG-modules and P and P_i be projective RG-modules, $i = 1, 2$. M_i are called (projectively) stably equivalent if for suitable choices of P_i, $M_1 \oplus P_1 \cong M_2 \oplus P_2$. Define the operation ω on the stable equivalence classes of RG-modules (well-defined up to projective equivalence by Schanuel's Lemma) via the short exact sequence $0 \to \omega(M) \to P \to M \to 0$ applied to a representative of the class. Denote stable equivalence class of M by $\langle M \rangle$ and $\omega(\langle M \rangle)$ by $\langle \omega(M) \rangle$. We can define the operation ω^{-1} for R-free RG-modules via $\omega^{-1}(M) \equiv \operatorname{Hom}_R(\omega(\operatorname{Hom}_R(M, R)), R)$. Inductively, $\omega^{n+1}(M) \equiv \omega(\omega^n(M))$, $n \in \mathbb{Z}$. In general, $\omega^{-1}(M)$ can be defined as $\omega^{-2}(\omega(M))$, using the R-free RG-module $\omega(M)$. Two RG-modules M_1 and M_2 are called ω-stably equivalent if $\langle \omega^r(M_1) \rangle = \langle \omega^s(M_2) \rangle$ for some $r, s \in \mathbb{Z}$.

We need the graded version of ω-stability also. Given a graded finitely generated RG-module $M^\bullet = \oplus_{i \geq 0} M^i$, choose $N \in \mathbb{Z}$ sufficiently large so that $M^i = 0$ for $i \geq N$. Define a sequence $\overline{M}^i$ as follows: $\overline{M}^0 = \omega^N M^0$ and the sequence $0 \to \overline{M}^i \to \overline{M}^{i+1} \to \omega^{N-(i+1)}(M^{i+1}) \to 0$ is exact. Thus, up to ω-stability, $\overline{M}^N$ is a composite extension with composition factors $\omega^{N-i}(M^i)$. The ω-stable class of $\overline{M}^N$ is well-defined and it is called an ω-composite extension of $M^\bullet$. The ω-stable class of $\overline{M}^N$ is denoted by $\mu(M^\bullet)$. Note that $\overline{M}^N$ depends on the choices of extensions. Hence, there are many possibilities for $\mu(M^\bullet)$ in general.

There are two closely-related notions of stability for G-spaces and G-complexes. Namely, two G-spaces X_i, $(i = 1, 2)$ are called freely equivalent, if there exists a G-space Y containing X_1 and X_2 as G-subspaces, and Y/X_i are compact G-spaces with free G-actions away from their base points corresponding to X_i. The corresponding notion for R-free RG-chain complexes are defined similarly: C_* and C'_* are freely equivalent if there is an RG-complex D_* containing both as RG-subcomplexes and such that D_*/C_* and D_*/C'_* are finitely generated and free over RG. Free equivalence is an equivalence relation.

Notation. For a homotopy action $\alpha : G \to \mathcal{E}(X)$, define $\eta(X, \alpha) = $ the set of ω-stable classes of ω-composite extensions for $H^*(X; R)$. $\eta(X, \alpha)$ incorporates all possible ω-stable classes which may arise from all possible topologically equivalent G-actions.

Next, we measure the effect of ω-stability of [4] on characteristic sheaves of Section Two above.

Definition. Let Λ^* be a graded R-algebra, and let F^* be a graded Λ^*-module. Let $S = \operatorname{Proj}(\Lambda^*)$ and $\mathcal{F}$ be the coherent $\mathcal{O}_S$-module corresponding to F^*. The shift of grading by d yields the Λ^*-module $F^*(d)$ defined by $F^*(d)^i = F^{d+i}$. The corresponding sheaf is denoted by $\mathcal{F}(d)$. It follows that $\mathcal{F}(d) \cong \mathcal{F} \otimes_{\mathcal{O}_S} \mathcal{O}_S(d)$. $\mathcal{O}_S(1)$ is called the Serre twisting sheaf, and $\mathcal{F}(d)$ is called a Serre twist of $\mathcal{F}$. Two $\mathcal{O}_S$-modules $\mathcal{F}$ and $\mathcal{F}'$ are called Serre-twist equivalent if $\mathcal{F}(d) \cong \mathcal{F}'(d')$ for some $d, d' \in \mathbb{Z}$.

3.1. Proposition. *Let $\mathcal{F}$ be an $\mathcal{O}_S$-module, and let $\mathcal{F}_\tau$ and $\mathcal{F}_\varphi$ be its torsion subsheaf and torsion-free quotient. Then:*

(a) *$\mathcal{F}(d)_\tau = \mathcal{F}_\tau(d)$. Hence $\mathrm{supp}(\mathcal{F})$ is invariant under Serre twists.*

(b) *$\mathcal{F}(d)_\varphi = \mathcal{F}_\varphi(d)$. Hence $\mathcal{F}_\varphi$ is locally-free iff $\mathcal{F}_\varphi(d)$ is locally free. Moreover, $\mathcal{F}_\varphi$ and $\mathcal{F}_\varphi(d)$ split into direct sums of sheaves of smaller rank in a similar manner.*

The proof follows from tensoring $0 \to \mathcal{F}_\tau \to \mathcal{F} \to \mathcal{F}_\varphi \to 0$ with $\mathcal{O}_S(d)$ over $\mathcal{O}_S$, and using the definitions.

Definition. Let $\alpha : G \to \mathcal{E}(X)$ be a homotopy action. The set $\mathcal{S}^\omega(X, \alpha) \equiv$ Serre-twist equivalence classes of characteristic sheaves $\mathcal{F}$ corresponding to one representative from each ω-stable class in $\eta(X, \alpha)$.

3.2. Proposition. *$\mathcal{S}^\omega(X, \alpha)$ is well-defined. In particular, the set $\mathrm{supp}(X, \alpha) = \{\mathrm{supp}(\mathcal{F}_\tau) \mid \mathcal{F}_\tau$ is a characteristic sheaf corresponding to a representative from an element of $\eta(X, \alpha)\}$ and $\mathrm{Fix}(X, \alpha) = \{$ Serre-twist equivalence classes of characteristic torsion-free sheaves corresponding to a representative from an element of $\eta(X, \alpha)\}$ are well-defined.*

Proof. With a slight abuse of notation, apply cohomology $H^*(G, -)$ to the exact sequence $0 \to \omega(M) \to P \to M \to 0$. It follows that $H^i(G; M) \cong H^{i+1}(G; \omega(M)) = H^*(G; \omega(M))(1)^i$. Hence, the characteristic sheaves of ω-stably equivalent RG-modules differ only by Serre twists. Note that two graded modules which are eventually isomorphic (i.e. isomorphic in all sufficiently large degrees) give rise to isomorphic sheaves upon Serre-Grothendieck construction. The conclusion follows from Proposition 3.2. $\qquad\square$

Remark. $\mathrm{Supp}(X, \alpha)$ and $\mathrm{Fix}(X, \alpha)$ incorporate the extensions in the ω-composite extension constructions on $H^*(X; R)$. They depend on R and the graded RG-representation $H^*(X; R)$. Otherwise, most other topological properties of X are weakened in the stability process.

3.3. Proposition.

(a) *Suppose X and Y are freely equivalent G-spaces. Then their characteristic sheaves are isomorphic.*

(b) *Assume further that $Y^G \neq \emptyset$ and Y is a Moore space with $H^n(Y; R) = M$ for some $n > 0$ and some RG-module M. Then the characteristic sheaves of X and M are Serre-twist equivalent. Similar statements hold for RG-complexes.*

Proof. It suffices to consider the case where Y contains X as a G-subspace, and $Y - X$ is a finite dimensional free G-space. Thus, $H_G^*(Y, X)$ vanishes in all degrees greater than $\dim(Y - X)$. The inclusion $j : X \to Y$ induces an eventual isomorphism $j^* : H_G^*(Y) \to H_G^*(X)$, hence an isomorphism after Serre-Grothendieck construction. To see (b), observe that $H^n(Y; R)$ is an ω-composite extension of $\oplus_i \overline{H}^i(X; R)$. Let $y_0 \in Y^G$. Then $\overline{H}_G^*(X) = \overline{H}_G^*(Y, y_0) \cong H^*(G; M)$ in all sufficiently high degrees and after a possible degree shift. Hence the characteristic sheaves differ at most by a Serre-twist. Similar comments apply to RG-complexes.   □

Remark. The condition $Y^G \neq \emptyset$ is not necessary if one is willing to replace G-spaces by suitable RG-chain complexes (e.g. the cellular chain complex if X and Y are G-CW complexes). Up to free equivalence, any RG-chain complex is equivalent to one with precisely one cohomology group M, which is a ω-composite extension of $\oplus_i H^i(X; R)$. In this situation, the statement of 2.4 remains true again.

4   Obstructions for topological replacements and applications

Throughout this section, we assume that $G = (\mathbb{Z}_p)^{n+1}$ and k is a field of characteristic p. Homology and cohomology groups have coefficients k.

4.1.   Theorem.   *Let (X, α) be a homotopy G-action on a connected CW complex X. Let $(S, \mathcal{O}_S)$ be The projective scheme $\mathrm{Proj}(H_G)$. Let $\eta(X, \alpha)$ be the set of ω-stable classes of ω-composite extensions and $\mathcal{S}^\omega(X, \alpha) =$ the set of Serre-twist equivalence classes of characteristic sheaves obtained from Serre-Grothendieck construction on representatives of $\eta(X, \alpha)$. If (X, α) is equivalent to a topological G-action, then there exists an $[\mathcal{E}] \in \mathcal{S}^\omega(X, \alpha)$ such that:*

(a) *ξ_φ is locally-free, and it splits into a direct sum of invertible $\mathcal{O}_S$-modules.*

(b) *$\mathrm{supp}[\xi_\tau]$ is a union of $\mathbb{F}_p$-rational linear subspaces of S.*

Outline of Proof: Suppose $\psi : G \times Y \to Y$ is a topological G-action replacing (X, α). Let $[\xi] \in \mathcal{S}^\omega(X, \alpha)$ be the Serre twist equivalence class of the characteristic sheaf of (Y, ψ) corresponding to $H_G^*(Y; k)$. As indicated before, we set $(S, \mathcal{O}_S)$ be $\mathrm{Proj}(H_G/\mathrm{Radical})$, so that $S = \mathbb{P}^n(k)$ and ξ is an $\mathcal{O}_S$-module. If $\dim Y < \infty$, then Theorem 2.5 implies that the support of ξ_τ consists of the essential stabilizers of the G-action on Y which are proper subgroups of G. Hence (b) follows in this case. Moreover, proof of 2.5 shows that $[\xi_\varphi]$ is the Serre twist equivalence class of the sheaf obtained from the graded H_G-module $H_G^*(X^G; k)$. The splitting of $H_G^*(X^G) = H^*(G) \otimes H^*(X; k) = H^*(G) \otimes (\oplus_d H^d(X; k) \cong \oplus_d H^*(G)(d)$ proves (a) for this case. In the genral case, we note that $\mathrm{supp}[\xi_\tau]$

is determined by the components of the closed subspace of $\mathbb{P}^n(k)$ defined by $\mathrm{ann}(u) \subset H_G$ for a suitable set of elements $u \in H_G^*(Y)$. On the other hand, $\mathrm{ann}(u)$ is invariant under the action of Steenrod algebra $\mathcal{A}_p$ [16] . A result of Serre [23] implies that these components are $\mathbb{F}_p$-rational linear subspaces of $\mathbb{P}^n(k)$, as required in (b). As for (a) in this case, we use the work of Lannes [17] [18]. Accordingly, the torsion-free quotient of $H_G^*(Y'; \mathbb{F}_p)$ is isomorphic in sufficiently high dimensions to $(H_G/\mathrm{Radical}) \otimes \tilde{T}_G(H_G^*(Y'; \mathbb{F}_p))$, where $\tilde{T}_G$ is a suitable version of Lannes' functor T_G and $Y' = \mathrm{Map}(EG, Y)$ with the diagonal G-action. Since $H_G(Y'; \mathbb{F}_p)$ is finitely generated over $H_G/\mathrm{Radical}$. $\tilde{T}_G(H_G^*(Y', \mathbb{F}_p))$ is a finite dimensional graded $\mathbb{F}_p$-vector space. On the other hand, $H_G^*(Y') = H_G^*(Y)$, so (a) follows just as in the previous case. $\qquad \square$

Remark. For a general finite group G, the Quillen stratification [19] of Spec H_G can be used to formulate a generalization of the above theorem. The proof above determines the behavior of $[\xi_\varphi]$ and $[\xi_\tau]$ on each stratum.

4.2. Corollary. *Suppose (X, α) is a homotopy action and keep the notation above. Assume that for all $[\xi] \in \mathcal{S}^\omega(X, \alpha)$, either $[\xi_\varphi]$ is not represented by a sum of invertible $\mathcal{O}_S$-module or $\mathrm{supp}[\xi_\tau]$ has components which are not $\mathbb{F}_p$-rational linear subspaces of $\mathbb{P}^n(k)$. Then (X, α) cannot be replaced by a topological action.*

4.3. Applications. In the late 1950's, Steenrod asked if every G-representation was obtained as the homology representation for a G-action on a finite Moore space. The answer is negative in general. A reformulation of this problem, called *The Steenrod Problem*, has been studied by several authors, cf. for example, [4] [5] [9] [15] [25] [26] [27] [28] and their references. The problem amounts to characterizing the homology representations of a group G acting on a Moore space, which may be approached via topological replacements for homotopy actions. In this case, $\mathcal{S}^\omega(X, \alpha)$ contains only one element, namely the equivalence class of the characteristic sheaf of the RG-module induced by α. Thus, Corollary 4.2 above provides new global obstructions for an RG-module to occur as a homology representation on a Moore space. The following examples show homotopy actions on somewhat more general spaces, and how they compare with the case of Moore spaces.

Example 1. Let G be an elementary abelian p-group of rank $n + 1$ and let K be a finite extension of $\mathbb{F}_p$ such that $[K : \mathbb{F}_p] > 1$. Let $P_i \in \mathbb{P}^n(k)$ be a set of points which are not $\mathbb{F}_p$-rational, $i = 1, \ldots, m$. Choose $(n + 1)$-vectors $x_i = (x_{i0}, \ldots, x_{in}) \in K^n$, and form the elements $u_i = 1 + x_{i0}(e_0 - 1) + \ldots \in KG$, where $\{e_0, \ldots, e_n\}$ is a basis for G regarded as an vector space. The subgroups $\langle u_i \rangle \subseteq KG$ are all isomorphic to $\mathbb{Z}_p$, and KG is a free $K\langle u_i \rangle$-module [8]. Define the $\mathbb{Z}$-torsion free modules M_i via exact sequence $0 \to M_i \to (\mathbb{Z}G)^S \to KG \otimes_{K\langle u_i \rangle} K \to 0$. Computation of $H^*(G, M_i)$ shows that its characteristic sheaf is a sky-scraper sheaf over $\mathbb{P}^n(k)$ with support $\{P_i\}$. On the other hand, consider a space X homotopy equivalent to a bouquet of spheres of dimensions d_i, $i =$

$1, \ldots, m + 1$. We assume that the d_i-th Betti number of X is the same as the $\mathbb{Z}$-rank of M_i for $1 \leq i \leq m$. Construct a homotopy action $\alpha : G \to \mathcal{E}(X)$ whose reduced homology representation is $\oplus_i M_i$. Here, we choose the trivial action on M_{m+1}. Suspending topological actions allows one to have a base point fixed by G. Such suspensions do not affect our arguments, so we shall consider reduced homology in the following. We leave out M_{m+1} for the moment, and consider the extensions $0 \to \overline{M}^i \to \overline{M}^{i+1} \to \omega^{N-(i+1)}(M^{i+1}) \to 0$ as in the definition and notation of Section 3. Here $H^*(X; k) = M^*$ and $i \geq 1$ in these extensions. The extensions are determined by an appropriate $\mathrm{Ext}^*_G(\overline{M}^i, M^{i+1})$ after dimension shifting, hence by a class in $H^*(G, \mathrm{Hom}(\overline{M}^i, M^{i+1}))$. On the other hand, a well-known result of Dade (called Dade's Lemma [7] [8]) and an argument similar to Mackey's formula can be used to show that $(KG \otimes_{K\langle u_i \rangle} K) \otimes (KG \otimes_{K\langle u_j \rangle} K)$ is a free KG-module if $i \neq j$. This type of argument and induction proves that the ω-composite extension corresponding to $M_1, \ldots, M_m$ is isomorphic to $\oplus_{i=1}^m \omega^{d_m - d_i}(M^i) = M'$. First, consider the case $M_{m+1} = 0$. Then, $S^\omega(X, \alpha)$ has only one element $[\xi]$, and $\xi = \xi_\tau$. It follows that $\mathrm{supp}[\xi_\tau] = \{P_1, \ldots, P_m\} \subset \mathbb{P}^n(k)$ satisfies the hypothesis of Corollary 4.2. Hence (X, α) does not have a topological realization. Next, if $M_{m+1} \neq 0$, $S^\omega(X, \alpha)$ can have more than one element. However, one can compute that for all $[\xi] \in S^\omega(X, \alpha)$, $\mathrm{supp}[\xi_\tau] \subset \{P_1, \ldots, P_m\}$. If we take $m > \dim H^{m+1}(X; k)$, then $\mathrm{supp}[\xi_\tau]$ is seen to be non-empty. Hence (X, α) is not equivalent to a topological action in this case either. Note that the approach through support and rank varieties in [4] will not detect these phenomena when $M_{m+1} \neq 0$.

Example 2. In this example, let P_1 be as above, and construct M_1 as before. Let $M_2 = \omega(M_1)$, and construct a homotopy action (X, α) as before with reduced homology $H_d(X) \cong M_1$ and $H_{d+1}(X) = M_2$. Again, we may consider reduced homology. In this case, then $S^\omega(X, \alpha)$ has only two classes. One corresponding to the case $\xi_\tau = 0$ and $\xi_\varphi = \mathcal{O}_S$ and another in which $\xi_\varphi = \mathcal{O}_S$ and ξ_τ is a sky-scraper sheaf with support $\{P_1\}$ and $H^i(S; \xi_\tau) \cong k^2$ for all sufficiently large i. In the first case, $[\xi]$ is represented by the characteristic sheaf of $\psi : G \times Y \to Y$, where Y is obtained as the mapping cone of $f : G_+ \wedge (S^d)^r \to G_+ \wedge (S^d)^r$. Such f_* in homology realizes the homomorphism ∂ in the sequence $0 \to \omega^2(M_1) \to (\mathbb{Z}G)^r \xrightarrow{\partial} (\mathbb{Z}G)^s \to M_1 \to 0$. Thus, this ω-composite extension corresponds to a semifree G-action on a finite dimensional space with a contractible fixed point set. In the second case, there is no topological action whose characteristic sheaf be equivalent to ξ. Hence this ξ does not correspond to a topological action.

Example 3. Let X be a closed-simply-connected topological 4-manifold, and let μ be its intersection form. Let $\mathrm{Aut}(H^2(X), \mu)$ denote the set of isometries of $H^2(X)$ with respect to the non-degenerate symmetric bilinear form μ. It is isomorphic to an arithmetic group of orthogonal type. According to Mike Freedman, every such orthogonal form μ on $\mathbb{Z}^n$ corresponds to $\mathrm{Aut}(H^2(X), \mu)$ for an appropriate simply-connected closed 4-manifold X as above, and at most there are two such X up to homeomorphism for each isomorphism class of μ. On the other hand, there is a surjection $\mathcal{E}(X) \to \mathrm{Aut}(H^2(X), \mu)$ with kernel isomorphic to $(\mathbb{Z}_2)^r$. Thus, a homotopy G-action $G \to \mathcal{E}(X)$ gives rise to a torsion-free orthogonal $\mathbb{Z}G$-module. Conversely, the results of Freedman-Quinn [11] imply that any homomorphism $G \to \mathrm{Aut}(H^2(X), \mu)$ lifts to $\mathcal{E}(X)$. Consequently, every torsion-free orthogonal $\mathbb{Z}G$-module corresponds to a homotopy G-action on a simply-connected closed 4-manifold (well-defined up to homotopy). In this case, the homotopy equivalences corresponding to action of $g \in G$ can be taken to be homeomorphisms. However, the group law in G is preserved only up to homotopy, so this homotopy action is still far from a topological G-action. Orthogonal versions of the above-mentioned examples can be constructed, showing that the characteristic torsion sheaves for these homotopy actions can detect new obstructions for topological realizability of "generic" homotopy actions on 4-manifolds. On the other hand, it is more interesting to treat these characteristic invariants for homotopy actions as substitutes for fixed point sets and inertia groups in order to study more flexible notions of symmetry that are dominant in this case.

References

[1] J. Alperin, Cohomology in representation theory, *Proc. Sympos. Pure Math.* **47** (1986), 3–11.

[2] J. Alperin and L. Evens, Varieties and elementary abelian groups, *J. Pure Appl. Algebra* **22** (1982), 1–9.

[3] A. Assadi, Extensions libres des actions des groupes finis, *Proc. Aarhus Top. Conf.* 1982, Springer LNM 1052 (1984).

[4] A. Assadi, Homotopy actions and cohomology of finite groups, *Proceedings of Conference on Transformation Groups*, Poznan, 1985, Lecture Notes in Mathematics, Vol. 1217, Springer-Verlag, Berlin/New York, 1986.

[5] A. Assadi, Algebraic invariants for finite group actions. I. Varieties, *Advances in Math.* **100** (1993) 232-261.

[6] A. Assadi and W. Browder, Construction of finite group actions on simply-connected manifolds (Manuscript).

[7] G. Avrunin and L. Scott, Quillent stratification for modules, *Invent. Math.* **66** (1982), 277-286.

[8] J. Carlson, The varieties and the cohomology ring of a module *J. Algebra* **85** (1983), 104-143.

[9] G. Carlsson, A counterexample to a conjecture of Steenrod, *Invent. Math.* **64** (1981), 171-174.

[10] G. Cooke, Replacing homotopy actions by topological actions, *Trans. AMS* **237** (1978), 391-406.

[11] M. Freedman and F. Quinn, *The Topology of 4-Manifolds*, Annals of Math. Studies, Princeton University Press, 1990.

[12] A. Grothendieck and J. Dieudonné, *Éléments de la Géometrie Algébrique I*, Springer-Verlag, Berlin/New York, 1971.

[13] A. Heller, Homological resolutions of complexes with operators, *Ann. Math.* **60** (1954), 283-303.

[14] W. Y. Hsiang, *Cohomological of Topological Transformation Groups*, Ergeb. Math., Springer-Verlag, Berlin, 1975.

[15] P. Kahn, Steenrod's Problem and k-invariants of certain classifying spaces, *Alg. K-theory*, Proc. Oberwolfach (1980), Springer LNM 967 (1982).

[16] P. Landweber and R. Stong, The depth of rings of invariants over finite fields, *Proceedings of Number Theory Conference*, Columbia Univ., Springer-Verlag, 1985.

[17] J. Lannes, Sur la cohomologie modulo p des groupes abelian élémentaires, *Proc. of Durham, Symposium in Homotopy Theory 1985*, Cambridge University Press (1987), 97-116.

[18] J. Lannes, Sur les espaces fonctionelles dont la source est la classifiant d'un p-groupe abélien élémentaire, (preprint) 1990.

[19] D. Quillen, The spectrum of an equivariant cohomology ring I and II, *Ann. of Math.* **94** (1971), 549-572 and 573-602.

[20] F. Quinn, Actions of finite abelian groups, *Proceedings Conf. Homotopy Theory*, LNM 658, Springer-Verlag, Berlin/New York, 1977.

[21] F. Quinn, Homotopy of modules and chain complexes over $\mathbb{Z}[\mathbb{Z}/p]$, (preprint 1976).

[22] R. Schwänzl and R. Vogt, Coherence in homotopy group actions, *LNM 1217*, Springer-Verlag (1986), 364-390.

[23] J. -P. Serre, Sur la dimension cohomologique des groupes profinis, *Topology* **3** (1965), 413-420.

[24] J. -P. Serre, Faiseaux algébrigue cohérents, Annals of Math. **61** (1955), 197-278.

[25] J. Smith, Topological realization of chain complexes. I. The general theory, *Topology Appl.* **22** (1986), 301-313.

[26] R. Swan, Invariant rational functions and a problem of Steenrod, *Inven. Math.* **7** (1969), 148-158.

[27] P. Vogel, On Steenrod's problem for non-abelian finite groups, *Proc. Alg. Top. Conf.*, Aarhus 1982, Springer LNM 1051 (1984).

[28] P. Vogel, A solution of the Steenrod problem for G-Moore spaces, *K-Theory*, No. 4 (1987), 325-335.

[29] A. Zabrodsky, On G. Cooke's theory of homotopy actions, *Proc. Topology Conf. London*, Ontario 1981, in New Trends in Algebraic Topology.

[30] A. Zabrodsky, Homotopy actions of nilpotent groups, *Proc. Topology Conf. Mexico 1981*, Contemporary Math. 1984.

Algebraic K-Theory of Local Number Fields: The Unramified Case

M. Bökstedt
I. Madsen

1 Introduction

For any unital ring A the cyclotomic trace of [BHM] gives a map, in the category of spectra,

$$\text{trc}: K(A) \to \text{TC}(A)$$

from Quillen's K-theory spectrum to the topological cyclic homology spectrum. The latter spectrum was introduced in [BHM], but we refer the reader to [HM2], for a detailed account of the construction.

If A is a finite dimensional algebra over the Witt vectors $W(k)$ of a perfect field k of characteristic $p \neq 0$, then the cyclotomic trace becomes a homotopy equivalence in non-negative degrees after completion at p, or equivalently,

$$\text{trc}_*: \pi_i(K(A); \mathbb{Z}_p) \cong \pi_i(\text{TC}(A); \mathbb{Z}_p), \quad i \geq 0 \tag{1.1}$$

by one of the main results from [HM2], based in part on a relative theorem due to R. McCarthy.

In this paper we evaluate the homotopy type of $\text{TC}(A)$ for certain discrete valuation rings, namely those which have characteristic zero, finite residue fields, and are absolutely unramified (A/pA is the residue field). These rings can be displayed as the Witt-vectors of finite fields with p^s elements, or equivalently as the ring of integers in local number fields which are unramified extensions of $\mathbb{Q}_p$, the field of p-adic numbers, cf. [Ser].

Let K denote the topological periodic K-theory spectrum whose $2n$'th space is $BU \times \mathbb{Z}$, and let $hF(\Psi^g)$ be the homotopy fiber of $\Psi^g - 1$,

$$hF(\Psi^g) \to K \to K,$$

i.e. the "homotopy fixed point spectrum" of the Adams operation Ψ^g, $g \in \mathbb{Z}$. This is a very well understood spectrum. In particular its homotopy groups are given by

$$\pi_i hF(\Psi^g) = \begin{cases} \mathbb{Z}/(g^n - 1) & \text{if } i = 2n - 1 \\ \mathbb{Z} & \text{if } i = 0, -1 \\ 0 & \text{otherwise.} \end{cases} \tag{1.2}$$

Given any spectrum T we let $T[n, \infty)$ denote its "$n - 1$-connected" cover; there is a map

$$f_n \colon T[n, \infty) \to T$$

with $\pi_i(f_n)$ an isomorphism for $i \geq n$ and $\pi_n T[n, \infty) = 0$ for $i < n$. Quillen proved in [Q1] that $hF(\Psi^g)$ is intimately related to the algebraic K-theory of the finite field $\mathbb{F}_g$ when $g = l^a$ is a prime power:

$$\begin{aligned} K(\mathbb{F}_g)^\wedge_p &\simeq hF(\Psi^g)^\wedge_p[0, \infty), \\ K(\mathbb{F}_g)^\wedge_l &\simeq H\mathbb{Z}_l \quad p \neq l. \end{aligned} \tag{1.3}$$

Here $(-)^\wedge_p$ indicates p-adic completion and $H\mathbb{Z}_l$ denotes the Eilenberg-Maclane spectrum with $\pi_0 H\mathbb{Z}_l = \mathbb{Z}_l$, the l-adic integers, and $\pi_i H\mathbb{Z}_l = 0$ for $i \neq 0$.

Let us now fix a prime p and write $A_s = W(\mathbb{F}_{p^s})$ for the Witt-vectors of the finite field with p^s elements. Our main result can be stated as follows.

Choose an integer g which generates the units $(\mathbb{Z}/p^2)^\times$ in the ring $\mathbb{Z}/p^2$ (or equivalently a topological generator of $\mathbb{Z}_p^\times$).

Theorem. *For an odd prime p, there is a homotopy equivalence of p-adic spectra,*

$$K(A_s)^\wedge_p \simeq hF(\Psi^g)^\wedge_p[0, \infty] \vee \Sigma\big(hF(\Psi^g)^\wedge_p[0, \infty)\big)$$

$$\vee \Sigma(K^\wedge_p[2, \infty)) \vee \bigvee^{s-1} \Sigma^{-1}(K^\wedge_p[2, \infty)),$$

where $\Sigma(-)$ denotes the suspension of the listed spectrum.

It is customary in topology to write $J_p = hF(\Psi^g)^\wedge_p$ and we shall adopt this notation in the rest of the paper. Since p is fixed throughout, we shall often drop the subscript p on J, $J = J_p$.

For readers which are not so familiar with the notion of spectra, let us rewrite the theorem above on the level of spaces as

$$(BGL(A_s)^+ \times \mathbb{Z})^\wedge_p \simeq J[0, \infty) \times B(J[0, \infty)) \times SU^\wedge_p \times \prod^{s-1} U^\wedge_p \tag{1.4}$$

where $(\ )^+$ is Quillen's plus construction, and $B(-)$ denotes the classifying space, $\Omega BX \simeq X$, and where SU resp. U are the infinite special unitary group, resp. unitary group. In particular, (1.4) implies the following results about the higher K-groups of A_s: for $i \geq 0$,

$$K_i(A_s; \mathbb{Z}_p) \cong \big(\pi_i(J) \oplus \pi_i(BJ) \oplus \pi_i(SU) \oplus \bigoplus^{s-1} \pi_i(U)\big) \otimes \mathbb{Z}_p. \tag{1.5}$$

Concretely, $\pi_0(J) = \mathbb{Z}_p$, $\pi_{2n}(J) = 0$ for $n > 0$ and

$$\pi_{2n-1}(J) = \begin{cases} \mathbb{Z}/p^{v_p(n+1)} & \text{if } n \equiv 0 \pmod{p-1} \\ 0 & \text{otherwise} \end{cases}$$

where $v_p(-)$ denotes the p-adic valuation, and

$$\pi_i(U) = \begin{cases} \mathbb{Z} & \text{for } i \text{ odd} \\ 0 & \text{for } i \text{ even} \end{cases}$$

while $\pi_i(SU) = \pi_i(U)$ for $i \neq 0$ and $\pi_1(SU) = 0$. Finally, $\pi_i(BJ) = \pi_{i-1}(J)$. The left-hand side of (1.5) are the higher K-groups with p-adic coefficients, i.e.

$$K_i(A_s; \mathbb{Z}_p) = \varprojlim \pi_i\big(BGL(A_s)^+ \times \mathbb{Z}; \mathbb{Z}/p^n\big).$$

The first two terms of (1.5) are p-torsion groups except for $i = 0, 1$. Rationally,

$$\dim\big(K_{2i-1}(A_s; \mathbb{Z}_p) \otimes_{\mathbb{Z}_p} \mathbb{Q}_p\big) = s \quad \text{for } i \geq 2. \tag{1.6}$$

This is in agreement with an old result of Wagoner [W], since one knows from [P] that $K_i(A_s; \mathbb{Z}_p) = \varprojlim K_i(A_s/p^k; \mathbb{Z}_p)$.

Suppose l is a prime different from p. Then

$$K_i(A_s; \mathbb{Z}_l) \simeq K_i(\mathbb{F}_{p^s}; \mathbb{Z}_l) = K_i(\mathbb{F}_{p^s}) \otimes \mathbb{Z}_l \tag{1.7}$$

by a result of Suslin, cf. [Sus], Theorem 2.8, and these groups we listed in (1.2), (1.3). Hence, in total, one now knows all higher K-groups of A_s with finite or adic coefficients. What is missing is the knowledge of the uniquely divisible part of $K_i(A_s)$; this is an entirely different story.

Let F_s be the quotient field. By localization, [Q2], there is a cofibration of spectra

$$K(A_s) \to K(F_s) \to \Sigma K(\mathbb{F}_{p^s})$$

where the cofiber is known by (1.3). Moreover, the homotopy exact sequence breaks up into short exact sequences

$$0 \to K_i(A_s; \mathbb{Z}_p) \to K_i(F_s; \mathbb{Z}_p) \to K_{i-1}(\mathbb{F}_{p^s}) \otimes \mathbb{Z}_l \to 0$$

cf. [Sou], so the above also calculates $K_i(F_s; \mathbb{Z}_l)$ for all l.

In the theorem we excluded $p = 2$. The expected result for $p = 2$, currently evolving, is similar but not identical to the case of odd primes. We refer the reader to sect. 6 below, where we also discuss the relations of our results to étale K-theory, and a possible attack on the ramified case.

In the rest of the paper p will denote an odd prime unless otherwise specified.

Table of contents.

This paper is intimately related to [BM], and consequently there are many cross references to that paper. We have strived, however to make the presentation relatively selfcontained. Finally it is a pleasure to acknowledge valuable conversations with L. Hesselholt; he pointed out a mistake in our initial formulation of the results.

2 Topological cyclic homology

Let A be a unital ring and let $T(A)$ denote the spectrum whose 0'th space is the topological Hochschild homology, THH(A). It is a connected equivariant S^1-spectrum, and moreover, a cyclotomic spectrum in the notion of [HM2]. In particular, there are maps

$$F, R: T(A)^{C_n} \to T(A)^{C_m}$$

of the fixed sets under the cyclic groups of order n and m, provided m divides n. The topological cyclic homology is defined to be

$$\mathrm{TC}(A) = \underset{F,R}{\mathrm{holim}}\, T(A)^{C_n}$$

cf. [HM2], sect. 3. We are in this paper only interested in its p-adic completion, where by [HM2], Theorem 3.1 one can simplify the construction by only using fixed sets under the groups $C_{p^n} \subset S^1$ of order p^n. Let us define

$$\mathrm{TF}(A) = \underset{F}{\mathrm{holim}}\, T(A)^{C_{p^n}}, \quad \mathrm{TR}(A) = \underset{R}{\mathrm{holim}}\, T(A)^{C_{p^n}}. \tag{2.1}$$

We note that

$$\pi_0 T(A)^{C_{p^n}} = W_{n+1}(A),$$

the Witt vectors of length $n + 1$, and that $\pi_0 F$, $\pi_0 R$ are the Frobenius and restriction maps of Witt vectors by [HM2], Theorem 2.3. One defines

$$\mathrm{TC}(A, p) = \mathrm{TF}(A)^{hR} = \mathrm{TR}(A)^{hF}. \tag{2.2}$$

Here hR and hF indicates homotopy fixed sets, i.e.

$$\mathrm{TF}(A)^{hR} \longrightarrow \mathrm{TF}(A) \xrightarrow{R-1} \mathrm{TF}(A)$$

is a cofibration of spectra. One has

$$\mathrm{TC}(A)_p^\wedge \simeq \mathrm{TC}(A, p)_p^\wedge. \tag{2.3}$$

Let Γ be a discrete group, and $\Lambda B\Gamma$ the free loop space of its classifying space. Then

$$T(A[\Gamma]) \simeq T(A) \wedge \Lambda B\Gamma_+$$

(in the category of spectra), and the maps R and F can be calculated as follows (on the level of prespectra)

$$R_{A[\Gamma]} = R \wedge \Delta_p^{-1}, \quad F_{A[G]} = F \wedge \text{incl} \tag{2.4}$$

as maps from $T(A)^{C_{p^n}} \wedge \Lambda B\Gamma_+^{C_{p^n}}$ to $T(A)^{C_{p^{n-1}}} \wedge \Lambda B\Gamma_+^{C_{p^{n-1}}}$. Here $\Delta_p \colon \Lambda B\Gamma^{C_{p^{n-1}}} \to \Lambda B\Gamma^{C_{p^n}}$ is the homeomorphism which maps a loop $\lambda(z)$ into the loop $\lambda(z^p)$. This follows from [HM2], Theorem 6.1 upon identifying the realization of the cyclic nerve $N^{cy}(\Gamma)$ with $\Lambda B\Gamma$, cf. [BHM], Proposition 2.6.

Suppose now that Γ is a finite group whose order is prime to p. Then

$$(\Lambda B\Gamma)_p^\wedge \simeq \pi_0 \Lambda B\Gamma = \Gamma/\text{conjugacy} \tag{2.5}$$

and Δ_p becomes $\Delta_p(\gamma) = \gamma^p$, [BHM], lemma 7.11. We decompose $\Gamma/\text{conjugacy}$ into its orbits under Δ_p, say

$$\Gamma/\text{conjugacy} = \coprod \Gamma_+^\alpha.$$

Since the C_{p^n}-action comes from a circle action

$$T(A[\Gamma])^{C_{p^n}} = \bigvee_\alpha T(A)^{C_{p^n}} \wedge \Gamma_+^\alpha,$$

and hence

$$\begin{aligned} \text{TF}(A[\Gamma]) &= \bigvee \text{TF}(A) \wedge \Gamma_+^\alpha \\ \text{TC}(A[\Gamma], p) &= \bigvee (\text{TF}(A) \wedge \Gamma_+^\alpha)^{hR}. \end{aligned}$$

Suppose Γ^α has cardinality $s(\alpha)$. Then

$$(\text{TF}(A) \wedge \Gamma_+^\alpha)^{hR^{s(\alpha)}} = (\text{TF}(A)^{hR^{s(\alpha)}} \wedge \Gamma_+^\alpha).$$

The rest of the R-action amounts to an action of the cyclic group $C_{s(\alpha)}$, which acts freely on Γ^α, by definition. Thus

$$\text{TC}(A, p) = \bigvee (\text{TF}(A)^{hR^{s(\alpha)}} \wedge \Gamma_+^\alpha)^{hC_{s(\alpha)}} = \bigvee \text{TF}(A)^{hR^{s(\alpha)}}. \tag{2.6}$$

We shall use (2.6) for $\Gamma = C_f$ with $(p, f) = 1$ and $A = \mathbb{Z}_p$. In this case the splitting (2.6) is analogous to the decomposition

$$\mathbb{Z}_p[C_f] = \prod_{d|f} \mathbb{Z}_p \otimes \mathbb{Z}[\zeta_d] = \prod_{d|f} \prod^{r(d)} \mathbb{Z}_p[\zeta_d] \tag{2.7}$$

with $r(d)$ the index of the local Galois group $G_d = \{p^i \mid i \in \mathbb{Z}\}$ in the global Galois group $\mathbb{Z}/d^\times$ of $\mathbb{Q}(\zeta_d)/\mathbb{Q}$. Recall from the introduction the notation $A_s =$

$W(\mathbb{F}_{p^s})$, the ring of integers in the unramified extension of $\mathbb{Q}_p$ of degree s. Then $A_s = \mathbb{Z}_p[\zeta_f]$ for suitable f (For given f, s is the minimal number with $p^s \equiv 1 \pmod{f}$).

Theorem 2.8. *There is a cofibration*

$$\mathrm{TC}(A_s)^\wedge_p \longrightarrow \mathrm{TF}(\mathbb{Z}_p)^\wedge_p \overset{R^s-1}{\longrightarrow} \mathrm{TF}(\mathbb{Z}_p)^\wedge_p.$$

Proof. Pick f so that $A_s = \mathbb{Z}_p[\zeta_f]$. We claim that $\mathrm{TF}(\mathbb{Z}_p \otimes \mathbb{Z}[\zeta_f])^\wedge_p$ is homotopy equivalent to the cofiber of

$$\bigvee \mathrm{TF}\left(\mathbb{Z}_p[C_d]\right)^\wedge_p \to \mathrm{TF}\left(\mathbb{Z}_p[C_f]\right)^\wedge_p \tag{*}$$

where the wedge runs over all divisors d of f with $d < f$. This follows for example from general induction theory: $\pi_*(\mathrm{TF}(\mathbb{Z}_p C_f); \mathbb{Z}_p)$ is a Mackey functor, hence a module over the Burnside ring $A(C_f) \otimes \mathbb{Z}_p$. The localization at the prime ideal $q(C_f) \subset A(C_f) \otimes \mathbb{Z}_p$, given by the C_f-sets X with $\mathrm{card}(X^{C_f}) \equiv 0 \pmod{p}$, is on the one hand equal to $\pi_*(\mathrm{TF}(\mathbb{Z}_p \otimes \mathbb{Z}[\zeta_f]); \mathbb{Z}_p)$ by [HaM], Proposition 6.17 and on the other hand

$$\pi_*(\mathrm{TF}(\mathbb{Z}_p[C_f]); \mathbb{Z}_p)_{q(C_f)} - p)$$
$$= \ker\left(\mathrm{Res}\colon \pi_*(\mathrm{TF}(\mathbb{Z}_p C_f); \mathbb{Z}_p) \to \bigoplus \pi_*(\mathrm{TF}(\mathbb{Z}_p C_d); \mathbb{Z}_p)\right)$$
$$= \mathrm{coker}\left(\mathrm{Ind}\colon \bigoplus \pi_*(\mathrm{TF}(\mathbb{Z}_p C_d); \mathbb{Z}_p) \to \pi_*(\mathrm{TF}(\mathbb{Z}_p C_f); \mathbb{Z}_d)\right)$$

The cofiber of $(*)$ is $\mathrm{TF}(\mathbb{Z}_p)^\wedge_p \wedge (\mathbb{Z}/f)^\times$ and thus

$$\mathrm{TC}(\mathbb{Z}_p \otimes \mathbb{Z}[\zeta_f])^\wedge_p = \overset{r(f)}{\bigvee} (\mathrm{TF}(\mathbb{Z}_p)^{hR^s})^\wedge_p.$$

Since $\mathbb{Z}_p \otimes \mathbb{Z}[\zeta_f] = \prod^{r(f)} \mathbb{Z}_p[\zeta_f]$ the result follows.   □

More generally, let $E \subset F$ be an unramified extension of local fields of degree s, and let $B \subset A$ be the corresponding extension of the integers in E and F. Then

$$\mathrm{TC}(A)^\wedge_p \to \mathrm{TF}(B)^\wedge_p \overset{R^s-1}{\longrightarrow} \mathrm{TF}(B)^\wedge_p \tag{2.9}$$

is a cofibration where p is the residue characteristic.

The general scheme for calculation, presented in the next three sections when $A = A_s$ and $B = \mathbb{Z}_p$, is to calculate $\pi_*(\mathrm{TF}(B); \mathbb{Z}_p)$ by comparing fixed sets to homotopy fixed sets. More precisely there is a cofibration diagram (cf. [HM2], Proposition 4.1)

$$
\begin{array}{ccccc}
T(B)_{hC_{p^n}} & \xrightarrow{N^h} & T(B)^{hC_{p^n}} & \xrightarrow{R^h} & \hat{\mathbb{H}}(C_{p^n}, T(B)) \\
\big\uparrow & & \big\uparrow{\scriptstyle \Gamma_n} & & \big\uparrow{\scriptstyle \hat{\Gamma}_n} \\
T(B)_{hC_{p^n}} & \xrightarrow{N} & T(B)^{C_{p^n}} & \xrightarrow{R} & T(B)^{C_{p^{n-1}}}
\end{array}
\tag{2.10}
$$

for each n. The upper sequence is the norm cofibration, valid for any equivariant spectrum. The lower sequence is the fundamental cofibration of cyclotomic spectra. One may take limits over n and observe that

$$\underleftarrow{\operatorname{holim}}\, T(B)^{hC_{p^n}} \simeq_p T(B)^{hS^1}$$

$$\underleftarrow{\operatorname{holim}}\, T(B)_{hC_{p^n}} \simeq_p \Sigma T(B)_{hS^1} \qquad (2.11)$$

$$\underleftarrow{\operatorname{holim}}\, \hat{\mathbb{H}}(C_{p^n}; T(B)) \simeq_p \hat{\mathbb{H}}(S^1, T(B))$$

where the notation is as follows: for any compact Lie group G, and any G-spectrum T

$$T_{hG} = T \wedge_G EG_+, \quad T^{hG} = \operatorname{Map}(EG_+, T),$$

$$\hat{\mathbb{H}}(G, T) = (\widetilde{EG} \wedge \operatorname{Map}(EG_+, T))^G$$

with EG the free contractible G-space, $\widetilde{EG} = \Sigma(EG)$; the smash products are taken in the category of G-spectra. The first equivalence in (2.11) is an easy consequence of the fact that the homotopy direct limit of BC_{p^n} is equivalent to BS^1 after p-adic completion. The second equivalence is proved in [BHM], lemma 5.15 when $T(B)$ has finite p-type, but is true in general. The third equivalence follows from the two first and the norm cofibration displayed in the upper line of (2.12).

The homotopy limit of diagram (2.10) then becomes the p-completion of

$$
\begin{array}{ccccc}
\Sigma T(B)_{hS^1} & \xrightarrow{\;N^h\;} & T(B)^{hS^1} & \xrightarrow{\;R^h\;} & \hat{\mathbb{H}}(S^1, T(B)) \\
\big\uparrow & & \big\uparrow{\scriptstyle\Gamma} & & \big\uparrow{\scriptstyle\hat{\Gamma}} \\
\Sigma T(B)_{hS^1} & \xrightarrow{\;N\;} & \mathrm{TF}(B) & \xrightarrow{\;R\;} & \mathrm{TF}(B)
\end{array}
\qquad (2.12)
$$

Finally one has the following conjecture similar to the affirmed Segal conjecture for the sphere spectrum:

Conjecture 2.13. *If B is the ring of integers in a local field with residue field of characteristic p then the p-completion of Γ and $\hat{\Gamma}$ in (2.12) defines homotopy equivalences onto the connected covers $T(B)^{hS^1}[0, \infty)$ and $\hat{\mathbb{H}}(S^1, T(B))[0, \infty)$.*

By a result of S. Tsalidis, explained in the next section, the conjecture follows from the statement that

$$\hat{\Gamma}_1 \colon T(B) \to \hat{\mathbb{H}}(C_p, T(B))[0, \infty) \qquad (2.14)$$

be a p-adic homotopy equivalence; (2.14) in turn was proved in [BM] for $B = \mathbb{Z}_p$, $p > 2$.

3 Spectral sequences

In [BM] the skeleton spectral sequences ([GM], [BM], sect. 2) with abutments $\pi_*(T(\mathbb{Z}_p)^{hC_{p^n}}; \mathbb{F}_p)$ and $\pi_*(\hat{\mathbb{H}}(C_{p^{n+1}}, T(\mathbb{Z}_p)); \mathbb{F}_p)$ were studied for odd primes p. The spectral sequences are of homology type, and

$$E^2_{p,q}(C_{p^n}) = H^{-p}(C_{p^n}; \pi_q(T(\mathbb{Z}_p); \mathbb{F}_p)) \Rightarrow \pi_{q-p}(T(\mathbb{Z}_p)^{hC_{p^n}}; \mathbb{F}_p)$$
$$\hat{E}^2_{p,q}(C_{p^n}) = \hat{H}^{-p}(C_{p^n}; \pi_q(T(\mathbb{Z}_p); \mathbb{F}_p)) \Rightarrow \pi_{q-p}(\hat{\mathbb{H}}(C_{p^n}, T(\mathbb{Z}_p)); \mathbb{F}_p)$$

where $\hat{H}$ denotes Tate cohomology. The E^2-terms are

$$E^2_{*,*}(C_{p^n}) = E\{u_n\} \otimes S\{t\} \otimes E\{e\} \otimes S\{f\}$$
$$\hat{E}^2_{*,*}(C_{p^n}) = E\{u_n\} \otimes S\{t, t^{-1}\} \otimes E\{e\} \otimes S\{f\}$$

with $E\{-\}$ and $S\{-\}$ denoting the exterior and symmetric algebras, respectively. The bi-degrees of the listed generators are

$$\deg u_n = (-1,0), \ \deg t = (-2,0), \ \deg e = (0, 2p-1), \ \deg f = (0, 2p).$$

Both spectral sequences are spectral sequences of algebras, that is, the differentials d^r are derivations. It was conjectured in [BM], (4.3) that the non-zero differentials are generated from the following statements:

(i) $\quad d^{2\mathfrak{p}(k+1)}(t^{p^k}) = \lambda_k t^{p^k + \mathfrak{p}(k+1)} e f^{\mathfrak{p}(k)} \quad$ for $k < n$, with $\lambda_k \in \mathbb{F}_p^\times$

(ii) $\quad d^{2\mathfrak{p}(n)+1}(u_n) = \mu_n t^{\mathfrak{p}(n)+1} f^{\mathfrak{p}(n-1)+1} \quad$ with $\mu_n \in \mathbb{F}_p^\times \qquad (3.1)$

(iii) $\quad d^r(tf) = 0, \ d^r(te) = 0, \ d^r(1) = 0$ for all r

where $\mathfrak{p}(k) = p(p^k - 1)/(p-1)$ with $\mathfrak{p}(0) = 1$.

Since the differentials are derivations, (i), (iii) imply that

$$d^{2\mathfrak{p}(k)}(t^{p^k i}) = i\lambda_{k-1} t^{p^k i + \mathfrak{p}(k+1)} e f^{\mathfrak{p}(k)}$$

for all i ($i \in \mathbb{Z}$ for $\hat{E}^{2\mathfrak{p}(k)}_{*,*}(n)$), and it is equivalent to give the differential structure for $E^r_{*,*}(C_{p^n})$ and $\hat{E}^r_{*,*}(C_{p^n})$.

In [BM], sect. 5 we showed

Lemma 3.2 ([BM]). *Conjecture 3.1 is true for $n = 1$, and 3.1 (iii) is true for all n.*

We shall not repeat the argument here. It is based upon a comparison with the corresponding spectral sequences for the sphere spectrum,

$$(S^0)^{hS^1} \to T(\mathbb{Z}_p)^{hS^1} \to T(\mathbb{Z}_p)^{hC_{p^n}}$$
$$\hat{\mathbb{H}}(S^1, S^0) \to \hat{\mathbb{H}}(S^1, T(\mathbb{Z}_p)) \to \hat{\mathbb{H}}(C_{p^n}, T(\mathbb{Z}_p)), \qquad (3.3)$$

the known structure in low degrees of the spectral sequence for $(S^0)^{hS^1}$, and upon the fact that S^0 is a so called split spectrum so that

$$(S^0)^{hS^1} \simeq \mathrm{Map}(BS^1_+, S^0).$$

The obvious map from S^0 to $\mathrm{Map}(BS^1_+, S^0)$ composed with the first map in (3.3) has the property that it takes the non-trivial elements

$$v_1 \in \pi_{2p-2}(S^0; \mathbb{F}_p), \quad \alpha_1 = \beta_1(v_1) \in \pi_{2p-3}(S^0; \mathbb{F}_p)$$

into tf and te, respectively.

For $n > 1$, conjecture 3.1 was derived from the assertion that there exists a factorization

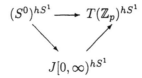

which we could not prove. The differentials in 3.1 (i), (ii) correspond to the differentials for $J[0, \infty)^{hS^1}$ upon substituting te, tf for α_1, v_1.

Lemma 3.4 ([BM]). *The map*

$$\hat{\Gamma}_1 \colon T(\mathbb{Z}_p)^\wedge_p \to \hat{\mathbb{H}}(C_p, T(\mathbb{Z}_p))[0, \infty)$$

is a homotopy equivalence.

Given (3.1) it is routine to evaluate the E^∞-terms, and observe that

$$\pi_i(T(\mathbb{Z}_p)^{hC_{p^n}}; \mathbb{F}_p) \cong \pi_i\big(\hat{\mathbb{H}}(C_{p^{n+1}}, T(\mathbb{Z}_p)); \mathbb{F}_p\big)$$

for $i \geq 0$. In fact we proved in [BM], that the following are homotopy equivalences

$$T(\mathbb{Z}_p)^{hC_{p^n}}[0, \infty) \xleftarrow[\simeq]{\Gamma_n} T(\mathbb{Z}_p)^{C_{p^n}} \xrightarrow[\simeq]{\hat{\Gamma}_{n+1}} \hat{\mathbb{H}}(C_{p^{n+1}}, T(\mathbb{Z}_p))[0, \infty)$$

In his 1994 thesis [T], S. Tsalidis proved by independent methods

Theorem 3.5 (Tsalidis). *If* $\Gamma_1 \colon T(\mathbb{Z}_p)^{C_p} \to T(\mathbb{Z}_p)^{hC_p}[0, \infty)$ *is a homotopy equivalence then so are*

$$\begin{aligned}
\Gamma_n &: T(\mathbb{Z}_p)^{C_{p^n}} \to T(\mathbb{Z}_p)^{hC_{p^n}}[0, \infty) \\
\hat{\Gamma}_{n+1} &: T(\mathbb{Z}_p)^{C_{p^n}} \to \hat{\mathbb{H}}(C_{p^{n+1}}, T(\mathbb{Z}_p))[0, \infty)
\end{aligned}$$

for all n.

Tsalidis' proof is modelled upon the corresponding statement for the sphere spectrum—known as (part of) the Segal conjecture. It does not however follow from the existing literature on the Segal conjecture, e.g. [C], [AHJM], [CMP], since $T(\mathbb{Z}_p)$ is not a split C_{p^n}-spectrum, but Tsalidis overcomes this difficulty. The start of the induction is provided by lemma 3.4 upon using diagram (2.10).

In [BM], (4.13) it was remarked (without proof) that the differentials listed in (3.1) above were the first non-zero differentials. More precisely:

Proposition 3.6. *Suppose the differentials (3.1), (i) and (3.1), (ii) are true for* $E^r_{*,*}(C_{p^n})$ *and* $\hat{E}^r_{*,*}(C_{p^n})$, *then the differentials for* $E^r_{*,*}(C_{p^{n+1}})$ *and* $\hat{E}^r_{*,*}(C_{p^{n+1}})$ *are as listed, except that* λ_n *and* μ_{n+1} *might be zero. If* λ_n *is non-zero, so is* μ_{n+1}

Based on this proposition Tsalidis proves that Theorem 3.5 implies (3.1) for all n. We owe a proof of Proposition 3.6; it did not seem that important at the time of writing [BM] as we had no way of proving that λ_n and μ_{n+1} were non-zero. See also remark 3.11 below.

The inductive proof is based upon comparison with the spectral sequences for $T(\mathbb{Z}_p)^{hS^1}$ and $\hat{\mathbb{H}}(S^1, T(\mathbb{Z}_p))$, and the obvious restriction maps

$$F: \hat{\mathbb{H}}(S^1, T(\mathbb{Z}_p)) \to \hat{\mathbb{H}}(C_{p^n}, T(\mathbb{Z}_p)).$$

The skeleton spectral sequence for $\pi_*(\hat{\mathbb{H}}(S^1, T(\mathbb{Z}_p)); \mathbb{F}_p)$ has E^2-term

$$\hat{E}^2_{*,*}(S^1) = S\{t, t^{-1}\} \otimes E\{e\} \otimes S\{f\}.$$

It is an upper half plane convergent spectral sequence, cf. [HM1]. We shall also make use of the transfer map

$$V_n: \hat{\mathbb{H}}(C_{p^n}, T(\mathbb{Z}_p)) \to \hat{\mathbb{H}}(C_{p^{n+1}}, T(\mathbb{Z}_p))$$

which works for all equivariant spectra, and in particular for the S^1/C_p-spectrum $\hat{\mathbb{H}}(C_p, T(\mathbb{Z}_p))$, cf. [HM2], Proposition 2.2.

Proof of Proposition 3.6. Suppose inductively that 3.1 (i) and 3.1 (ii) hold for $\hat{E}^r_{*,*}(C_{p^n})$. The restriction map F induces a map of spectral sequences, which on the E^2-term is the obvious injection. The terms te and tf in $\hat{E}^2_{*,*}(S^1)$ are permanent cycles by lemma 3.2. It follows from the injectivity of F on $\hat{E}^2_{*,*} = \hat{E}^{2p}_{*,*}$ that $d^{2p}(t) = t^{p+1}e$ in $E^2_{*,*}(S^1)$, and hence that $d^{2p}(t^{i+j}f^j) = it^{i+p+j}ef^j$ for $i \in \mathbb{Z}$ and $j \geq 0$. Thus

$$\hat{E}^{2p+1}_{*,*}(S^1) = \mathbb{F}_p\{t^{pi+j}f^j \mid i \in \mathbb{Z}, \, j \geq 0\} \oplus \mathbb{F}_p\{t^{pi+j}ef^j \mid i \in \mathbb{Z}, \, j \geq 0\}$$

where $\mathbb{F}_p\{\cdots\}$ is the $\mathbb{F}_p$ vector space generated by the listed elements. Similarly we calculate $\hat{E}^{2p+1}_{*,*}(C_{p^n})$. Since $d^{2p}(u_n) = 0$ we get

$$\hat{E}^{2p+1}_{*,*}(C_{p^n}) = E\{u_n\} \otimes \hat{E}^{2p+1}_{*,*}(S^1)$$

and F is injective on $\hat{E}^{2p+1}$. By our induction hypothesis, $\hat{E}^{2p+1}_{*,*}(C_{p^n}) = \hat{E}^{2p(2)}_{*,*}(C_{p^n})$ when $n > 1$, and

$$d^{2p(2)}(t^{pi+j}f^j) = it^{pi+j+p(2)}ef^{j+p(1)}$$

which makes it possible to calculate $\hat{E}^{2p(2)+1}_{*,*}(S^1)$ and to check it injects into $\hat{E}^{2p(2)+1}_{*,*}(C_{p^n})$ provided $n \geq 2$. In this fashion we obtain

$$\hat{E}^{2p(n)+1}_{*,*}(S^1) = \mathbb{F}_p\{t^{p^n i+j}f^j \mid i \in \mathbb{Z},\ j \geq 0\} \oplus \mathbb{F}_p\{t^{p^n i+j}ef^j \mid i \in \mathbb{Z},\ j \geq 0\} \oplus$$
$$\bigoplus_{k=1}^{n-1}\mathbb{F}_p\{t^{p^k i+j}ef^j \mid i \in \mathbb{Z},\ v_p(i) < n-k,\ p(k-1) \leq j < p(k)\} \tag{3.7}$$

where $v_p(i)$ denotes the p-adic valuation and $\hat{E}^{2p(n)+1}_{*,*}(C_{p^n}) = \hat{E}^{2p(n)+1}_{*,*}(S^1) \otimes E\{u_n\}$, so that F is still an injection on $\hat{E}^{2p(n)+1}_{*,*}$.

We next show that $\hat{E}^{2p(n)+1}_{*,*}(S^1) = \hat{E}^{2p(n+1)}_{*,*}(S^1)$. Let us look in total degree $2p^n - 1$. We have the elements $z(k-1)$, $k = 1,\ldots,n-1$, in the last $n-1$ summands of (3.7),

$$z(k-1) = t^{p^k(1-p^{n-k})+p(k-1)}ef^{p(k-1)},$$

and in addition the elements

$$z(n-1,l) = t^{p^{n+1}l+p(n-1)}ef^{p(n-1)+p^n l},\ l \geq 0$$

in the second summand of (3.7). There are no further elements in degree $2p^n - 1$ in $\hat{E}^{2p(n)+1}_{*,*}(S^1)$. One knows in advance that all elements of (3.7) involving e are infinite cycles. Indeed we can compare with the spectral sequence converging to p-integral homotopy with E^2-term

$$\hat{H}^{-p}(S^1; \pi_*T(\mathbb{Z}_p)) = S\{t,t^{-1}\} \otimes \pi_*T(\mathbb{Z}_p)$$

and abutment $\pi_*\hat{\mathbb{H}}(S^1, T(\mathbb{Z}_p))$. The homotopy groups of $T(\mathbb{Z}_p)$ are concentrated in odd degrees, and in degree zero,

$$\pi_{2n-1}T(\mathbb{Z}_p) = \mathbb{Z}/n \otimes \mathbb{Z}_p$$

and the reduction map

$$\pi_{2n-1}T(\mathbb{Z}_p) \to \pi_{2n-1}(T(\mathbb{Z}_p); \mathbb{F}_p)$$

is surjective, [B]. Thus in positive fiber degrees the p-integral spectral sequence is concentrated in odd total degrees, and the reduction map from the integral spectral sequence to the mod p spectral sequence is injective on the E^2-level with image consisting of the elements of the form $t^i ef^j$, which must therefore be permanent cycles.

In particular the elements $z(k-1)$, $z(n-1,l)$ are all infinite cycles in the mod p spectral sequence $\hat{E}^r_{*,*}(S^1)$. Since d^r raises fiber degrees by $r-1$ and since the spectral sequence lies in the upper half plane, $z(k-1)$, $k = 1, \dots, n-1$ and $z(n-1)$ cannot be d^r-boundaries when $r \geq 2p(n)+1$. However $z(n-1,1) = t^{p^{n+1}+p(n-1)} e f^{p^n+p(n-1)}$ could be a boundary, in that

$$d^{2p(n+1)} t^{-p^n} = \lambda_n t^{p^{n+1}+p(n-1)} e f^{p^n+p(n-1)}, \quad \lambda_n \in \mathbb{F}_p.$$

Since $p^n + p(n-1) = p(n)$ and $p^{n+1} + p(n-1) = p(n+1) - p^n$ this is equivalent to the statement

$$d^{2p(n+1)} t^{p^n} = \lambda_n t^{p^n+p(n+1)} e f^{p(n)}, \quad \lambda_n \in \mathbb{F}_p \tag{3.8}$$

in $\hat{E}^{2p(n+1)}_{*,*}(S^1)$. Note also that the argument implies that $\hat{E}^{2p(n)+1}_{*,*}(S^1) = \hat{E}^{2p(n+1)}_{*,*}(S^1)$ as postulated.

Consider now the maps

$$F_n \colon \hat{\mathbb{H}}(C_{p^{n+1}}, T(\mathbb{Z}_p)) \to \hat{\mathbb{H}}(C_{p^n}, T(\mathbb{Z}_p))$$
$$V_n \colon \hat{\mathbb{H}}(C_{p^n}, T(\mathbb{Z}_p)) \to \hat{\mathbb{H}}(C_{p^{n+1}}, T(\mathbb{Z}_p))$$

where F_n is the inclusion of fixed sets and V_n is the transfer. On the level of homotopy (and in the spectral sequences) V_n is F_n-multiplicative in the sense that

$$V_n(x F_n(y)) = V_n(x) y \tag{3.9}$$

for $x \in \hat{E}^r_{*,*}(C_{p^n})$, $y \in \hat{E}^r_{*,*}(C_{p^{n+1}})$, cf. [HM2], (2.2). On $\hat{E}^2_{*,*}(C_{p^n})$, $V_n(u_n) = u_{n+1}$. This is just the usual statement that the transfer or induction from $\hat{H}^1(C_{p^n}; \mathbb{F}_p)$ to $\hat{H}^1(C_{p^{n+1}}; \mathbb{F}_p)$ is an isomorphism. Since the elements $t^i e f^j$ and $t^i f^j$ are in the image of F_n, it follows from (3.9) that the elements

$$u_{n+1} t^i e f^j, u_{n+1} t^i f^j \in \hat{E}^2_{*,*}(C_{p^{n+1}})$$

are in the image of V_n, and inductively that they survive to $\hat{E}^r_{*,*}(C_{p^{n+1}})$ if they survive to $\hat{E}^r_{*,*}(C_{p^n})$. We conclude that

$$\hat{E}^{2p(n)+1}_{*,*}(C_{p^{n+1}}) \cong \hat{E}^{2p(n)+1}_{*,*}(S^1) \otimes E\{u_{n+1}\} \cong \hat{E}^{2p(n)+1}_{*,*}(C_{p^n})$$

with the last isomorphism induced by F_n. In total degree $2p^{n+1} - 2$ and fiber degree larger than $2p(n+1)$ there are the elements

$$t^{p^{n+1}l+p(n)+1} f^{p^n l+p(n)+1}, \quad u_{n+1} e t^{p^{n+1}l+p(n)} f^{p^n l+p(n)}.$$

Using $F \colon \hat{E}^r_{*,*}(S^1) \to \hat{E}^r_{*,*}(C_{p^{n+1}})$ and (3.8) we conclude that $\hat{E}^{2p(n)+1}_{*,*}(C_{p^{n+1}}) = \hat{E}^{2p(n+1)}_{*,*}(C_{p^{n+1}})$, and that $d^{2p(n+1)}$ is determined by

$$d^{2p(n+1)}(t^{-p^n}) = \lambda_n t^{p(n)} e f^{p(n)}$$
$$d^{2p(n+1)}(u_{n+1} t^{-p^{n+1}}) = \nu_{n+1} u_{n+1} e t^{p(n)} f^{p(n)} \tag{3.10}$$

with $\lambda_n, \nu_{n+1} \in \mathbb{F}_p$. This proves the first part of Proposition 3.6. To prove the second part we must show that $\nu_{n+1} = 0$. Here is an argument based on Tsalidis' Theorem 3.5.

With the induction hypothesis we can evaluate the spectral sequence

$$E_{*,*}^r(C_{p^n}) \Rightarrow \pi_*(T(\mathbb{Z}_p)^{hC_{p^n}}; \mathbb{F}_p).$$

The result we need (cf. [BM], (4.4)) is that $E_{i,j}^{2\mathfrak{p}(n)+2}(C_{p^n}) = E_{i,j}^{\infty}(C_{p^n})$ for $j - i \geq 0$ is concentrated in the strip $-2\mathfrak{p}(n) \leq p \leq 0$, and that

$$0 \neq (tf)^{\mathfrak{p}(n)} \in E_{*,*}^{2\mathfrak{p}(n)+2}(C_{p^n})$$

but $(tf)^{\mathfrak{p}(n)+1} = 0$. As mentioned above, in the paragraph following lemma 3.2, the unit $S^0 \to T(\mathbb{Z}_p)^{hC_{p^n}}$ maps $v_1 \in \pi_{2p-2}(S^0; \mathbb{F}_p)$ into tf (modulo higher filtration, of course), so $v_1^{\mathfrak{p}(n)} \neq 0$ in $\pi_*(T(\mathbb{Z}_p)^{hC_{p^n}}; \mathbb{F}_p)$. More importantly, $v_1^{\mathfrak{p}(n)+1} = 0$. This is true modulo filtration by the vanishing of $(tf)^{\mathfrak{p}(n)+1}$ in $E_{*,*}^{2\mathfrak{p}(n)+2}(C_{p^n})$. Moreover, as $E_{i,j}^{2\mathfrak{p}(n)+2}(C_{p^n}) = 0$ for $i < -2\mathfrak{p}(n)$ when $j - i = 2(p-1)\mathfrak{p}(n) \geq 0$, the vanishing of $v_1^{\mathfrak{p}(n)+1}$ in $\pi_*(T(\mathbb{Z}_p)^{hC_{p^n}}; \mathbb{F}_p)$ follows.

By Theorem 3.5 we can conclude that $v_1^{\mathfrak{p}(n)+1} = 0$ in $\pi_*(\hat{H}(C_{p^{n+1}}; T(\mathbb{Z}_p)); \mathbb{F}_p)$. Suppose now in (3.10) that $\nu_{n+1} \neq 0$. Then $\hat{E}_{2p^{n+1}-1,0}^{2\mathfrak{p}(n+1)+1} = 0$ and consequently

$$d^{2\mathfrak{p}(n+1)+1} \colon \hat{E}_{2p^{n+1}-1,0}^{2\mathfrak{p}(n+1)+1} \to \hat{E}_{-2\mathfrak{p}(n)-2,2\mathfrak{p}(n+1)}^{2\mathfrak{p}(n+1)+1}$$

vanishes. The range contains $(tf)^{\mathfrak{p}(n)+1}$ which must then survive to $\hat{E}^{\infty}$ contradicting the vanishing of $v_1^{\mathfrak{p}(n)+1}$. We conclude that $\nu_{n+1} = 0$, so that the first possible differential on $u_{n+1} t^{-p^{n+1}}$ is

$$d^{2\mathfrak{p}(n+1)+1}(u_{n+1} t^{-p^{n+1}}) = \mu_{n+1} (tf)^{\mathfrak{p}(n)+1}.$$

Since $t^{p^{n+1}i}$ is a $d^{2\mathfrak{p}(n+1)+1}$-cycle in $\hat{E}_{*,*}^r(S^1)$ it is a $d^{2\mathfrak{p}(n+1)+1}$-cycle in $\hat{E}_{*,*}^r(C_{p^{n+1}})$ and the $d^{2\mathfrak{p}(n+1)+1}$ differential on $u_n = (u_n t^{-p^{n+1}}) t^{p^{n+1}}$ is as claimed. $\square$

Remark 3.11. We must admit that in [BM] we overlooked the possibility that $d^{2\mathfrak{p}(n+1)}(u_{n+1}) \neq 0$ corresponding to $\nu_{n+1} \neq 0$ in the argument above.

Theorem 3.12 (Tsalidis). *The differentials in the spectral sequences*

$$E_{*,*}^r(C_{p^{n+1}}) \Rightarrow \pi_*(T(\mathbb{Z}_p)^{hC_{p^{n+1}}}; \mathbb{F}_p)$$
$$\hat{E}_{*,*}^r(C_{p^{n+1}}) \Rightarrow \pi_*(\hat{\mathbb{H}}(C_{p^{n+1}}, T(\mathbb{Z}_p)); \mathbb{F}_p)$$

are given by (3.1).

Proof. The case $n = 0$ and 3.1 (iii) is covered in lemma 3.2. By Proposition 3.6 it suffices to establish that $\lambda_n \neq 0$ and $\mu_{n+1} \neq 0$. Tsalidis proceeds by induction. Given the statement for $n-1$, the argument of Proposition 3.6 or [BM], (4.4) shows that

$$\dim_{\mathbb{F}_p} \pi_{2p^n-1}(T(\mathbb{Z}_p)^{hC_{p^n}}; \mathbb{F}_p) = n$$

(generated by $ef^{p^{n-1}-1}, t^p ef^{p^{n-1}}, t^{p(2)} ef^{p^{n-1}+p(1)}, \dots, t^{p(n-1)} ef^{p^{n-1}+p(n-2)}$). By Theorem 3.5, $\dim_{\mathbb{F}_p} \pi_{2p^n-1}(\hat{\mathbb{H}}(C_{p^n}, T(\mathbb{Z}_p)); \mathbb{F}_p) = n$. Given Proposition 3.6,

$$\hat{E}_{*,*}^{2p(n+1)}(C_{p^{n+1}}) = \mathbb{F}_p\{u_{n+1}^\epsilon t^{p^n i+j} f^j \mid i \in \mathbb{Z}, \ j \geq 0, \ \epsilon = 0,1\}$$
$$\oplus \mathbb{F}_p\{u_{n+1}^\epsilon t^{p^n i+j} ef^j \mid i \in \mathbb{Z}, \ j \geq 0, \ \epsilon = 0,1\} \oplus$$
$$\bigoplus_{k=1}^{n-1} \mathbb{F}_p\{u_{n+1}^\epsilon t^{p^k i+j} ef^j \mid v_p(i) < n-k, \ p(k-1) \leq j < p(k), \ \epsilon = 0,1\}.$$

In total degree $2p^n - 1$ we have the classes

$$y(k) = t^{p(k)-p^n} ef^{p(k-1)}, \quad k = 1, \dots, n$$

which must all survive to $\hat{E}_{*,*}^\infty(C_{p^{n+1}})$, since they are infinite cycles and have fiber degree less that $2p(n+1) - 1$. The class

$$y(n+1) = t^{p(n+1)-p^n} ef^{p(n)} \in \hat{E}_{*,*}^{2p(n+1)}$$

has precisely fiber degree $2p(n+1) - 1$. It is an infinite cycle (cf. the proof of Proposition 3.6), and

$$d^{2p(n+1)}(t^{-p^n}) = \lambda_n y(n+1).$$

If $\lambda_n = 0$ then $y(n+1)$ will survive to $\hat{E}_{*,*}^\infty(C_{p^{n+1}})$ and would violate that $\dim_{\mathbb{F}_p} \hat{\mathbb{H}}(C_{p^{n+1}}, T(\mathbb{Z}_p)) = n$. Tsalidis' argument for $d^{2p(n+1)+1}(u_{n+1}) \neq 0$ is a similar counting argument, but in fact our proof of Proposition 3.6 already shows this. □

Corollary 3.13. *The spectral sequences* $E_{*,*}^r(S^1)$ *and* $\hat{E}_{*,*}^r(S^1)$ *converging to* $\pi_*(T(\mathbb{Z}_p)^{hS^1}; \mathbb{F}_p)$ *and* $\pi_*(\hat{\mathbb{H}}(S^1, T(\mathbb{Z}_p)); \mathbb{F}_p)$ *has* E^∞-*terms*

$$E_{*,*}^\infty(S^1) = E\{e\} \otimes S\{tf\} \oplus \prod_{k=0}^\infty \mathbb{F}_p\{t^i ef^j \mid p(k) \leq i < p(k+1), \ v_p(i-j) \geq k\}$$

$$\hat{E}_{*,*}^\infty(S^1) = E\{e\} \otimes S\{tf\} \oplus \prod_{k=0}^\infty \mathbb{F}_p\{t^{p^{k+1}i+j} ef^j \mid i \in \mathbb{Z}, \ p(k) \leq j < p(k+1)\}.$$

Proof. The restriction homomorphisms induce isomorphisms

$$\pi_*(T(\mathbb{Z}_p)^{hS^1}; \mathbb{F}_p) \xrightarrow{\cong} \varprojlim \pi_*(T(\mathbb{Z}_p)^{hC_{p^n}}; \mathbb{F}_p)$$

$$\pi_*(\hat{\mathbb{H}}(S^1, T(\mathbb{Z}_p)); \mathbb{F}_p) \xrightarrow{\cong} \varprojlim \pi_*(\hat{\mathbb{H}}(C_{p^n}, T(\mathbb{Z}_p)); \mathbb{F}_p)$$

and using the above theorem we find that the differentials in the spectral sequences are determined from (3.1), (i) and (3.1), (iii). A routine calculation calculates the E^∞-terms.   □

In even degrees we have only the powers of tf, and the unit

$$S^0 \to T(\mathbb{Z}_p)^{hS^1} \xrightarrow{R^h} \hat{\mathbb{H}}(S^1, T(\mathbb{Z}_p))$$

maps the periodicity element $v_1 \in \pi_{2p-2}(S^0; \mathbb{F}_p)$ into tf. In odd degrees $2r - 1$, one can conveniently list the generators as follows. Write

$$(r - p)/(p - 1) = a_0 + a_1 p + a_2 p^2 + \cdots \tag{3.14}$$

with $0 \le a_0 < p$ and $0 < a_i \le p$ for $i > 0$. We call it the p-expansion of $(r - p)/(p - 1)$. Define

$$x_{2r-1}(k) = t^{a_0 + \cdots + a_k p^k} e f^{(r - p + a_0 + \cdots + a_k p^k)/p} \in E^\infty_{*,*}(S^1)$$
$$\hat{x}_{2r-1}(k) = t^{p - r + a_0 p + \cdots + a_k p^{k+1}} e f^{a_0 + a_1 p + \cdots + a_k p^k} \in \hat{E}^\infty_{*,*}(S^1). \tag{3.15}$$

Then in total degree $2r - 1$,

$$E^\infty \pi_{2r-1}(T(\mathbb{Z}_p)^{hS^1}; \mathbb{F}_p) = \prod_{k=1}^\infty \mathbb{F}_p\{x_{2r-1}(k)\}$$

$$E^\infty \pi_{2r-1}(\hat{\mathbb{H}}(S^1, T(\mathbb{Z}_p)); \mathbb{F}_p) = \prod_{k=1}^\infty \mathbb{F}_p\{\hat{x}_{2r-1}(k)\}. \tag{3.16}$$

In (3.14) we let $l = l(r)$ be the minimal number so that $a_k = a_{k+1}$ for $k > l$, i.e.

$$(r - p)/(p - i) = \sum_{k=0}^l a_k p^k + a_\infty p^{l+1}/(1 - p) \in \mathbb{Z}_p \tag{3.17}$$

with $a_l \ne a_\infty$.

Remark 3.18. The ordinary p-adic expansion

$$r/(p - 1) = \sum_{i=0}^\infty b_i p^i, \quad 0 \le b_i < p - 1$$

is related to the p-expansion of (3.14) by $a_0 = b_0$ and $a_i = b_i + 1$ for $i \ge 1$. If we write $r/(p - 1) = x + y/(p - 1)$ with x an integer and $y = a_\infty - 1$ we see by performing the subtraction (in base p) $r/(p-1) - y/(p-1)$ that $r > 0$ precisely when $a_l > a_\infty$.

One can use (3.15) to evaluate

$$E^\infty(R^h): E^\infty \pi_{2r-1}(T(\mathbb{Z}_p)^{hS^1}; \mathbb{F}_p) \to E^\infty \pi_{2r-1}(\hat{\mathbb{H}}(S^1, T(\mathbb{Z}_p)); \mathbb{F}_p).$$

Indeed, one just checks incidence relations $x_{2r-1}(i) = \hat{x}_{2r-1}(j)$. The result is (cf. [BM], sect. 7): The classes $x_{2r-1}(i)$ with $E^\infty(R^h)(x_{2r-1}(i)) \neq 0$ are as follows:

(i) $E^\infty(R^h)(x_{2r-1}(k)) = \hat{x}_{2r-1}(k-1)$ for $k > l$

(ii) $E^\infty(R^h)(x_{2r-1}(l-1)) = \hat{x}_{2r-1}(l-1)$ if $r \equiv 1 \pmod{p-1}, r > 1$

(iii) $E^\infty(R^h)(x_{2r-1}(\kappa)) = \hat{x}_{2r-1}(l-1)$ if $r = p + p^{\kappa+\nu+2} + (p-1)a$

   with $\mathfrak{p}(\kappa) \leq a < \mathfrak{p}(\kappa+1)$, $\nu > 0$ and $\kappa \geq 0$.   (3.19)

In exceptional case (ii), $(r-p)/(p-1) = \sum_{n=0}^{l-1} a_k p^k$, $a_l = p$ and $a_\infty = p-1$.
In the exceptional case (iii), $r \equiv 2 \pmod{p-1}$, and (3.14) takes the form:

$$(r-p)/(p-1) = \sum_{k=0}^{\kappa} a_k p^k + p \cdot p^{\kappa+1} + (p-1) \cdot p^{\kappa+2} + \cdots$$
$$+ (p-1)p^{\kappa+\nu+1} + (p-2)p^{\kappa+\nu+2} + \cdots.$$

So $l = \kappa + \nu + 1$ and $a = \sum_{k=0}^{\kappa} a_k p^k$.
   We close this section by listing the E^∞-terms of the spectral sequences converging to $\pi^*(T(\mathbb{Z}_p)^{hC_{p^n}}; \mathbb{F}_n)$ and $\pi_*(\hat{\mathbb{H}}(C_{p^{n+1}}, T(\mathbb{Z}_p)); \mathbb{F}_p)$ in positive total degrees. The calculations are routine, given Theorem 3.12; in fact most of the calculations were done in the course of proving Proposition 3.6. We only need the result in odd positive degrees. With the notation of (3.15) we have for $r > 0$:

$$E^\infty \pi_{2r-1}(T(\mathbb{Z}_p)^{hC_{p^n}}; \mathbb{F}_p) = \begin{cases} \mathbb{F}_p\{x_{2r-1}(0), \ldots, x_{2r-1}(n-1)\}, & n \geq v_p(r) \\ \mathbb{F}_p\{x_{2r-1}(0), \ldots, x_{2r-1}(n)\}, & n < v_p(r) \end{cases} \quad (3.20)$$

$$E^\infty \pi_{2r-1}(\hat{\mathbb{H}}(C_{p^{n+1}}, T(\mathbb{Z}_p)); \mathbb{F}_p) = \begin{cases} \mathbb{F}_p\{\hat{x}_{2r-1}(0), \ldots, \hat{x}_{2r-1}(n-1)\}, & n \geq v_p(r) \\ \mathbb{F}_p\{\hat{x}_{2r-1}(0), \ldots, \hat{x}_{2r-1}(n)\}, & n < v_p(r). \end{cases}$$

Moreover, the restriction maps

$$F: E^\infty \pi_{2r-1}(T(\mathbb{Z}_p)^{hS^1}; \mathbb{F}_p) \to E^\infty \pi_{2r-1}(T(\mathbb{Z}_p)^{hC_{p^n}}; \mathbb{F}_p) \quad (3.21)$$
$$F: E^\infty \pi_{2r-1}(\hat{\mathbb{H}}(S^1, T(\mathbb{Z}_p)); \mathbb{F}_p) \to E^\infty \pi_{2r-1}(\hat{\mathbb{H}}(C_{p^{n+1}}, T(\mathbb{Z}_p)); \mathbb{F}_p)$$

are the obvious surjections which annihilates the classes $x_{2r-1}(i)$ and $\hat{x}_{2r-1}(i)$ not present in the target. Indeed, F preserve filtrations, and the classes from (3.15) not present on the right hand side in (3.21) lic in filtration degrees where the target is zero.

4 The map $R: \mathrm{TF}(z_p) \to \mathrm{TF}(z_p)$

In this section we shall use the diagram

$$
\begin{array}{ccc}
T(\mathbb{Z}_p)^{hS^1} & \xrightarrow{\ R^h\ } & \hat{\mathbb{H}}(S^1, T(\mathbb{Z}_p)) \\[4pt]
\Big\uparrow{\scriptstyle \Gamma} & & \Big\uparrow{\scriptstyle \hat{\Gamma}} \\[4pt]
\mathrm{TF}(\mathbb{Z}_p) & \xrightarrow{\ R\ } & \mathrm{TF}(\mathbb{Z}_p)
\end{array}
\qquad (4.1)
$$

and the calculations of sect. 3 to study the endomorphism induced by R on mod p homotopy. It follows from Theorem 3.5 that Γ and $\hat{\Gamma}$ define homotopy equivalences, onto the connective covers $T(\mathbb{Z}_p)^{hS^1}[0, \infty)$ and $\hat{\mathbb{H}}(S^1, T(\mathbb{Z}_p))[0, \infty)$, respectively. Our main result is the following theorem where $\mathrm{TF}_*(\mathbb{Z}_p; \mathbb{F}_p) = \pi_*(\mathrm{TF}(\mathbb{Z}_p); \mathbb{F}_p)$. With the notation of (3.17) we have

Theorem 4.2. (i) *There are classes* $\xi_{2r-1}(i) \in \mathrm{TF}_{2r-1}(\mathbb{Z}_p; \mathbb{F}_p)$ *so that* $\mathrm{TF}_{2r-1}(\mathbb{Z}_p; \mathbb{F}_p) = \prod_{i=0}^{\infty} \mathbb{F}_p\{\xi_{2r-1}(i)\}$ *and such that*

$$
\begin{aligned}
&R(\xi_{2r-1}(i)) = \xi_{2r-1}(i-1) \quad \text{for } i > l, \text{ with } l = l(r) \text{ defined in } (3.17) \\
&R(\xi_{2r-1}(l-1)) = \xi_{2r-1}(l-1) \quad \text{if } r > 1 \text{ and } r \equiv 1 \ (\mathrm{mod}\, p-1) \\
&R(\xi_{2r-1}(\kappa)) = \xi_{2r-1}(l-1) \quad \text{if } r = p + p^{\kappa + \nu + 1} + (p-1)a \\
&\quad\quad \text{with } \mathfrak{p}(\kappa) \le a < \mathfrak{p}(\kappa+1) \text{ and } \nu > 0 \\
&R(\xi_{2r-1}(i)) = 0 \quad \text{in all other cases.}
\end{aligned}
$$

(ii) $\mathrm{TF}_{2r}(\mathbb{Z}_p; \mathbb{F}_p) = \mathbb{F}_p\{v_1^{r/p-1}\}$ *if* $r \equiv 0 \ (\mathrm{mod}\, p-1)$ *and is zero otherwise, and* R *is the identity in even degrees.*

The proof of this result will occupy the main part of the section, but first let us derive the structure of the homotopy fiber $hF(R^s)$ in

$$
hF(R^s) \xrightarrow{\ S\ } \mathrm{TF}(\mathbb{Z}_p) \xrightarrow{\ R^s - 1\ } \mathrm{TF}(\mathbb{Z}_p).
$$

This gives the exact sequences of $\mathbb{F}_p$ vector spaces

$$
0 \to \mathrm{TF}_{2r}(\mathbb{Z}_p; \mathbb{F}_p) \xrightarrow{\partial_*} \pi_{2r-1}(hF(R^s); \mathbb{F}_p) \to \ker(R^s - 1)_{2r-1} \to 0
$$

$$
0 \to \mathrm{cok}(R^s - 1)_{2r-1} \xrightarrow{\partial_*} \pi_{2r-2}(hF(R^s); \mathbb{F}_p) \to \mathrm{TF}_{2r-2}(\mathbb{Z}_p; \mathbb{F}_p) \to 0 \qquad (4.3)
$$

upon taking into considerations that $R^s = 1$ on $\mathrm{TF}_{2*}(\mathbb{Z}_p; \mathbb{F}_p)$. Since $R(\xi_{2r-1}(l)) = 0$ we have the exact diagram

$$
\begin{array}{ccccccccc}
0 & \to & \prod_{i=l}^{\infty} \mathbb{F}_p\{\xi_{2r-1}(i)\} & \to & \mathrm{TF}_{2r-1}(\mathbb{Z}_p; \mathbb{F}_p) & \to & \prod_{i=0}^{l-1} \mathbb{F}_p\{\xi_{2r-1}(i)\} & \to & 0 \\[4pt]
& & \Big\downarrow{\scriptstyle R^s-1} & & \Big\downarrow{\scriptstyle R^s-1} & & \Big\downarrow{\scriptstyle R^s-1} & & \\[4pt]
0 & \to & \prod_{i=l}^{\infty} \mathbb{F}_p\{\xi_{2r-1}(i)\} & \to & \mathrm{TF}_{2r-1}(\mathbb{Z}_p; \mathbb{F}_p) & \to & \prod_{i=0}^{l-1} \mathbb{F}_p\{\xi_{2r-1}(i)\} & \to & 0
\end{array}
$$

The right-hand vertical arrow has kernel and cokernel $\mathbb{F}_p\{\xi_{2r-1}(l-1)\}$ when $r \equiv 1 \pmod{p-1}$ and $r > 1$. (If $r = 1$ then $l = 0$ and $\prod_{i=0}^{l-1} = 0$), and is otherwise an isomorphism by Theorem 4.2 (i). The left hand vertical map is surjective, and its kernel is an s-dimensional vector space with generators

$$\zeta_{2r-1}^{(k)} = \lim_{\leftarrow}\{\xi_{2r-1}(i) \mid i \equiv k \pmod{s}, \ i \geq l\} = \sum_{i \equiv k \pmod{s}} \xi_{2r-1}(i), \quad k = 1, \dots, s.$$

$$(4.4)$$

Indeed, $R^s\xi_{2r-1}(i) = \xi_{2r-1}(i-s)$ for $i \geq l+s$ and $R^s\xi_{2r-1}(i) = 0$ if $i < l+s$. We have proved

Corollary 4.5. $\pi_i(hF(R^s); \mathbb{F}_p) = 0$ for $i < -1$ and

(i) $\quad \pi_{2r-2}(hF(R^s); \mathbb{F}_p) = \begin{cases} \mathbb{F}_p \oplus \mathbb{F}_p & \text{if } r \equiv 1 \pmod{p-1}, \ r > 1 \\ \mathbb{F}_p & \text{if } r = 1 \\ 0 & \text{otherwise} \end{cases}$

(ii) $\quad \pi_{2r-1}(hF(R^s); \mathbb{F}_p) = \begin{cases} \mathbb{F}_p^{\oplus(s+1)} & \text{if } r \equiv 1 \pmod{p-1}, \ r > 1 \\ \mathbb{F}_p^{\oplus s} & \text{if } r = 1 \\ \mathbb{F}_p^{\oplus(s+1)} & \text{if } r \equiv 0 \pmod{p-1}, \ r > 0 \\ \mathbb{F}_p^{\oplus s} & \text{if } r = 0 \\ \mathbb{F}_p^{\oplus s} & \text{otherwise}, \ r > 0. \end{cases}$

In short there is an (abstract) isomorphism

$$\pi_*(hF(R^s); \mathbb{F}_p) \cong \pi_{*-1}(J; \mathbb{F}_p) \oplus \pi_*(J; \mathbb{F}_p) \oplus \pi_*(SU \times U^{(s-1)}; \mathbb{F}_p)$$

for $* \geq 0$, where J is the connective image of J space.

We are now ready to begin the proof of Theorem 4.2. Let us first introduce a filtration on the homotopy groups of $T(\mathbb{Z}_p)^{hS^1}$, $\hat{\mathbb{H}}(S^1, T(\mathbb{Z}_p))$ and $\mathrm{TF}(\mathbb{Z}_p)$ by setting

$$W_*^{n+1} = \ker\left(\pi_*(T(\mathbb{Z}_p)^{hS^1}; \mathbb{F}_p) \to \pi_*(T(\mathbb{Z}_p)^{hC_{p^n}}; \mathbb{F}_p)\right)$$
$$\hat{W}_*^{n+1} = \ker\left(\pi_*(\hat{\mathbb{H}}(S^1, T(\mathbb{Z}_p)); \mathbb{F}_p) \to \pi_*(\hat{\mathbb{H}}(C_{p^{n+1}}, T(\mathbb{Z}_p)); \mathbb{F}_p)\right) \quad (4.6)$$
$$V_*^{n+1} = \ker\left(\pi_*(\mathrm{TF}(\mathbb{Z}_p); \mathbb{F}_p) \to \pi_*(T(\mathbb{Z}_p)^{C_{p^n}}; \mathbb{F}_p)\right)$$

where the maps are induced from the restriction maps F which includes S^1 fixed sets into the C_{p^n} or $C_{p^{n+1}}$ fixed set in the first two cases, and the restriction map $\mathrm{holim}\, T(\mathbb{Z}_p)^{C_{p^n}} \to T(\mathbb{Z}_p)^{C_{p^n}}$ in the last case. By convention, W_*^0, $\hat{W}_*^0$ and V_*^0 are the entire homotopy group in each of the three cases.

Since there are p-adic homotopy equivalences (cf. (2.11)):

$$T(\mathbb{Z}_p)^{hS^1} \to \mathrm{holim}\, T(\mathbb{Z}_p)^{hC_{p^n}}$$
$$\hat{\mathbb{H}}(S^1, T(\mathbb{Z}_p)) \to \mathrm{holim}\, \hat{\mathbb{H}}(C_{p^{n+1}}; T(\mathbb{Z}_p))$$

we get from Theorem 3.5 *filtered* isomorphisms

$$\pi_*(\hat{\mathbb{H}}(S^1, T(\mathbb{Z}_p)); \mathbb{F}_p) \xleftarrow{\hat{\Gamma}} \pi_*(\mathrm{TF}(\mathbb{Z}_p); \mathbb{F}_p) \xrightarrow{\Gamma} \pi_*(T(\mathbb{Z}_p)^{hS^1}; \mathbb{F}_p). \qquad (4.7)$$

The isomorphic filtration quotients in odd degrees

$$\hat{W}_{2r-1}^k / \hat{W}_{2r-1}^{k+1} \cong V_{2r-1}^k / V_{2r-1}^{k+1} \cong W_{2r-1}^k / W_{2r-1}^{k+1} \qquad (4.8)$$

are easily calculated from (3.12) and (3.21) to be

$$W_{2r-1}^k / W_{2r-1}^{k+1} = \begin{cases} \mathbb{F}_p & \text{if } k \neq v_p(r) \\ 0 & \text{if } k = v_p(r) \end{cases} \qquad (4.9)$$

and the $\mathbb{F}_p$ is generated by $x_{2r-1}(k-1)$ if $k > v_p(r)$ and by $x_{2r-1}(k)$ if $k < v_p(r)$.

We point out that the filtrations W_{2r-1}^k and $\hat{W}_{2r-1}^k$ (for $k \neq v_p(r)$) simply amounts to a renumbering of the skeleton filtrations associated to the spectral sequences which converges to $\pi_{2r-1}(T(\mathbb{Z}_p)^{hS^1}; \mathbb{F}_p)$ and $\pi_{2r-1}(\hat{\mathbb{H}}(S^1; T(\mathbb{Z}_p)); \mathbb{F}_p)$. The precise renumbering however depends on r or more precisely on the p-expansion of $(r-1)/(p-1)$ in (3.14). Let $E^0\mathrm{TF}_{2r-1}(\mathbb{Z}_p; \mathbb{F}_p)$ denote the associated graded vector space in the V-filtration. Then (4.1), (4.8), (4.9) and (3.19) calculates

$$E^0 R: E^0\mathrm{TF}_{2r-1}(\mathbb{Z}_p; \mathbb{F}_p) \to E^0\mathrm{TF}_{2r-1}(\mathbb{Z}_p; \mathbb{F}_p)$$

to be

$$E^0 R(x_{2r-1}(k)) = \begin{cases} x_{2r-1}(k-1) & \text{if } k > l \\ x_{2r-1}(k) & \text{if } k = l-1 \text{ and } r \equiv 1 \ (\mathrm{mod}\, p-1), r > 1 \\ x_{2r-1}(l-1) & \text{if } k = \kappa \text{ and } r = p + p^{\kappa+\nu+1} + a(p-1) \\ & \qquad \text{with } \mathfrak{p}(\kappa) \leq a < \mathfrak{p}(\kappa+1), \nu > 1 \\ 0 & \text{otherwise.} \end{cases}$$

$$(4.10)$$

Theorem 4.2 simply asserts that we can lift this formula to $\mathrm{TF}_{2r-1}(\mathbb{Z}_p; \mathbb{F}_p)$ for suitable choices of classes $\xi_{2r-1}(k) \in \mathrm{TF}_{2r-1}(\mathbb{Z}_p; \mathbb{F}_p)$ lifting $x_{2r-1}(k)$.

Proof of Theorem 4.2. There are two main points: one needs to lift classes in the kernel of $E^0 R$ to classes in the kernel of R, and when $r \equiv 1 \ (\mathrm{mod}\, p-1)$, $r > 1$ one needs to exhibit $\xi_{2r-1}(l-1)$ which is invariant under R. If we can solve these two points then the rest follows algebraically.

The first problem of lifting elements in the kernel of $E^0 R$ to elements in the kernel of R was treated in [BM], Theorem 2.15. It is based on the pleasant fact that the structure of the norm cofibration in homotopy

$$\Sigma T(\mathbb{Z}_p)_{hS^1} \xrightarrow{N^h} T(\mathbb{Z}_p)^{hS^1} \xrightarrow{R^h} \hat{\mathbb{H}}(S^1, T(\mathbb{Z}_p))$$

is intimately related to the structure of the spectral sequence $\hat{E}^r_{*,*}(S^1)$ which converges to $\hat{\mathbb{H}}(S^1, T(\mathbb{Z}_p))$. Indeed, the usual spectral skeleton spectral sequence for $T(\mathbb{Z}_p)_{hS^1}$ is contained in $\hat{E}^r_{*,*}(S^1)$:

$$E^2_{i,j}(T(\mathbb{Z}_p)_{hS^1}) = \hat{E}^2_{i+2,j}(S^1), \quad i \geq 0.$$

The differentials in $\hat{E}^r_{i,j}(S^1)$ inside the range $i \geq 2$ corresponds to differentials in $E^r_{*,*}(T(\mathbb{Z}_p)_{hS^1})$. The differentials

$$d^r: \hat{E}^r_{i,j}(S^1) \to \hat{E}_{i-r,j+r-1}(S^1)$$

with $i \geq 2$ and $i \leq r$ on the other hand detects the norm map

$$N^h: \pi_*(\Sigma T(\mathbb{Z}_p)_{hS^1}; \mathbb{F}_p) \to \pi_*(\hat{\mathbb{H}}(S^1, T(\mathbb{Z}_p)); \mathbb{F}_p).$$

See sect. 2 of [BM] for more details. In our case the classes $x_{2r-1}(k) \in E^\infty_{*,*}(S^1)$ which are annihilated by $E^\infty R^h$ are the classes which are killed by differentials in $\hat{E}^r_{*,*}(S^1)$. Concretely, let us write

$$(r-p)/(p-1) = a + p^s b/(p-1), \quad a = \sum_{i=0}^{k} a_i p^i$$

with $b = 0$ or $v_p(b) = 0$. From remark 3.18 and a minor computation we see that $k \leq l(r)$ and $r > 0$ implies that $b > 0$. It is easy to see that $b = 0$ if and only if $x_{2r-1}(k)$ belongs to case 3.19 (ii), and that $x_{2r-1}(k)$ belongs to case 3.19 (iii) if and only if $b = 1$ and $s \geq k + 3$. Thus $E^0 R(x_{2r-1}(k)) = 0$ implies that either $b \geq 2$ or $b = 1$ and $s \leq k + 2$. In both these cases $a + p^{s-1} b \geq \mathfrak{p}(s-1)$ and hence $(r - p + a)/p \geq \mathfrak{p}(s-1)$. But this means that there exists a class

$$u = t^{a-\mathfrak{p}(s)} f^{(r-p+a)/p-\mathfrak{p}(s-1)} \in E^2(\hat{\mathbb{H}}(S^1, T(\mathbb{Z}_p)); \mathbb{F}_p)$$

and using that $v_p(b) = 0$ we have

$$d^{2\mathfrak{p}(s)} u = \lambda x_{2r-1}(k), \quad \lambda \in \mathbb{F}_p^\times.$$

In this situation there is a representative for $x_{2r-1}(k)$ which lies in the image of N^h namely N^h applied to a class in $\pi_*(\Sigma T(\mathbb{Z}_p)^{hS^1}; \mathbb{F}_p)$ represented by u. Since $R^h \circ N^h$ is trivial, this ensures a representative $y_{2r-1}(k)$ for $x_{2r-1}(k)$ which is annihilated by R^h, and we can use $\xi_{2r-1}(k) = \Gamma^{-1}(y_{2r-1}(k)) \in \mathrm{TF}_{2r-1}(\mathbb{Z}_p; \mathbb{F}_p)$. This represents $x_{2r-1}(k)$ and lies in the kernel of R.

The second problem to find a representative for $x_{2r-1}(l-1)$ invariant under R was solved in sect. 7, and in the appendix, of [BM]. It is based on the diagram of spectra

$$
\begin{array}{ccccc}
K(\mathbb{Z}_p) & \xrightarrow{\mathrm{Trc}} & TC(\mathbb{Z}_p) & \xrightarrow{S} & T(\mathbb{Z}_p)^{hS^1} \\
\downarrow{\scriptstyle tr} & & \downarrow{\scriptstyle S} & & \downarrow{\scriptstyle id} \\
T(\mathbb{Z}_p) & \xleftarrow{F} & TF(\mathbb{Z}_p) & \xrightarrow{\Gamma} & T(\mathbb{Z}_p)^{hS^1}
\end{array}
$$

and the fact that in dimension $2p - 1$,

$$\mathrm{tr}: K_{2p-1}(\mathbb{Z}_p; \mathbb{F}_p) \to \pi_{2p-1}(T(\mathbb{Z}_p); \mathbb{F}_p)$$

is surjective. Let $\mathrm{tr}(e_K) = e$. Since $K_*(\mathbb{Z}_p; \mathbb{F}_p)$ is a $\pi_*(S^0; \mathbb{F}_p)$ module, we have the element $e_K v_1^{(r-1)/(p-1)} \in K_{2r-1}(\mathbb{Z}_p; \mathbb{F}_p)$ and it maps to

$$e(tf)^{(r-p)/(p-1)} \in E^\infty \pi_{2r-1}(T(\mathbb{Z}_p)^{hS^1}; \mathbb{F}_p).$$

But this is precisely $x_{2r-1}(l-1)$. Thus we set

$$\xi_{2r-1}(l-1) = S \circ \mathrm{Trc}\big(e_K v_1^{(r-p)/(p-1)}\big).$$

Since $(R - \mathrm{id})S$ is null homotopic, $\xi_{2r-1}(l-1)$ is invariant under R.

Finally, since $E^0 R \colon V_{2r-1}^{k+1}/V_{2r-1}^k \to V_{2r-1}^k/V_{2r-1}^{k-1}$ is an isomorphism when $k \geq l$ we can always find representatives $\xi_{2r-1}(k)$ for $k > l$ with the wanted property, and also a representative $\xi_{2r-1}(\kappa)$ with $R\xi_{2r-1}(\kappa) = \xi_{2r-1}(l-1)$ in the exceptional case $r \equiv 2 \pmod{p-1}$. $\qquad\square$

5   The homotopy types

In this section we examine the homotopy type of the fiber $hF(R^s)$ in the cofibration

$$hF(R^s) \longrightarrow \mathrm{TF}(\mathbb{Z}_p) \overset{R^s-1}{\longrightarrow} \mathrm{TF}(\mathbb{Z}_p).$$

Every spectrum is a module over the sphere spectrum, so $v_1 \in \pi_{2p-2}(S^0; \mathbb{F}_p)$ acts on the mod p homotopy groups, and in particular on $\pi_*(hF(R^s); \mathbb{F}_p)$. We shall examine this action and begin by the action on

$$E^0 \pi_*(\mathrm{TF}(\mathbb{Z}_p); \mathbb{F}_p) = \prod_{k=0}^{\infty} V_*^k/V_*^{k+1}$$

where $V_*^0 \supseteq V_*^1 \supseteq V_*^2 \supseteq \cdots$ is the filtration introduced in (4.6). We recall from (4.8) and (4.9) that $V_{2r-1}^k/V_{2r-1}^{k+1}$ is zero for $k = v_p(r)$ and otherwise is a single copy of $\mathbb{F}_p$, generated by $x_{2r-1}(k-1)$ resp. $x_{2r-1}(k)$ when $k > v_p(r)$ resp. $k < v_p(r)$.

Let $(a_0, a_1, \ldots)$ be the coefficients which appears in the p-expansion (3.14) of $(r-p)/(p-1)$.

Lemma 5.1. In $E^0 \pi_{2r-1}(\mathrm{TF}(\mathbb{Z}_p); \mathbb{F}_p)$,

$$v_1 \cdot x_{2r-1}(k) = \begin{cases} x_{2s-1}(k) & \text{if } (a_0, a_1, \ldots, a_k) < (p-1, p, \ldots, p) \\ x_{2s-1}(k+1) & \text{if } (a_0, a_1, \ldots, a_k, a_{k+1}) = (p-1, p, \ldots, p, p) \\ 0 & \text{otherwise} \end{cases}$$

in the left lexicographic ordering; $s = r + p - 1$.

Proof. We apply Γ and calculate instead in $E^0 \pi_*(T(\mathbb{Z}_p)^{hS^1}; \mathbb{F}_p) \cong E^\infty_{*,*}(S^1)$ where by (3.15),

$$x_{2r-1}(k) = t^{a(k)} e f^{1/p(r-p+a(k))}, \quad a(k) = a_0 + a_1 p + \cdots + a_k p^k$$

and where $v_1 \cdot x_{2r-1}(k) = t f x_{2r-1}(k)$. This is non-zero if $t^{a(k)+1} e f^{1/p(r-1+a(k)+1)} \neq 0$, i.e. if $a(k)+1$ appears as a partial sum in the p-expansion of $(s-1)/p-1$. This easily gives the claimed result. □

Example 5.2. Consider $r = p + p^{n+2}$. The p-expansion of $(r-p)/(p-1)$ has coefficients $(0, p, (p-1), \ldots, (p-1), (p-2), (p-2), \ldots)$ with n entries $(p-1)$, and $x_{2r-1}(0) = e f p^{n+1}$. It follows that

$$v_1^{p(n+2)-1} x_{2r-1}(0) \neq 0 \text{ and } v_1^{p(n+2)} x_{2r-1}(0) = 0$$

(in $E^0 \pi_*(\mathrm{TF}(\mathbb{Z}_p); \mathbb{F}_p)$). From (the graded version of) Theorem 4.2,

$$R(x_{2r-1}(0)) = x_{2r-1}(l-1) = v_1^{p^{n+1}} e f p^{n}$$

and hence

$$R(v_1^a x_{2r-1}(0)) = v_1^{p^{n+1}+a} e f p^{n} \qquad (*)$$

which is non-zero for $a < p(n)$. Now, in the notation of (4.2),

$$v_1^a x_{2r-1}(0) = x_{2t-1}(\kappa), \quad t = p + p^{n+2} + a(p-1),$$

and $(*)$ is precisely the (exceptional) formula $R x_{2t-1}(\kappa) = x_{2t-1}(l-1)$ in (3.19) (iii).

Proposition 5.3. *For every $s \geq 1$ and $i \geq 0$,*

$$v_1 \colon \pi_i(hF(R^s); \mathbb{F}_p) \to \pi_{i+2p-2}(hF(R^s); \mathbb{F}_p)$$

is injective. It is an isomorphism for $i \neq 1$.

Proof. This follows from the exact sequences (4.3) and lemma 5.1. Indeed, the action of v_1 in $\mathrm{TF}_{2r}(\mathbb{Z}_p; \mathbb{F}_p)$ is an isomorphism because $\mathrm{TF}_{2*}(\mathbb{Z}_p; \mathbb{F}_p) = S\{v_1\}$, so we must check how v_1 acts on $\ker(R^s - 1)_{2r-1}$ and on $\mathrm{cok}(R^s - 1)_{2r-1}$. We have already calculated these two modules in the course of proving corollary 4.5, e.g.

$$\ker(R^s - 1)_{2r-1} = \begin{cases} \mathbb{F}_p^{\oplus(s+1)} & \text{if } r \equiv 1 \ (\mathrm{mod}\, p-1), \ r > 1 \\ \mathbb{F}_p^{\oplus s} & \text{otherwise.} \end{cases}$$

The generators are $\xi_{2r-1}(l-1) = e_K \cdot v_1^{(r-p)/(p-1)}$ with $e_K \in K_{2p-1}(\mathbb{Z}_p; \mathbb{F}_p)$ and

$$\zeta_{2r-1}^{(k)} = \varprojlim \{\xi_{2r-1}(i) \mid i \equiv k \ (\mathrm{mod}\, s), \ i \geq l\}.$$

We must check that $v_1 \zeta_{2r-1}^{(k)} = \zeta_{2r+2p-3}^{(k)}$. This is a consequence of lemma 5.1 as follows: In the p-expansion

$$(r-p)/(p-1) = \sum_{k=0}^{l} a_k p^k + a_\infty p^{l+1} + a_\infty p^{l+2} + \cdots$$

the constant terms $a_\infty \leq p - 1$ by remark 3.18. Thus

$$(a_0, \ldots, a_i) < (p - 1, p, \ldots, p, p) \text{ when } i \geq l + 1$$

and hence that $v_1 \cdot x_{2r-1}(i) = x_{2r+2p-3}(i)$. Since $\xi_{2r-1}(l + 1)$ corresponds to $x_{2r-1}(l + 1)$ modulo higher filtration, it follows that

$$v_1 \xi_{2r-1}(i) = \xi_{2r+2p-3}(i) \in E^0 \pi_*(\mathrm{TF}(\mathbb{Z}_p); \mathbb{F})$$

for large i. Thus $v_1 \zeta_{2r-1}^{(k)} = \zeta_{2r+2p-3}^{(k)}$ in $E^0 \pi_*(\mathrm{TF}(\mathbb{Z}_p); \mathbb{F}_p)$ and v_1 maps the s-dimensional vector space spanned by the $\zeta_{2r-1}^{(k)}$ isomorphically to the s-dimensional vector space spanned by the $\zeta_{2r+2p-3}^{(k)}$. Hence

$$v_1 : \ker(R^s - 1)_{2r-1} \to \ker(R^s - 1)_{2r+2p-3}$$

is an isomorphism.

For the cokernel,

$$\mathrm{cok}(R^s - 1)_{2r-1} = \begin{cases} \mathbb{F}_p & \text{if } r \equiv 1 \ (\mathrm{mod}\, p - 1), \ r > 1 \\ 0 & \text{otherwise} \end{cases}$$

the generator of $\mathbb{F}_p$ is $e_K v_1^{r-p/p-1}$ and these generators are visibly stable under multiplication by v_1. $\qquad\square$

Before we continue it is in order to remind the reader about L_1-localization (= localization at p-local topological K-theory, p a fixed odd prime), cf. [Bou], [Rav]. The L_1-localization of the sphere spectrum fits into a cofibration

$$L_1(S^0) \to J \to \Sigma^{-1} H\mathbb{Q},$$

where J is the p-local periodic J spectrum as in sect. 1. Moreover for any spectrum E,

$$L_1(E) \simeq E \wedge L_1(S^0).$$

In particular, if E has finite type then

$$L_1(E)_p^\wedge \simeq E \wedge J_p^\wedge.$$

Let E/p be the cofiber of $p \colon E \to E$, so that S^0/p is the Moore spectrum. The element $v_1 \in \pi_{2p-2}(S^0/p) = \pi_{2p-2}(S^0; \mathbb{F}_p)$ factors to define a map $v_1 \colon \Sigma^{2p-2}(S^0/p) \to S^0/p$, [A1] and we let $S^p/p[1/v_1]$ be the homotopy colimit of the desuspended iterates of v_1. It is a fundamental fact of homotopy theory that

$$S^0/p[1/v_1] \simeq L_1(S^0)/p,$$

and hence more generally that $E/p[1/v_1] = L_1(E)/p$, [Mi]. Thus multiplication by v_1 defines an isomorphism on $\pi_*(L_1(E); \mathbb{F}_p)$, and moreover:

Lemma 5.4. *Let E be any spectrum (of finite type) such that $v_1: \pi_i(E; \mathbb{F}_p) \to \pi_{i+2p-2}(E; \mathbb{F}_p)$ is an isomorphism for $i \geq N$. Then the connected covers $E[N,\infty)$ and $L_1(E)[N, \infty)$ agree after completion at p.* □

In particular, this applies to $hF(R^s)$ to give

$$hF(R^s)[2, \infty) \simeq_p L_1(hF(R^s))[2, \infty). \tag{5.5}$$

Let $bu = K[2, \infty)$ be the spectrum with 0'th space BU, and spectrum structure induced from Bott periodicity. Theorem 2.8 and the following proves the theorem of the introduction.

Theorem 5.6. *There is a p-adic homotopy equivalence*

$$hF(R^s)[0, \infty) \simeq_p J[0, \infty) \vee \Sigma J[0, \infty) \vee \Sigma bu \vee \bigvee^{s-1} \Sigma^{-1} bu,$$

in agreement with the abstract isomorphism of homotopy groups from corollary 4.5.

Proof. Let us first comment on the case $s = 1$ which was treated in [BM], Theorem 9.17, where

$$hF(R) \simeq TC(\mathbb{Z}_p) \simeq K(\mathbb{Z}_p).$$

Denote by

$$f: S^0 \to TC(\mathbb{Z}_p), \quad g: \Sigma(S^0) \to TC(\mathbb{Z}_p)$$

the unit map and the map which represents $1 - p \in \pi_1 K(\mathbb{Z}_p) = \mathbb{Z}_p^\times$. If $S: hF(R) \to TF(\mathbb{Z}_p)$ is the inclusion of the fiber, then keeping the notation of (4.4)

$$(Sf)_*(v_1) = v_1 \in TF_{2p-2}(\mathbb{Z}_p; \mathbb{F}_p)$$
$$(Sg)_*(\Sigma v_1) = v_1 \zeta_1^{(1)} \in TF_{2p-1}(\mathbb{Z}_p; \mathbb{F}_p).$$

Moreover, $v_1 \zeta_1^{(1)} = \zeta_{2p-1}^{(1)}$ since this is true modulo filtration and since the elements lie in the R-invariant part of $TF_{2p-1}(\mathbb{Z}_p; \mathbb{F}_p)$, generated by e and $\zeta_{2p-1}^{(1)}$.

We claim that the mod p Bockstein operator is non-trivial on $f_*(v_1)$ and $g_*(v_1)$. This is proved in [BM], lemma 9.8. We shall not repeat the proof here, but just mention that it uses comparison with topological cyclic homology in the Waldhausen setting where TC splits off the spectrum $S^0 \vee \Sigma(S^0)$.

In the notation of the previous section, and with $\alpha_1 = \beta_1 v_1$

$$(Sf)_*(\alpha_1) = \zeta_{2p-1}^{(1)}, \quad g_*(\Sigma \alpha_1) = \partial_*(e)$$

cf. (4.3), (4.4). It is a consequence of (5.5) that $f \vee g$ factors as

$$
\begin{array}{ccc}
S^0 \vee \Sigma S^0 & \xrightarrow{\ f \vee g\ } & hF(R) \\
\downarrow & & \downarrow{\scriptstyle id} \\
J[0, \infty) \vee \Sigma J[0, \infty) & \xrightarrow{\ \bar{f} \vee \bar{g}\ } & hF(R)
\end{array}
$$

Now $\pi_*(J[0,\infty);\mathbb{F}_p) = E\{\alpha_1\} \otimes S\{v_1\}$ and since $v_1^k \cdot \zeta_{2p-1}^{(0)} = \zeta_{2p+2(p-1)k-1} \neq 0$ and $v_1^k e \neq 0$, $\pi_*(\bar{f};\mathbb{F}_p)$ and $\pi_*(\bar{g};\mathbb{F}_p)$ are both injective. (The higher Bockstein structure: $\beta_n(v_1^{p^n}) = v_1^{p^n-1}\alpha_1$ shows that the same is the case on $\mathbb{Z}_p$-integral homotopy groups).

We can use the diagram

$$
\begin{array}{ccccc}
hF(R) & \longrightarrow & \mathrm{TF}(\mathbb{Z}_p) & \xrightarrow{R-1} & \mathrm{TF}(\mathbb{Z}_p) \\
\downarrow{\scriptstyle i} & & \downarrow{\scriptstyle id} & & \downarrow{\scriptstyle (R^s-1)/(R-1)} \\
hF(R^s) & \longrightarrow & \mathrm{TF}(\mathbb{Z}_p) & \xrightarrow{R^s-1} & \mathrm{TF}(\mathbb{Z}_p)
\end{array}
$$

to conclude that $i \circ (\bar{f} \vee \bar{g})$ is also injective on homotopy groups; indeed $i_*(\zeta_{2r-1}^{(1)})$ $= \sum_{k=1}^{s} \zeta_{2r-1}^{(k)}$ by the definition of the classes.

We now look in the cofiber

$$J[0,\infty) \vee \Sigma J[0,\infty) \to hF(R^s) \to \overline{hF}(R^s).$$

The mod p homotopy groups of $\overline{hF}(R^s)$ are concentrated in odd degrees, and by corollary 4.5 abstractly given by

$$\pi_*(\overline{hF}(R^s);\mathbb{F}_p) \cong \pi_*(\Sigma bu \vee \bigvee^{s-1} \Sigma^{-1}bu;\mathbb{F}_p).$$

Moreover,

$$v_1: \pi_{2r-1}(\overline{hF}(R^s);\mathbb{F}_p) \to \pi_{2r+2p-3}(\overline{hF}(R^s);\mathbb{F}_p)$$

is an isomorphism for $r \neq 1$, and injective for $r = 1$. It follows from [Rog1] that

$$\overline{hF}(R^s) \simeq_p \Sigma bu \vee \bigvee^{s-1} \Sigma^{-1}bu$$

giving the cofibration

$$J[0,\infty) \vee \Sigma J[0,\infty) \to hF(R^s) \to \Sigma bu \vee \bigvee^{s-1} \Sigma^{-1}bu. \qquad (*)$$

By [MST], [Rog1] or [A2], this cofibration is necessarily split:

$$[bu, J[0,\infty)_p^\wedge] = [bu, \Sigma^2 J[0,\infty)_p^\wedge] = 0,$$

and the homomorphisms

$$[bu, \Sigma J[0,\infty)_p^\wedge] \to \mathrm{Hom}(\pi_*(bu_p^\wedge), \pi^*(\Sigma J[0,\infty)_p^\wedge))$$

$$[bu, \Sigma^3 J[0,\infty)_p^\wedge] \to \mathrm{Hom}(\pi_*(bu_p^\wedge), \pi_*(\Sigma^3 J[0,\infty)_p^\wedge))$$

are injective. Since the exact homotopy sequence of $(*)$ is short exact, $(*)$ must split. $\qquad\square$

6 Concluding remarks

One would like of course to extend the results of this paper to cover the discrete valuation rings which are ramified. Let us discuss the most obvious case, namely the ring $\mathbb{Z}_p[\zeta_p]$. There are at least two ways to attack the calculation of $TC(\mathbb{Z}_p[\zeta_p])$.

One could follow the general scheme explained at the end of sect. 2, and begin by calculating $T(\mathbb{Z}_p[\zeta_p])$, then its homotopy fixed sets and Tate theory for C_{p^n} and S^1, and verify (2.14). This would give $\pi_*(TF(\mathbb{Z}_p[\zeta_p]); \mathbb{F}_p)$ and a calculation of $\pi_* R$ would then lead to $TC_*(\mathbb{Z}_p[\zeta_p])$.

An alternative approach, which we find promising, is to compare $\mathbb{Z}_p[\zeta_p]$ with $\mathbb{Z}_p[C_p]$ and use the general decomposition of $T(\mathbb{Z}_p[G])$. This leads to the following cofibration, cf. [M]:

$$TC(\mathbb{Z}_p) \wedge BC_{p+} \to TC(\mathbb{Z}_p[C_p]) \to \bigvee^{p-1} \Sigma(T(\mathbb{Z}_p) \wedge_{S_1} \rho^*(BC_p)_+) \qquad (6.1)$$

where $\rho^*(BC_p)$ denotes BC_p with a free S^1-action (i.e. $\rho^*(ES^1/C_p)$ with $\rho: S^1 \to S^1/C_p$ the p'th root isomorphism). One knows the homotopy groups of the left-hand term by our main theorem and standard facts from algebraic topology, and it is not so hard to work out entirely the homotopy groups of the right-hand wedge terms.

The real difficulty is in the comparison of the group ring with its maximal order, i.e. the calculation of the relative term in

$$TC(\mathbb{Z}_p[C_p] \to \mathcal{M}_p) \to TC(\mathbb{Z}_p[C_p]) \to TC(\mathcal{M}_p)$$

where $\mathcal{M}_p = \mathbb{Z}_p \times \mathbb{Z}_p[\zeta_p]$.

Let us compare with the localization theorem in algebraic K-theory which imply the cofibrations

$$K(\mathbb{Z}_p[C_p] \to \mathbb{Q}_p[C_p]) \to K(\mathbb{Z}_p[C_p]) \to K(\mathbb{Q}_p[C_p])$$
$$K(\mathbb{F}_p) \times K(\mathbb{F}_p) \to K(\mathcal{M}_p) \to K(\mathbb{Q}_p[C_p])$$

since the relative K-theory of the integers in a local field is equivalent to the K-theory of the residue field. Moreover, as $K(\mathbb{F}_p)_p^\wedge \simeq H\mathbb{Z}_p$ the equality of TC and K for orders give

$$TC(\mathbb{Z}_p[C_p] \to \mathcal{M}_p)_p^\wedge \to K(\mathbb{Z}_p[C_p] \to \mathbb{Q}_p[C_p])_p^\wedge \to H\mathbb{Z}_p \vee H\mathbb{Z}_p. \qquad (6.2)$$

It is a standard conjecture in algebraic K-theory that $K(A)_p^\wedge$ is essentially equal to the completion of its L_1-localization for integers in (local) fields, i.e. that

$$K(A)_p^\wedge \to L_1 K(A)_p^\wedge \qquad (6.3)$$

is injective on homotopy, and induces a homotopy equivalence of 1-connected covers: $K(A)_p^\wedge[2, \infty) \simeq L_1 K(A)_p^\wedge[2, \infty)$. This statement is sometimes called the (local) Lichtenbaum-Quillen conjecture.

Let us compare with (6.1). The right-hand term in (6.1) has trivial L_1-localization after completion since this is the case for $T(\mathbb{Z}_p)$, and since localization commutes with smash products. The localization of the left-hand term in (6.1) can be determined from our main result to be

$$(J \vee \Sigma J \vee \bigvee^p \Sigma K) \wedge BC_{p_+}.$$

Using these results one may work out that

$$\mathrm{TC}(\mathbb{Z}_p[C_p])^\wedge_p \to L_1\mathrm{TC}(\mathbb{Z}_p[C_p])^\wedge_p$$

is not injective on homotopy groups, not even in high dimensions. Thus $K(\mathbb{Z}_p[C_p])^\wedge_p$ is not "L_1-local". This however in itself does not contradict (6.3). Indeed one expects $K(\mathbb{Z}_p[C_p] \to \mathcal{M}_p)^\wedge_p$ to be related to $K(\mathbb{F}_p[C_p])^\wedge_p$ which has trivial L_1-localization; it seems a good working hypothesis that

$$L_1(K(\mathbb{Z}_p[C_p]) \to \mathcal{M}_p)^\wedge_p = 0$$

and (6.3) would then be equivalent to the statement that

$$\mathrm{TC}(\mathbb{Z}_p[C_p] \to \mathcal{M}_p)^\wedge_p[2, \infty) \to \mathrm{TC}(\mathbb{Z}_p[C_p])^\wedge_p[2, \infty) \to L_1\mathrm{TC}(\mathbb{Z}_2[C_p])^\wedge_p[2, \infty)$$
$$(6.4)$$

be a cofibration. We plan to return to these questions in a future paper.

For any local number field F, that is finite extension F of $\mathbb{Q}_p$, Dwyer and Mitchell has in [DH], Theorem 13.3 evaluated $L_1K(F)^\wedge_p$. With our notions the result is

$$L_1K(F)^\wedge_p \simeq hF(\Psi^q)^\wedge_p \vee \Sigma hF(\Psi^q)^\wedge_p \vee \bigvee^r \Sigma K^\wedge_p, \quad r = |F : \mathbb{Q}_p| \qquad (6.5)$$

where q is a topological generator of a certain subgroup $\Gamma_F \subseteq \mathbb{Z}_p^\times$. To describe Γ_F, let $F_0 = F[\zeta_p]$ have degree d_F over F and let a_F be the maximal a such that F_0 contains p^a'th roots of one. Then $\Gamma_F \subseteq \mathbb{Z}_p^\times$ is the subgroup generated by $1 + p^{a_F}\mathbb{Z}_p \subseteq \mathbb{Z}_p^\times$ and the d_F'th roots of one. If F is unramified over $\mathbb{Q}_p$ then $d_F = p - 1$ and $a_F = 1$, so (6.5) is in agreement with our theorem since $L_1K(F)^\wedge_p \simeq L_1K(A)^\wedge_p$, where A is the integers in F, and since $L_1K[n, \infty)^\wedge_p \simeq K^\wedge_p$ for all n. Given (6.1) it seems somewhat surprising that the result only depends on the degree and Γ_F, in particular when one, as we do, believes in (6.3). But it is not impossible as far as we can judge at present.

In the proof of (6.5), Dwyer and Mitchell uses that

$$L_1K(F)^\wedge_p[2, \infty) \simeq K^{et}(F)^\wedge_p[2, \infty) \qquad (6.6)$$

where the right-hand side is the étale K-theory of F. By definition, étale K-theory has the homotopy limit property:

$$K^{et}(F)^G \simeq K^{et}(F)^{hG} \qquad (6.7)$$

for any subgroup G of the Galois group of $F/\mathbb{Q}_p$. The same equation then holds on connective covers. Returning to the unramified case one then gets

$$K(A_s)^G[2,\infty]_p^\wedge \simeq K(A_s)^{hG}[2,\infty)_p^\wedge$$

and one wonders what the Galois action might be on the right-hand side of

$$K(A_s) \simeq_p J[0,\infty) \times \Sigma J[0,\infty) \times \Sigma(K[2,\infty)) \times \prod^{r-1} \Sigma^{-1}(K(2,\infty)).$$

The first guess that it be trivial on the J-factors cannot be true since

$$J^{hG} = \mathrm{Map}(BG_+, J)$$

is more than a single copy of J when $(p,|G|) \neq 0$.

Let us finally comment on $p = 2$ which was excluded in our theorem. Rognes has in [Rog2] calculated $\hat{\mathbb{H}}(C_2, T(\mathbb{Z}_2))$ and has verified (2.14), hence conjecture (2.13). It appears that one can use the argument of sect. 3 to evaluate the spectral sequences $\hat{E}^r_{*,*}(C_{2^n})$ and $\hat{E}^r_{*,*}(S^1)$ at 2, and that one obtains an answer very similar to the answer for odd primes. In particular multiplication by the 8-dimensional Adams periodicity operator at 2 is injective on mod 2 homotopy group. In the basic case of $K(\mathbb{Z}_2) = \mathrm{TC}(\mathbb{Z}_2)$ it seems entirely likely that there are cofibrations

$$\Sigma J[0,\infty) \to K(\mathbb{Z}_2)_2^\wedge \to X$$
$$SU_2^\wedge \to X \to J[0,\infty) \qquad (6.8)$$

where as above J is the *complex* image of J space, and *not* its real analogue. In contrast to our results at odd primes, we do not think that both cofibrations in (6.8) are split.

References

[A1]   J. F. Adams, *J(X) IV*, Topology vol **5** (1966), 21–71.

[A2]   J. F. Adams, *Infinite loop spaces,* Ann. Math. Studies, no **90**, Princeton University Press (1978).

[AHJM] J. F. Adams, J.-P. Haeberly, S. Jackowski, J. P. May, *A generalisation of the Atiyah-Segal completion theorem,* Topology **27** (1988), 1–6 and 7–21.

[B]   M. Bökstedt, *Topological Hochschild homology of $\mathbb{Z}$ and $\mathbb{Z}/p$*, (preprint), Bielefeld.

[BM]   M. Bökstedt, I. Madsen, *Topological cyclic homology of the integers,* Preprint Aarhus University 1993, no **6** and Asterisque (to appear).

[BHM]   M. Bökstedt, W. C. Hsiang, I. Madsen, *The cyclotomic trace and al-gebraic K-theory of spaces*, Invent. Math. (1993), 465–540.

[Bou]   A. K. Bousfield, *The localization of spectra with respect to homology*, Topology **18** (1979), 257–281.

[C]   G. Carlsson, *Segal's Burside ring conjecture and the homotopy limit problem in Homotopy Theory*, edited by E. Rees and J. D. S. Jones, Cambridge University Press, 1987.

[CMP]   J. Caruso, J. P. May, S. B. Priddy, *The Segal Conjecture for elementory abelian p-groups*, Topology **26** (1987), 413–433.

[DM]   W. Dwyer, S. Mitchell, *On the K-theory spectrum of a ring of algebraic integers*, (preprint)

[GM]   J. P. C. Greenlecs, J. P. May, *Generalized Tate cohomology*, Memoirs AMS (to appear).

[HaM]   I. Hambleton, I. Madsen, *Actions of finite group on $\mathbb{R}^{n+k}$ with fixed set $\mathbb{R}^k$*, Canadian J. Math. vol XXXVIII, no **4** (1986), 781–860

[HM1]   L. Hesselholt, I. Madsen, *The S^1-Tate spectrum for J*, Bol. Soc. Math. Mexicana, memorial issue for J. Adem (to appear).

[HM2]   L. Hesselholt, I. Madsen, *Topological cyclic homology of perfect fields and their dual numbers*, Preprint, Institut Mittag Leffler 1993/94 no **21**.

[M]   I. Madsen, *The cokernel of the assembly map in algebraic K-theory*, (in preparation).

[Mi]   H. Miller, *On relations between Adams spectral sequences, with an ap-plication to the stable homotopy of Moore spaces*, J. Pure and Appl. Alg. **20** (1981), 287–312

[MST]   I. Madsen, V. Snaith, J. Tornehave, *Infinite loop maps in geometric topology*, Proc. Camb. Phil. Soc. **81** (1977), 399–429.

[P]   I. A. Panin, *On a theorem of Hurewicz and K-theory of complete dis-crete valuation rings*, Math USSR Izvestiya **29** (1987), 119–131.

[Q1]   D. Quillen, *On the cohomology and K-theory of general linear groups over finite fields*, Ann. of Math vol **96** (1972), 552–586.

[Q2]   D. Quillen, *Higher algebraic K-theory I*, In: H. Bass (ed.) Algebraic K-theory I, LNM vol **341**, pp. 85–147, Springer 1973.

[Rav]   D. Ravenel, *Localization with respect to certain periodic homology theories,* Amer. J. Math. vol **106** (1984), 351–414.

[Rog1] J. Rognes, *Two lemmas on BU,* Proc. Camb. Math. Soc. (to appear).

[Rog2] J. Rognes, *The homotopy limit problem for $T(\mathbb{Z})$ at 2,* preprint, Oslo University.

[Ser]   J.-P. Serre, *Local fields,* GTM no **67** (1979), Springer.

[Sou]   C. Soulé, *K-théorie des anneaux d'entiers de corps de nombres et cohomologie étale,* Invent. Math. vol **55** (1979), 239–262.

[Sus]   A. A. Suslin, *Algebraic K-theory of fields,* Proc. ICM, Berkely 1986, 222–244.

[T]   S. Tsalidis, *The equivariant structure of topological Hochschild homology and the topological cyclic homology of the integers,* thesis, Brown University 1994.

[W]   J. Wagoner, *Algebraic K-theory,* Evanston, LNM **551**, Springer (1976), 241–248.

The Mapping Cone and Cylinder of a Stratified Map

Sylvain E. Cappell
Julius L. Shaneson

In previous papers we studied the behavior of invariants and characteristic classes of stratified spaces in general and algebraic varieties in particular under maps

$$f : X \longrightarrow Z .$$

The main applications up to now have involved trying to use the existence and properties of the map to understand something about the spaces, for example to compute the Todd classes of toric varieties as in [CS3] or the L-classes of hypersurfaces [CS1]. In this paper we will show that the mapping cone and cylinder of a stratified map are manifold homotopy stratified spaces as defined by Quinn. Hopefully this result will be useful in applying results from the theory of stratified spaces to the study of local and global properties of maps.

Let X and Z be Whitney stratified subsets of smooth manifolds M and N, respectively. Then a stratified map

$$f : X \longrightarrow Z$$

is by definition a proper map that is the restriction of a smooth map from M to N and that has the property that for each stratum W of Z, the inverse image $f^{-1}(W)$ is a union of strata of X, each of which is mapped submersively onto W (see [GM]).

The mapping cone and cylinder constructions of algebraic topology provide standard methods for converting problems about a map f to problems about a space. By definition, the cone $C(f)$ is obtained from the disjoint union of the cone

$$cX = (X \times [0,1])/X \times \{1\}$$

and Z by identifying $x \in X = X \times \{0\} \subset cX$ with $f(x) \in Z$. Alternatively, $C(f)$ is obtained from the mapping cylinder

$$M(f) = \frac{X \times [0,1] \cup Z}{\{(x,0) \sim f(x)\}}$$

by collapsing $X \times \{1\}$ to a point. For $\dim X = \dim Z$, let

$$\hat{C}(f) = C(f) \cup_Z c(Z)$$

be obtained by attaching the cone on Z to $Z \subset C(f)$. Let $\mathfrak{C}(f) = C(f)$ for $\dim X > \dim Z$ and $\hat{C}(f)$ for $\dim X = \dim Z$.

The natural choice for the set of strata of a stratification of $\mathfrak{C}(f)$ seems clear: the cone point (or points), $V \times (0,1) \subset cX$ for V a stratum of X, $W \subset Z \subset \mathfrak{C}(f)$, for W a stratum of Z, and, in the equidimensional case, $W \times (0,1) \subset cZ$. The reader can modify this appropriately for $M(f)$ as a stratified space with boundary, and these collections will be called the *natural strata* for $\mathfrak{C}(f)$ and $M(f)$.

Under the extra assumption that f satisfies Thom regularity, it is not difficult to show that $\mathfrak{C}(f)$ is indeed a stratified space with the natural strata, and that $M(f)$ a stratified space with boundary. However, the example in the second section labeled 4° in [T1] can be used to show that, with the natural strata, $\mathfrak{C}(f)$ and $M(f)$ will not in general have the structure of a stratified space in which the strata have locally trivial bundle neighborhoods (e.g. a Whitney stratified space). In this note it will be shown that $\mathfrak{C}(f)$ and $M(f)$ are manifold homotopy stratified spaces as defined by Quinn [Q1,2], with the natural strata. Thus Quinn's homotopy stratified spaces arise naturally even if one is interested only in stratified objects or even just complex algebraic varieties and their morphisms.

A bit more will actually be shown. A manifold homotopy link-stratified space dimension n is defined by induction on dimension as a space, filtered by closed subspaces,

$$\phi = Z_{-1} \subset Z_0 \subset Z_1 \subset \cdots \subset Z_{n-2} \subset Y_{n-1} \subset Y ,$$

such that the strata $Z_i - Z_{i-1}$ satisfy the following properties:

1. $Z - Z_{n-1}$ is an n-dimensional manifold and is dense in X.

2. $Z_i - Z_{i-1}$ is an i-dimensional manifold (or $\emptyset$).

3. For $x \in Z_i - Z_{i-1}$, there exists an open neighborhood U of x in $Z_i - Z_{i-1}$, a compact manifold link-stratified space L_x of dimension $n - i - 1$, and a map
$$\varphi_x : U \times cL_x \longrightarrow Y ,$$
with $\varphi_x(z, c) = z$ for $z \in U$ that is a stratum preserving homotopy equivalence near U, relative U.

The final condition is a slightly strengthened version of the conclusion of Theorem 2 of [Q1]. Thus, as in [Q2], there is an open neighborhood V of U in Y and a stratum preserving map

$$\psi : V \longrightarrow U \times cL_x$$

that is the identity on U, such that the compositions $\varphi_x\psi$ and $\psi\varphi_x$, where defined, are stratum preserving homotopic relative U to the inclusions.

The present notion is apparently more restrictive than that of [Q1,2] in that the same structure is imposed on the links. This may be useful in applying the methods of the theory of stratified spaces and calculations inductively.

Theorem. *Let f be a stratified map of Whitney stratified subsets of smooth manifolds. Then $\mathfrak{C}(f)$ is a manifold homotopy link-stratified space, and $M(f)$ a manifold homotopy stratified space with boundary, both with the natural strata.*

As mentioned above, if f satisfies Thom regularity, then $\mathfrak{C}(f)$ is actually a stratified space. Hence, if

$$f : M \longrightarrow N$$

is a topologically structurally stable smooth map, then $\mathfrak{C}(f)$ will be homeomorphic (but not with the natural strata) to a stratified space. It would be interesting to study the precise relation between the topological type of f and the obstructions to imposing a stratified structure on $\mathfrak{C}(f)$. More generally, given the powerful methods available for the study and classification of stratified spaces [W], it seems natural to ask how much information about the local or global topological type of f is captured by the local or global homeomorphism type of the mapping cone or cylinder. It is not even entirely clear if the cone or cylinder can always be stratified, with non-natural strata.

The proof of the theorem will also provide a description of the links. For example, for $\mathfrak{C}(f)$ we have:

Corollary (Of the proof). *Assume above the hypotheses.*
Let $\{T_W, \pi_W, \rho_W\}$ be a system of "control data" for the strata of Z with $[0, \epsilon] =$ image ρ_W (see [Gi] [GM] [V] [Ma]). For W a singular stratum of $Z, w \in W$, and $\delta \in (0, \epsilon)$, let

$$S_{w,\delta} = Z \cap (\pi_W, \rho_W)^{-1}(w, \delta)$$

and let

$$D_{w,\delta} = Z \cap (\pi_W, \rho_W)^{-1}(\{w\} \times [0, \delta]) \cong cS_{w,\delta} .$$

(Thus $D_{w,\delta}$ is the fiber of a locally trivial cone-bundle neighborhood of W in Z.)
Let

$$g_{w,\delta} : f^{-1}S_{w,\delta} \longrightarrow S_{w,\delta}$$

be the restriction of f. Then, for $\dim X > \dim Z$, we may take, in 3. in the definition,

$$L_w = M(g_{w,\delta}) \cup_{f^{-1}S_{w,\delta}} f^{-1}D_{w,\delta} ,$$

where $f^{-1}S_{w,\delta}$ is identified with $f^{-1}S_{w,\delta} \times \{1\}$ in the mapping cylinder. For $\dim X = \dim Z$,

$$L_w = D_{w,\delta} \cup_{S_{w,\delta}} M(g_{w,\delta}) \cup_{f^{-1}S_{w,\delta}} f^{-1}D_{w,\delta} .$$

For W the non-singular stratum of Z,

$$L_w = f^{-1}(w) , \quad \dim X > \dim Z ,$$

and

$$L_w = f^{-1}(w) \cup \{pt\}$$

is the union of $f^{-1}(w)$ with a disjoint point for $\dim X = \dim Z$.

This result implies that the stratum preserving homotopy type of L_w is independent of the choices and, within a component of W, of w. If f is assumed to satisfy Thom regularity, then the topological type of L_w is independent of the choices and of w.

We now turn to the proof of the Theorem, concentrating on the mapping cone and leaving the mapping cylinder to the reader. We will first show that $Y = \mathfrak{C}(f)$ is a manifold weakly stratified set in the sense of [Q1]. (In [Q2] the terminology "manifold homotopy stratified set" is used.) It clearly suffices to verify this in the neighborhood of Y. From the statement labeled "CONVERSE" on page 240 of [Q1], it is clearly more than sufficient to show that for each pair consisting of a stratum W of Y and a higher dimensional stratum T of Z whose closure in Z meets W, the subspace $T \cup W$ is a stratified space, with strata T and W, and links that are not assumed to be compact.

If T is itself a stratum of Z, this follows immediately from the fact that Z is stratified. The case (for $\dim Z = \dim X$) in which T is a stratum of the interior of cY is also easy and is left to the reader. It remains to consider the case $T = V \times (0, 1)$, where W' is a stratum of Z containing W in its closure and V is a stratum of W with $V \subset f^{-1}(W')$.

Suppose first that $W' = W$. Then the smooth submersion $f|V : V \longrightarrow W$ will be a smooth fiber bundle, and $T \cup W$ will be its mapping cylinder. Hence, locally near a point $x \in T$, $(T \cup W, W)$ is homeomorphic to $U \times (cF, \{c\})$, U a neighborhood of x in W and F the fibre of $f|V$. Thus $T \cup W$ is a stratified pseudomanifold in this case. Note that the first isotopy lemma of Thom actually implies that the entire mapping cylinder of the restriction

$$f|f^{-1}W : f^{-1}W \longrightarrow W$$

is a stratified space near W. In particular, Y is a stratified space near the non-singular stratum of Z, with link of the fibre of f.

Finally, suppose $W' \neq W$. Then, as part of the "control data" for Z (see [Gi] [GM] [V] [Ma]) W has a tubular neighborhood T_W in N, together with a tubular projection

$$\pi_W : T_W \longrightarrow W$$

and a tubular distance function

$$\rho_W : T_W \longrightarrow [0, \epsilon] \,,$$

with $W = \rho_W^{-1}(0)$, such that

$$(\pi_W, \rho_W)|W' \cap T_W : W' \cap T_W \longrightarrow W \times (0, \epsilon)$$

is a smooth submersion. By the second isotopy lemma of Thom, the smooth submersion

$$f|V \cap f^{-1}(T_W) : V \cap f^{-1}(T_W) \longrightarrow W' \cap T_W$$

is locally trivial over

$$(\pi_W, \rho_W) \,| W' \cap T_W : W' \cap T_W \longrightarrow W \times (0, \epsilon) .$$

It follows that each point of W has a neighborhood U such that, near U, $T \cup W' \cup W$ is stratum preserving homeomorphic the "open" mapping cylinder of the composite

$$U \times (0, \epsilon) \times G \xrightarrow{\varphi} U \times (0, \epsilon) \times F \subset U \times c^\circ F ,$$

where F is a fiber of $(\pi_W, \rho_W) \,| W' \cap T_W$, $G = f^{-1}(F)$,

$$c^\circ F = F \times [0, \epsilon)/F \times \{0\}$$

is the open cone, and φ is the product of the identity of $U \times (0, 1)$ and the smooth submersion of manifolds

$$f|G : G \longrightarrow F .$$

Here the strata of the open mapping cylinder

$$\frac{U \times (0, \epsilon) \times G \times (0, 1] \cup U \times c^\circ F}{\{(u, t, g, 0) \sim (u, t, f(g))\}}$$

are $U \times (0, \epsilon) \times G \times (0, 1)$, $U \times (0, \epsilon) \times F$, and $U = U \times \{c\}$. We leave to the reader the (easy) exercise to show that this mapping cylinder is a stratified space. For example, in the neighborhood of U it looks like $U \times c(M^\circ(f|G))$, $M^\circ(f|G)$ the open mapping cylinder of $f|G$. Thus $T \cup W' \cup W$, and hence $T \cup W$, is a stratified space. This completes the proof that Y is a manifold weakly stratified set in the sense of [Q1].

Hence, by Theorem 2 of [Q1] (see also the footnote on page 445 of [Q2]), given $w \in W$, W a stratum of Y, there exist a neighborhood U of w in W, a homotopy stratified set K_w, and a map

$$U \times cK_w \longrightarrow Z$$

that is a stratum preserving homotopy equivalence near U. More precisely, let $h_s(Y, W)$ be the stratified homotopy link of W in Z. By definition [Q1], this is the space of paths

$$\theta : [0, 1] \longrightarrow Z$$

with $\{0\} = \theta^{-1}(W)$ and with $\theta(0, 1)$ contained in a single stratum. It has the structure of a homotopy stratified set with strata the homotopy links $h(T \cup W, W) = h_s(T \cup W, W)$. Let

$$p : h_s(Y, W) \longrightarrow W$$

be evaluation $p(\theta) = \theta(0)$. Then the methods of [Q1], as modified in [Q2], lead to the following conclusions:

(i) The map evaluation $(\theta, t) \mapsto \theta(t)$ induces a map

$$M(p) \longrightarrow Z ,$$

$M(p)$ the mapping cylinder of p, that is a stratum preserving homotopy equivalence near W.

(ii) p restricts to a fibration on each stratum and there exist homotopy control data on $h_s(Y, W)$ (i.e., compatible nearly stratum preserving deformation retractions of neighborhoods of strata) that are fiberwise with respect to p. In particular, p is locally stratum preserving homotopy equivalent to a trivial fibration.

Thus, it remains to show that $K_w = p^{-1}(w)$ is stratum preserving homotopy equivalent to a compact manifold homotopy link-stratified space.

(Remark on (ii): In [Q2] it is shown that manifold homotopically stratified spaces are isotopically homogeneous along components of strata not incident to any stratum of dimension below six.)

We consider the case $\dim X > \dim Y$ and leave it to the reader to make the necessary modifications for the equivimensional case. Fix $\delta < \epsilon$, and let

$$T_\delta = Y \cap \rho_W^{-1}[0, \delta]$$

and let f_δ be the stratified map

$$f_\delta = f | f^{-1}(T_\delta) : f^{-1}T_\delta \longrightarrow T_\delta .$$

The mapping cylinder

$$M(f_\delta) = \frac{\{f^{-1}(T_\delta) \times [0, \delta]\} \cup T_\delta}{\{(x, 0) \sim f(x)\}} \subset C(f)$$

will be a closed neighborhood of W in Y. Let $S_\delta = \rho_W^{-1}(\delta) \cap Z$. The restriction of π_W to S_δ is a locally trivial map with fiber at x the link of W in y at x. Then the frontier $\partial M(f_\delta)$ of this neighborhood is the union

$$\partial M(f_\delta) = M(g_\delta) \cup f^{-1}(T_\delta) \times \{\delta\} ;$$
$$g_\delta : f^{-1}S_\delta \longrightarrow S_\delta$$

the restriction of f. We stratify $\partial M(f_\delta)$ to have strata that are unions of the intersections of strata of the mapping cone of g_δ with $M(g_\delta)$ and of strata of $X \times \{\delta\}$ with $f^{-1}(T_\delta)$. Thus the strata of $\partial M(f_g)$ are the intersections of the strata of Z with $\partial M(f_\delta)$. Let $M(f_\delta)$ be stratified similarly by intersection with strata of Y.

We claim that $\partial M(f_\delta)$ is a manifold weakly stratified set. In view of what has already been proven, applied to g_δ, it more than suffices to show that $f^{-1}(T_\delta)$ is a stratified pseudomanifold with boundary $f^{-1}S_\delta$, in the usual sense, i.e., with a boundary collar. This follows from the first isotopy lemma, applied to the composite

$$f^{-1}(Z \cap T_W - W) \xrightarrow{f|} Z \cap T_W - W \xrightarrow{\rho_W} (0, \epsilon) .$$

Thus, $\partial M(f_\delta)$ is actually a stratified pseudomanifold except perhaps near S_δ.

Further, $\partial M(f_\delta)$ is homotopically transverse [Q2, 3.4] to the strata of Y. As noted in [Q2, 3.4], to check this it is more than sufficient to see that for T and W_1 strata of Y with W_1 contained in the frontier of T, $\partial M(f_\delta) \cap (T \cup W_1)$ is transverse to $W_1 \subset T \cup W_1$. (For this intersection to be non-empty, W must be in the frontier of W_1.) The argument is quite similar to the proof that Y is a weakly stratified space. For example, the (hardest) case $T = V \times (0, 1)$, V a stratum of $f^{-1}W'$, where W' is a stratum of Y with W_1 in its frontier, is handled by exhibiting actual (smooth) transversality by application of the second isotopy lemma to trivialize

$$f|V \cap f^{-1}T_W : V \cap f^{-1}T_W \longrightarrow W' \cap T_W$$

over

$$\rho_W : W' \cap T_W \longrightarrow (0, \epsilon) .$$

We leave the details to the reader. In particular, it follows that $(M(f_\delta), \partial M(f_\delta))$ is a manifold weakly stratified set with boundary.

Let $w \in W$ and set

$$S_{w,\delta} = (\pi_W|S_\delta)^{-1}(w) ,$$

and

$$D_{w,\delta} = cS_{w,\delta} = (\pi_W|T_\delta)^{-1}(w) .$$

Let $f_{w,\delta}$ be the stratified map

$$f_{w,\delta} = f|f^{-1}D_{w,\delta} : f^{-1}D_{w,\delta} \longrightarrow D_{w,\delta} ,$$

and let

$$g_{w,\delta} = g_\delta|f^{-1}S_{w,\delta} : f^{-1}S_{w,\delta} \longrightarrow S_{w,\delta} .$$

Let

$$L_w = M(g_{w,\delta}) \cup_{f^{-1}S_{w,\delta} \times \{\delta\}} f^{-1}D_{w,\delta} \times \{\delta\} = \partial M(f_{w,\delta}) .$$

From the preceeding, applied to $f_{w,\delta}$, $(M(f_{w,\delta}), L_w)$ is a manifold weakly stratified set with boundary, and L_w is a stratified pseudomanifold except perhaps near $S_{w,\delta}$. Near $S_{w,\delta}$, L_w is identical to $Y(g_{w,\delta})$. Hence, by induction on dimension, it may be assumed that L_w is a homotopy stratified pseudomanifold.

Next we observe that the inclusion

$$(M(f_{w,\delta}), \partial M(f_{w,\delta})) \subset (M(f_\delta), \partial M(f_\delta))$$

is homotopically transverse to the strata of $M(f_\delta)$. Again, one considers intersection with a pair of strata, and proves smooth transversality using the first or second isotopy lemmas, this time with respect to submersions over W; we omit the details.

One could also define homotopy transversality, as in [Q2, 3.4], but using the stratified homotopy link. However, from the equation

$$h_s(B, A) = h(B, A) \cap h_s(D, C) ,$$

for $(B, A) \subset (D, C)$ with strata of B the intersection of B with strata of D, it follows readily that homotopy transversality as defined in [Q2] implies the stratified version. (The converse holds for the condition "homotopically transverse to the skeleta," since locally — in the sense of the second paragraph of 3.4 of [Q2] — the homotopy link and the stratified homotopy link are the same.) Up to stratum link preserving fiberwise equivalence, the homotopy link of W depends only on the neighborhood of W; hence by transversality there is a stratum preserving homotopy equivalence

$$K_w = p^{-1}(w) \cong h_s\left(M\left(f_{w,\delta}\right), \{w\}\right) .$$

It is not hard to see from the definition of the mapping cylinder and the local triviality of

$$(\pi_W, \rho_W) | Y \cap T_W - W : Y \cap T_W - W \longrightarrow W \times (0, \epsilon)$$

that the inclusion

$$L_w = \partial M\left(f_{w,\delta}\right) \subset M\left(f_{w,\delta}\right) - \{w\}$$

is a homotopy equivalence on each stratum. In fact, the strata of $M\left(f_{w,\delta}\right) - \{w\}$ are homeomorphic to the products of the strata of the boundary with $(0, \delta]$. By the Corollary on p. 494 of [Q2] and neglecting low dimensional difficulties (e.g. if there are no strata of dimension less than five), there is a stratum preserving homeomorphism

$$M\left(f_{w,\delta}\right) - \{w\} \cong L_w \times (0, \delta] .$$

Hence there is a stratum preserving homeomorphism

$$\left(M\left(f_{w,\delta}\right), \{w\}\right) \cong \left(cL_w, \{c\}\right) .$$

The cone structure provides an obvious stratum preserving homotopy equivalence of $h_s\left(M\left(f_{w,\delta}\right), \{w\}\right)$ and L_w; hence K_w is stratum preserving homotopy equivalent to the homotopy stratified pseudomanifold L_w.

To handle the low dimensional difficulties, one notes that the proof of the Corollary in [Q2] produces a stratum preserving proper homotopy equivalence

$$M\left(f_{w,\delta}\right) - \{w\} \simeq L_w \times (0, \delta] ;$$

the low dimensional difficulties are the usual ones in imposing an actual product structure. Clearly a proper homotopy equivalence will suffice in the present situation. Alternatively, cross with a Euclidean space and apply a proper version of the Corollary. This completes the proof.

References

[CS1] S.E. Cappell and J.L. Shaneson, Singular spaces, characteristic classes, and intersection homology, *Annals of Math.* **134** (1991), 274–325.

[CS2] S.E. Cappell and J.L. Shaneson, Stratifiable maps and topological invariants, *J. Amer. Math. Soc.* **4** (1991), 521–551.

[CS3] S.E. Cappell, Genera of algebraic varieties and counting of lattice points, *Bull. Amer. Math. Soc.* **30** (1994), 62–69.

[Gi] C.G. Gobson, K. Wirthmuller, A.A. duPlessis, and E.J.N. Looijenga, *Topological Stability of Smooth Mappings*, Springer Lecture Notes in Mathematics **552**, Springer-Verlag, New York, 1976.

[GM] M. Goresky and R. MacPherson, Stratified Morse Theory, *Ergebnisse der Mathematik und ihrer Grengebiete*, Vol. 14, Springer-Verlag, New York, 1988.

[Ma] J. Mather, Stratifications and mappings, in: *Dynamical Systems* (M.M. Peixota, ed.), Academic Press, New York, 1973.

[Q1] F. Quinn, Homotopically stratified spaces, *J. of the Amer. Math. Soc.* **1** (1988), 441–499.

[Q2] F. Quinn, Intrinsic skeleta and intersection homology of weakly stratified sets, *Lecture Notes in Pure and Appl. Math.* **105**, Marcel Dekker, New York, 1987, 233–249.

[Sh] J.L. Shaneson, Characteristic classes, lattice points, and Euler-MacLaurin formulae, to appear in: *Proceedings ICM 1994*.

[T1] R. Thom, La stabilite des applications polynomiales, *L'Enseignement Mathematique II* **8** (1962), 24–33.

[T2] R. Thom, Ensembles et morphismes stratifiees, *Bull. Amer. Math. Soc.* **75** (1969), 240-284.

[V] A. Verona, Stratified Mappings — Structure and Transversality, *Lecture Notes in Math.* **1102**, Springer-Verlag, New York, 1984.

[W] S. Weinberger, *Classifications of Stratified Spaces*, University of Chicago Press, 1994.

Replacement of Fixed Sets and of their Normal Representations in Transformation Groups of Manifolds

Sylvain Cappell
Shmuel Weinberger

Abstract

This paper gives general results on two problems in transformation groups acting on manifolds. The first is: *When can a component of the fixed set be replaced by a given (simple) homotopy equivalent manifold as a corresponding component of the fixed set of an action of the same group on the same manifold?* The main results show that this can usually be achieved for PL locally linear actions of odd order groups without changing the other components of fixed points. However, for even order groups and compact Lie groups there are obstructions to replacement of fixed points; these are studied and some geometrical consequences are drawn. The second problem is: *When can we change a PL locally linear group action on a manifold to produce a new PL locally linear action of the same group on the same manifold with the same fixed points but with the normal linear representation to one component of the fixed points replaced by another inequalivalent representation without changing the action near the other components of the fixed points?* Classical rigidity results of Atiyah-Bott, Bredon, Sanchez, etc. and the Atiyah-Singer *G*-signature theorem place severe restrictions on such results, particularly for odd prime power order groups. However, the new phenomena constructed here show that such normal representation replacement does occur, even for (free) representations of both odd and even prime-power order cyclic groups, when the component of the fixed points is simply connected and *not a point*. All these replacement results for fixed points (respectively: normal representations) use an investigation, presented here, of the *base (resp: fiber) change problem for fibrations*. The geometrical results use applications of this to the stratified fibration formed by a new construction, the "bubble quotient" of a group action.

Both authors are supported by NSF grants.

Dedication

A distinguishing feature of Bill Browder's research contributions is that they are so often based more on creative new ideas and perspectives than on computation or technique. They have entered into the foundational structure and language in which Topology is expressed. Some of those contributions and results that grew out of them are drawn upon in the present effort, e.g., simply-connected surgery and his embedding theorems and use of Poincaré embeddings, his early work with Levine on fibering over a circle, the Browder-Quinn development of a "transverse isovariant" surgery theory (applied here to non-transverse constructions), Browder's unpublished work on the relation between equivariant and isovariant surgery under the large gap condition, etc. We offer this work in tribute to him upon his sixtieth birthday.

Table of Contents

Introduction

It is obvious that equivariantly (simple) homotopy equivalent group actions have (simple) homotopy equivalent fixed sets; the restriction of the given map and its inverse to the respective fixed sets produces the desired homotopy equivalence.

The converse, when a manifold homotopically equivalent to the fixed points of a (locally linear) group action on a manifold arises as the fixed points of another (locally linear) action of the same group on the same, or a homotopy equivalent, manifold is a classical question in transformation groups. Deep results on this problem have previously been obtained largely in the context of smooth actions on particular smooth manifolds, where, moreover, such replacement of fixed points is often highly obstructed; see the surveys of [DPS] [CS4]. In this paper we study this converse for PL locally linear (or, equivalently, locally smoothable) actions where the methods developed here are quite different and results apply to all groups and have a general character: These general results on this replacement problem for fixed points of group actions of all odd order groups are presented in Theorem 0.1 of this Introduction. The results on fixed point replacement for actions of even order groups are less general and involve a characteristic class obstruction; they are given for semifree actions in Theorem 0.2 of this Introduction. Both results are proved in Section 2.1.

We then present below results on the problem of changing the normal linear representation to one component of the fixed points of a PL locally linear group action without changing the group, the manifold, the fixed points, and the action near the other components of fixed points, and thus, in particular, the other normal representations. These results on normal linear representation replacement around a single component of the fixed points display some new phenomena even for odd prime power cyclic transformations with fixed points of dimension greater than zero. This contrasts with the classical rigidity results of Atiyah-Bott [AB] associated to isolated fixed points. Those results used the Atiyah Singer G-signature theorem [AS] to relate the representations around different fixed points. In contrast, here we show that for certain pairs of linear representations, normal linear representation replacement is possible around a simply connected component of fixed points F if and only if F is not a point. For certain other pairs of representations, the normal replacement is possible only when the Euler characteristic of F is even. We give a general result in in Theorem 0.4 and exhibit examples for $\mathbb{Z}_p r$ in Theorem 0.5 of this Introduction and prove them in Sections 2.2. and 3.2.

Theorem 0.1. *Suppose that G is an odd order group acting PL locally linearly on a closed manifold M, and smoothly on a neighborhood of a 1-skeleton of the fixed point sets, with no codimension two gaps in the dimensions of fixed point sets of subgroups. Then any compact boundary-less PL manifold simple homotopy equivalent to the fixed set is the fixed set of another PL locally linear G action on M that is equivariantly simple homotopy equivalent to the initial action. Moreover, this can be achieved with the normal linear representation of G to the fixed points unchanged.*

Addendum. *The theorem remains valid for G odd abelian even if the smoothness assumption on the 1-skeleton is dropped. This assumption can also be dropped for arbitrary odd order G if the normal representation to the fixed points satisfies the "large gap hypothesis".[1]*

It is conceivable that the smoothness assumption on the 1-skeleton can al-

---

[1]Earlier uses of the "large gap hypothesis" condition in other problems in transformation groups are reviewed in [DPS].

ways be dropped for odd order groups. The present hypothesis on the absence of codimension two gaps (i.e., that $\dim M^H \cap M^K) \neq \dim(M^H) - 2$ for all pairs of components of fixed sets of subgroups of G) is, however, required for the theorem to be true. Codimension two gaps lead to rigidity statements for submanifolds consisting of fixed points of locally linear actions.[2]

We study, in addition the parallel question for even order groups, and have some, though less complete, results. Here the results are more complicated, and the replacement of the fixed set is not always possible. To proceed we need a definition.

Definition. A *strong replacement* by F' of the fixed set F of a G action on M consists of another G-action on M with the following:

　　1) an isovariant simple homotopy equivalence ϕ to the original action,

　　2) the restriction of ϕ to the complement of an equivariant regular neighborhood of the fixed set, an isovariant homeomorphism, and the restriction of ϕ maps this regular neighborhood of the fixed set to an equivariant regular neighborhood of F.

　　3) a (unequivariant) homotopy, relative to this decomposition of the G-manifolds and the identification of complements, of this ϕ to a homeomorphism, and 4) the new action has fixed set F'.

One can specify the simple homotopy equivalence $F' \to F$ and insist that ϕ restrict to this map. In Theorem 0.1 it is moreover the case, and this is what we will prove, that the fixed set can, in fact, be changed by strong replacements with the normal linear representation of G to the fixed points unchanged as well.

For even order groups we present the following partial result on locally linear semifree (that is, free on the complement of the fixed points) actions, as a contrast:

Theorem 0.2. *If G is even order and acts semifreely PL locally linearly on a closed PL manifold with a component of the fixed points of codimension greater than two and simply connected, then it is possible to strongly replace this component of the fixed set by a given (simple) homotopy equivalent closed PL manifold iff the Kervaire classes of the homotopy equivalence vanish. Moreover, when this happens, the strong replacement can be done without changing the normal linear representation of G to the fixed points.*

Recall that in Sullivan's reformulation of Browder-Novikov surgery theory, a homotopy equivalence between closed PL manifolds $F' \to F$ gives rise to a classifying map $F \to \mathbf{F/PL}$, called a normal invariant; here $\mathbf{F/PL}$ is the classifying space for PL block bundles equipped with a homotopy trivialization. Sullivan's analysis [Su], see [MM], of this space includes the construction of

---

[2]However, codimension two fixed point set replacement may occur frequently for nonlocally linear actions where the codimension two fixed sets are not even demanded to be locally flat (cf. [CS3]).

characteristic classes $k^{4i+2} \in H^{4i+2}(\mathbf{F/PL}; \mathbb{Z}_2)$ which detect certain Kervaire invariants and are called Kervaire classes. The condition of Theorem 0.2 for the strong replacement of F, a component of the fixed points, by F' is that the pullbacks of these Kervaire classes to F is trivial.

For more general actions of groups of even order we do not have even a conjectured analogue to Theorem 0.1. It seems likely that for semifree actions but with F not simply connected but with $\pi_1(F)$ satisfying a form of the Borel conjecture, there is an extension of Theorem 0.2 with the given Kervaire class condition replaced by the vanishing of generalized characteristic classes lying in $H^{4i+2}(B\pi_1(F), F; \mathbb{Z}_2)$. We conjecture that even for actions of even order groups which are not semifree, when F is simply connected the vanishing condition on the Kervaire classes still suffices for replacement, but this seems difficult at present.

In addition, we shall also study the case of $G = S^1$ or $SU(2)$, but there replacement is much rarer. Indeed, in some situations, a strong replacement $F' \to F$ is never possible for any $F' \neq F$ (see the rigidity Theorem 2.4 below).

For many applications to classical examples, e.g. linear actions on lens spaces, one needs a replacement result analogous to Theorem 0.1 generalized to homotopy equivalences $F' \to F$ that are not necessarily simple. For this extension, there is, in general, a further obstruction, the image of Whitehead torsion in a usually finite quotient group of the Whitehead group whose computation involves subtle number theory, see Theorem 3.1 of Section 3. Fortunately, in many cases of interest this quotient group, or at least the relevant element in it, can be seen to vanish as in 3.2 and, for involutions, 3.3. For example, for applications to (linear) actions on lens spaces, we have:

Corollary 0.3. *Suppose given a cyclic group* $\mathbb{Z}_n$ *acting PL locally linearly on a manifold* M *without codimension two gaps and with a connected component of the fixed points a homotopy lens space* F *of order* m $(= |\pi_1(F)|)$ *prime to* n. *If* F' *is a PL manifold homotopy equivalent to* F *and if*

(i) $n = 2$

or (ii) n *is prime to* $\sigma(m)$

or (iii) F *and* F' *are linear lens spaces, and if* n *is even the action is semifree,*[3]

then there is another PL locally linear isovariantly homotopy equivalent action of $\mathbb{Z}_n$ *on* M *with* F' *replacing* F *in the fixed points set. Moreover, this can be achieved with the normal linear representation to the fixed points unchanged.*

In (ii), $\sigma(m)$ is the order of the finite quotient of $\text{Wh}(\mathbb{Z}_m)$ by the "subgroup of the cyclotomic units", i.e., the subgroup of Whitehead torsions of homotopy equivalences of lens spaces; of course these torsions can be read off from the quotients of their Reidemeister torsions. Thus, for m prime $\sigma_m = h_+$, the

---

[3]This semifree hypothesis for n even could, with more work, be dropped.

second factor of the class number h of the cyclotomic field of m-th roots of unity.

Each of these three criteria in Corollary 0.3, and others in § 3, do yield examples of semifree homotopy linear PL locally linear actions of prime order cyclic groups on (linear) lens spaces with (linear) lens space fixed points whose fixed points violate relations on Reidemeister torsions that hold for linear actions. On the other hand, there are examples of linear lens spaces F and homotopy equivalent smooth nonlinear lens spaces F' for which these conditions don't hold and, in fact, for which computations involving their Reidemeister torsions show that, in some situations, replacement is not possible (even for semifree linear actions and $\pi_1(F)$ and G being cyclic groups of relatively prime odd prime order).

It is possible to study weaker forms of the above replacement questions. For instance, one might not insist that the new action realizing F' as fixed points be on the same manifold but just on an equivariantly homotopy equivalent manifold. For G even, this weaker replacement is more often possible, and, we shall see that there is no obstruction to this for certain orientation preserving involutions, although for other group actions there is the same Kervaire type obstruction. Similar phenomena also occur for certain circle and $SU(2)$ actions; see Theorem 2.5. This will be further analyzed in a future paper by the present authors together with Min Yan using more techniques in the theory of stratified spaces.

One could study as well the question of obtaining the action on M, but with the complements of the fixed sets not necessarily equivariantly PL homeomorphic. As far as we could tell, the solution to this problem is not uniform, and can best be analyzed by comparing the present strong replacement obstructions with the surgery theory of the complement.

After treating the problem of replacing fixed points in locally linear PL actions, we turn to the problem of replacing the linear normal representation about a component of the fixed points by different linear representation. For a PL locally linear G action on M, cognate to the notion of strong replacement of fixed points above, we will formulate and study when we can "h-strongly normally replace" the normal linear representation γ of G around a component F of the fixed points by another linear representation γ' of G. This notion of h-strong normal replacement, made precise in section 2.2, will produce a new PL locally linear action of G on the same manifold M without changing the fixed points, the action outside a regular neighborhood of F and the isovariant (and, a fortiori, the equivariant) homotopy type of the action. Problems of changing normal representations have been studied previously only in rather special contexts, particularly for M a sphere and F a point in the context of the Smith conjecture; see e.g. [CS1], [Pe]. As we shall see, that case does not display all the characteristic features of the present general problem. Indeed, using the following result of § 2.2, we construct examples of pairs of free linear representations of a group G for which normal representations replacement will occcur for all simply connected components of fixed points F except F a point.

Theorem 0.4. *Suppose that G is a finite group acting by a PL locally linear semifree action on M and that F is a simply connected component of the fixed points and let γ denote the normal linear representation of G at points of F. Let γ' be another linear representation of G and let $S(\gamma)$, $S(\gamma')$ denote the unit spheres of the representation spaces. Then there is a PL locally linear G action on M with γ h-strongly replaced by γ' if and only if*

(i) *F is not a point,*

(ii) *the action of G on $S(\gamma')$ is free and there is a homotopy equivalence $h\colon S(\gamma')/G \to S(\gamma)/G$,*

(iii) *the extended normal invariant of h, $(S(\gamma)/G) \cup e^{\dim(\gamma)} \to \mathbf{F}/\mathbf{PL}$, vanishes*

(iv) *the Euler characteristic $\chi(F)$ is even, or $\chi(F)$ odd and the Whitehead torsion of h represents 0 in $H^*(\mathbb{Z}_2; Wh(G))$.*

These conditions are readily turned into computations and verified in many cases of interest. Condition (ii) on homotopy equivalence of quotients of free actions on spheres is classical and is equivalent to the equality of the easily calculated associated k-invariants in $(\mathbb{Z}/|G|)^\times$; see, for example [M], [Co]. The extended normal invariant of (iii) and methods of computing it are discussed in detail in sections 2.2 and 3.2; it suffices here to remark that it depends only upon the restriction of the G action on the representation spheres $S(\gamma)$ and $S(\gamma')$ to the Sylow subgroups; for example, it trivially vanishes when γ and γ' agree on the Sylow subgroups. Condition (iv) is, of course, always satisfied for $\chi(F)$ even; in any case, when $SK_1(G) = 0$, e.g. when G is cyclic, the Whitehead torsion of h is given by the quotients of the Reidemeister torsions, $(\tau(S(\gamma)/G)/\tau(S(\gamma')/G))$ and $H^*(\mathbb{Z}_2; Wh(G)) = Wh(G)/(2Wh(G))$. Indeed, when $\chi(F)$ is odd and G cyclic, the condition (iv) follows from divisibility of $(\gamma - \gamma')$ by a suitable power of 2 in the representation theory of G.

Examples of nontrivial normal linear representation replacement about fixed points obtained using Theorem 0.4 for some G of composite order for all simply connected F, F not a point, are given in section 2.2. With some further number theoretic computations, we obtain in Section 3.2 the following examples of nontrivial normal representation replacement for PL locally linear actions of $\mathbb{Z}_m$, for both $m = 2^r$, $r \geq 3$ and for $m = p^r$, p an odd prime and $r \geq 2$. Let t^k denote the underlying real 2-dimensional representation of $\mathbb{Z}_n$ of the complex 1-dimensional representation sending the obvious generator to $(e^{2\pi i k/n})$; the real representations t^k and $t^{k'}$ are thus equivalent if and only if $k \equiv \pm k' (\mathrm{mod}\ p^r)$.

Theorem 0.5. *Let $\mathbb{Z}_m$, $m = p^r$, p prime, act by a PL locally linear action on the closed PL manifold M and let F be one component of the fixed points of the action. Let $\gamma = 2p(t^k)$, k prime to p, denote the normal linear representation of $\mathbb{Z}_m$ to F and let $\gamma' = 2pt^{k'}$ be another linear representation of $\mathbb{Z}_m$ with $k \equiv k' (\mathrm{mod}\ p^{r-1})$. Then if F is simply connected and not a point, there exists* *another (isovariantly homotopy equivalent) PL locally linear action of $\mathbb{Z}_m$ on the same manifold M with the same fixed points and with the action unchanged on $M - F$ (so that, in particular, the normal linear representations to all components of the fixed points other than F are unchanged) but with γ' now being the normal linear representation to F.*

In this proposition, from the description of γ and γ' the actions are readily seen to be semifree near F and this replacement of γ by γ' is contructed in Section 3.2 to be h-strong. On the other hand, we show that for $p = 2$, $r \geq 3$, F simply connected, $\gamma = 6t^k$, k odd, and $\gamma' = 6t^{k+2^{r-1}}$, γ' can h-strong normally replace γ if and only if the Euler characteristic $\chi(F)$ is even.

Many of the results on normal representation replacement, including Theorems 0.4 and 0.5, have extensions when $\pi_1(F)$ is of odd order with F not a point; this can be applied, for example, to linear actions on lens spaces to change normal representations. On the other hand, for F a $K(\pi, 1)$-manifold, by employing some results related to the Novikov conjecture [FRW], [FW1, 2] the present methods can often be used to show that nontrivial normal replacement does not occur for odd order groups; see the example at the end of Section 2.2.

Even for actions that are not semifree the methods of section 2.2 do reduce the normal representation replacement problems to calculations. In the most general case these are, however, difficult at this time as they would require more understanding, to begin with, of the isovariant homotopy theory of linear group actions on spheres. We do compute, for certain nonfree representations, around which connected components F of the fixed point sets normal replacement is possible; see section 3.2. In particular, the work of Cappell and Shaneson which gave counterexamples to one of the Smith conjectures for G even [CS1] can be used to yield some examples where the unit spheres of the representations are equivariantly PL h-cobordant but not free and, correspondingly, h-strong normal replacement is possible for these non-semifree actions for any F. This contrasts with the results on semifree actions, where, as noted above, normal replacement is not possible for F a point or for many $K(\pi, 1)$; see Section 3.2.

Later papers will combine the present methods with [CW2,3, We2] to deal with the replacement problems of equivariant maps that are only unequivariant homotopy equivalences.

As the present results on fixed point replacement involve, in particular, the solution of certain embedding problems, one might hope to approach them with the conventional paradigm; e.g., first do work on equivariant bundle theory, to provide candidate neighborhoods for the putative fixed point set and then trying to apply manifold surgery theory to glue the boundary of this neighborhood onto the boundary of the complement, and then study whether one has obtained the original manifold. Indeed, this is the method used classically for proving the unequivariant embedding theorem of Browder-Casson-Haefliger-Sullivan-Wall [Wa1], that under the assumptions that F PL embeds in M in codimension at least three, so does any PL manifold F' homotopy equivalent to F; it has been used as well for many theorems in the theory of transformation groups. However, for the topological problems considered here the precise integral information that would be required is quite difficult to get, and the equivariant bundle theory that would be necessary is unstable. Moreover, we do not know how to deal with the glueing problem for the neighborhood of the fixed points to the complement by this head on approach. Instead, the present results are obtained by an integrated geometrical analysis of neighborhoods and blocked surgery; classifying spaces for bundles never enter[4], and the glueing issue is directly visible and resolved using an associated stratified space, the "bubble quotient" of a group action, introduced in section 2.

For applications to the present results on group actions, we will consider in § 1.1 the general obstruction to deforming a map $W \to B$ to a block bundle or a block fibration. This theory was first developed by Casson [Ca] for the case where B is a sphere and fibers are manifolds, and for general B by Quinn [Q]. The book [BLR] gives a careful exposition of Quinn's ideas in a special case. Here, using in addition the work of Browder and Quinn [BQ] on transverse isovariant surgery, we describe the solution of the general block fibration problem, needed below, when the fiber is not necessarily a manifold. We present a useful reformulation of the theory and develop some methods of calculation, new even for the case of manifold fibers. These results are applied in § 1.2 in studying the base change problem in fibrations, that is, when can the base of a block fibration be replaced by a homotopy equivalent manifold without changing the total space or fiber. Our analysis yields in 1.7 a result applicable to many fibrations with trivial monodromy.

Theorem 0.6. *Given a block fibration* $W \xrightarrow{f} B$ *which is fiber homotopically trivial over the 1-skeleton of the PL manifold* B, *with a compatible splitting of* $\pi_1(W) \xrightarrow{f_*} \pi_1(B)$, *and with the manifold fiber (algebraic[5]) Poincaré nullcobordant over its fundamental group. If* B' *is a PL manifold and* $B' \xrightarrow{g} B$ *is a simple homotopy equivalence, then the composite map* $g^{-1} \circ f \colon W \to B'$ *is homotopic to a block fibration with the same fiber as* f.

Such results on the problem of when the base of a fibration can be changed, for nonmanifold fibers, are applied in Section 2.1 to "bubble quotient fibrations" in obtaining results on replacement of fixed points of group actions.

Similarly, results are obtained in § 1.3 on the fiber-change problem, when can we change the (possibly nonmanifold) fiber of a fibration, without changing the base or total space. These results are applied in § 2.2 to "bubble quotient fibrations" in obtaining results on the normal replacement of the linear representations about a component of the fixed points of a group action on a manifold.

---

[4]However one should be able to obtain some results on (unstable) equivariant bundle classifying spaces as a consequence of the result proven here.

[5]E.g., the Ranicki-Mischenko symmetric signature ([R]) vanishes.

A future paper will discuss the question of analogues of the present results on PL locally linear actions for PL semilinear, topological locally linear and topological semilinear actions.

While all of this work on fixed point replacement and on normal representation replacement is previously unpublished, some of these results on fixed point replacement were obtained by us a number of years ago in the case of simply connected manifolds; the first author lectured on these in his 1987 Conference Board of the Mathematical Sciences lectures at Virginia Polytechnic University. In the intervening years, we had increased the range of applicability one class of fundamental groups at a time, by calculation. Moreover, some of the ideas contained herein helped motivate some of our intervening and current works on other problems, e.g. [CW4, CSW, We1]. Finally, we decided to systemize and write down our tentative results and, in so doing, discovered a geometrical method (see the proof of Theorem 1.7) that at once radically simplifies the argument, eliminates extensive computations and removes all fundamental group hypotheses from Theorem 0.1 on fixed point replacement. We should remark that this method is a descendant of one of the ideas in Edwards's shrinking theorem [Da], and a variant is also used in [BFMW] for a very different application.

1   The block fibration, base charge and fiber charge problems

1.1   Block fibration and trivializing bundles of surgery spectra

Recall that a block bundle over a triangulated base space consists of a total space W which is decomposed into pieces that correspond to the simplices of the base. Over vertices we have copies of the fiber F. Over 1-simplices we have s-cobordisms that are isomorphic to $F \times I$. Over n-simplices Δ we have a space isomorphic to $\Delta \times F$. All of the face relations of the base are to be mimicked in the total space.

There is also a variant of the notion of a block bundle in which all of the blocks are only supposed homotopy equivalent to the model $\Delta \times F$. These have a similar theory.

One reason for the utility of block bundles is that the regular neighborhood of one PL manifold M locally flatly embedded in another has the structure of block bundle over M with fiber a ball $(D, \partial D)$ ([RS]). The following is proved in identical fashion:

Proposition 1.1. *An equivariant regular neighborhood of the fixed set of a PL locally linear G action has the structure of an equivariant block bundle, that is all the blocks have G action preserving all face relations, and all of the structure* *equivalences to* $\Delta \times (D, \partial D)$ *are equivariant.* $\square$

The solution to the block fibering problem requires the introduction of *surgery spaces* or *surgery spectra* as homotopy functors of a space X. These are Δ-sets (see [Q]) defined by a variant of chapter 9 of [Wa] as surgery problems equipped with maps to X.

Definition of $\mathbf{L}_n(X)$. 0-simplices are n-dimensional objects in the sense of [Wa1, Chapter 9]. That is, we have an n-dimensional Poincare complex with boundary mapping into X, and a degree one normal invariant, relative to the boundary, of this Poincare space. 1-simplices are cobordisms between such objects. Observe that the boundary is not fixed during the course of a cobordism, but that if it varies, it does so in the same homotopical fashion in domain and range of the cobordism. Higher simplices are defined similarly.

Remarks. 1) There are variant notions of **L**-spectra according to whether or not one assumes that these homotopy equivalence maps on the boundaries are simple; these spectra are decorated according to the usual corresponding convention for *L*-groups.

2) The surgery obstruction induces an obvious map $\pi_j(\mathbf{L}_n(X)) \to L_{n+i}(X)$. A generalization of the argument from [Wa1, Chapter 9] shows that it is an isomorphism.

3) This obvious map is a special case of a more general *assembly map:*

$$\text{Maps}[M\colon \mathbf{L}_n(X)] \to \mathbf{L}_{n+m}(X \times M) \ .$$

Here, m is the dimension of M. Yet more generally, if $W \to M$, is a block fibration with fiber F, then there is an associated fibration over M, $E(\mathbf{L}_n(\mathbf{F}) \downarrow M) \to M$, with fiber $\mathbf{L}_n(F)$, and then a further assembly map defined on the space of sections of this associated fibration,

$$\text{Sect}[M\colon E(\mathbf{L}_n(F) \downarrow M)] \to \mathbf{L}_{n+m}(W) \ .$$

4. Finally, all of these spaces and maps are infinite loop spaces and maps. This is because $\times \mathbb{C}P^2$ induces a periodicity homotopy equivalence $\mathbf{L}_n \to \mathbf{L}_{n+4}$, and this map commutes with all of these constructions.

We will find it very convenient to have an extension of this formalism to stratified spaces; we will use the setting of Browder and Quinn [BQ].

Definitions. X is a *strongly stratified space* if it is given a stratification $X = X_n \supset X_{N-1} \supset X_{n-2} \supset \cdots \supset X_0 \supset X_{-1} = \emptyset$, such that each stratum is closed, the differences between consecutive strata are open manifolds (called *open* or *pure strata*), and there are block bundle neighborhoods around neighborhoods around the pure strata in the higher strata. A map between stratified spaces is said to be stratified if it preserves pure strata. (Browder and Quinn use the word isovariant for stratified, but we reserve isovariant for maps between G-spaces.)

It is transverse if it furthermore pulls pack these block bundles from the range to the structure block bundles in the domain.

Browder and Quinn developed a surgery theory for transverse stratified classification [BQ]. (Some of our classification results for stratified spaces without transversality restrictions were given in Cappell's 1987 CBMS lectures on transformation groups and in [CW4]. The second author removed the transversality condition in a general stratified surgery theory [We1].) The idea is as follows. By [Wa1, Chap. 9] the key point that one must establish to have a surgery theory is the so-called $\pi - \pi$ theorem on the triviality of surgery obstructions in the appropriate relative situation. For strongly stratified spaces, one proves the π-π theorem inductively; for the bottom stratum, it holds by the result for manifolds. Then one uses the bundle structure for a neighborhood and the transversality to give a solution in a neighborhood of the bottom stratum. On removing this neighborhood one has a relative problem on a space with few strata that is still (stratified) π-π.

Now, as we have seen, these considerations from Chapter 9 of [Wa1] are also all that is necessary to define **L**-spaces and spectra. We therefore can define $\mathbf{L}^{BQ}(X)$ for a stratified space X. Then the analogues of all of the remarks made above hold here as well. For example, its homotopy groups are the Browder-Quinn obstruction groups to transverse stratified surgery.

It is interesting to observe that while the Browder-Quinn transverse isovariant surgery theory is adequate for the blocked surgery problems encountered in the present paper, the group actions constructed by our fixed set repalcement results and normal representation replacement results are not transverse isovariantly modeled on the original actions. One may also note that the techniques developed in this section for studying the fibration, base change and fiber change problems, e.g. the results on and consequences of our trivialization of **L**-space bundles, should be applicable to other settings for surgery theories besides that of [BQ].

To have any hope of deforming a map between manifolds to the projection of a block bundle, a necessary condition is that the homotopy fiber be given the structure of a simple Poincare complex, and that the space is simple homotopy equivalent to the total space of the associated fibration. There is an entirely analogous, but more elaborate, condition that can be formulated to apply to stratified spaces. On the bottom stratum it is this condition, and then one uses the bundle structures to require that the remaining strata are simple stratified Poincare objects of a one lower sort. We call this condition on a (stratified) map that it be a *(stratified) homotopy bundle*. Given this data, we can describe the obstruction to blocking a map (see [Q], [BLR]):

Blocking Theorem 1.2. *Let $p : V \to M$ be a map from a stratified space into a manifold. Suppose that p is a stratified homotopy bundle with fiber Q. Suppose, in addition, that all strata of Q have dimension at least five. Then one can associate to p an element of the set of components of the homotopy fiber of* *the map:*

$$Sect[M: E(\mathbf{L}^{BQ}(Q) \downarrow M)] \rightarrow \mathbf{L}^{BQ}(V)$$

which vanishes iff the map p can be homotoped to a blocked fibration. □

Remarks. 1. This map is an assembly map as above.

2. To produce the section, one makes p transverse to a triangulation of M, and associates to each simplex of the simplex of $\mathbf{L}^{BQ}(Q_\delta)$ that lies over it, where Q_δ is a copy of Q lying over the barycenter δ of a simplex. (These are matched together for different simplices according to the data of the fibration.) These assemble to the identity element $V \rightarrow V$, so that associated to p is an element of the fiber.

3. Blocked (stratified) surgery establishes the theorem. In fact, it shows that homotopy fibrations modulo block bundles are naturally homotopy equivalent to this homotopy fiber.

4. Of course, the necessity of the vanishing condition of (1.2) is valid without the dimension condition on strata of Q. For sufficiency, one can sometimes remove by ad hoc arguments the condition on the dimension of the fiber. For instance, the classical surgery exact sequence corresponds to deforming a simple homotopy equivalence to a homeomorphism. Here the fiber is zero dimensional. It is precisely in the ad hoc low dimensional surgery that Rochlin's theorem enters, and one obtains that the genuine PL obstruction involves $\mathbf{F}/\mathbf{PL}$ as opposed to $\mathbf{L}(e) = \mathbf{F}/\mathbf{Top} \times \mathbf{Z}$. We will sometimes need to do similar low dimensional arguments when we deal with submanifolds of low codimension.

5. As a sample application of this theorem we point out a simple proof of the theorem of Browder, Casson, Haefliger, Sullivan, and Wall on the existence of a unique PL embedding in codimension three corresponding to a Poincare embedding. Using the elementary π-π case of codimension one splitting [Wa1] one obtains a putative regular neighborhood of the potential submanifold. The obstruction to making this into a block bundle over the submanifold lies in the homotopy fiber of two contractible spaces! The sections are into an $\mathbf{L}(D, \partial D)$-bundle, where D is the normal disk, which is contractible by the $\pi - \pi$ theorem ($\pi = e$) and one maps into the global $\mathbf{L}$-space of this neighborhood, which also vanishes by π-π. Having made the neighborhood into a block bundle over M, it is a simple matter to embed M into it using Zeeman unknotting in codimension greater than 2.

Note that this argument works exactly as stated if the codimension is at least six, but that it is not that difficult to resolve by hand the problems that must be solved to get the result through codimension three.

In general it is not as easy to get information about the homotopy fiber in Theorem 1.2. As is customary with assembly fibrations in surgery theory, we can give a two-part treatment here by viewing the section piece as the *primary obstruction* to fiberring, and the remaining L-piece as the *secondary obstruction*. Since the primary obstruction lies in a space of sections of a bundle, it is important to know when this bundle is trivial; when it is trivial, we will be able to replace the space of sections by an easier to analyze space of maps. Our next goal is to give criteria for this:

Proposition 1.3. *Let $p : V \to M$ be a manifold homotopy fibration over M with fiber Q. Then the associated fibration of $\mathbf{L}$-spectra is trivial if the homotopy sequence of the fibration ends in a split exact sequence:*

$$0 \to \pi_1 Q \to \pi_1 V \to \pi_1 M \to 0 .$$

For the stratified situation we do not have as pleasant a criterion. However, we do have the following, more special result:

Proposition 1.4. *Let $p : V \to M$ be a stratified homotopy fibration over M. Then the associated fibration of $\mathbf{L}$-spectra is trivial if the inclusion of the stratified homotopy fiber induces an isomorphism on the fundamental group of every stratum of the fiber to the corresponding stratum of V.*

Notice that the hypothesis of 1.4 on fundamental groups is satisfied if M is 2-connected. A similar result which can often be applied when M is not simply connected is:

Proposition 1.5. *Let $p: V \to M$ be a stratified homotopy fibration over M. Assume that there are splittings of the induced maps of fundamental groups of each stratum of V to M and that these splittings are compatible with a homotopy trivialization of p over a 1-skeleton of M. Then the associated fibration of L-spectrum over M is trivial.*

Proof of 1.3, 1.4 and 1.5. In each case, letting Q_δ denote the fibre over a point δ of the base, there is a natural map from $\mathbf{L}(Q_\delta)$ to $\mathbf{L}$(something) which is an isomorphism, and therefore trivializes the $\mathbf{L}$-fibration. In the first case one uses the map to V and then use the projection to $\mathbf{L}(\pi_1 Q)$ provided by the hypothesis. In the second case $\mathbf{L}^{BQ}(V)$ itself is isomorphic to an iterated loop space of $\mathbf{L}^{BQ}(Q)$, and we make use of this. (More precisely, we have labelled Browder-Quinn $\mathbf{L}$ spectra by the stratified space, while we have labelled ordinary $\mathbf{L}$-spectra by the dimension of the manifolds used in defining the 0-simplices. If we should label Browder-Quinn spectra by the dimensions of the lowest stratum manifold used for the 0-simplices, instead of using the dimension given by the target space, then there would be none of these confusing loops.) In the last case, using the trivialization over the 1-skeleton of M, regarded as a bouquet of circles, to attach to V copies of $D^2 \times$ fiber to each circle $\times$ fiber in V to obtain a space Y, we can proceed as before to trivialize the fibration using $\mathbf{L}(Q_\delta) \to \mathbf{L}(V) \to \mathbf{L}(Y)$.   □

Remark. Analogous to all these results on trivialization of bundles of $\mathbf{L}$-spectra, there are much easier corresponding results on trivialization of bundles of "Whitehead spaces". These spaces classify block bundles of h-cobordisms of stratified spaces; their homotopy groups are the Whitehead groups of such spaces. (They are quite distinct from the Whitehead spaces that arise in pseu-doisotopy theory.) These are easy because, using the PL structure to remove tubular neighborhoods of strata, the Whitehead spaces in the Browder-Quinn theory can be decomposed into standard Whitehead spaces, which "spacify" classical Whitehead groups.

1.2 The base change problem for fibrations

We now use the present results on these **L**-spectra bundles in an analysis of the problem of base change. These will be used in Sections 2 and 3 in obtaining results on replacement of fixed points.

Definition. The *base change problem* is the problem of deciding for a block bundle projection $W \to M$ and a simple homotopy equivalence of manifolds $M' \to M$, whether or not the obvious map $W \to M'$ is homotopic to a block bundle projection.

Having provided conditions for when the corresponding fibration of **L**-spectra trivializes, we now are in better shape for killing the obstruction. What follows are two analyses: the first applies to the primary obstruction when the fiber is a manifold. It is based on well known results on **L**-spectra. The second analysis involves some new geometrical constructions and methods. It is more useful for Theorem 0.1 as it deals with both obstructions (i.e., the total obstruction of Theorem 1.2) simultaneously. However, the first analysis is used in determining the obstruction to fixed point replacement for G of even order in Theorem 0.2.

Proposition 1.6. *Let $p : W \to M$ be a manifold block bundle over M with the fundamental group situation split as in Proposition 1.3. If the action of the fundamental group on the homotopy of the fiber is trivial, with the trivialization compatible with the splitting of $\pi_1 W \to \pi_1 M$, then the primary obstruction to base change can be identified with the map induced by crossing with the fiber Q:*

$$[M: \mathbf{L}(e)] \to [M; \mathbf{L}(F)] .$$

Theorem 1.7 (Monodromy-Signature Criterion for Base Change). *If the block fibration $W \to M$ with stratified fiber Q is stratified fiber homotopi-cally trivial over the 1-skeleton of the manifold base M, with the trivialization compatible with a splitting of the map from π_1 of each stratum of the total space to π_1 of the base, then the obstruction to base change using the simple homo-topy equivalence $M' \to M$ can be identified with the product by Q of the total obstruction to homotopy of $M' \to M$ to a PL-homeomorphism. In particular, if the fiber Q is a manifold which is (algebraic) Poincare nullcobordant (see [R]) over its fundamental group then the total obstruction to base change vanishes. Moreover, in this case the base change is achieved without changing the fiber.*

Remarks. Here the algebraic Poincare cobordism class of the fibre is its (Miscenko-Ranicki) symmetric signature. There are situations somewhat more

general than vanishing of the symmetric signature for which the conclusion of
Theorem 1.7 holds. Indeed, it will be necessary to employ its proof in Sections
2.1 and 3.1. One important point is that there are many Q of interest in transformation groups (such as $\mathbb{R}P^{4k+3}$) for which, while the symmetric signature is
nonzero, yet all products with it yield vanishing quadratic surgery obstructions
(see [WY2]).

In the applications of (1.7) in § 2 below, the needed splitting of fundamental
groups follows easily on each stratum using the group action and covering space
theory.

Proof of Theorem 1.6. From the point of view of the theory above, one sees
directly that the obstruction to base change is associated to the transfer that
maps:

$$\text{Fiber}([M; \mathbf{L}(e)] \to \mathbf{L}(M)] \to \text{Fiber}(\text{Sect}[M: E(\mathbf{L}_n(Q) \downarrow M)] \to \mathbf{L}_{n+m}(W))$$

In light of the remark, we will concentrate on the primary obstruction. We
will see that many transfers can be described purely in terms of their monodromy, so that not only is this space of sections turned into a function space,
but the transfer map into it can be computed by taking a product.

All of the spectra involved are modules over the Thom spectrum MSO and
even over the natural symmetric signature spectrum $\mathbf{L}^*(\mathbb{Z})$, where $\pi_i(\mathbf{L}^*(\mathbb{Z})) =
L^i(\mathbb{Z})$ is Ranicki's group of symmetric (Witt) forms; this is seen simply by cross-
ing defining manifold problems by oriented closed algebraic Poincare complexes.
(Strictly speaking, for this last module structure to make sense, one needs algebraic analogues of the definitions of the geometric $\mathbf{L}$-spectra. Ranicki has done
this, but only in unpublished work; however, it follows easily from the formality
of the process, as described above, and the main theorem of [R] describing L
groups as bordism of algebraic Poincare complexes.)

Now, $\mathbf{L}^*(\mathbb{Z})$-modules spectra have remarkable properties (see [Su, TW].).
Their homotopy types are functors of their homotopy groups, and maps between
them (respecting the $\mathbf{L}^*(\mathbb{Z})$-module structure) are determined by their induced
maps on homotopy with coefficients $\pi_*(; \mathbb{Z}_n)$.

Geometrically, the way one can recognize an induced map on $\pi_*(; \mathbb{Z}_n)$ for
such modules is by mapping in $\mathbb{Z}_n$-manifolds and computing the surgery-with-
coefficient obstruction for these singular manifolds. (See [MS] for a more elaborate discussion.) As a consequence, to prove the result regarding the identification of the surgery transfers needed for the primary obstruction, it suffices to
prove the following lemma:

Lemma 1.8 (transfers as products). *If $p : E \to B$ is a block bundle over
a ($\mathbb{Z}_n$-) manifold with fiber Q and which is trivial as a fibration over the 1-
skeleton, with this trivialization compatible with a splitting $\pi_1 Q \times \pi_1 B \to \pi_1 E$,
then the transfer of a surgery problem for B is identified with the product with
the symmetric signature of Q.*

In light of this lemma, the theorem is proven. This lemma can be proven geometrically as follows. Since the fibration is trivial around the one skeleton, with a trivialization compatible with the given splitting of fundamental groups, we can do surgery on circles representing the relations in the fundamental group to produce a new fibration, with a new base, but that has the property that there is an embedded 2-complex on which the fibration is trivial, and which has isomorphic fundamental group.

Using the π-π theorem one can perform surgeries on the normal invariant of this base, to make the map an equivalence outside this neighborhood of the 2 complex, so that the transfer is now seen to be simply product with Q. $\square$

Remark. This lemma was independently proven by W. Luck and E. Pederson.

Proof of Theorem 1.7. We shall use some ideas similar to the proof of the lemma on transfers. The use of $\mathbf{L}^*(\mathbb{Z})$ module structure is replaced by an explicit use of geometric periodicity, namely $\times \mathbb{C}P^2$, taking products with this complex projective space.

We will first give the argument assuming that the fibration is trivial on the 2-skeleton.

The obstruction to base change is (up to sign) a transfer. We can cross with $\mathbb{C}P^2$, without changing the obstruction. Now consider a fixed fine triangulation of $M \times \mathbb{C}P^2 \to M$. We can surger the structure of M to be a blocked (over M) homotopy equivalence outside a neighborhood of the 2-skeleton of $M \times CP^2$ by using repeated application of the π-π theorem. At the end of the surgery one has a homotopy equivalence $V \to M \times CP^2$ which is blocked over M outside a regular neighborhood of the 2-skeleton of the target. Over this neighborhood, we have a trivialization of the Q bundle, so the transfer is product with the symmetric signature, which by assumption vanishes.

Now for the general case. To simplify discussion we introduce a couple of terms: A *3/2 skeleton* for a space X is a 2-complex $K \to X$ which is an isomorphism on π_1. The second term, *homotopy equivalence exchange* is the following process: If we have a normal invariant to $(W; A, B)$, with A and B codimension 0 submanifolds (or stratified spaces) intersecting along their mutual boundary, and everything is π-π, and the map is already a homotopy equivalence to A, then one can normally cobord (rel *other* boundary components) so that instead the map is a homotopy equivalence to B. (This follows immediately from the π-π theorem as in Chapter 11 of [Wa1].)

Remarks. In a sense, homotopy equivalence exchange is a red herring as one does not need the condition on A to achieve the homotopy equivalence on B; we are emphasizing in this terminology that one loses what one has on A to obtain the equivalence on B.

We take the map to $M \times \mathbb{C}P^2$ produced above and make it, in addition, blocked split to a regular neighborhood of the 1-skeleton. This is possible by the π-π case of codimension one splitting. Thus, all of the nontriviality is now taking place in a region between a regular neighborhood of the 1-skeleton and the 2-skeleton. We can even arrange this to be a blocked homeomorphism in this region, if we wished, because structure spaces of thickened 1-complexes are trivial, but, in any case, we do not need this.

Inside the 1-neighborhood we have a trivialization of the Q-fibration compatible with the given splittings of fundamental groups. We now can form in the obvious way a 3/2 skeleton for the block and attach it back onto this one skeleton. The 3/2 skeleton is just obtained by attaching cells to kill the additional fundamental group; we extend the Q bundle trivially to the new cells. This produces a 3/2 skeleton within the region bounded by the boundary of the regular neighborhood of the original 2-skeleton.

Now perform a homotopy equivalence exchange (blockwise), to obtain the equivalence to be blocked homotopy outside the 3/2-neighborhood. In addition, the global assembly of the blocked exchanges can be made into a homotopy equivalence, by the π-π theorem, if you like.

Now on this 3/2 skeleton the bundle is trivial, so the argument finishes as before: the transfer is simply a product, which vanishes when we assume that the fiber Q bounds in a suitable sense.

This completes the argument to achieve the needed block fibration. We could moreover have arranged to keep the fiber unchanged by dealing with the base change problem relative to a solution over fixed discs in M' and M for which $M' \to M$ restricts to a homotopy equivalence of (M'-disc) to (M-disc). □

1.3   The fiber change problem for fibrations

Now we consider the cognate problem of changing the fiber of a block fibration. This is technically easier in that it does not involve the two stage difficulties of the base change problem of the previous section. These results are applied in sections 2.2 and 3.2 to solve problems in transformation groups concerning change of normal representations to fixed points. The main technical result is:

Theorem 1.9. *Let $f\colon V \to B$ be a PL block fibration of the PL stratified space V over a PL manifold base B. Assume that this fibration is trivial over the 1-skeleton of B and that there are compatable splittings of the fundamental group of each stratum of V to the fundamental group of the corresponding stratum of the stratified fiber Q. Then if the dimension of each stratum of the fiber is ≥ 5, a stratified PL space Q' is the fiber of a map $f'\colon V \to B$ homotopic to f if and only if:*

(i) *there is a transverse stratified homotopy equivalence $g\colon Q' \to Q$ for which the transverse stratified normal invariant $\nu(g) \in [Q; \mathbf{F/PL}]$ is zero; and*

(ii) *the Whitehead torsion $\tau(g)$ of g satisfies $\tau(g) \times \chi(B) = 0$ in $Wh(Q)$, for $\chi(B)$ denoting the Euler characteristic of B; and*

(iii) *letting $\alpha \in L_{*+1}^{\chi(B)}(Q)$ denote the surgery obstruction of a normal cobordism of g to the identity of Q, evaluated in the surgery obstruction group in which Whitehead torsions are evaluated in $Wh(Q)/\chi(B) \cdot Wh(Q)$,*

$$(\alpha \times \sigma^*(B)) \in (Im\ A_0) \ in\ L_{*+1}(Q \times B) \ .$$

Here $\sigma^*(B)$ denotes the Ranicki (or Ranicki-Weiss) symmetric signature of B evaluated in the theory of chain complexes with Euler-characteristic divisible by $\chi(B)$. The map A_0 denotes the composite of

$$[B, (\mathbf{L}_1(Q))_0] \rightarrow [B, (\mathbf{L}_1(Q))] \overset{A}{\rightarrow} L_{*+1}(Q \times B)$$

where $(\mathbf{L}_1(Q))_0$ denotes the connected component of the base point of the spectrum $\mathbf{L}_1(F)$ given by the appropriate simple homotopy equivalence surgery groups and A denotes the "partial assembly" (or, assembly with coefficients) of L-theory.

In some cases of interest, condition (iii) can be dropped.

Corollary 1.10. *Let $f: V \rightarrow B$ be a PL block fibration of the PL stratified space V over a 2-connected PL manifold base B, and B is not a point. Then if the dimension of each stratum of the stratified fiber Q is ≥ 5, a stratified PL space Q' is the fiber of a block fibration $f': V \rightarrow B$ homotopic to f if and only if conditions (i) and (ii) of Theorem 1.9 are satisfied.*

Examples. i) As a consequence of Corollary 1.10, fiber change by a normally cobordant space is often feasible over a simply connected base. On the other extreme, if B is a $K(\pi, 1)$ space, generalizations of the Novikov conjecture assert the injectivity of the map

$$A: [B; \mathbf{L}(Q)] \rightarrow L_*(B \times Q)$$

away from the prime 2, and even integrally when the relevant algebraic K-theoretic groups of Q vanish. When this is true, even when $Q' \rightarrow Q$ is a transverse simple homotopy equivalence condition (iii) will imply that $Q' = Q$ and so nontrivial fiber change is *not* possible. This injectivity hypothesis on A is satisfied for B a torus, when the L groups and spaces are decorated by the appropriate algebraic K-theory symbols (s, h, etc.) using the methods of [Sh] and more generally, as noted in [We4], it also follows for a class of fundamental groups considered in [Ca] using methods similar to that paper. Moreover, from these considerations, other cases in which the Novikov conjecture has been verified, e.g. [FW1], [FW2], yield similar examples in which nontrivial fiber change is impossible. (On the other hand, note that when Q' is h-cobordant to Q, even if $Q' \neq Q$ (whence, in particular $Wh(Q) \neq 0$), $S^1 \times Q$ does fiber over S^1 with fiber Q'.)

ii) An example of interest in group actions is when B is a (homotopy) lens space with π_1 of odd order. In that case $\chi(B) = 0$ and also $\sigma^*(B) = 0$, as it is in the image of "reduced" bordism of $\mathbb{Z}_n$ in $L^{odd}(\mathbb{Z}_{odd})$ which, as it agrees with L_{odd} away from 2, is 2-torsion. Thus in this base a lens space case, a fiber can be changed to precisely the stratified transverse normally cobordant homotopy equivalent spaces. In Section 3.2 this is applied to changing normal representations around (homotopy) lens spaces.

Remarks. 1) The condition of triviality over the 1-skeleton could be weakened to stratified homotopy triviality over the 1-skeleton, or equivalently, homotopy triviality of the stratified monodromy.

2) In fact, we could just view the surgery obstruction α of a normal cobordism of Q to Q' as lying in $L_1(Q)$ for $\chi(B)$ odd and in $L_1^{(2^a)}(F)$ for $\chi(B) = 2^a \cdot b$, b odd.

Similarly $\sigma^*(B)$ could be viewed as an element of the usual Ranicki (or Ranicki-Weiss) symmetric signature groups for $\chi(B)$ odd, and as an element of the "round" versions of these groups, for $\chi(B)$ even; recall that elements of the round groups are represented by chain complexes of Euler characteristic zero [HRT].

3) Of course, even when the dimensions of strata of Q are less than 5, the given conditions are still necessary for obtaining fibrations. Furthermore, one could often obtain sufficient conditions under much weaker dimension conditions; see, e.g. Remark 4 to Theorem 1.2 above.

4) Condition (i) is necessary for the fibration of f' with fiber Q' even without the hypothesis on π_1 or monodromy.

5) The condition in Corollary 1.10 on B being 2-connected and not a point could be weakened to the condition that B is not a point and that for each stratum of Q, π_1 is the same as that of the corresponding stratum of V.

6) The surgery group $L(Q \times B)$ could be viewed as a functor of Q as a stratified space and of $\pi_1(B)$ and the orientation character $\pi_1(B) \to \{\pm 1\}$ of the manifold B.

Proof of Theorem 1.9. The homotopy equivalence and normal cobordism of the fibers of the homotopic maps f and f' are by standard exercises in homotopy theory and transversality verified by making a homotopy of f to f' transverse to a point of B (which is possible because B is a manifold). This shows the necessity of condition (i).

Now, as shown in § 1.1 above, the bundle of **L** spaces over B, $E(\mathbf{L}(Q) \downarrow B)$ with fibre $\mathbf{L}(Q)$, induced by $f\colon V \to B$, is trivialized using the given hypothesis on π_1 and the 1-skeleton. Moreover, as remarked at the end of § 1.1, similar but much easier considerations apply to the bundle over B with fiber the spectrum $\mathbf{Wh}(Q)$. Then, comparing $f\colon V \to B$ with the fibration $f'\colon V \to B$, and using standard additivity for Whitehead torsions over a triangulation of B, we get that $\chi(B) \times \tau(g))$ represents $\tau(\text{identity of } V) = 0$ in $\mathrm{Wh}(V) = \mathrm{Wh}(V \times \pi_1(B))$. But as $\chi(B) \times \tau(g)$ represents an element of $\mathrm{Wh}(V)$, which maps split injectively to Wh(V), we conclude that

$$\chi(B) \times \tau(g) = 0 \quad \text{in} \quad \text{Wh}(Q)$$

To see the necessity of condition (iii), let $h\colon V \times I \to B$ be a homotopy of $f\colon V = V \times 0 \to B$ to $f'\colon V = V \times 1 \to B$. Let $p \in B$, so that we make the identifications $F = f^{-1}(p)$, $F' = {f'}^{-1}(p)$. We may assume that h is transverse to p so that $h^{-1}(p)$ is a transverse isovariant normal cobordism of 1_F, the identity of F to a homotopy equivalence $g\colon Q' \to Q$. As noted above, this normal cobordism represents an element α of $L_1^{\chi(B)}(Q) = \pi_0(\mathbf{L}_1^{\chi(B)}(Q))$. Using the trivialization in § 1.1 of the bundle $E(\mathbf{L}(Q) \downarrow B)$ as $B \times \mathbf{L}(Q)$, we can construct a normal cobordism blocked over B, representing $B \to \{\text{a point}\} \overset{a}{\to} \mathbf{L}_1^{\chi(Q)}(F)$ where the image of a lies in the component corresponding to $(-\alpha)$, yielding a cobordism Y of $f'\colon V \to B$ to a blocked map $f''\colon V' \to B$. In fact, if we do this construction of a blocked over B normal cobordism of V with some further care, by constructing it inductively over simplices of B, beginning over 0-simplices with the given normal cobordism of Q to Q', so that the torsion of the given homotopy equivalence over each simplex of B from the inverse image in V' to that in V is constantly $\tau(g\colon Q' \to Q)$, we would get that the given blocked map $f''\colon V' \to B$ is a blocked fibration with fiber Q'. (In a less geometrical and more abstract language, this could be rephrased as doing a cobordism construction representing a map from B to the fiber over the component corresponding to $\tau(Q' \to Q)$ of the "spacified" version of the Rothenberg map [Sh],

$$L_*^h(Q) \to \mathbb{H}^*(\mathbb{Z}_2; \text{Wh}(Q)) \;.$$

Of course, this fiber could be identified with the usual description of the fiber over the connected component of the identity, $\mathbf{L}_*(Q)$, the simple surgery spectrum.) Consider now the blocked cobordism over $W = V \times I \cup_{V \times 1} Y \to B$. Regarded as a normal cobordism from $V \to V$ to $V' \to V$, this has surgery obstruction given by, letting s denote surgery obstructions, $s(W) = s(V \times I) + s(Y) = 0 + \sigma^*(B) \times (-\alpha) = -\sigma^*(B) \times \alpha$. Moreover, from the construction, the inverse image of p in W is a blocked normal cobordism representing $\alpha + (-\alpha) = 0$. Thus, W can be viewed as representing an element of $[B, (\mathbf{L}_1(Q))_0]$ and thus $s(W) \in \text{Im } A$, for A the assembly map

$$[B, \mathbf{L}(Q)] \overset{A}{\to} L(B \times Q)$$

Thus, $-\sigma^*(B) \times \alpha \in \text{Im}(A)$.

Conversely, arguing as above, given a transverse stratified normal cobordism from Q to Q' with surgery obstruction $\alpha \in \mathbf{L}_*^{\chi(Q)}(Q)$ satisfying (i), (ii) and (iii), and using the trivialization of the fibration of $E(\mathbf{L}(Q)) \downarrow B)$ over B, we can construct a normal cobordism with surgery obstruction $\sigma^*(B) \times (\alpha)$ of V to a simple homotopy equivalent V' blocked over B with the inverse image of a basepoint $p \in B$ being Q'. Then, using $-(\sigma^*(B) \times (\alpha)) \in \text{Im}(A_0)$, we can construct a further normal cobordism, represented by $(B, p) \to \mathbf{L}(Q)$, to a blocked fibration V'' over B with the inverse image of the basepoint p still Q'. Moreover, we have arranged that the composed normal cobordism from V to V'' has zero surgery obstruction. Thus $V = V''$ and the construction is completed. $\qquad\square$

Proof of Corollary 1.10. It is a familiar and easy fact that the maps from B, an oriented manifold of dim $n > 0$, to $\mathbf{F/TOP}$ satisfy $[B, \mathbf{F/TOP}] = [B - D^n, \mathbf{F/TOP}] \times [D^n \text{ rel } \partial, \mathbf{F/TOP}]$; this is obtained using the map induced from the pinch map $B \to S^n$ (or by using $\mathbf{L}^*(\mathbb{Z})$ orientations) to split the surgery obstruction assembly map. The same methods readily yield the generalization that $[B, \mathbf{L}(Q)] = [B - D^n, \mathbf{L}(Q)] \times [D^n \text{ rel } \partial; \mathbf{L}(Q)]$ where for $D^n = n$-disc, $[D^n \text{ rel } \partial, \mathbf{L}_*(Q)] = L_{*+n}(Q)$. Using this last fact, it is easy to see that in this case A_0 is surjective, yielding condition (iii) automatically for $\pi_1 B = 0$. $\qquad\square$

In the applications to normal representation replacement in transformation groups in § 2.2 below, we can often make do with slightly weaker forms of fiber change than those of Theorem 1.9 and Corollary 1.10. We give sample results in which we allow the total space of the fibration to vary by an h-cobordism and get just two conditions instead of the three of Theorem 1.9.

Theorem 1.11. *Hypothesis as in 1.9. Then if the dimension of each stratum of the fiber is ≥ 5, a stratified PL space Q' is the fiber of a block fibration $f': V' \to B$, where V' is (transverse) stratified h-cobordant to V and f' is homotopic to the composite map $V' \to V \to B$ if and only if*

(i) *there is a transverse stratified homoptopy equivalence $g: Q' \to Q$ for which the transverse stratified normal invariant $\nu(g) \in [Q; \mathbf{F/PL}]$ is zero;*

(ii) *letting $\alpha \in L^h_*(Q)$ denote the h-surgery obstruction of a normal cobordism of g to the identity of Q,*

$$\alpha \times \sigma^*(B) \in Im(A_0^h) \text{ in } L^h_{*+1}(Q \times B) \ .$$

Here $\sigma^*(B)$ denotes the Ranicki (or Ranicki-Weiss) symmetric signature of B and A^h is the composite

$$[B, (\mathbf{L}(Q))_0] \xrightarrow{A} L_*(Q \times B) \to L^h_*(Q \times B)$$

Remarks. 1) We emphasize that here undecorated L-groups or spaces continue to denote the obstruction groups or spaces to simple homotopy equivalence.

2) Note that condition (ii) implies that for $\chi(B) \not\equiv 0 \pmod 2$ $\tau(g)$ represents 0 in $H^*(\mathbb{Z}_2; \text{Wh}(Q))$. This is because $\tau(g) \times \chi(B)$ represents the image of $(\alpha \times \sigma^*(B))$ under the map in the Rothenberg exact sequence map $L^h_*(Q \times B) \to H^*(\mathbb{Z}_2; \text{Wh}(Q \times B))$

The proof of Theorem 1.11 and its Corollary 1.12 are similar to those of 1.9 and 1.10 and so are omitted. $\qquad\square$

Corollary 1.12. *Let* $f: V \to B$ *be a PL block fibration of the PL stratified space* V *over a 2-connected PL-manifold base* B, *and* B *is not a point. Then if the dimension of each stratum of the fiber* Q *is* ≥ 5, *a stratified PL space* Q' *is the fiber of a block fibration* $f': V' \to B$ *where* V' *is (transverse) stratified h-cobordant to* V *and* f' *is homotopic to the composite* $V' \to V \to B$ *if and only if*

    (i) *there is a transverse stratified homotopy equivalence* $g: Q' \to Q$ *for which the transverse stratified normal invariant* $\nu(g) \in [Q; \mathbf{F}/\mathbf{PL}]$ *is zero; and*

    (ii) $\chi(B)$ *is even, or* $\chi(B)$ *is odd and* $\tau(g)$ *represents zero in* $H^*(\mathbb{Z}_2;\ Wh(F))$.

    The remarks 1, 3, 4, 5 and 6 after Corollary 1.10 apply to Theorem 1.11 and Corollary 1.12 as well.   □

Remark. The simplifications of the conditions of Theorems 1.9 and 1.11 for a simply connected base B in Corollaries 1.10 and 1.12 also apply when $\pi_1(B)$ is of odd order and the actions of $\pi_1(B)$ on the fiber is stratified homotopically trivial. The proof is similar to that of the corollaries using that $\sigma^*(B)$ lies in the image of $L^*(0)$; this is because, as in Example (ii) after 1.10, its image in the reduced L^*-group, which has no odd torsion for finite π_1, is in the image of a reduced bordism group of $\pi_1(B)$ and so is of odd order.

2   Bubble quotients and transformation groups

2.1   Fixed set replacement

A basic idea for proving the results mentioned in the introduction, and the others to be found below, is to view the boundary of a neighborhood of the fixed set of a given action as an equivariant block sphere bundle over the fixed set. Then to replace the fixed set we might try to use results of Section 1.2 to change the base of the block fibration to the new candidate for submanifold of fixed points. If we can do this, we will achieve at least a *weak replacement*, that is a homotopy equivalent manifold with the desired submanifold as fixed set of an equivariantly homotopy equivalent action on it. In order to also control the PL homeomorphism type of this manifold we will perform a change of base to a more subtle block bundle: the "bubble quotient" neighborhood.

Definitions. The *bubble quotient* of a G action on M consists of the space $M/\sim$, where $m \sim m'$ iff either $m = m'$ or m and m' both lie on the same G orbit which does not pass within a small regular neighborhood of the fixed set.

    The picture for the nontrivial involution on $\mathbb{R}^1$ explains the name: it looks like the usual quotient except for a "bubble" near the image of the fixed set. The *bubble quotient neighborhood* is the image of the chosen small invariant regular neighborhood in the bubble quotient. For instance if the fixed set F has codimension c for a locally linear involution, then the bubble quotient neighborhood has the structure of an $\mathbb{R}P^c$ block bundle over F

In fact, the bubble quotient is naturally a stratified space with three strata (in the semifree case). The smallest stratum is the quotient of the boundary of the regular neighborhood. Attached to it are two other strata — the quotient of the complement of this neighborhood, and the neighborhood itself. Restricting to the bubble quotient neighborhood, we see that it has a natural stratification as a space with two strata. In the example for an involution, this would be a block fibration over the fixed points with fiber the pair $(\mathbb{R}P^c, \mathbb{R}P^{c-1})$.

Proposition 2.1. *A necessary and sufficient condition for performing a strong replacement for a component of the fixed point set of an action is to be able to perform the corresponding base change on the bubble quotient neighborhood block fibration.*

The proof is almost by definition. The definition of strong replacement gives a homeomorphism of the complements, and in particular, asserts that the boundary of a regular neighborhood of F block fibers over F'. The statement that unequivariantly the result of glueing in the associated disk block bundle over F' is homeomorphic to the result of glueing in the one over F is precisely the extension of this over the "bubble". The converse is a similar disentangling of the definitions.   □

In view of this, to obtain results on replacement of fixed points, we will use the results of Section 1.2 on the base change problem. These are applicable by the following:

Proposition 2.2. *Let G be a finite group acting PL and locally linearly on a PL manifold G, and smoothable in a neighborhood of a 1-skeleton of a component F of the fixed point set. Then the associated bubble quotient bundle over F satisfies the hypothesis of Proposition 1.5. Hence, the associated bundle of $\mathbf{L}$-spectra over F is trivial.*

Proof. If the action is smooth near the 1-skeleton, then the monodromy is readily seen to be trivial. This is because the structure group for such equivariant bundles is a product of unstable unitary groups, and hence, connected — see e.g. [LR].   □

Remark. Note that, in particular, from this assumption of smoothness near the 1-skeleton of the fixed points and G odd, the monodromy transformation of the equivariant normal disc around a loop in the fixed points must be orientation-preserving. Moreover, this conclusion remains valid under the hypotheses of the Addendum to Theorem 0.1 and of Theorem 0.2. This follows, for example, from the fact that nontrivial linear fixed point free spheres for odd order groups or general free actions cannot have equivariant orientation reversing homeomorphisms; indeed, the ρ-invariants of cyclic odd or free linear actions are non-trivial but reverse sign under change of orientation.

Now we can prove the theorems on replacing fixed points asserted in the introduction.

Proof of Theorems 0.1 and 0.2. Let us first concentrate on the semifree case. In that case the bundles have fiber the local bubble pair $B = (K \cup D, K)$ where K is a spaceform covered by the sphere ∂D and D is a disk, and where we glue K to the boundary via the covering map of the sphere to the space form. The Browder-Quinn $\mathbf{L}$ space for this situation is the fiber of the transfer map $\mathbf{L}(G) \to \mathbf{L}(e)$. This induces a homotopy exact sequence:

$$\to L_c(G) \to L_c(e) \to L(B) \to L_{c-1}(G) \to L_{c-1}(e)$$

(where c denotes the codimension of the fixed set, and is assumed at least three). We note the following features:

Lemma 2.3. *For G of odd order, $L(B)$ is torsion free, and detected by the G-signature of the singular set, as a split summand of $L_{c-1}(G)$. For G of even order there is always a split surjection $L_c(B) \to \mathbb{Z}_2$, which is given either by the Kervaire invariant of the singular set, or when $G = \mathbb{Z}_2$, and c is even, the Kervaire invariant of the forgetful map of the singular structure.*

The map for $G = \mathbb{Z}_2$ forgetting the singular structure is simply the map that recalls that in this case the bubble fiber is a genuine manifold, namely $\mathbb{R}P^c$. This theorem follows immediately from the calculations in [Wa2] and the exact sequence above.   □

A first step in proving the replacement theorems for fixed points is then to compute the primary obstructions to base change in the associated bubble quotient neighborhood bundle. As seen in Section 1.2 above, this primary obstruction is given by composing a normal invariant with a map of $\mathbf{L}$-spectra given by taking products with B. As in the discussion of maps of $\mathbf{L}$ spectra there, we have to compute the result of forming a product with $\mathbb{Z}_n$-manifolds, i.e. compute the product $\times B$:

$$L_i(e; \mathbb{Z}_n) \to L_{i+b}(B; \mathbb{Z}_n) , \qquad b = \dim B .$$

Now, L-theory with coefficients satisfies a Bockstein long exact sequence. Thus for G of odd order, it is easy to see that either the domain or range is trivial except in the case $L_3(e; \mathbb{Z}_n) \to L_{3+b}(B; \mathbb{Z}_n)$, n even. Here the domain is detected by the Bockstein, but the range is detected by restricting to the bubble singularity, i.e. the map is detected by

$$L_3(e; \mathbb{Z}_n) \to L_{3+c}(K; \mathbb{Z}_n)$$

under the map of crossing by K. Now this map is zero because K has vanishing symmetric signature. This vanishing can be seen by first combining the results of [R] on the difference between quadratic and symmetric L-theory being 2-torsion

and the computations of [Wa2] that for a finite group the torsion subgroup
of the quadratic L group has only 2-torsion to yield the same conclusion for a
symmetric L group. Now, compare this with the fact that a space form with odd
order fundamental group is odd torsion even in the *geometric* bordism group.

For G of even order, the calculation is somewhat different; the Kervaire
elements do not vanish after crossing with B, as in the calculation for real
projective space in [Wa1]. The product of B with terms like $L_0(e; \mathbb{Z}_n)$ vanish
by the calculations in [St].

To deal with the secondary obstructions to this base change problem, we use
the following trick: We puncture F' and replace for the punctured action on M.
Doing this changes the situation from a closed problem to a π-π (*e-e*) problem,
so that there is no secondary obstruction at all.

It remains to fill in this remaining puncture with a locally linear action on a
disc; the difficulty is to see linearity on the boundary sphere. However, one can
certainly cone the boundary to obtain a group action with locally flat fixed set
but the resulting action is not necessarily locally linear at the cone point. The
remaining obstruction to concording this to a locally linear action was analyzed
completely in [CW3] and lies, in this case, in the Tate cohomology of Wh(G).
However, since the construction is performed here with simple homotopy equivalences, this obstruction vanishes.

Theorem 0.2 and the simply connected case of Theorem 0.1 are now proven
in the semifree case. Our original approach to Theorem 0.1 continued along
similar lines using the stratified theory. The propositions of the previous section still apply to trivialize the bundle of **L**-spectra. Thus, we are reduced to
analyzing similar product formulae for the more seriously singular spaces that
arise, and more precisely to demonstrating the same vanishing theorem for all
these products.

Now we will apply the above monodromy-signature criterion to prove the
theorem. Thus, we would only have to check that the symmetric signature
vanishes. However, explication of this vanishing of signature is a little prob-
lematic because there is no available treatment of stratified (or equivariant)
symmetric L-theory in the literature. The natural possibility of using Weiss's
visible L-theory [Ws] (which has a transfer and can be therefore amplified into a
stratified symmetric analogue of Browder-Quinn theory, see [We1]) makes sense;
however, it would not be directly applicable to the present problem, as in this
theory the symmetric signature of the quotient of a representation on a sphere
is a nontrivial element.

Instead, we observe that the transverse isovariant π-π theorem in quadratic
L-theory gives a canonical reason why finding G-manifolds (bubbled) whose
boundary is the given representation (bubbled) will trivialize products. Thus it
will suffice to find such "coboundaries" for our representations.

Away from the prime 2 it is quite straightforward to use symmetric signatures to kill the products (see e.g. [CSW] and [LM II]), so we localize at
2. The argument in [LM II] shows that an odd multiple of the representation

bounds in (essentially) a π-π fashion using that $MU(G)$ is odd torsion in odd dimensions. Moreover, this argument easily extends to bubble quotients; this is seen by regarding them as representing elements in the fiber of the transfer or "forgetful" map $MU(G) \longrightarrow MU(\text{point})$, which is split surjective at $(1/|G|)$. This completes the proof of Theorem 0.1.

To prove the addendum, one observes further that assuming the gap hypothesis, isovariant monodromy can be computed from the equivariant monodromy (from Browder's theorem comparing isovariant and equivariant classifications under gap hypothesis, see [Do] for a similar result), and the latter vanishes because of the orientability of all fixed point sets and the above proof. If G is odd order abelian, even without the smoothness on the 1-skeleton the normal representation to the fixed points is complex and some multiple W is a "periodicity representation" in the sense of Yan [Y] by the main result of [WY1]. Then there is a map

$$\mathbf{S}_G(M) \to \mathbf{S}_G(M \times D(W) \text{ rel } \partial) ,$$

which is an isomorphism, and which commutes with the forgetful map to be unequivariantly the Siebenmann periodicity [KS], [CW1] and also restricts to commute with Siebenmann periodicity on the fixed sets. Using this we stabilize, strongly replace, and then invert the periodicity to replace on the original manifold. □

Remarks. 1) Using the present methods and the thesis [Y] one can show that for arbitrary odd order groups, there are representations at the fixed set that enable replacement. For instance, if all possible isotropy occurs near the fixed set then one can cross with $4k\times$ regular representation and repeat the same argument.

2) An argument similar to our proof above using a "multiply bubbled quotient" space (constructed by the device of glueing on many more bubbles; one for each conjugacy class of subgroup of isotropy; that is, for each fixed sphere S^H, we attach a disk of the relevant dimension) actually guarantees that all of the fixed sets of all subgroups, other than the fixed set of the whole group, are unchanged by the replacement.

3) By working relative to fixed discs in F' and F, one can readily arrange that the normal representations of G to F and F' are the same.

As promised in the Introduction, we also describe some variants and extensions. First, we have the following result for circle actions:

Theorem 2.4. *A simple homotopy equivalence in dimension not 3 or 4 can be strongly replaced for a PL semifree circle (or SU(2)) action iff it is homotopic to a PL homeomorphism.*

Addendum. This is also valid for dimension 3 or 4 fixed points of codimension not 2 with the PL homeomorphism condition replaced by stable (after $\times \mathbb{C}P^2$) PL homeomorphism.

Proof. The exact same analysis performed above applies here, except that now the bubble fiber always contains an even dimensional complex (or quaternionic) projective space either as the singular set or as the nonsingular total space. Of course, multiplication by such a projective space induces the periodicity isomorphism.   □

Remarks. This result is closely related to our geometrical interpretation of Siebenmann periodicity phenomena, see [CW1]. More precise information can be obtained using the results of [CW4].

For even order groups we will use a weak replacement construction to produce group actions in some cases that we have shown do not satisfy strong replacement. To achieve weak replacement, consider a variant application of our base change method; we simply forget the bubble quotient and consider instead just the easier base change problem for the block fibration with total space the quotient of the equivariant normal sphere bundle to the component of the fixed points and base this component of the fixed points.

We will illustrate how the obstructions decrease for weak replacement in the case of semifree actions. Recall that a group acting semifreely on a manifold has either cyclic or quaternionic 2-Sylow subgroup; this is because the normal representation to the fixed points is then free, i.e., the action on its unit sphere is free.

Theorem 2.5. *If the 2-Sylow subgroup of G is cyclic then for any orientation preserving free representation, one can weakly replace. If the 2-Sylow subgroup is Quaternionic, then the vanishing of the Kervaire classes condition is still necessary. If $G = S^1$ acts semifreely with the codimension of the fixed point component to be replaced divisible by 4, then one can weakly replace.*

The proof is exactly the same as described for strong replacement, except the calculations work out a little differently. In the case of the circle, one observes that all of those representations extend to representations of $SU(2)$; one can hence use the associated Hopf fibrations to still make the representation spheres equivariant π-π boundaries.   □

Remarks. In [CWY] we will show that for arbitrary locally linear actions of abelian groups it possible to perform weak replacement of fixed sets whose normal representations are twice a complex representation. For $G = S^1$ the condition on a normal representation that it extend to $SU(2)$ will be relaxed to just it being twice a complex representation.

2.2   Normal representation replacement

In this section we study the problem of changing the normal representation to one component F of the fixed points of a PL locally linear G action on a manifold M. The methods developed will achieve this, for certain reprsentations and certain F, without changing G, M, F, the isovariant homotopy type of the action and the group action outside a closed neighborhood of F. Just as the results on replacing fixed points of section 3.1 uses the results on the base change problem in fibration of section 1.2 applied to the "bubble quotients" of group actions introduced above, the present results on replacing normal representations will use the results on the fiber change problem of Section 1.3 applied to these bubble quotients.

Analogous to our formulation in the Introduction of the notion of strong replacement for fixed points, we formulate a cognate notion of strong normal replacement for the normal linear representations to fixed points. For technical reasons, related to Whitehead torsion, it will also be convenient to have available a slightly weaker notion of h-strong normal replacement.

Definition. A *strong normal replacement* of the normal linear representation γ to a component F of the fixed points of a locally linear PL action of a group G on M, by a linear representation γ' of G is another PL locally linear action of G on M with the following:

1) an isovariant homotopy equivalence ϕ to the original action and ϕ restricts to a PL homeomorphism on $\phi^{-1}(F) \to F$;

2) the restriction of ϕ to the complement of an equivariant regular neighborhood of $\phi^{-1}(F)$ an isovariant homeomorphism, and the restriction of ϕ maps this regular neighborhood of $\phi^{-1}(F)$ to an equivariant regular neighborhood of F;

3) an unequivariant homotopy, relative to this decomposition of the G-manifolds and the identification of complements, of this ϕ to a homeomorphism; and

4) the new action has normal representation γ' to the fixed set instead of γ.

An *h-strong normal replacement* of γ to γ' is defined just as a strong normal replacement, except that in condition 2, the condition on an "equivariant regular neighborhood" is slightly weakened to a "G-invariant regular neighborhood."

An example that illustrates the difference between G-invariant and equivariant regular neighborhoods is given by the following. Let G act linearly on a disc D, with the action free on the boundary sphere S, and we let W be a nontrivial equivariant h-cobordism with boundary $S \cup S'$; then in $D \cup_S W$ both D and $D \cup_S W$ are G-invariant regular neighborhoods of the fixed point at the origin, but only the former is an equivariant regular neighborhood.

Thus, as an h-strong normal replacement also does not alter G, M, the component F of the fixed point set, the isovariant homotopy type of the action and the action outside a G-invariant regular neighborhood of F, but does change the normal linear representation from γ to γ', it is powerful enough to construct the new G-actions on M we will need.

For γ a linear representation, we let the bubble quotient of γ denote the bubble quotient of a disc neighborhood of the origin; that is the bubble quotient of the orthogonal representation γ of G is the stratified quotient space of the unit disc D of the representation space $D/\{x \sim gx \mid g \in G, x \in S$, the boundary of $D\}$. This bubble quotient, for a nontrivial representation, is never a manifold.

Cognate to Proposition 2.1 and with analogous proof is then the following result:

Proposition 2.6. *A necessary and sufficient condition for performing a strong normal replacement for the normal linear representation at a component of the fixed points of an action is to be able to perform the corresponding fiber change on the bubble quotient neighborhood block fibration with fibers the corresponding bubble quotients of linear representations.* □

Remark. There is an analogous result for h-strong normal replacement in which the corresponding fiber change can be achieved on a space h-cobordant to the bubble quotient neighborhood, where the torsion of the h-cobordism transfers trivially to the Whitehead group of the base; this condition on Whitehead torsion can be reformulated by requiring the image of this element under Wh(bubble quotient) $\rightarrow$ Wh(π_1(fixed points)) vanishing. Here the map on the PL Whitehead group is the split surjection given by restricting to the complement of a regular open neighborhood of the singular set, which is the quotient of the sphere bundle in the bubble quotient.

Combining this with Theorem 1.9 yields the following result:

Corollary 2.7. *Let G act by a PL locally linear action on a PL manifold M with F a component of the fixed points and γ the normal G-representation to F with the action smoothable near a 1-skeleton of F. Assume that the quotient of the unit sphere of the representation space $S(\gamma)/G$ has no strata of dimension 3 or 4. Then the representation γ can be strongly normally replaced by γ' if and only if*

(i) *there is a transverse stratified homotopy equivalence of the unit spheres of the representation spaces $g: S(\gamma')/G \rightarrow S(\gamma)/G$ determining the trivial normal invariant $B(\gamma) \rightarrow \mathbf{F/PL}$, for $B(\gamma)$ the bubble quotient of γ.*

(ii) *The Euler characteristic of the fixed points, $\chi(B) = 0$.*

(iii) *is the same condition on surgery obstructions as (iii) of Theorem 1.9.*

Remarks. Notice that $B(\gamma) = (S(\gamma)/G) \cup e^{\dim(\gamma)}$ so that the normal invariant of condition (i) is that of $S(\gamma)/G$ extended to one more cell. This was called the extended normal invariant in Theorem 0.4.

Proof. This follows from Theorem 1.9 and Proposition 2.6 with the following simplifications. Using the radial direction, $g: S(\gamma')/G \rightarrow S(\sigma)/G$ induces a map of bubble quotients $\bar{g}: B(\gamma') \rightarrow B(\gamma)$; if g is a transverse stratified homotopy equivalence, so is $\bar{g}$. The condition $\chi(F) \times \tau(\bar{g}) = 0$ implies that $\chi(B) = 0$ as the Whitehead torsion of a homotopy equivalence of unequivalent linear actions on spheres is not a torsion element. This is classical for lens spaces and follows in general as a linear representation is determined by its restriction to cyclic groups. □

When the fixed points do not have Euler characteristic zero, it is more convenient to change normal representations to fixed points by using instead the following:

Corollary 2.8. *Let G act by a PL locally linear action on a PL manifold M with F a component of the fixed points and γ the normal representation to F and with the action smoothable near a 1-skeleton of F. Assume that the quotient of the unit sphere of the representation space $S(\gamma)/G$ has no strata of dimension 3 or 4. Then the representation γ can be h-strongly normally replaced by γ' if and only if*

(i) *is as condition (i) on vanishing of extended normal invariants of Corollary 2.7; and*

(ii) *is the same condition on surgery obstructions as (iii) of Corollary 2.7 (i.e., of Theorem 1.9).*

Proof. Similarly to the proof of Corollary 2.7 above, the proof follows by combining Proposition 2.6 and Corollary 1.11. Here there is the simplification that the Whitehead group of the disc is 0. □

For many fixed points, particularly for simply connected spaces *not a point*, the condition on surgery obstructions in (ii) of Corollary 2.8 can be radically simplified.

Corollary 2.9. *Let G act by a PL locally linear action on a manifold M with F a component of the fixed points and F simply connected but not a point. Assume that the quotient of the unit sphere of the representation $S(\gamma)/G$ has no strata of dimension 3 or 4. Then the representation γ can be h-strongly normally replaced by γ' if and only if*

(i) *is as condition (i) on vanishing of extended normal invariants of Corollary 2.7; and*

(ii) $\chi(K)$ *is even or $\tau(g)$ represents 0 in $H^*(\mathbb{Z}_2;\ Wh(S(\gamma)/G))$.*

Addendum. *There are similar results when the condition on F being connected is generalized to $\pi_1(F)$ of odd order and the action is smoothable near a 1-skeleton of F.*

Proof. Combine the above results with Corollary 1.12. Of course, Corollary 1.12 used a stronger 2-connected hypothesis on F, but the condition on π_2 was only used in the proof to obtain a computation of π_1 of the total space of fibrations; this is easily obtained directly for the quotients of strata of equivariant sphere bundles. □

Remarks. 1) Note that for G cyclic, $\tau(S(\gamma')/G \to S(\gamma)/G)$ can be readily computed from the difference of the Reidemeister torsions of $S(\gamma')/g$ and $S(\gamma)/G$. This is, of course, classical; see the exposition in [M], [Co].

2) If the G action is semifree, so that the action on $S(\gamma)$ is free, the condition in (i) of the above corollaries on the quotients of the spheres being transverse isovariantly homotopy equivalent just reduces to their quotients being homotopy equivalent. Moreover, this is classically equivalent to the associated and easily computed k-invariants in $(\mathbb{Z}/|G|)^\times$ of these space-forms being equal; see [M], [Co].

Verification of the conditions on extended normal invariants of this section is facilitated by the following two propositions, which can often be effectively used together.

Proposition 2.10. *Let γ and γ' be linear repesentations of a finite group G for which the associated unit spheres are transverse isovariantly homotopy equivalent. Then the corresponding extended normal invariant of the bubble quotient, $B \to \mathbf{F}/\mathbf{PL}$, is trivial if and only if this is true for the bubble quotients for the Sylow subgroups of G.*

Moreover, we note that for cyclic groups, the determination of when these extended normal invariants of bubble quotients of group representations vanish can be turned into a more conventional problem concerning the vanishing of the usual normal invariant of a quotient of a stabilization of the spheres of the representations.

Proposition 2.11. *Let γ and γ' be representations of the cyclic group $G = \mathbb{Z}_m$ and with $S(\gamma)/G$ and $S(\gamma')/G$ transverse isovariant homotopy equivalent; let δ be a free representation of G. Then the extended normal invariant of the bubble quotients $B(\gamma)$ vanishes if and only if the normal invariant $S(\gamma \oplus \delta)/G \to \mathbf{F}/\mathbf{PL}$, determined by the homotopy equivalence $S(\gamma' \oplus \delta)/G \to S(\gamma \oplus \delta)/G$, vanishes through dimension $dim(\gamma) + 1$.*

There is also an analogous principle for any group G equipped with a free representation δ. For G cyclic, this result is most easily applied by taking δ to be a 1-dimensional free complex representation. In that case, as $\pi_{2k+1}(\mathbf{F}/\mathbf{PL}) = 0$, the criteria reduces to just the vanishing of the normal invariant $S(\gamma \oplus \delta)/G \to \mathbf{F}/\mathbf{PL}$.

Proof of Proposition 2.10. It is a standard (and easy) fact from homotopy theory that a normal invariant, that is a map to $\mathbf{F}/\mathbf{PL}$, vanishes if and only if all its localizations do. Moreover, letting G_p denote the p-Sylow subgroup of G, the map of bubble quotients $B(\gamma|G_p) \to B(\gamma)$ is a p-local homotopy equivalence. Combining these two facts, the result follows. $\square$

Proof of Proposition 2.11. In fact, we can identify B with the skeleton of $S(\gamma \oplus \delta)$ and thus identify the normal invariants. $\square$

In view of Corollary 2.9, when F is simply connected, not a point, and with $\chi(F)$ odd, construction of normal replacement requires some Whitehead torsion, modulo 2× Whitehead group, calculations. This can often be readily seen to vanish. For example, we have the following result, applicable to doubles of representations:

Proposition 2.12. *Let γ and γ' be linear representations of the same dimensions of the cyclic group G which are free actions on the unit spheres with $\gamma - \gamma' = 2(\alpha - \beta)$ in the representation theory of G where α and β hvae equal k-invariants in $(\mathbb{Z}/|G|)^\times$. Then, letting $g\colon S(\gamma)/G \to S(\gamma')/G$ be the homotopy equivalence, its Whitehead torsion $\tau(g)$ represents 0 in $H^*(\mathbb{Z}_2; Wh(G))$.*

Proof. Under the hypothesis on $r - r'$ being a double, g is obtained as an equivariant join of the form $g = f * h * h$, where f is an identity map on a linear lens space. To see that the difference of the Whitehead torsions of g and of f is then twice that of h, we use the standard decomposition of joins and the addition and multiplication formulae for Whitehead torsions of unions and products. □

Example. (i): Let $G = \mathbb{Z}_{pq}$, a cyclic group, where p and q are relatively prime integers, $p, q \geq 5$. Pick a, b prime to p, $1 \leq a < b \leq p/2$, and pick c, d prime to q, $1 \leq c < d \leq q/2$. Let s_1, s_2, s_3, s_4 be the integers modulo (pq) given by

$$s_1 \equiv s_3 \equiv a(\bmod\ p)\ , \quad s_2 \equiv s_4 \equiv b(\bmod\ p)$$
$$\text{and } s_1 \equiv s_4 \equiv c(\bmod\ q)\ , \quad s_2 \equiv s_3 \equiv d(\bmod\ q).$$

Then letting t^ℓ denote the complex 1-dimensional representation of G sending the generator to $e^{2\pi i(\ell/pq)}$, the real representations $2(t^{s_1} \oplus t^{s_2})$ and $2(t^{s_3} \oplus t^{s_4})$ can (h-strongly) normally replace each other around a simply connected fixed point component F if and only if F is not a point.

(ii) On the other hand, the argument in example (i) following Corollary 1.10 above can be used to show that for G of odd order normal replacement is never possible when the fixed point component is $K(\pi, 1)$ space for π satisfying a sufficiently strong form of the Novikov conjecture, e.g. for F a torus and the other examples discussed above. Of course, this is related to the dichotomy in normal representation replacement between a point and all other simply-connected fixed point components.

3 Computations for examples

3.1 Fixed points and Whitehead torsion

So far in our treatment of fixed point replacement in Section 2.1, we have only considered the question of replacing by simple homotopy equivalent submanifolds. In some of the natural examples, e.g. for linear group actions on lens

spaces with fixed points sub-lens spaces, the homotopy equivalences of interest are often not simple. We describe briefly the modifications in the theory
necessary to analyze fixed point replacement by homotopy equivalence.

The analysis for this case is somewhat similar to that of § 1.2 and § 2.1, but
relies on the notion of round-L-theory to define the relevant transfers. (See [R3].)
(We still strive for equivariant homeomorphisms of complements.) However, the
bubble now gives some new difficulty because the Euler characteristic of a disk
is 1 and thus contributes Whitehead torsions in transfers.

The strong replacement method of § 2.1 then yields an equivariantly homotopy equivalent action (equivariantly homeomorphic outside regular neighborhoods of fixed sets) on manifolds whose torsions are the same as the original
torsions between the fixed sets, i.e. for $i: F \to M$ the given inclusion of fixed
points,

$$\tau(M' \to M) = i_* \tau(F' \to F) .$$

At this point one can apply one's favorite equivariant surgery theory to study
to what extent one can change the complement to get rid of this torsion. For
instance, let $tr : \ker\big(\text{Sym}\ \big(Wh^G(M - N(F))\big) \xrightarrow{r} L_*^{S,G}(M)\big) \longrightarrow Wh(M)$ be
given by the "forget equivariance" transfer map on the given subgroup of the
symmetric elements in the equivariant Whitehead group of the complement of
an open neighborhood $N(F)$ of F, and r the map of an equivariant Rothenberg
sequence. One then obtains:

Theorem 3.1. *Let G be an odd order group acting PL locally linearly on M
with fixed set F, satisfying the no codmension two gap hypothesis and smoothness on the 1-skeleton or abelian (or large gap) hypotheses of the Introduction
and $F' \to F$ a homotopy equivalence of PL manifolds. Then there is an equivariantly homotopy equivalent PL locally linear action on M with fixed set F' if the
image of the Whitehead torsion, $i_* \tau(F' \to F)$ satisfies $i_* \tau(F' \to F) \in \text{Im}\,(tr)$.
Moreover, in that case the action on M with fixed set F' can be constructed to
have the same normal representation of G as F.* □

Addendum. *Similarly, there is an analogue of Theorem 0.2 of the introduction
for semifree actions of even order groups when $\pi_1(F)$ is of odd order and $F' \to F$
a homotopy equivalence (that is not necessarily simple), in terms of the vanishing
condition on the Kervaire classes of 0.2 and the extra torsion obstruction as in
Theorem 3.1.*

Here the Rothenberg map is the map in the Rothenberg exact sequence [Sh]
from Tate cohomology of Whitehead groups to L-groups. The addendum uses
also the Remark at the end of § 1. □

As an example, consider the case when $\pi_1(F) = \mathbb{Z}_q$ is cyclic and $G = \mathbb{Z}_n$, a
prime to q, has a semifree action on M with F a component of the fixed points.
Using standard calculations with the L-theory of cyclic groups and the relations
to torsions, we see that this Whitehead torsion obstruction to replacing fixed

points vanishes in even dimensions and in odd dimensions is the image of $\tau(F' \to F)$ in the cokernel of the transfer homomorphism $Wh(Z_n \times Z_q) \to Wh(Z_q)$; transfer is the same as forgetting equivariance.

This transfer for $G = Z_n$, n prime to q is, of course, closely related to the norm map for units in cyclotomic fields. Now let $B_q =$ the subgroup of "cyclotomic units" in $Wh(Z_q)$. Geometrically, this can be described as the subgroup generated by Whitehead torsions of homotopy equivalent linear lens spaces with fundamental group Z_q; these are given by the quotients of their Reidemeister torsions. Algebraically, B_q has the same rank as the group $Wh(Z_q)$. Indeed, the finite cokernel $A_q = Wh(Z_q)/B_q$ is closely related to class number considerations; let $\sigma(q)$ denote its order. Now, it is easy to see geometrically that the transfer homomorphism, for n prime to q, from $B_{n \times q}$ to B_q is onto. Hence in studying such replacement for spaces with fundamental group Z_q, we can perform our transfer and torsion calculations in A_q, yielding:

Proposition 3.2. *Under the hypotheses of 3.1 and its addendum for $\pi_1(F) = Z_q$ and $G = Z_n$, n prime to q, the torsion obstruction to fixed point set replacement of F by a homotopy equivalent manifold F',*

(i) *for dim F even, vanishes*

(ii) *for dim F odd, is the image of $\tau(F' \to F)$ in the cokernel of the map of finite groups $A_{nq} \to A_q$ (or, equivalently, in $Coker(A_{nq} \otimes Z_n \to A_q \otimes Z_n))$.*

$\square$

Another case for which the torsion obstruction always vanishes is given by:

Proposition 3.3. *Suppose that F is a component of the fixed set of a PL locally linear involution on a manifold M with F of codimension greater than 2 and with $\pi_1(F)$ finite of odd order. Then for any homotopy equivalence $F' \to F$ with vanishing Kervaire classes, there is an equivariantly homotopy equivalent PL locally linear involution on M with fixed points set in which F' replaces F.*

Here the vanishing of the Kervaire classes is actually a necessary and sufficient condition for replacement with the appropriate modification of condition 2 in the definition of strong replacement. Combining 2.5 and 3.3, for an involution with $\pi_1(F)$ of odd order there is no obstruction for weak replacement by a homotopy equivalence. Proposition 3.3 follows from the above considerations and the following easy fact which we believe is previously known but which we include for the reader's convenience:

Lemma 3.4. *For G of odd order, the transfer map induced from the inclusion $G \to Z_2 \times G$,*

$$Wh(Z_2 \times G) \to Wh(G) ,$$

is surjective.

Proof. Clearly this is true for Wh(G) localized away from 2. We now check it for localization at 2 as well by using the pullback diagram of rings:

$$\begin{array}{ccc}
\mathbb{Z}[\mathbb{Z}_2 \times G] & \to & \mathbb{Z}[G] \\
\downarrow & & \downarrow \\
\mathbb{Z}[G] & \to & \mathbb{Z}_2[G]
\end{array}$$

Now, as $\mathbb{Z}_2[G]$ is a product of matrix rings over 2-fields, whose units are of odd order, its K_1 is here irrelevant. Hence, it suffices just to check that the diagonal inclusion of $\mathbb{Z}[G]$ in $\mathbb{Z}[G] \times \mathbb{Z}[G]$ induces under the transfer a surjection. But this transfer is easily seen to send the element $(\lambda_1, \lambda_2) \in \text{Wh}(\mathbb{Z}[G]) \times \text{Wh}(\mathbb{Z}[G])$ to $\lambda_1 \lambda_2$. □

For the study of replacement in (linear) lens spaces with a (linear) lens space of fixed points, we can use Corollary 0.3 of the introduction to yield examples where the Reidemeister torsions are not related as in the linear actions.

Proof of Corollary 0.3. Note that quite generally for a homotopy equivalence of even order linear lens spaces, by passing to the real projective covering spaces, the Kervaire classes are seen to vanish. In any case, for case (iii) the torsion of $F' \to F$ in this case clearly lies in B_q and so we can apply Theorem 3.1 or its addendum. For case (ii), note further that the transfer map transfer: $\text{Wh}(\mathbb{Z}_q \times \mathbb{Z}_n) \to \text{Wh}(\mathbb{Z}_q) \to A_q$ certainly has, using the inclusion $\text{Wh}(\mathbb{Z}_q) \subset \text{Wh}(\mathbb{Z}_q \times \mathbb{Z}_n)$, image including $n \cdot A_q$. For case (i), use Proposition 3.3. □

In view of this reduction in 3.2 ii to an algebraic calculation in $A_q \otimes \mathbb{Z}_n$, it is often quite a delicate number theoretic calculation to determine from the Reidemeister torsions when one lens space can replace another as the fixed set of a homotopy linear action on a lens space. However, it is obvious that for any relatively prime q and n, taking repeated equivariant joins of the lens spaces produces many homotopy equivalences whose torsions vanish in $A_q \otimes \mathbb{Z}_n$. Thus, Theorem 3.2 yields PL locally linear homotopy linear actions on lens spaces with lens space fixed set which violate relations on Reidemeister torsions that hold in the linear case for any nontrivial n and q besides those of Corollary 0.3.

3.2　Normal representation replacement for p-groups

In this section we use the results of Section 2.2 and some computations to yield several kinds of examples of nontrivial normal representation replacement for cyclic p-groups both for $p = 2$ and for odd primes.

We begin by considering semifree G actions for G cyclic. For a simply connected component F of the fixed points of such an action, in view of Corollary 2.9 and the remarks following it, and Propositions 2.10 and 2.11, we are reduced to studying when certain associated prime-power order lens spaces are PL normally cobordant and, for $\chi(F)$ odd, a computation involving Reidemeister torsions modulo squares.

Remark. When studying PL normal cobordism problems at odd primes, it is often useful to recall that, away from the prime 2, homotopy equivalences of PL manifolds are normally cobordant if and only if they are just oriented PL cobordant as maps. This follows readily from Sullivan's $\mathbf{F}/\mathbf{PL} \otimes \mathbb{Z}[1/2]$ orientation of PL manifolds, regarded as a map of the PL Thom Spectrum $\mathbf{MSPL} \to (\mathbf{F}/\mathbf{PL}) \otimes \mathbb{Z}[1/2]$. Hence, the normal cobordism issue is reduced, at least at odd primes, to more familiar cobordism issues. In the smooth category, these are classically computed from characteristic numbers of maps; such smooth invariants are not always well defined in the PL category, but they certainly would yield sufficient conditions. (On the other hand, for the present application to prime-power order lens spaces, for all primes above some number determined by the dimension, the PL and smooth conditions coincide, making this characteristic class condition also necessary in these cases.)

We will now prove Theorem 0.5 of the Introduction, which gave examples of nontrivial h-strong replacement of normal representations for $G = \mathbb{Z}_p r$, p odd and $r \geq 2$, or $p = 2$ and $r \geq 3$.

Proof of Theorem 0.5. We first consider the case p odd. The case p even is dealt with in the example following the statement of Theorem 3.5 below.

In view of Corollary 2.9 and Proposition 2.11, the key point is to check the vanishing of the normal invariant (and even of the extended normal invariant discussed above) of the homotopy equivalence of the lens spaces L and L' corresponding to the quotients of the unit spheres of the representations γ and γ' by $G = \mathbb{Z}_p r$. Indeed, the extended normal invariant has already been identified with the usual normal invariant of the homotopy equivalence of L_+ and L'_+, where these are associated to the representations $\gamma \oplus t^1$ and $\gamma' \oplus t^1$. Of course, this normal invariant is detected by its localization at p and from the results of [Wal, Chapter 14E] on odd lens spaces, vanishes if and only if the difference of ρ-invariants (i.e., signature defects) of L_+ and L'_+ is integral, localized at p. To simplify notation, we may take, for example, $\gamma = 2pt^1$ and $\gamma' = 2pt^k$, $k \equiv 1 \pmod{p^{r-1}}$.

$$\text{Now, } \rho(L_+) = \left(\frac{1+x}{1-x}\right)^{2p}\left(\frac{1+x}{1-x}\right)$$

$$\rho(L'_+) = \left(\frac{1+x^d}{1-x^d}\right)\left(\frac{1+x}{1-x}\right), \qquad kd \equiv 1 \pmod{p^r}$$

$$\text{and so } \rho(L'_+) - \rho(L_+) = \left(\left(\frac{1+x^d}{1-x^d}\right)^{2p} - \left(\frac{1+x}{1-x}\right)^{2p}\right)\left(\frac{1+x}{1-x}\right)$$

in the ring $\mathbb{Z}[1/p][x]/(1 + x + x^2 \ldots + x^{p^r-1})$. This ring decomposes into a product of number rings and the given difference vanishes in all factors other than $\mathbb{Z}[1/p](\xi)$, ξ a primitive p^r-th root of unity. Factoring, we must then check the integrality of

$$\left(\prod_{b=0}^{p^r-1} V_b\right)\left(\frac{1+\xi}{1-\xi}\right),$$

$$\text{where } V_b = \left(\frac{1+\xi^d}{1-\xi^d}\right) - \xi^b\left(\frac{1+\xi}{1-\xi}\right).$$

Now each V_b, and hence their product, is easily seen to be integral, but we need the stronger fact that this product is divisible by $(1-\xi)$. This can be checked by showing that at least one V_b vanishes in $\mathbb{Z}[\xi]/(1-\xi) \cong \mathbb{Z}_p$. For this, we note that the differences,

$$(V_b - V_0) = (\xi^b - 1)\left(\frac{1+\xi}{1-\xi}\right) = -(1 + \xi + \xi^2 + \ldots + \xi^{b-1})(1+\xi)$$

$$= -2b \text{ in } \mathbb{Z}_p.$$

Hence, for some choice of b, V_b is indeed divisible by $(1-\xi)$, as needed.

Lastly, when $\chi(F)$ is odd we must check the Reidemeister torsion condition of the remark to Corollary 2.9, which is easily verified using Proposition 2.12.□

Remarks. (i) The above result for $\mathbb{Z}_p r$, p odd, on normal representation replacement can be generalized to $\gamma = 2pt^k + \eta$, $\gamma' = 2pt^{k'} + \eta$, k and k' as before but with η any free representation below some dimension depending on p^r. On the other hand, characteristic class computations, as suggested near the beginning of Section 3.2, will show that such h-strong normal replacement will not exist if η is of sufficiently high dimension.

(ii) It is possible to relax the condition that F is simply connected to just that $\pi_1(F)$ is of odd order and its monodromy action on F is isovariantly homotopically trivial; this last condition follows from smoothness of the action near a 1-skeleton of F. See the Example (ii) after Corollary 1.10. In particular, with the above calculations, this can be applied to achieve normal representation replacement to the fixed points of some linear actions of $\mathbb{Z}_p r$ on lens spaces.

(iii) It is possible to write down examples of h-strong normal replacement for $\mathbb{Z}_p$, p an odd prime, when F is simply connected and $\chi(F)$ is even. To determine if there are examples when $\chi(F)$ is odd, but F is not a point, would require further number theoretic calculations which we have not done.

Next we consider normal representation replacement for semifree actions of $G = \mathbb{Z}_{2^r}$. In this case, it is possible to give, at least for a simply connected component of the fixed points, a complete reduction to numerical calculations of when normal replacement is possible by combining the above discussion with the "undertrick" method of computing PL normal invariants of lens spaces at the prime 2 of Cappell and Shaneson in [CS1] recalled below.

Theorem 3.5. *Let $G = \mathbb{Z}_{2^r}$ act by a semifree PL locally linear action on a manifold M. Denote by F a component of the fixed point set and let α denote the* *normal representation of G to F. Assume F is simply connected. let β denote another free representation of G. Then β can h-strongly normally replace α if writing $\alpha = \sum_a t^{i_a}$, $\beta = \sum_b t^{i_b}$, $c = \dim M - \dim F = \dim \alpha = \dim \beta$.*

(i) $\Pi i_a \equiv \Pi i_b (mod\ 2^r)$,

(ii) $\Pi(1 = i_a^2 x^2) + (1 + x^2)^{(c/2)-1}(\Pi i_b^2 - \Pi i_a^2) = \Pi(1 + i_b^2 x^2)$ *in* $\mathbb{Z}_{2^{r+3}}[x]/(x^{c/2+1})$,

(iii) $\sum i_a^2 + \prod i_b^2 \equiv \sum i_b^2 + \prod i_a^2$ *(mod* 2^{k+4}*), and*

(iv) $\chi(F)$ *is even, or* $\chi(F)$ *is odd and F is not a point and* $\alpha - \beta = 2 \cdot \sum_d m_d t^{i_d}$ *where* $\prod i_d^{m_d} \equiv 1 (mod\ 2^r)$.

Example. Let $G = \mathbb{Z}_{2^r}$, $r \geq 3$. Then in a semifree G action with F a simply connected component of the fixed points, $4t^k$ and $4t^{(k+2^{r-1})}$ can h-strongly normally replace each other if (and only if) F is not a point. However, $6t^k$ and $6t^{(k+2^{r-1})}$, k odd, can h-strongly normally replace each other for F simply connected if (and only if) $\chi(F)$ is even.

Proof of Theorem 3.5. The proof is just a computation of the conditions of Corollary 2.9 and we briefly indicate how this is done. Of course, condition (i) is just a rewriting of the usual (oriented) homotopy equivalence condition for the associated lens spaces. Conditions (ii) and (iii) can be seen to be equivalent to the triviality of the extended normal invariant. But, a normal invariant is described at the prime 2 by the Morgan-Sullivan $\mathcal{L}$ class and the Kervaire class [MS]. Now, Kervaire classes here are seen to be trivial, as in the proof of Corollary 0.3, by restricting to $\mathbb{Z}_2$. The $8 \times \mathcal{L}$ class can be computed from the $\mathcal{L}$-class and to remove this factor of 8 we use the "undertrick" of [CS1, 3]; that is, we write down a homotopy equivalence of lens spaces with $\pi_1 = \mathbb{Z}_m$, $m = 8 \times 2^r = 2^{r+3}$, which the given homotopy equivalence covers and then compute the $\mathcal{L}$-class from the L-class of these new lens spaces. To carry this out, we use the lens spaces associated to the representations

$$\left(\sum_a t^{i_a}\right) + t^{n_b} \quad \text{and} \quad \left(\sum_b t^{i_b}\right) + t^{n_a}, \quad n_b = \prod i_b, \quad n_a = \Pi i_a,$$

of $\mathbb{Z}_{2^{r+3}}$. Using the standard formulae for the Pontrjagen classes of lens spaces and applying the L-polynomials to these to calculate the L-classes then yields necessary and sufficient numerical conditions for such normal cobordisms. To obtain some simpler sufficient conditions, as the L-classes have odd denominators, we could work instead just with the easier to use total Pontrjagen classes.[6] But this is computed for these lens spaces in conditions (ii) and (iii), where (iii) is a refinement of the condition on the class in dimension 4 as we are working in

---

[6] We are grateful to Ian Hambleton for bringing to our attention that, as pointed out in the 1995 M.Sc. thesis of Carmen Young at McMaster University written under his supervision, because of the presence of some powers of 2 in the numerators of the L-polynomials, this later condition on the Pontrjagen classes of $\mathbb{Z}_{2^r}$ lens spaces is not always necessary for the above condition on L-classes to hold.

F/PL rather than **F/Top**. Condition (iv), when $\chi(F)$ is odd, is the Reidemeister torsion condition. It is easy to see, using the oddness of the class number of 2^r, that the relevant quotient of Reidemeister torsions must be of the form u^2, when $\alpha - \beta$ is divisible by 2. Then, using the usual Rim diagram,

$$
\begin{array}{ccc}
\mathbb{Z}[\mathbb{Z}_{2^r}] & \rightarrow & \mathbb{Z}[\mathbb{Z}_{2^r}]/(1 + x + \cdots + x^{2^r-1}) \\
\downarrow & & \downarrow \\
\mathbb{Z} & \rightarrow & \mathbb{Z}_{2^r}
\end{array}
$$

we see that the condition $\Pi i_d^{m_d} \equiv 1 \pmod{2^r}$ just says that u is in the image of the units, $(\mathbb{Z}[\mathbb{Z}_{2^r}])^\times \rightarrow (\mathbb{Z}[\mathbb{Z}_{2^r}]/(1 + x + \cdots x^{2^r-1}))^\times$. □

Last, we point out that for certain groups of order divisible by 4 there are examples of pairs of linear representations, though not for semifree actions, where h-strong normal replacement is possible even for a fixed point component which is just an isolated point. This was done in the construction of counterexamples to the Smith conjecture in [CS1]. Moreover, in the present context by applying Corollary 2.8 to the computations of normal invariants and surgery obstructions of [CS1] (or, by directly using the equivariant PL h-cobordism constructed here) we obtain the following:

Example. Let $G = \mathbb{Z}_{4m}$, $m > 1$, act by a PL locally linear action on a manifold M with F a component of the fixed points with normal representation $\delta + 4t^k$ where δ is the nontrivial 1-dimensional representation and with the action smoothable near a 1-skeleton of F, k prime to $4m$. (The action is thus orientation reversing and not semifree.) Then this representation can be h-strongly normally repalced by $\delta + 4t^{2m+k}$; in particular, there is another PL locally linear action of G on M which is the same outside a regular neighborhood of F but with normal representation $\delta + 4t^{2m+k}$.

References

[AB]   Atiyah, M.F., and Bott, R., A Lefschetz fixed point formula for elliptic complexes: II. Applications. Ann. of Math. 88 (1968), 451–491.

[AS]   Atiyah, M.F., and Singer, I.M., The index of elliptic operators III. Ann. of Math. 87 (1968), 546–604.

[BFMW]   Bryant, J., Ferry, S., Mio, W., and Weinberger, S., The topology of homology manifolds, Bull. Amer. Math. Soc. 28 (1993), 324–328.

[BQ]   Browder, W., and Quinn, F., A surgery theory for G-spaces and stratified sets, Manifolds Tokyo 1973, Univ. of Tokyo Press, 1975, 27-36.

[BLR]   Burghelea, Lashof, R., and Rothenberg, M., Automorphisms of manifolds, Springer-Verlag Lecture Notes in Math 473 (1975)

[Ca]   Cappell, S.E., On homotopy invariance of higher signatures. Invent. Math. 33 (1976), No. 2, 171–179.

[CS1]   Cappell, S.E., and Shaneson, Periodic differentiable maps, Invent. Math. 68 (1982), 1–20.

[CS2]   ——, Piecewise linear embeddings and their singularities, Ann. of Math. 103 (1976), 163-228.

[CS3]   ——, Branched cyclic converings, in Knots, groups and 3-manifolds, Ann. of Math. Studies, No. 84, Princeton University Press, Princeton, N.J., 165–173.

[CS4]   Cappell, S.E., and Shaneson, Representations at fixed points. Proc. Amer. Math. Soc. Res. Conf. on Group Actions, Contem. Math. 36 (1985), 151–158.

[CSW]   Cappell, S., Shaneson, and Weinberger, S., Classes topologiques caracteristiques pour les actions de groupes sur les espaces singuliers, Comptes Rendus Acad. Sci., Paris, Serie I. Math. 313 (1991), 293–295.

[CW1]   Cappell, S. and Weinberger, S., A geometric interpretation of Siebenmann's Periodicity Phenomenon, Proc 1986 Georgia Topology Conference (McCrory and Shiffrin editors) 47–52.

[CW2]   ——, Homology propagation of group actions, Comm. Pure and Applied Math. (1987) No. 6, 723–744.

[CW3]   ——, A simple construction of Atiyah Singer classes and PL group actions. J. Diff. Geo. 33 (1991), 731–742.

[CW4]   ——, Classification de certains espaces stratifies. Comptes Rendus Acad. Sc. Ser. I. Math. 313 (1991), No. 6, 399–401.

[CWY]   ——, and Yan, Equivariant functoriality of isovariant structure sets (in preparation).

[C]   Casson, A., Fibrations over spheres, Topology 6 (1967), 489–499.

[Co]   Cohen, M. A coarse in simple-homotopy theory. Graduate Texts in Mathematics, Vol. 10, Springer-Verlag, New York-Berlin.

[Da]   Daverman, R., Decompositions of Manifolds, Academic Press 1986.

[Do]   Dovermann, K.H., Almost isovariant normal maps, Amer. J of Math. 111 (1989), 851–904.

[DPS] Dovermann, K.H., Petrie, T. and Schultz, R., Transformation groups and fixed point data, Proc. Amer. Math. Soc. Res. Conf. on Group Actions (in Boulder, Colorado in 1983), Contem. Math. 36 (1985), 161–191.

[DS] Dovermann, K. H. and Schultz, R., Equivariant Surgery Theories and their Periodicity Properties, Lecture Notes in Math. 1443, Springer (1991).

[FRW] Ferry, S., Weinberger, S. and Rosenberg, J. Phenomenes de rigidite topologique equivariante, Comptes Rendus Acad. Sci. Paris Ser. I Math. 306 (1988), No. 19, 777–782.

[FW1] Ferry, S., Weinberger S. Curvature, tangentiality, and controlled topology, Invent. Math. (Inventiones Mathematique) 105 (1991), No. 2, 401–414.

[FW2] ——, A coarse approach to the Novikov conjecture, Proc. 1993 Oberwolfach conference on the Novikov Conjecture, to appear.

[HRT] Hambleton, I., Ranicki, A., Taylor, L. Round L-theory. J. Pure Appl. Algebra 47 (1987), No. 2, 131–154.

[KS] Kirby, R., and Siebenmann, L., Foundational Essays on Topological Manifolds, Smoothings, and Triangulations, Princeton University Press 1977.

[LR] Lashof, R., and Rothenberg, M., G-smoothing theory, Proc. Symp. Pure Math. 32 (1978), 211–266.

[LM] Luck, W., and Madsen, I., Equivariant L-theory I, II Math. Zeit 203 (1990), 503-526, 204 (1990), 253-268.

[M] Milnor, J., Whitehead torsion, Bull. Amer. Math. Soc. 72 (1966), 358–426.

[MM] Madsen, I., and Milgram, J., The Classifying spaces for surgery and cobordism of manifolds, Annals of Math. Studies, No. 92 Princeton University Press.

[MS] Morgan, J., and Sullivan, D., The transversality characteristic class and linking cycles in surgery theory, Ann. of Math. 99 (1974), 384–463.

[Pe] Petrie, T., Smith equivalence of representations, Proc. Camb. Phil. Soc. 94 (1983), 61–99.

[Q] Quinn, F., A geometric formulation of surgery, 1969 Princeton PhD thesis

[R]   Ranicki, A., The algebraic theory of surgery I,II, Proc LMS 40 (1980), 87–192, 193–283.   .

[R2]   ———, Localisation in quadratic L-theory, Springer-Verlag Lecture Notes in Math. 741 (1979), 102–157.

[RW]   Rosenberg, J., Weinberger, S., An equivariant Novikov conjecture. *K*-Theory 4 (1990), No. 1, 29–53.

[RS]   Rourke, C.P., and Sanderson, B.J., Block bundles, I, II, III. Ann. of Math. 87 (1968), 1–28, 256–278, 431–483.

[Sh]   Shaneson, J., Wall's surgery obstruction groups for $G \times \mathbb{Z}$, Annals of Math. 90 (1969), 293–334.

[St]   Stein, E., Surgery on products with finite fundamental group. Topology 16 (1977), No. 4, 473–493.

[Su]   Sullivan, D., Geometric topology seminar notes (unpublished notes, Princeton 1968).

[TW]   Taylor, L., and Williams, B., Surgery spaces: structure and formulae, LNM 741 (1979), 170–195

[Wa1]   Wall, C.T.C., *Surgery on Compact Manifolds*, Academic Press, 1969.

[Wa2]   ———, Classification of Hermitian Forms VI, Ann. of Math. 103, 1–80.

[We1]   Weinberger, S., *The topological classification of stratified spaces*, University of Chicago Press, 1994.

[We2]   ———, Semifree PL locally linear actions on the sphere. Israel J. of Math. 66 (989), 351–363.

[We3]   ———, Aspects of the Novikov conjecture. Contemp. Math., 105, Amer. Math. Soc., Providence, R.I. 281–297.

[WY1]   Weinberger, S., and Yan, M., Exotic products and equivariant periodicity for abelian group actions. (in preparation)

[WY2]   ———, Some examples in symmetric *L*-theory (in preparation)

[Ws]   Weiss, M., Visible *L*-theory. Forum Mathematicum 4 (1992), No. 5, 465–498.

[Y]   Yan, M., The periodicity in stable equivariant surgery. Comm. Pure Appl. Math. 46, (1993), No. 7, 1013–1040.

Finite $K(\pi,1)$s for Artin Groups

Ruth Charney
Michael W. Davis

Introduction

It is not so widely known that there is a beautiful and simple description of a certain finite CW-complex Z which is a $K(\pi,1)$ for the braid group on $n+1$ strands. The complex Z is obtained by identifying certain faces of an n-dimensional convex polytope called a "permutohedron". The resulting CW complex has exactly one k-cell for each k-element subset of $\{1,\ldots,n\}$. We are not sure who first discovered this complex, but we first heard it described by C. Squirer in the mid 1980's and later by K. Tatsuoka [T]. It has also been known to J. Milgram for some time (a closely related construction appears already in [Mi]). Other references include [FSV], [P], and [S1]. The purpose of this paper is to give the details of the construction of this complex and its generalizations to other Artin groups.

It is a well-known fact that an Eilenberg-MacLane space for the pure braid group on $n+1$ strands is the set of points in $\mathbb{C}^{n+1}$ with all coordinates distinct. This space is the complement of an arrangement of hyperplanes in $\mathbb{C}^{n+1}$ associated to the action of the symmetric group on $n+1$ letters.

In [S], Salvetti showed how any hyperplane complement, obtained by complexifying a real arrangement, was homotopy equivalent to a certain cell complex. In the case at hand, this complex is a union of n-dimensional permutahedra. The symmetric group acts naturally and freely on it. Taking the quotient by the symmetric group, we obtain the complex Z.

There are two ingredients in the above program. The first is the fact that the hyperplane complement in $\mathbb{C}^{n+1}$ is a $K(\pi,1)$. The second is Salvetti's identification of the hyperplane complement with a certain cell complex.

Suppose that W is a Coxeter group with fundamental generating set S. In the same way as the braid groups on $n+1$ strands is associated to the symmetric group on $n+1$ letters, one can associate to (W,S) an "Artin group" (or "generalized braid group") A_W. We would like to carry out a similar program to obtain a nice $K(\pi,1)$-complex for A_W.

As in [V] one can always find a representation of W as a linear reflection group on a real vector space V so that W acts properly on a certain nonempty, W-stable, convex, open set I (the interior of the "Tits cone"). When W is finite, $I = V$. Inside $V \otimes \mathbb{C}$ one has the domain $V + iI$. Consider the hyperplane

Both authors were partially supported by NSF Grant DMS 9208071.

complement

$$M = [(V + iI) - \cup \text{ reflection hyperplanes}]/W.$$

As for the first ingredient in the program, it is known that $\pi_1(M) = A_W$ (see [CD1; Corollary 3.2.4] or [L]). In the case where W is finite, Deligne proved in [D] that M is a $K(A_W, 1)$. In [CD1] the authors proved that the same result holds for many infinite Coxeter groups. (A precise statement can be found in §1.4, below.) Thus, this paper is a continuation of [CD1].

As explained in §1 and §2, the second ingredient (Salvetti's complex) works for all W. Specifically, we show in Theorem 1.4.1 and Corollary 2.2.3, below, that M is homotopy equivalent to a certain finite CW-complex Z_W. The complex Z_W has one cell of dimension k for each k-element subset T of S such that the subgroup generated by T is finite. From this it is easy to compute the cohomological dimension of A_W (Corollary 1.4.2) and its Euler characteristic (Corollary 2.2.5) at least in the cases where M is known to be a $K(\pi, 1)$-space.

In §3, we consider the case where W is "right-angled", that is, the product of any two generators has order either 2 or ∞. The first ingredient holds in this case. In Theorem 3.1.1, we show that Z_W with its natural piecewise Euclidean metric is nonpositively curved. Thus, in the right-angled case, A_W is a "semihy-perbolic group" in the sense of [AB]. (This was previously proved by Hermiller and Meier in [HM] using combinatorial methods.) In this case, it is an easy matter to calculate the cohomology ring of A_W and we do so in Theorem 3.2.4.

After completing this paper we learned of Salvetti's recent paper [S2], essentially on the same topic as this paper, at least in the case where W is finite.

1 Basic Definitions and Constructions

1.1 Coxeter groups and Artin groups

A *Coxeter matrix* on a finite set S is an S by S symmetric matrix $(m_{ss'})$ such that each entry is a positive integer or ∞ and such that $m_{ss'} = 1$ if $s = s'$ and $m_{ss'} \geq 2$ if $s \neq s'$. Associated to $(m_{ss'})$ there is a group W defined by the presentation:

$$W = \langle S \mid (ss')^{m_{ss'}} = 1 \rangle.$$

The natural map $S \rightarrow W$ is an injection ([B,Ch. V, §4.3, Prop. 4]) and hence-forth, we identify S with its image in W. The pair (W, S) is a *Coxeter system* and W is a *Coxeter group*. If T is any subset of S then the subgroup W_T, gener-ated by T, is called a *special subgroup*. It is known ([B; ch. IV, §1.8, Théoréme 2(i)]) that (W_T, T) is the Coxeter system associated to the restriction of the Coxeter matrix to T.

For each $s \in S$ introduce a symbol x_s and let $\mathcal{X} = \{x_s\}_{s \in S}$. Associated to a Coxeter matrix $(m_{ss'})$ there is also an *Artin group*, denoted A_W, and defined by the presentation

$$A_W = \langle \mathcal{X} \mid \mathrm{prod}\,(x_s, x_{s'}, m_{ss'}) = \mathrm{prod}\,(x_{s'}, x_s; m_{ss'}) \rangle$$

where $\mathrm{prod}\,(x, y; m)$ denotes the word $xyx \ldots$ of length m. We say that A_W is of *finite type* if W is a finite group.

If W is the symmetric group on n letters, then A_W is the braid groups on n strands. Thus, Artin groups are sometimes called "generalized braid groups".

1.2   Posets and simplicial complexes associated to a Coxeter system

Let (W, S) be a Coxeter system. Associated to (W, S) there are the following three posets:

$$\mathcal{S}^f = \{T \subseteq S \mid W_T \text{ is finite}\}$$
$$W\mathcal{S}^f = \{wW_T \mid w \in W, T \in \mathcal{S}^f\},$$
$$W \times \mathcal{S}^f.$$

The sets $\mathcal{S}^f$ and $W\mathcal{S}^f$ are partially ordered by inclusion. The partial order on $W\mathcal{S}^f$ is given explicitly as follows. If wW_T and $w'W_{T'}$ are elements of $W\mathcal{S}^f$, then $wW_T < w'W_{T'}$ if and only if the following two conditions hold:

(1) $T < T'$,
(2) $w^{-1}w' \in W_{T'}$.

The partial ordering on $W \times \mathcal{S}^f$ is defined as follows.

If (w, T) and (w', T') are elements of $W \times \mathcal{S}^f$, then $(w, T) < (w', T')$ if and only if conditions (1), (2) and the following hold:

(3)   for all $t \in T$, $\ell(w^{-1}w') < \ell(tw^{-1}w')$.

(Here ℓ denotes word length with respect to the generating set S.) The geometric meaning of (3) will be given in §1. The natural map $W \times \mathcal{S}^f \to W\mathcal{S}^f$ defined by $(w, T) \to wW_T$ is obviously order-preserving.

The posets $\mathcal{S}^f$ and $W\mathcal{S}^f$ appear in [CD1]. The poset $W \times \mathcal{S}^f$ is one of the main objects of study in this paper.

For the moment, suppose that $\mathcal{P}$ is an arbitrary poset. Given $p \in \mathcal{P}$, define a sub-poset, $\mathcal{P}_{\geq p} = \{x \in \mathcal{P} \mid x \geq p\}$. The sub-posets $\mathcal{P}_{\leq p}$, $\mathcal{P}_{>p}$ and $\mathcal{P}_{<p}$ are defined similarly. The poset $\mathcal{P}^{op}$, called the *dual* of $\mathcal{P}$, is equal to $\mathcal{P}$ as a set but with the order relations reversed. The *derived complex* of $\mathcal{P}$, denoted by $\mathcal{P}'$, is the set of finite chains in $\mathcal{P}$, partially ordered by inclusion. It is an abstract simplicial complex.

The geometric realizations of $(\mathcal{S}^f)'$, $(W\mathcal{S}^f)'$, and $(W \times \mathcal{S}^f)'$ are denoted, respectively, by

$$K_W = |(\mathcal{S}^f)'|$$
$$\Sigma_W = |(W\mathcal{S}^f)'|$$
$$\tilde{\Sigma}_W = |(W \times \mathcal{S}^f)'|$$

When there is no ambiguity, we shall omit the subscript "W" from our notation. Following [CD1] we call Σ the *modified Coxeter complex* of W. Here $\tilde{\Sigma}$ will be called the *Salvetti complex* of W. We note that W acts naturally and simplicially on Σ and on $\tilde{\Sigma}$.

There is a projection map $\pi: W\mathcal{S}^f \to \mathcal{S}^f$ defined by $wW_T \to T$ and an embedding $i: \mathcal{S}^f \to W\mathcal{S}^f$ defined by $T \to W_T$. These induce simplicial maps $\Sigma \to K$ and $K \to \Sigma$ which we again denote by π and i. We can identify K with its image $i(K)$. The map π is then a retraction; moreover, it induces a simplicial isomorphism

$$\Sigma/W \cong K.$$

The group W acts freely on $\tilde{\Sigma}$. The orbit space will be denoted by Z_W (or simply by Z), i.e.,

$$\tilde{\Sigma}/W = Z_W.$$

It is our primary object of study.

1.3 Sectors

Let R denote the set of conjugates of S in W. An element of R is called a *reflection*. Given $r \in R$, its fixed points set on Σ is denoted by Σ^r and called a *wall* of Σ. The space $\Sigma - \Sigma^r$ has two connected components, which are interchanged by r. Each such component is called a *half-space*. The half-space containing the interior of K (where K is regarded as a subcomplex of Σ) is called *positive* and denoted H_+^r.

Let T be a subset of S. Put

$$R_T = R \cap W_T.$$

The components of

$$\Sigma - \bigcup_{r \in R_T} \Sigma^r$$

are permuted freely and transitively by W_T. Each such component is called a W_T-*sector* of Σ. Such a sector is an intersection of half-spaces. The W_T-sector containing the interior of K is called *positive*. It is the intersection of the positive half-spaces H_+^t, $t \in T$.

If Λ is a W_T-sector then we shall also want to call its translate $w\Lambda$ by an element w of W a "sector". We note that $w\Lambda$ is a component of

$$\Sigma - \bigcup_{r \in wR_Tw^{-1}} \Sigma^r.$$

We shall call it a wW_Tw^{-1}-*sector*.

1.4   The spaces Q_W and Y_W

Put

$$Y_W = \Sigma \times \Sigma - \bigcup_{r \in R} \Sigma^r \times \Sigma^r$$

and

$$Q_W = Y_W / W .$$

As usual, we shall often omit the subscript W from our notation.

The following facts are proved in [CD1].

(i) Y is W-equivariantly homotopy equivalent to the complex hyperplane complement associated to any representation of W as a real linear reflection group ([CD1, Corollary 2.2.5]).

(ii) $\pi_1(Q) = A_W$ ([CD1, Corollary 3.2.4]).

(iii) $\pi_1(Y) = PA_W$, where PA_W denotes the kernel of the natural map $A_W \rightarrow W$. (PA_W is the "pure Artin group".)

The main conjecture of [CD1], then becomes the following.

Main Conjecture. *Q_W is an Eilenberg-MacLane space $K(A_W, 1)$.*

The poset $S_{>\phi}^f$ is an abstract simplicial complex. Let K_0 denote its geometric realization. (Thus, K is the cone on the barycentric subdivision of K_0.)

A simplicial complex is said to be a *flag complex* if any set of vertices which are pairwise joined by edges span a simplex. In [CD1], it is proved that the Main Conjecture holds under either of the following two hypotheses:

(A) K_0 is a flag complex, or

(B) $\dim K_0 \leq 1$.

For example, (A) holds if W is finite.

The main result of this section is the following:

Theorem 1.4.1. *Q_W is homotopy equivalent to Z_W.*

Before giving a proof in the next subsection, let us deduce the corollary below.

Put $n_W = \max\{ \operatorname{Card}(T) \mid T \in S^f \}$. In other words, n_W is the common dimension of the simplicial complexes, K, Σ, $\widetilde{\Sigma}$, and Z.

Corollary 1.4.2. *Suppose the Main Conjecture holds for (W, S). Set $n = n_W$. Then the following statements are true.*

(i)   *A_W has a finite, n-dimensional $K(\pi, 1)$-space, namely Z_W.*

(ii)   *A_W is of type FP.*

(iii)   *The cohomological dimension of A_W is n.*

Proof. If the Main Conjecture holds, (i) follows immediately from Theorem 1.4.1 and (ii) follows immediately from (i).

Also by (i) the cohomological dimension of A_W is $\leq n$. Suppose $T \in S^f$ has n elements. By [Br], $H^n(PA_{W_T}; \mathbb{Z}) \neq 0$; hence, the cohomological dimension of A_W is $\geq n$ so (iii) holds.   □

1.5 An open cover of Y

The vertices of K are naturally indexed by the elements of $\mathcal{S}^f$. For each $T \in \mathcal{S}^f$ let v_T denote the corresponding vertex of K. Similarly, the vertices of Σ are naturally indexed by $W\mathcal{S}^f$. For each $wW_T \in W\mathcal{S}^f$, let wv_T denote the corresponding vertex of Σ. Let Star (wv_T) denote the open star of wv_T in Σ.

Consider the open cover of an arbitrary simplicial complex X by open stars of vertices. Of course, the nerve of this open cover is just X. Applying this fact in the case $X = \Sigma$ gives the following lemma.

Lemma 1.5.1. *Suppose $w_i W_{T_i}$, $0 \leq i \leq k$, are distinct elements in $W\mathcal{S}^f$. Then*

$$\text{Star}\,(w_0 v_{T_0}) \cap \ldots \cap \text{Star}\,(w_k v_{T_k}) \neq \emptyset$$

if and only if $\{w_0 W_{T_0}, \ldots, w_k W_{T_k}\}$ is a chain in $W\mathcal{S}^f$.

For each $(w, T) \in W \times \mathcal{S}^f$, let Sec (w, T) denotes the open $wW_T w^{-1}$-sector of Σ which contains the interior of wK. Define a subset $U(w, T)$ of $\Sigma \times \Sigma$ by

$$U(w, T) = \text{Sec}\,(w, T) \times \text{Star}\,(wv_T).$$

Clearly, $\Sigma^r \cap \text{Star}\,(wv_T) \neq \emptyset$ if and only if $r \in wR_T w^{-1}$. Also, for all $r \in wR_T w^{-1}$, we have Sec $(w, T) \cap \Sigma^r = \emptyset$. Hence, $U(w, T) \cap (\Sigma^r \times \Sigma^r) = \emptyset$ for all r in R. That is to say, $U(w, T)$ is an open subset of Y. Moreover, $\mathcal{U} = \{U(w, T)\}_{(w,T) \in W \times \mathcal{S}^f}$ covers Y.

Lemma 1.5.2.
(i) *If $U(w, T) = U(w', T')$, then $(w, T) = (w', T')$.*
(ii) *Suppose (w_i, T_i), $0 \leq i \leq k$, are distinct elements in $W \times \mathcal{S}^f$. Then $U(w_0, T_0) \cap \ldots \cap U(w_k, T_k) \neq \emptyset$ if and only if $\{(w_0, T_0), \ldots (w_k, T_k)\}$ is a chain in $W \times \mathcal{S}^f$.*

Proof. (i) Suppose $U(w, T) = U(w', T')$. Then Star $(wv_T) = $ Star $(w'v_{T'})$ and hence, $wW_T = w'W_{T'}$. It follows that $T = T'$ and that $w^{-1}w' \in W_T$. The condition that Sec $(w, T) = $ Sec (w', T) is equivalent to Sec $(1, T) = $ Sec $(w^{-1}w', T)$. Thus, $w^{-1}w'$ lies in W_T and the interior of $w^{-1}w'K$ lies in the positive W_T-sector. Since W_T acts freely on the set of W_T-sectors, this forces, $w^{-1}w' = 1$, i.e., $w' = w$.

(ii) Suppose (w, T) and (w', T') are such that $wW_T < w'W_{T'}$. We claim that the following statements are then equivalent:
(a) $(w, T) < (w', T')$,
(b) for all $t \in T$, $\ell(w^{-1}w') < \ell(tw^{-1}w')$,
(c) Sec $(w, T) \supset$ Sec (w', T'),
(d) Sec $(w, T) \cap$ Sec $(w', T') \neq \emptyset$.
Condition (b) is just condition (3) of §1.2. Thus, (a) and (b) are equivalent. Given $r \in R$ and $u \in W$, the condition that $\ell(u) < \ell(ru)$ means that the interior of uK lies in the positive half-space for r. Thus, (b) means that the interior

of $w^{-1}w'K$ lies in the positive W_T-sector. By hypothesis, $T < T'$ so every $W_{T'}$-sector is contained some W_T-sector. Since $w^{-1}w' \in W_{T'}$, we see that (b) is equivalent to the condition that Sec $(1,T) \supset$ Sec $(w^{-1}w',T')$ which is equivalent to (c). Obviously (c)$\Rightarrow$(d). On the other hand, since we are assuming $T < T'$ and $w^{-1}w' \in W_{T'}$, we see that the intersection Sec $(1,T) \cap$ Sec $(w^{-1}w',T')$ can be nonempty only if Sec $(w^{-1}w',T') \subset$ Sec $(1,T)$. Thus, (d)$\Rightarrow$(c). Statement

(ii) of the lemma now follows from the previous lemma together with the equivalence of (a) and (d).

Proof of Theorem 1.4.1. Consider the open cover $\mathcal{U} = \{U(w,T)\}$, of Y, indexed by the elements of $W \times S^f$. Let $\sigma = \{(w_0,T_0),\dots,(w_k,T_k)\}$ be a set of distinct elements of $W \times S^f$, and put $U_\sigma = U(w_0,T_0) \cap \dots U(w_k,T_k)$. By the previous lemma, U_σ is nonempty if and only if σ is a simplex in the derived complex $(W \times S^f)'$. Moreover, if this is the case, then after renumbering we may assume that $(w_0,T_0) < \dots < (w_k,T_k)$. It follows from the proof of the previous lemma (the equivalence of (c) and (d)) that

$$U_\sigma = \text{Sec}\,(w_k,T_k) \times \text{Star}\,(p(\sigma))$$

where $p \colon \tilde{\Sigma} \to \Sigma$ is the natural projection. It is shown in Lemma 2.2.6 of [CD1] that for $T \in S^f$, each W_T-sector of Σ is homotopy equivalent to K_{W_T}. In particular, U_σ is contractible. Hence, the nerve of $\mathcal{U}$ is $\tilde{\Sigma}$ and each nonempty intersection is contractible. It follows that the spaces $\tilde{\Sigma}$ and Y are homotopy equivalent. The group W acts freely on both spaces. We leave it as an exercise for the reader to construct a W-equivariant embedding $\tilde{\Sigma} \to Y$ such that Y equivariantly deformation retracts onto $\tilde{\Sigma}$. Taking quotients by W we get the desired result: Z is homotopy equivalent to Q. $\square$

2   A cell structure on Z

2.1   Coxeter cells

In this subsection only, the Coxeter group W will be required to be finite.

Thus, S^f will be the poset of all subsets of S. We shall describe a certain convex polytope P_W, called a "Coxeter cell", such that the poset of faces of P_W is isomorphic to WS^f. Some of this material is also described in [CD2; §6].

Associated to (W,S) there is an (essentially unique) representation of W on $\mathbb{R}^n$, $n = \text{Card}\,(S)$, as an orthogonal linear reflection group ([B; Ch. V, §4]). Each element of S acts on $\mathbb{R}^n$ as an orthogonal reflection across a hyperplane. A "fundamental chamber" C is a simplicial cone bounded by the hyperplanes corresponding to the elements of S. To each function $x \colon S \to (0,\infty)$ there is a unique point, which we will also denote by x, in the interior of C such that the distance from x to the hyperplane fixed by s is $x(s)$. Explicitly, if u_s is the outward-pointing unit normal vector to the hyperplane fixed by s, then x is the point defined by: $x \cdot u_s = -x(s)$, $s \in S$.

Definition 2.1.1. The *Coxeter cell* P_W associated to (W, S) and x is the convex hull of the W-orbit of x.

Examples 2.1.2. (i) If $W = \mathbb{Z}/2$, then P_W is an interval $[-x(s), x(s)]$.
    (ii) If W is the dihedral group of order $2m$, then P_W is a $2m$-gon.
    (iii) If (W, S) is a direct product of two Coxeter systems, $(W, S) = (W_1 \times W_2, S_1 \coprod S_2)$, then P_W is isometric to $P_{W_1} \times P_{W_2}$.
    (iv) In particular, if $W \cong (\mathbb{Z}/2)^n$ and x is the constant function, then P_W is an n-cube.

Lemma 2.1.3. *Suppose W is a finite Coxeter group. The poset of faces of the Coxeter cell P_W is isomorphic to $W\mathcal{S}^f$. In particular, for $T \subset S$, the convex hull of the W_T-orbit of x is a face of P_W and this face is isomorphic to P_{W_T}.*

Proof. Let e_s be a vector on the extremal ray of C which is opposite to the face fixed by s. Consider the linear form φ or $\mathbb{R}^n$ defined by $v \to v \cdot e_s$. Put $c = x \cdot e_s$. We claim that c is the maximum value of φ on P_W. Indeed, since C is a Dirichlet fundamental domain for the action of W on $\mathbb{R}^n$ any point of C is at least as close to x as it is to any other point in the orbit of x. In particular, $|wx - e_s|^2 \geq |x - e_s|^2$ for all $w \in W$. But this implies $wx \cdot e_s \leq x \cdot e_s$ for all $w \in W$ and hence, that the maximum value of φ is attained at x. Let $T = S - \{s\}$. The affine hyperplane $\varphi(v) = c$ contains the orbit of x under the subgroup W_T and is spanned by this orbit. It follows that $\varphi(v) = c$ is a supporting hyperplane of P_W and that the convex hull of $W_T x$ is a codimension one face of P_W.. Letting s vary over S we obtain all supporting hyperplanes containing the vertex x. Replacing x by wx and e_s by we_s we obtain in this way a description of all the supporting hyperplanes of P_W. The lemma follows easily. $\square$

Remarks 2.1.4. (i) Associated to any convex polytope there is a dual polytope. The boundary complex of the dual polytope to P_W is called the *Coxeter complex* of W. The Coxeter complex is an $(n-1)$-dimensional simplicial complex. It is combinatorially isomorphic to the triangulation of S^{n-1} whose spherical $(n-1)$-simplices are the intersections $S^{n-1} \cap wC$, $w \in W$.
    (ii) Since Σ_W is the geometric realization of $(W\mathcal{S}^f)'$, we see that Σ_W can be identified with the barycentric subdivision of P_W.

2.2 Cellulations of Σ and $\tilde{\tilde{\Sigma}}$

We return to the general situation where W can be infinite.
    For each $wW_T \in W\mathcal{S}^f$ we have

$$(W\mathcal{S}^f)_{\leq wW_T} \cong W_T \mathcal{S}^f_{\leq T} .$$

Hence, the geometric realization of the derived complex of $(W\mathcal{S}^f)_{\leq wW_T}$ is a subcomplex of Σ_W isomorphic to Σ_{W_T}. By Lemma 2.1.3 and Remark 2.1.4(ii) we can identify this subcomplex with the Coxeter cell P_{W_T}. Thus, Σ_W is naturally cellulated by Coxeter cells. This gives Σ_W the structure of a convex cell complex: the associated poset of cells is $W\mathcal{S}^f$.

Similarly, for each $(w, T) \in W \times \mathcal{S}^f$, we have

$$(W \times \mathcal{S}^f)_{\leq(w,T)} \cong *(W_T \times \mathcal{S}^f_{\leq T})_{\leq(1,T)}$$
$$\cong *W_T \mathcal{S}^f_{\leq T}$$

where the second isomorphism is basically the observation that the projection $W \times \mathcal{S}^f \to W\mathcal{S}^f$ restricts to an isomorphism

$$(W \times \mathcal{S}^f)_{\leq(w,T)} \cong (W\mathcal{S}^f)_{\leq wW_T}.$$

Hence, $\tilde{\Sigma}_W$ also has a cellulation by Coxeter cells which projects to the cellulation on Σ_W described in the previous paragraph. We state this as the following lemma.

Lemma 2.2.1. *$\tilde{\Sigma}_W$ is cellulated by Coxeter cells: there is one cell for each element of $W \times \mathcal{S}^f$.*

Remark 2.2.2. This cellulation of $\tilde{\Sigma}$ does not give it the structure of a convex cell complex in the strictest sense: the intersection of two cells need not be a common face of both, rather it is a union of such faces. For example, if $W = \mathbb{Z}/2$, then $\tilde{\Sigma}$ is a circle, cellulated into two intervals.

Since the W-action on $\tilde{\Sigma}_W$ is obviously cellular, we get the following corollary, which can be considered the main result of this paper.

Corollary 2.2.3. *Z_W has the structure of a CW-complex: there is one cell of dimension $\operatorname{Card}(T)$ for each $T \in \mathcal{S}^f$.*

Corollary 2.2.4. *The Euler characteristic, $\chi(Z_W)$ is given by the formula:*

$$\chi(Z_W) = \sum_{T \in \mathcal{S}^f} (-1)^{\operatorname{Card}(T)}$$
$$= 1 - \chi(K_0).$$

(Here K_0 is as in §1.4: it is the geometric realization of the simplicial complex $\mathcal{S}^f_{>\emptyset}$.)

Proof. The first equation is immediate from the previous corollary. To see the second, note that the dimension of the simplex of K_0 corresponding to $T \in \mathcal{S}^f_{>\emptyset}$ is $\operatorname{Card}(T) - 1$. Hence

$$\chi(K_0) = -\sum_{T \in \mathcal{S}^f_{>\emptyset}} (-1)^{\operatorname{Card}(T)}. \qquad \square$$

Corollary 2.2.5. *If the Main Conjecture holds for (W, S), then the Euler characteristic of the Artin group A_W is given by the same formula:*

$$\chi(A_W) = 1 - \chi(K_0).$$

Lemma 2.2.6. *Suppose that (W, S) is the direct product of two Coxeter systems: $(W, S) = (W_1 \times W_2, S_1 \coprod S_2)$. Then*
  (i) $A_W = A_{W_1} \times A_{W_2}$,
  (ii) $\Sigma_W = \Sigma_{W_1} \times \Sigma_{W_2}$,
  (iii) $\widetilde{\Sigma}_W = \widetilde{\Sigma}_{W_1} \times \widetilde{\Sigma}_{W_2}$,
  (iv) $Z_W = Z_{W_1} \times Z_{W_2}$.

Proof. Clear.

Corollary 2.2.7. *Suppose $W = (\mathbb{Z}/2)^n$. Then*
  (i) $A_W \cong \mathbb{Z}^n$,
  (ii) Σ_W *(= P_W) is an n-cube,*
  (iii) $\widetilde{\Sigma}_W$ *is an n torus (cellulated by 2^n n-cubes), and*
  (iv) Z_W *is an n-torus (formed by identifying opposite faces of a single n-cube in the standard fashion).*

2.3   Links

We have just explained how the space $\widetilde{\Sigma}$ is cellulated by Coxeter cells. The barycentric subdivision of this cell structure gives $\widetilde{\Sigma}$ its natural simplicial structure discussed in §1.

Let $(w, T) \in W \times S^f$. Then (w, T) corresponds to a vertex v in the simplicial structure on $\widetilde{\Sigma}$. This vertex is the barycenter of a unique Coxeter cell σ (of dimension $\text{Card}(T)$). Any top-dimensional simplex in the barycentric subdivision of σ has v as its maximal vertex. The link of such a simplex in the simplicial structure on $\widetilde{\Sigma}$ is the geometric realization of the derived complex of $(W \times S^f)_{>(w,T)}$. This can also be thought of as the barycentric subdivision of the link of σ in $\widetilde{\Sigma}$ which we denote $Lk(\sigma, \widetilde{\Sigma})$. The underlying poset of cells in $Lk(\sigma, \widetilde{\Sigma})$ is $(W \times S^f)_{>(w,T)}$.

Each Coxeter cell is a "simple" polytope. This means that for each pair (σ, τ) where τ is a Coxeter cell and σ is a face of τ, that $Lk(\sigma, \tau)$ is a simplex. It follows that $Lk(\sigma, \widetilde{\Sigma})$ is a "simplicial cell complex", in the sense that all its cells are simplices. (We use this term even though it is not a convex cell complex in the strict sense of Remark 2.2.2. We reserve the term "simplicial complex" when the strict property of Remark 2.2.2 is satisfied.)

Example 2.3.1. Let v be the 0-cell in $\widetilde{\Sigma}_W$ corresponding to the element $(1, \emptyset)$ in $W \times S^f$. We compute $Lk(v)$ $(= Lk(v, \widetilde{\Sigma}))$ in some simple cases.

(i) If $W = \mathbb{Z}/2$, then $\tilde{\Sigma}_W$ is a circle and $Lk(v) = S^0$.

(ii) If $W = (\mathbb{Z}/2)^n$, then $\tilde{\Sigma}_W$ is a Cartesian product of n circles and $Lk(v)$ is the n-fold join $S^0 * \cdots * S^0$. In other words, $Lk(v)$ is the boundary complex of an n-dimensional octahedron.

(iii) Suppose that W is a dihedral group of order $2m$ and that $S = \{s_1, s_2\}$. Thus, $\tilde{\Sigma}_W$ is a 2-complex cellulated by $2m$ $2m$-gons. The complex $Lk(v)$ is 1-dimensional. Its 0-cells correspond to the elements of

$$\left(W_{\{s_1\}} \times \{s_1\}\right) \cup \left(W_{\{s_2\}} \times \{s_2\}\right).$$

Hence, there are four 0-cells. There are $2m$ edges corresponding to the elements of W. In fact, $Lk(v)$ is as pictured below.

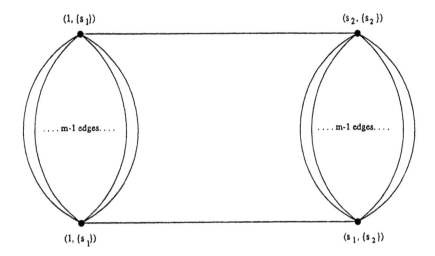

Figure 1.

Remark 2.3.2. Suppose that W is the dihedral group of order $2m$ and that $m > 2$. Then the link pictured above has cycles consisting of two edges. Thus, there is no way to assign lengths in $(0, \pi)$ to these edges so that this link is large (that is, so it does not contain cycles of length $< 2\pi$). It follows from this that any piecewise Euclidean structure on $\tilde{\Sigma}_W$ which is compatible with the given cell structure and with the W-action cannot satisfy CAT (0) (see [G] for the definition of the "CAT" inequalities). On the other hand, if $m = 2$, then $\tilde{\Sigma}$ is a flat 2-torus.

3   The right-angled case

A Coxeter system (W, S) is *right-angled* if given any two distinct elements s and s' of S, the order of ss' is either 2 or ∞ (i.e., if each off-diagonal entry of the Coxeter matrix is either 2 or ∞).

Artin groups associated to right-angled Coxeter groups are known as *graph groups*; they are "graph products" of groups where each vertex groups is infinite cyclic. (See [C], [HM].)

We will say that a simplicial cell complex is a *flag complex* if it is a genuine simplicial complex and if given any set of vertices which are pairwise joined by edges, then this set actually spans a simplex. If W is right-angled, then it follows immediately from the definitions that the geometric realization K_0 of $S_{>\emptyset}^f$ is a flag complex. Hence, as explained in §1.4, the Main Conjecture holds in this case.

3.1   Nonpositive curvature

If W is finite and right-angled, then $W \cong (\mathbb{Z}/2)^n$ for some n and the Coxeter cell P_W can be taken to be isometric to a regular Euclidean n-cube.

It then follows from §2.2 that if W is an arbitrary right-angled Coxeter group, then both the Salvetti complex $\widetilde{\Sigma}_W$ and its quotient Z_W naturally have the structure of piecewise Euclidean complexes in which each cell is a regular Euclidean cube.

Such a piecewise Euclidean complex is *nonpositively curved* if, locally, all geodesic triangles satisfy Gromov's CAT (0)-inequalities (cf. [G, p. 119]). Moreover, if each cell in the complex is a regular cube, then nonpositive curvature is equivalent to the condition that the link of each vertex is a flag complex (cf. [G, p. 122] and [M]).

The main result of this subsection is the following.

Theorem 3.1.1. *Suppose W is right-angled. Then Z_W with its natural piecewise Euclidean, cubical structure is nonpositively curved.*

Remark 3.1.2. Any nonpositively curved, piecewise Euclidean, finite complex is an Eilenberg-MacLane space ([G, p. 119]). Hence, in the case where W is right-angled, the above theorem gives an alternate proof of the main result of [CD1]. The fact that Z_W is aspherical is claimed as Theorem 10 in [KR]; the proof given there is incorrect as it is based on a false lemma ([KR, Lemma 9]). However, there is a simple and direct argument for this along the lines indicated in the remark following Lemma 4.3.7 of [CD1].

According to the remarks preceding the theorem, the statement that Z_W is nonpositively curved is equivalent to the statement that the link of its 0-cell is a flag complex. Equivalently, we can consider a 0-cell in $\widetilde{\Sigma}_W$. Thus, the theorem is an immediate consequence of the following.

Lemma 3.1.3. *Suppose W is right-angled and that v is the 0-cell of $\widetilde{\Sigma}_W$ corresponding to $(1, \emptyset) \in W \times S^f$. Then $Lk(v, \widetilde{\Sigma}_W)$ is a flag complex.*

Proof. We first show that $Lk(v, \widetilde{\Sigma}_W)$ is a simplicial complex. For this, we must show that $(W \times S^f)_{>(1,\emptyset)}$ is isomorphic to the poset of simplices in a simplicial complex. Given $(w, T) \in (W \times S^f)_{>(1,\emptyset)}$, let $(W \times S^f)_{((1,\emptyset),(w,T)]}$ denote the "half-open interval" between $(1, \emptyset)$ and (w, T), i.e., it is

$$\{(w', T') \in W \times S^f \mid (1, \emptyset) < (w', T') \le (w, T)\}.$$

We must show that $(W \times S^f)_{((1,\emptyset),(w,T)]}$ is isomorphic to the poset of all nonempty subsets of T. Since $(w, T) > (1, \emptyset)$, $w \in W_T$. Let $t_1, \ldots, t_n$ be the elements of T. Since $W_T \cong (\mathbb{Z}/2)^n$, any $w \in W_T$ can be put in the form

$$w = t_1^{\varepsilon_1} \ldots t_n^{\varepsilon_n} \tag{*}$$

where each ε_i is either 0 or 1. Then $(w', T') < (w, T)$ if and only if $T' < T$ and the expression for w' is obtained from $(*)$ by deleting those t_i which lie in $T - T'$. In other words, given (w, T) and T' with $(1, \emptyset) < (w, T)$, and $T' \le T$, there is a unique element w' such that $(1, \emptyset) < (w', T') \le (w, T)$. Thus, $(W \times S^f)_{>(1,\emptyset)}$ is an abstract simplicial complex as claimed.

The vertices of $Lk(v, \widetilde{\Sigma}_W)$ correspond to those elements (w, T) in $(W \times S^f)_{>(1,\emptyset)}$ such that T is a singleton. Hence, a vertex corresponds to an element $(w, \{t\})$ where $w = t$ or $w = 1$. Suppose $\{(w_0, \{t_0\}), \ldots, (w_k, \{t_k\})\}$ corresponds to a set of distinct vertices which are pairwise joined by edges. Put $T = \{t_0, \ldots, t_k\}$. The condition that $(w_i, \{t_i\})$ is joined by an edge to $(w_j, \{t_j\})$ means that $t_i t_j$ has order 2 and hence, that $W_T = (\mathbb{Z}/2)^{k+1}$. Thus, $T \in S^f$. Let $w = w_1 w_2 \ldots w_k \in W_T$. By the discussion above, (w, T) is a k-simplex of $(W \times S^f)_{>(1,\emptyset)}$ whose vertices are $(w_0, \{t_0\}), \ldots, (w_k, \{t_k\})$. Thus, $Lk(v, \widetilde{\Sigma}_W)$ is a flag complex. $\qquad\square$

3.2   Cohomology

Proposition 3.2.1. *Suppose W is right-angled. The CW structure on Z_W is "perfect" in the sense of Morse theory. That is to say, in the cellular chain complex, $C_*(Z_W)$, all boundary maps are 0.*

Proof. As pointed out in [KR, p. 180] the space Z_W can be identified with a subcomplex of the torus $(S^1)^S$ with its standard cubical cell structure: $Z_W = \{(x_s)_{s \in S} \in (S^1)^S \mid$ if s and t do not commute, then either $x_s = 1$ or $x_t = 1\}$. In other words, the i-cell corresponding to T, $T \subset S$ and $\text{Card}(T) = i$, belongs to Z_W if and only if $T \in S^f$. It follows that the cellular chain complex for Z_W injects into that of the torus. In the cellular chain complex of a torus, all boundary maps are 0; hence, the same is true in Z_W. $\qquad\square$

Corollary 3.2.2. *Suppose W is right-angled. Then $H_k(A_W)$ $(= H_k(Z_W))$ is free abelian. Its rank is the number of elements T in S^f such that $\text{Card}(T) = k$.*

Remark 3.2.3 The homology of Z_W was calculated by Kim and Roush in [KR, Cor. 11].

Still supposing W is right-angled, we turn now to the calculation of the ring structure of $H^*(A_W)$. The k-cells of Z_W are in one-to-one correspondence with those subsets T of S such that $T \in \mathcal{S}^f$ and $\mathrm{Card}\,(T) = k$. Order the elements of S, $s_1, \ldots, s_n$. Let e_i denote the 1-cell corresponding to $\{s_i\}$. Choose an orientation for e_i. If $T = \{s_{i_1}, \ldots, s_{i_k}\}$, with $i_1 < \cdots < i_k$, is an element of $\mathcal{S}^f$, then the oriented k-cell e_T corresponding to T is the Cartesian product: $e_T = e_{i_1} \times \cdots \times e_{i_k}$. The group of cellular 1-chains $C_1(Z_W)$ is the free abelian group on $\{e_1, \ldots, e_n\}$. Let $z_1, \ldots, z_n$ be the dual basis for $C^1(Z_W)$, the 1-cochains. By Proposition 3.2.1 all boundary maps and coboundary maps are zero; hence we can safely blur the distinctions between cochains, cocycles and cohomology classes. If $T = \{s_{i_1}, \ldots, s_{i_k}\}$, then put

$$z_T = z_{i_1} \cup \ldots \cup z_{i_k}\,.$$

Then $z_T(e_T) = 1$ and $z_T(e_{T'}) = 0$ if $T' \neq T$. Thus, $\{e_T\}$ and $\{z_T\}$, $T \in \mathcal{S}^f$, $\mathrm{Card}\,(T) = k$, are dual bases for $C_k(Z_W)$ and $C^k(Z_W)$. From the above remarks we can easily deduce the following theorem.

Theorem 3.2.4. *Suppose (W, S) is right-angled. Let $n = \mathrm{Card}\,(S)$ and let $\Lambda[y_1, \ldots, y_n]$ be the exterior algebra (over $\mathbb{Z}$) on indeterminates $y_1, \ldots, y_n$. Let I be the ideal generated by all products $y_i y_j$ such that $s_i s_j$ has infinite order in W. Then the map $y_i \to z_i$ defines an isomorphism of graded rings $\varphi \colon \Lambda[y_1, \ldots, y_n]/I \to H^*(Z_W)$. Thus,*

$$H^*(A_W) \cong \Lambda[y_1, \ldots, y_n]/I\,.$$

References

[AB]   J. Alonzo and M. Bridson, Semihyperbolic groups, *Proc. London Math. Soc. (3)* **70** (1995), 56-114.

[B]   N. Bourbaki, *Groupes et Algebres de Lie*, Masson, Paris (1981), Chapters IV–VI.

[Br]   E. Brieskorn, Sur les groupes des tresses *Séminaire Bourkai* **401** (1971).

[CD1]   R. Charney and M. Davis, The $K(\pi, 1)$-problem for hyperplane complements associated to infinite reflection groups *J. Amer. Math. Soc.*, to appear.

[CD2]   R. Charney and M. Davis, Euler characteristic of a nonpositively curved, piecewise Euclidean manifold, *Pac. J. Math.*, to appear.

[C]   I. M. Chiswell, The Euler characteristic of graph products and of Coxeter groups in: *Discrete Groups and Geometry*, W. J. Harvey and C. Machachlan (eds.), *London Math. Soc. Lecture Note Series* **173** (1992), 36–46.

[D]   P. Deligne, Les immeubles des groupes de tresses généralisés *Invent. Math.* **17** (1972), 273–302.

[FSV]   Z. Fiedorowicz, R. Schwänzl and R. Vogt, Iterated moniodal categories, preprint.

[G]   M. Gromov, Hyperbolic groups in: *Essays in Group Theory*, S. M. Gersten (ed.), *M.S.R.I. Publ.* **8** Springer-Verlag: New York (1987), 75–264.

[HM]   S. Hermiller and J. Meier, Algorithms and geometry for graph products of groups *Journal of Algebra*, to appear.

[KR]   K. H. Kim and F. W. Roush, Homology of certain algebras defined by graphs *Journal of Pure and Applied Algebra* **17** (1980), 179–186.

[L]   H. van der Lek, The homotopy type of complex hyperplane complements, Ph.D. Thesis, Univeristy of Nijmegan, 1983.

[Mi]   R. J. Milgram, Iterated loop spaces *Ann. of Math.* **84** (1966), 386–403.

[M]   G. Moussong, Hyperbolic Coxeter groups, Ph.D. thesis, The Ohio State University, 1988.

[P]   L. Paris, Universal cover of Salvetti's complex and topology of simplicial arrangements of hyperplanes *Trans. Amer. Math. Soc.* **340** (1993), 149–178.

[S1]   M. Salvetti, Topology of the complement of real hyperplanes in $\mathbb{C}^n$ *Invent. Math* .**88** (1987), 603–618.

[S2]   M. Salvetti, The homotopy type of Artin groups *Math. Res. Lett.* **1** (1994), 565–577.

[T]   K. Tatsuoka, A finite $K(\pi, 1)$ for Artin groups of finite type, preprint.

[V]   E. B. Vinberg, Discrete linear groups generated by reflection *Math. USSR Izvestija* **5** (1971), 1083–1119.

Double Point Manifolds of Immersions of Spheres in Euclidean Space

Peter J. Eccles

Abstract

Anyone who has been intrigued by the relationship between homotopy theory and differential topology will have been inspired by the work of Bill Browder. This note contains an example of the power of these interconnections.

We prove that, in the metastable range, the double point manifold of a self-transverse immersion $S^n \looparrowright \mathbb{R}^{n+k}$ is either a boundary or bordant to the real projective space $\mathbb{R}P^{n-k}$. The values of n and k for which non-trivial double point manifolds arise are determined.

1 Introduction

Given a self-transverse immersion $f: S^n \looparrowright \mathbb{R}^{n+k}$, the *r-fold intersection set* $I_r(f)$ is defined as follows:

$$I_r(f) = \{\, f(x_1) = f(x_2) = \ldots = f(x_r) \mid x_i \in S^n,\ i \neq j \Rightarrow x_i \neq x_j \,\}.$$

The self-transversality of f implies that this subset of $\mathbb{R}^{n+k}$ is itself the image of an immersion (not necessarily self-transverse)

$$\theta_r(f): L^{n-k(r-1)} \looparrowright \mathbb{R}^{n+k}$$

of a manifold L of dimension $n - k(r-1)$ called the *r-fold intersection manifold of f*. It is natural to ask for a given manifold L whether it can arise as an intersection manifold of an immersed sphere and if so for which dimensions. Alternatively we can consider simply the bordism class of L.

In the stable range $n < k$ the map f is necessarily an embedding as $I_r(f)$ is empty for $r \geqslant 2$. In this note we consider immersions in the metastable range, $n < 2k$. In these cases the intersection manifolds are empty for $r \geqslant 3$ and so the double point manifold is embedded by $\theta_2(f)$.

Theorem 1.1 *In the metastable range, $k \leqslant n < 2k$,*

(a) *if $n-k$ is odd then the double point manifold of a self-transverse immersion $S^n \looparrowright \mathbb{R}^{n+k}$ is a boundary;*

This work began during a visit to the Departments of Analysis and of Geometry of Eötvös University, Budapest. The author is grateful to these departments and András Szűcs for their invitation and hospitality, and to the British Council for support with travel costs.

(b) *if $n - k$ is even then the double point manifold of a self-transverse im-*
mersion $S^n \looparrowright \mathbb{R}^{n+k}$ is either a boundary or bordant to the real projective
space $\mathbb{R}P^{n-k}$.

It should be recalled that odd dimensional real projective spaces are boundaries.

Theorem 1.2 *In the metastable range, $0 \leqslant 4p < n$, there exists an immersion*
$S^n \looparrowright \mathbb{R}^{2n-2p}$ with double point manifold bordant to $\mathbb{R}P^{2p}$ if and only if $n \equiv$
$2^q - 1 \bmod 2^q$, where $q = p$ if $p \equiv 0$ or 3 mod 4 and $q = p+1$ if $p \equiv 1$ or 2 mod 4.

In fact, it is known ([17]) that for $p \geqslant 5$ there is an immersion with double
point manifold homeomorphic to $\mathbb{R}P^{2p}$ in the dimensions given by this theorem. I am grateful to András Szűcs for drawing this reference to my attention.
In addition, it was my efforts to understand his result that there exists an immersion $M^n \looparrowright \mathbb{R}^{2n-2}$ with double point manifold bordant to the projective
plane if and only if $n \equiv 3$ mod 4 ([14]) which led to the present work. Theorem 1.2 shows that in this result the manifold M can always be taken to be a
sphere.

The proof of these theorems is an application of the general method introduced in [3]. It may be summarized as follows.

Step 1. Write $\mathbb{R}P_k^\infty$ for the truncated real projective space $\mathbb{R}P^\infty/\mathbb{R}P^{k-1}$.
Then to an element $\alpha \in \pi_n \mathbb{R}P_k^\infty$ we can associate an immersion $i_\alpha \colon S^n \looparrowright \mathbb{R}^{n+k}$
well-defined up to regular homotopy. In the metastable range every regular
homotopy class of such immersions arises in this way.

Step 2. Let $MO(k)$ be the Thom space of the universal $O(k)$-bundle so that
each element of $\pi_n MO(k)$ represents a bordism class of $(n - k)$-dimensional
manifolds. Then there is a map

$$Mk\eta \colon \mathbb{R}P_k^\infty \to MO(k)$$

of Thom complexes, induced by the bundle $k\eta$ over $\mathbb{R}P^\infty$, the Whitney sum of
k copies of the Hopf line bundle. The element $(Mk\eta)_*\alpha \in \pi_n MO(k)$ represents
the bordism class of the double point manifold of the immersion i_α.

Step 3. Since the bordism classes of a manifold is determined by its Stiefel-
Whitney numbers we apply the Hurewicz map in $\mathbb{Z}/2$-homology to the class of
the double point manifold. The following diagram commutes by naturality.

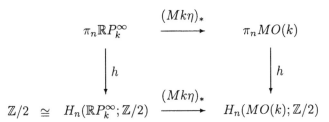

Results about the S-reducability of truncated projective spaces ([1]) imply that, for $k \leqslant n < 2k$, the left hand Hurewicz homomorphism is zero if and only if $k < n + 1 - \rho(n + 1)$. Here $\rho(n + 1)$ is the Hurwitz-Radon-Eckmann number which has the property that $\rho(n + 1) - 1$ is the maximum number of linearly independent tangent vector fields on S^n. Thus, in this case, the double point manifold is a boundary. The complete solution is obtained by evaluating the map $(Mk\eta)_*$ in homology.

This paper is organized as follows. In §2 we discuss the generalized J-homomorphism which leads to the first step of the above proof. In §3 we consider the relationship between this homomorphism and the Hopf invariant. The relationship between the Hopf invariant and the double point manifold described in §4 then leads to the second step. The calculations required for the third step are described in §5.

2   The generalized J-homomorphism

Let $G(m)$ be a closed subgroup of the orthogonal group $O(m)$ with inclusion map $i: G(m) \to O(m)$. Then, if ξ is an m-dimensional vector bundle on S^n represented by $\xi \in \pi_n BO(m)$, a $G(m)$-structure on ξ is a choice of element $\bar{\xi} \in \pi_n BG(m)$ such that $i_* \bar{\xi} = \xi$.

The normal bundle ν of the standard embedding $S^n \hookrightarrow \mathbb{R}^{n+m}$ is trivial and so each element of $\pi_n O(m)/G(m)$ determines a $G(m)$-structure on ν. With this structure, the embedded sphere represents an element of $\pi_{n+m} MG(m)$ by the Pontrjagin-Thom construction. This process defines the *generalized J-homomorphism*

$$J: \pi_n O(m)/G(m) \to \pi_{n+m} MG(m)$$

introduced by Bruno Harris ([4]).

The image of this map J consists of those elements which may be represented by the standard embedding of S^n in $\mathbb{R}^{n+m}$ with some $G(m)$-structure. When $G(m)$ is the trivial group, this reduces to the classical J-homomorphism $\pi_n O(m) \to \pi_{n+m} S^m$.

We now consider the case of $G(m) = O(k)$, for $k \leqslant m$, embedded in the standard way. In this case $O(m)/O(k)$ is the real Stiefel manifold $V_{m-k}(\mathbb{R}^m)$ and $MG(m)$ is the suspension $\Sigma^{m-k} MO(k)$ so that we have the following map:

$$J: \pi_n V_{m-k}(\mathbb{R}^m) \to \pi_{n+m} \Sigma^{m-k} MO(k).$$

As an aside, we now describe a map of spaces inducing this J-homomorphism which will be needed later. Recall that, as the classifying space $BO(k)$, we can take the infinite Grassmannian $G_k(\mathbb{R}^\infty)$ of k-dimensional linear subspaces of $\mathbb{R}^\infty$ (see for example [9]). In this case, the total space $EO(k)$ of the universal bundle is given by

$$EO(k) = \{\, (u, U) \mid u \in U,\ U \in G_k(\mathbb{R}^\infty \,\}.$$

Now, let $v = (v_1, \ldots, v_{m-k}) \in V_{m-k}(\mathbb{R}^m)$ be an orthogonal $(m-k)$-frame. Write $U = \langle v_1, \ldots, v_{m-k}\rangle^\perp \subseteq \mathbb{R}^m \subseteq \mathbb{R}^\infty$, the subspace of $\mathbb{R}^m$ orthogonal to each v_i. Then a point of $\mathbb{R}^m$ may be written uniquely as $u + t_1 v_1 + \ldots + t_{m-k} v_{m-k}$ where $t_i \in \mathbb{R}$ and $u \in U$. So we may define a continuous map $J(v): \mathbb{R}^m \to EO(k) \times \mathbb{R}^{m-k}$ by

$$J(v)(u + t_1 v_1 + \ldots + t_{m-k} v_{m-k}) = ((u, U), t_1, \ldots, t_{m-k}).$$

This induces a map $J(v): S^m \to MO(k) \wedge S^{m-k}$, i.e. $J(v) \in \Omega^m \Sigma^{m-k} MO(k)$. Checking the definitions carefully gives the following result.

Proposition 2.1 *The continuous map $J: V_{m-k}(\mathbb{R}^m) \to \Omega^m \Sigma^{m-k} MO(k)$ induces the generalized J-homomorphism*

$$J: \pi_n V_{m-k}(\mathbb{R}^m) \to \pi_n \Omega^m \Sigma^{m-k} MO(k) \cong \pi_{n+m} \Sigma^{m-k} MO(k).$$

Returning to the differential topology, recall ([16]) that the bordism group of immersed manifolds $M^n \looparrowright \mathbb{R}^{n+k}$ is isomorphic to the stable homotopy group $\pi_n^S MO(k)$. An n-manifold embedded in $\mathbb{R}^{n+m}$ with an $O(k)$-structure, as above, has $(m-k)$ linearly independent normal vector fields and by [5] these lead to a regular homotopy of the embedding to an immersion in $\mathbb{R}^{n+k} \subseteq \mathbb{R}^{n+m}$. This process corresponds to the stabilization map

$$\pi_{n+m} \Sigma^{m-k} MO(k) \to \pi_{n+k}^S MO(k).$$

Composing this with J gives a stable J-homomorphism

$$J^S: \pi_n V_{m-k}(\mathbb{R}^m) \to \pi_{n+k}^S MO(k)$$

whose image consists of classes represented by immersed spheres, and all such classes if $m > n$ since then the stabilization map is an epimorphism. Recall the classical result of Stephen Smale ([10]) that for $m > n+1$ the homotopy group $\pi_n V_{m-k}(\mathbb{R}^m)$ represents regular homotopy classes of immersed spheres $S^n \looparrowright \mathbb{R}^{n+k}$; the stable J-homomorphism maps regular homotopy classes to bordism classes.

Finally, recall that hyperplane reflection defines a $(2k-1)$-equivalence $\lambda: \mathbb{R}P_k^{m-1} \to V_{m-k}(\mathbb{R}^m)$. The composition

$$J^S \circ \lambda_*: \pi_n \mathbb{R}P_k^{m-1} \to \pi_{n+k}^S MO(k)$$

provides the map $\alpha \mapsto [i_\alpha: S^n \looparrowright \mathbb{R}^{n+k}]$ claimed in Step 1 of the introduction. From the above discussion, all bordism classes represented by immersed spheres lie in the image of this map if $m > n$ and $2k > n$.

3 The Hopf invariant and the *J*-homomorphism

In this section we demonstrate that the multiple suspension Hopf invariant is closely related to the generalized *J*-homomorphism described in the previous section.

To describe this invariant we need a preliminary definition. The *quadratic construction* on a pointed space X is defined to be

$$D_2 X = X \wedge X \times_{\mathbb{Z}/2} S^\infty = X \wedge X \times_{\mathbb{Z}/2} S^\infty / \{*\} \times_{\mathbb{Z}/2} S^\infty,$$

where the non-trivial element of the group $\mathbb{Z}/2$ acts on $X \wedge X$ by permuting the coordinates and on the infinite sphere S^∞ by the antipodal action. This space has a natural filtration given by $D_2^i X = X \wedge X \times_{\mathbb{Z}/2} S^i$. The following homeomorphisms may be obtained directly from this definition.

Proposition 3.1 *For a pointed space X,*

(a) $D_2^0 X \cong X \wedge X$;

(b) *for $i \geqslant 1$,* $D_2^i X / D_2^{i-1} X \cong \Sigma^i (X \wedge X)$;

(c) *for $i, j \geqslant 1$,* $D_2^i \Sigma^j X \cong \Sigma^j (D_2^{i+j} X / D_2^{j-1} X)$;

(d) *in particular, for $i, j \geqslant 1$,* $D_2^i S^j \cong \Sigma^j \mathbb{R}P_j^{i+j}$.

We can now describe the basic properties of the multiple suspension *James-Hopf invariants* h_2^i. Write QX for the direct limit $\Omega^\infty \Sigma^\infty X = \lim \Omega^n \Sigma^n X$. For a connected space X and $i \geqslant 0$, there is a natural transformation (see [11])

$$h_2^i = h_2^i(X) : \Omega^{i+1} \Sigma^{i+1} X \to Q D_2^i X.$$

In the case $i = 0$ this is simply the stabilization of the classical James-Hopf invariant:

$$\Omega \Sigma X \to \Omega \Sigma (X \wedge X) \to Q(X \wedge X).$$

The main property of these maps is a generalization of the classical *EHP*-sequence.

Theorem 3.2 ([8]) *Let X be a $(k-1)$-connected space where $k \geqslant 1$. Then, in the metastable range $j < 3k$, there is an exact sequence as follows:*

$$\pi_j X \xrightarrow{i_*} \pi_j \Omega^{i+1} \Sigma^{i+1} X \xrightarrow{h_2^i} \pi_j Q D_2^i X \longrightarrow \pi_{j-1} X.$$

Using adjointness and stability we can rewrite this as follows:

$$\pi_j X \xrightarrow{\Sigma^{i+1}} \pi_{i+j+1} \Sigma^{i+1} X \xrightarrow{h_2^i} \pi_j D_2^i X \longrightarrow \pi_{j-1} X.$$

We also need to record the relationships between these invariants for different values of i. These follow directly from their definitions.

Proposition 3.3 (a) *For each $i \geqslant 0$ the following diagram is commutative.*

$$\begin{array}{ccc} \Omega^{i+1}\Sigma^{i+1}X & \xrightarrow{h_2^i} & QD_2^i X \\ \downarrow & & \downarrow \\ \Omega^{i+2}\Sigma^{i+2}X & \xrightarrow{h_2^{i+1}} & QD_2^{i+1}X \end{array}$$

(b) *For each $i, j \geqslant 0$, the following diagram is commutative.*

$$\begin{array}{ccc} \Omega^{i+j+1}\Sigma^{i+j+1}X & \xrightarrow{\quad h_2^{i+j}(X) \quad} & QD_2^{i+j}X \\ {\scriptstyle 1}\downarrow & & \downarrow \\ \Omega^j\Omega^{i+1}\Sigma^{i+1}\Sigma^j X & \xrightarrow{\Omega^j h_2^i(\Sigma^j X)} \Omega^j QD_2^i\Sigma^j X & \cong \quad Q(D_2^{i+j}X/D_2^{j-1}X) \end{array}$$

The first part of this proposition means that the invariants h_2^i combine to form the *stable James-Hopf invariant* $h_2^S \colon QX \to QD_2X$.

We can now state the main result relating the Hopf invariant and the J-homomorphism.

Theorem 3.4 *For $1 \leqslant k < m$, the following diagram is commutative.*

$$\begin{array}{ccccc} \pi_n\mathbb{R}P_k^{m-1} & \xrightarrow{\lambda_*} & \pi_n V_{m-k}(\mathbb{R}^m) & \xrightarrow{J} & \pi_{n+m}\Sigma^{m-k}MO(k) \\ & & & & \downarrow{\scriptstyle \cong} \\ {\scriptstyle \Sigma^k}\downarrow & & & & \pi_{n+k}\Omega^{m-k}\Sigma^{m-k}MO(k) \\ & & & & \downarrow{\scriptstyle h_2^{m-k-1}} \\ \pi_{n+k}^S\Sigma^k\mathbb{R}P_k^{m-1} & \xrightarrow[\cong]{} \pi_{n+k}^S D_2^{m-k-1}S^k & \xrightarrow{i_*} & \pi_{n+k}^S D_2^{m-k-1}MO(k) \end{array}$$

Here the isomorphism in the bottom row comes from the homeomorphism of Proposition 3.1(d) and the map i is induced by the map $S^k \to MO(k)$ arising from the inclusion of a fibre in the universal bundle.

This result may be proved by making use of the map inducing the J-homomorphism introduced in the previous section beginning with a lemma which corresponds to the $k = 0$ version of the theorem.

Lemma 3.5 *For $m \geqslant 2$, the following diagram is homotopy commutative.*

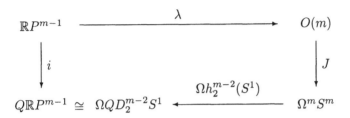

Proof. The proof is by induction on m. For $m = 2$, the result is a formulation of the statement that the Hopf map $S^3 \to S^2$ has Hopf invariant 1.

The inductive step is based on the following commutative diagram.

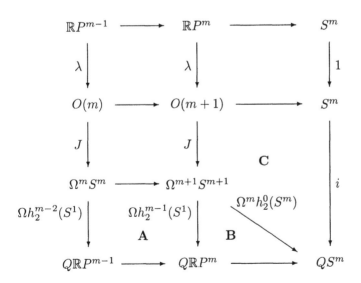

In this diagram, **A** is commutative by Proposition 3.3(a), **B** is commutative by Proposition 3.3(b) and **C** is commutative by [6]. Adjoining the vertical maps in the diagram gives the following diagram of stable maps.

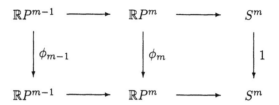

The inductive hypothesis is that $\phi_{m-1} \simeq 1$. Since the rows are cofibre sequences this implies that the stable map $\phi_m \simeq 1$ as required to complete the inductive step.   □

Proof of Theorem 3.4. Consider the following commutative diagram.

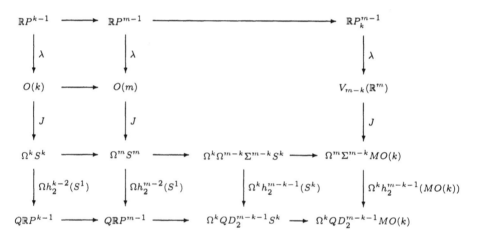

In this case, adjointing the vertical maps and using $D_2^{m-k-1}S^k \cong \Sigma^k \mathbb{R}P_k^{m-1}$ we obtain the following commutative diagram of stable maps.

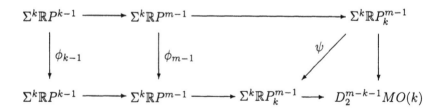

Here the stable map ψ exists by the basic properties of cofibre sequences. By the lemma, $\phi_{k-1} \simeq 1$ and $\phi_{m-1} \simeq 1$. It follows that the stable map $\psi \simeq 1$. Now applying π_n to the unstable adjoint of the right hand triangle gives the theorem.   □

4   The Hopf invariant and the double point manifold

The final ingredient required to enable us to read off the bordism class of the double point manifold of an immersed sphere is the observation that this is determined by the stable Hopf invariant $h_2^S \colon \pi_{n+k}^S MO(k) \to \pi_{n+k}^S D_2 MO(k)$ in the following sense. If the self-transverse immersion $f \colon M^n \looparrowright \mathbb{R}^{n+k}$ represents the element $\alpha \in \pi_{n+k}^S MO(k)$, then the double point manifold $\theta_2(f) \colon L^{n-k} \looparrowright \mathbb{R}^{n+k}$ respresents the element $h_2^S(\alpha)$. This has been proved independently by Pierre Vogel ([15]), by András Szűcs ([12], [13]), and by Ulrich Koschorke and Brian Sanderson ([7]).

Notice that $\theta_2(f)$ represents an element in the stable homotopy of $D_2 MO(k)$ because the immersion of the double point manifold automatically acquires some additional structure on its normal bundle, namely that at each point $f(x_1) = f(x_2)$, the normal $2k$-dimensional space may be decomposed as the direct sum of the two (unordered) k-dimensional normal spaces of f at the points x_1 and x_2. The universal bundle for this structure is

$$EO(k) \times EO(k) \times_{\mathbb{Z}/2} S^\infty \to BO(k) \times BO(k) \times_{\mathbb{Z}/2} S^\infty$$

which has the Thom complex $D_2 MO(k)$.

In the case of immersed spheres in the metastable range, Theorem 3.4 shows that the double point manifold has a more refined structure corresponding to the restriction of the above universal bundle to $\{*\} \times \{*\} \times_{\mathbb{Z}/2} S^\infty$:

$$\mathbb{R}^k \times \mathbb{R}^k \times_{\mathbb{Z}/2} S^\infty \to \mathbb{R}P^\infty.$$

As before, the involution on $\mathbb{R}^k \times \mathbb{R}^k$ is obtained by interchanging coordinates, but in this case by a change of basis it can be written as $(x, y) \mapsto (-x, y)$. This demonstrates that this bundle is simply the Whitney sum $k\eta \oplus k$ where η is the Hopf bundle. This is another way of proving that the Thom complex $D_2 S^k$ is homeomorphic to the suspended truncated real projective space $\Sigma^k \mathbb{R}P_k^\infty$ (using [2]).

In fact we are here not interested in this additional structure but merely in the unoriented bordism class of the double point manifold. Forgetting the additional structure induces a map of Thom complexes $\zeta \colon D_2 MO(k) \to MO(2k)$. The bordism class of the double point manifold is then given by the composition

$$\pi_{n+k}^S D_2 MO(k) \xrightarrow{\zeta_*} \pi_{n+k}^S MO(2k) \to \pi_{n-k} MO$$

where the final stabilization map to the homotopy of the MO-spectrum corresponds to forgetting the immersion. The above identification of the restricted universal bundle imples that the composition $\zeta_* i_* \colon \pi_{n+k}^S \Sigma^k \mathbb{R}P_k^\infty \cong \pi_{n+k}^S D_2 S^k \to \pi_{n+k}^S D_2 MO(k) \to \pi_{n+k}^S MO(2k)$ is given by $\Sigma^k (Mk\eta)_*$.

We can sum this up in the following result.

Theorem 4.1 *Let $\theta_2: \pi^S_{n+k}MO(k) \to \pi_{n-k}MO$ be the map defined geometrically by the double point manifold of a self-transverse immersion. Then the following diagram is commutative.*

$$
\begin{array}{ccccccc}
\pi^S_{n+k}MO(k) & \xrightarrow{\ h^S_2\ } & \pi^S_{n+k}D_2MO(k) & \longleftarrow & \pi^S_{n+k}D_2S^k & \overset{\cong}{\longleftarrow} & \pi_{n+k}\Sigma^k\mathbb{R}P^\infty_k \\
\Big\downarrow{\scriptstyle \theta_2} & & \Big\downarrow{\scriptstyle \zeta_*} & & & & \Big\downarrow{\scriptstyle \cong} \\
\pi_{n-k}MO & \longleftarrow & \pi^S_{n+k}MO(2k) & \longleftarrow & \pi^S_n MO(k) & \overset{(Mk\eta)_*}{\longleftarrow} & \pi^S_n\mathbb{R}P^\infty_k
\end{array}
$$

Some geometric comments

It is easy to see directly that the double point manifold of an immersed sphere has the refined structure discussed above.

Let $\tilde{L}$ be the double cover of the double point manifold L so that there is the following commutative diagram.

$$
\begin{array}{ccccc}
\tilde{L}^{n-k} & \longrightarrow & S^n & \xrightarrow{\ \nu\ } & BO(n+k) \\
\Big\downarrow & & \Big\downarrow{\scriptstyle f} & & \\
& {\scriptstyle \theta_2(f)} & & & \\
L^{n-k} & \longrightarrow & \mathbb{R}^{n+k} & &
\end{array}
$$

Then $\nu|\tilde{L}^{n-k}$ is trivial since $n - k < n$. A choice of trivialization induces a $\mathbb{Z}/2$-structure on the normal bundle of L^{n-k} corresponding to the universal bundle discussed above with Thom complex $D_2S^k \cong \Sigma^k\mathbb{R}P^\infty_k$.

Notice that this construction applies to any immersion of a sphere and suggests that a stable retraction of $V_{m-k}(\mathbb{R}^m)$ onto $\mathbb{R}P^{m-1}_k$ can be included in the diagram of Theorem 3.4. This ought to lead to an extension of the results of this paper beyond the metastable range to the general case.

5 Determining the double point manifold

In this section we prove Theorems 1.1 and 1.2. The calculation is based upon the following immediate corollary of Theorems 3.4 and 4.1.

Proposition 5.1 *For $k \leqslant n < 2k$, the following diagram is commutative.*

$$\pi_n \mathbb{R}P_k^{m-1} \xrightarrow{\;i_*\;} \pi_n \mathbb{R}P_k^\infty \xrightarrow{\;(Mk\eta)_*\;} \pi_n MO(k)$$

$$\Big\downarrow J^S \circ \lambda_* \qquad\qquad\qquad\qquad\qquad\qquad \Big\downarrow \sigma$$

$$\pi_{n+k}^S MO(k) \xrightarrow{\qquad\qquad\theta_2\qquad\qquad} \pi_{n-k} MO$$

Thus, given $\alpha \in \pi_n \mathbb{R}P_k^{m-1}$, if $i_\alpha \colon S^n \hookrightarrow \mathbb{R}^{n+k}$ is a self-transverse immersion representing $J^S \circ \lambda_*(\alpha)$ then the bordism class of the double point manifold of i_α is represented by $(Mk\eta)_* i_*(\alpha) \in \pi_n MO(k)$.

For $m > n + 1$ the above map i_* is an isomorphism. We may complete the proof by applying the Hurewicz homomorphism in $\mathbb{Z}/2$-homology to $Mk\eta$ because this Hurewicz homomorphism for MO is a monomorphism ([9]).

$$\pi_n \mathbb{R}P_k^\infty \xrightarrow{\;(Mk\eta)_*\;} \pi_n MO(k) \xrightarrow{\;\sigma\;} \pi_{n-k} MO$$

$$\Big\downarrow h \qquad\qquad\qquad \Big\downarrow h \qquad\qquad\qquad \Big\downarrow h$$

$$\mathbb{Z}/2 \cong H_n \mathbb{R}P_k^\infty \xrightarrow{\;(Mk\eta)_*\;} H_n MO(k) \xrightarrow{\;\sigma\;} H_{n-k} MO$$

Here and from now on we write $H_n X$ for $H_n(X; \mathbb{Z}/2)$.

It is well-known ([1]) that, for $k \leqslant n < 2k$, the left hand Hurewicz homomorphism is zero if and only if $k < n + 1 - \rho(n+1)$. So in this case $\sigma \circ (Mk\eta)_* = 0 \colon \pi_n \mathbb{R}P_k^\infty \to \pi_{n-k} MO$ and the double point manifold is a boundary.

Proposition 5.2 *For $k \leqslant n < 2k$, if $k \geqslant n + 1 - \rho(n+1)$, let $\alpha \in \pi_n \mathbb{R}P_k^\infty$ be such that $h(\alpha) \neq 0$. Let L^{n-k} be the double point manifold of a corresponding immersion $i_\alpha \colon S^n \hookrightarrow \mathbb{R}^{n+k}$. Then the (normal) Stiefel-Whitney numbers of L are given by*

$$w_I[L^{n-k}] = \binom{k}{i_1}\binom{k}{i_2}\cdots\binom{k}{i_t}$$

for partitions $I = (i_1, i_2, \ldots, i_t)$ such that $i_1 + i_2 + \ldots + i_t = n - k$.

Proof. The Thom complex $MO(k)$ is homotopy equivalent to the quotient space $BO(k)/BO(k-1)$ so that $H^* MO(k) \cong w_k \mathbb{Z}/2[w_1, \ldots, w_k]$ with the Thom isomorphism $\tilde{H}^{k+i} MO(k) \cong H^i BO(k)$ given by $w_I w_k \leftrightarrow w_I$. Hence the Stiefel-Whitney number

$$w_I[L] = \langle h(Mk\eta)_*(\alpha), w_I w_k \rangle \quad \text{by the Thom isomorphism}$$
$$= \langle h(\alpha), w_I(k\eta) a^k \rangle \quad \text{by naturality}$$

where $a \in H^1\mathbb{R}P^\infty$ is the generator so that $a^k \in H^k\mathbb{R}P^\infty_k$ is the Thom class.

Now the total Stiefel-Whitney class $w(k\eta) = (1 + a)^k \in H^*\mathbb{R}P^\infty$ so that $w_i(k\eta) = \binom{k}{i}a^i$. The result follows.   □

To complete the calculation we confirm that these are the normal Stiefel-Whitney numbers of $\mathbb{R}P^{n-k}$.

Proposition 5.3 *If* $k \geqslant n + 1 - \rho(n + 1)$, *then the (normal) Stiefel-Whitney numbers of* $\mathbb{R}P^{n-k}$ *are given by*

$$w_I[\mathbb{R}P^{n-k}] = \binom{k}{i_1}\binom{k}{i_2}\cdots\binom{k}{i_t}.$$

Proof. Put $n + 1 = (2a + 1)2^b$. It is clear from the definition of ρ that $\rho(n + 1) \leqslant 2^b$ so that $k \geqslant n + 1 - \rho(n + 1) \Rightarrow n - k \leqslant \rho(n + 1) - 1 < 2^b$.

The total tangent Stiefel-Whitney class of $\mathbb{R}P^{n-k}$ is given by $(1 + a)^{n-k+1}$ since $\tau \oplus 1 \cong (n - k + 1)\eta$. Hence the total normal class is given by $w(\mathbb{R}P^{n-k}) = (1 + a)^{-n-1+k} \in H^*\mathbb{R}P^{n-k}$. Hence, for $i \leqslant n - k$, $w_i(\mathbb{R}P^{n-k}) = \binom{-n-1+k}{i}a^i = \binom{k}{i}a^i$ since, by the above, $i \leqslant n - k \Rightarrow i < 2^b$ and $n + 1$ is a multiple of 2^b. The result follows.   □

This completes the proof of Theorem 1.1.

To complete the proof of Theorem 1.2 we simply examine the definition of the function ρ. We have proved that in the case of even $n - k$, say $2p$, there exists an immersion $S^n \looparrowright \mathbb{R}^{n+k}$ with double point manifold bordant to $\mathbb{R}P^{n-k}$ if and only if $k \geqslant n + 1 - \rho(n + 1)$, i.e. $\rho(n + 1) \geqslant 2p + 1$.

Recall that, for $n + 1 = (2a + 1)2^b$ where $b = c + 4d$ for $0 \leqslant c \leqslant 3$ and $d \geqslant 0$, the value of ρ is given by $\rho(n + 1) = 2^c + 8d$. It follows that the least value of b for which $\rho(n + 1) \geqslant 2p + 1$ is given by

for $p \equiv 0 \bmod 4$: $c = 0$ and $8d = 2p$, i.e. $b = p$;
for $p \equiv 1 \bmod 4$: $c = 2$ and $8d = 2(p - 1)$, i.e. $b = p + 1$;
for $p \equiv 2 \bmod 4$: $c = 3$ and $8d = 2(p - 2)$, i.e. $b = p + 1$;
for $p \equiv 3 \bmod 4$: $c = 3$ and $8d = 2(p - 3)$, i.e. $b = p$.

This completes the proof of Theorem 1.2.

September 1994

References

[1]   J.F. Adams, *Vector fields on spheres*, Ann. of Math. **75** (1962) 603–632.

[2]   M.F. Atiyah, *Thom complexes*, Proc. London Math. Soc. (3) **11** (1961) 291–310.

[3]   P.J. Eccles, *Characteristic numbers of immersions and self-intersection manifolds,* to appear in the Proceedings of the Colloquium in Topology, Szekszárd, Hungary, August 1993.

[4]   B. Harris, *J-homomorphisms and cobordism groups,* Inv. Math. **7** (1969) 313–320.

[5]   M.W. Hirsch, *Immersions of manifolds,* Trans. Amer. Math. Soc., **93** (1959) 242–276.

[6]   M. Kervaire, *An interpretation of G. Whitehead's generalization of H. Hopf's invariant,* Ann. of Math. (2) **69** (1959) 345–365.

[7]   U. Koschorke and B. Sanderson, *Self intersections and higher Hopf invariants,* Topology **17** (1978) 283–290.

[8]   R.J. Milgram, *Unstable homotopy from the stable point of view,* Lecture Notes in Mathematics **368** (Springer, 1974).

[9]   J.W. Milnor and J.D. Stasheff, *Characteristic classes,* Ann. of Math. Studies **76** (Princeton University Press, 1974).

[10]   S. Smale, *The classification of immersions of spheres in Euclidean space,* Ann. of Math. (2) **69** (1959) 327–344.

[11]   V.P. Snaith, *A stable decomposition of $\Omega^n S^n X$,* J. London Math. Soc. (2) **7** (1974) 577–583.

[12]   A. Szűcs, *Gruppy kobordizmov l-pogruženiĭ I,* Acta Math. Acad. Sci. Hungar. **27** (1976) 343–358,

[13]   A. Szűcs, *Gruppy kobordizmov l-pogruženiĭ II,* Acta Math. Acad. Sci. Hungar. **28** (1976) 93–102.

[14]   A. Szűcs, *Double point surfaces of smooth immersions $M^n \to \mathbb{R}^{2n-2}$,* Math. Proc. Cambridge Philos. Soc. **113** (1993) 601–613.

[15]   P. Vogel, *Cobordisme d'immersions,* Ann. Sci. Ecole Norm. Sup. (4) **7** (1974), 317–358.

[16]   R. Wells, *Cobordism groups of immersions,* Topology **5** (1966) 281–294.

[17]   R. Wells, *Double covers and metastable immersions of spheres,* Can. J. Math. **26** (1974) 145–176.

Controlled Linear Algebra

Michael H. Freedman
Zhenghan Wang

1 Introduction

These notes derive from lectures given at UCSD in 1990. The purpose of those lectures was to elucidate the two main theorems of controlled linear algebra: the vanishing theorem for Whitehead and K_0-type obstructions when only "simply connected directions" are left uncontrolled and the squeezing principle−that once a threshold level of geometric control is obtained any finer amount of control is also available. These notes present in detail (and with perhaps some new estimates on the optimal relations between ϵ, δ and dimension) the vanishing theorem of Frank Quinn's Section 8 [Q1] in the context of Whitehead torsion. Quinn's Section 8 is brilliant and difficult. In our efforts to understand the details we developed−with copious advice from Frank−the argument presented here. We have neglected all the important applications to savor the intricacy of the central argument.

The idea that controlled linear algebra might exist and be useful to topology goes back to Connell and Hollingsworth [CH] in the late 1960s. By 1977, Ferry [F1] and Chapman [C1] had obtained by geometric arguments theorems which strongly suggested the algebraic unification reached in Quinn's papers [Q1][Q2]. These effectively realized Connell and Hollingsworth's program. Since that time controlled algebraic topology and its cousin bounded algebraic topology have thrived (see e.g. [FP],[RY] and [FHP]). The theorem of Bryant-Ferry-Mio-Weinberger [BFMW] on the existence of "non-euclidean manifolds"−perhaps the most striking result at the wild end of topology in a decade−is evidence of the subject's vigor. Controlled linear algebra also has found applications outside of topology. The interested reader may consult [FrP] for an application to signal processing.

Although these notes do not cover the squeezing principle, the basic techniques of pulling back modules and truncating "flanges" by running the "standard deformation" of $A \oplus A^{-1}$ to id can be applied there as well. The reader who wishes to follow the approach of these notes to reach the squeezing principle may appreciate a hint: given any constant $c > 1$, the identity map on any manifold control space is homotopic to a composition f_c of folding maps which stretch all sufficiently small distances by a factor $\geq c$. For ϵ sufficiently small, the maps f_c can be used (in the place of handle-wise torus tricks) to deform an ϵ-isomorphism to an arbitrarily small ϵ'-isomorphism ($0 < \epsilon' < \epsilon$).

Michael Freedman was supported in part by NSF grant DMS-8901412.

Acknowledgement: The authors would like to thank Steve Ferry, Bruce Hughes, Andrew Ranicki and Richard Stong for their valuable suggestions on the presentation of this material.

2 Controlled linear algebra

Suppose X is a Riemannian manifold with the metric d, and π is a fixed group. Let $Z[\pi]$ be the group ring of π. A *controlled module* M over X is defined to be a free $Z[\pi]$-module with a preferred basis $S_M = \{e_i\}_{i=1}^{+\infty}$ and a proper map of the basis $\varphi : S_M \longrightarrow X$. We require that S_M is countable and endowed with the discrete topology. X is called the *control space* of M and φ the *control map* of M. The controlled module is denoted by a triple (M, X, φ). But in the following, only one or two symbols are given if the others can be understood without confusion. From the definition, for each compact subset of X, there are only finitely many basis elements mapping to it. It is convenient to identify the basis elements with their images in the control space. We will do this whenever no confusion will result.

Suppose two controlled modules (M, X, φ), (N, X, ϕ) are given with preferred bases $S_M = \{e_i\}_{i=1}^{+\infty}$ and $S_N = \{f_j\}_{j=1}^{+\infty}$, respectively.

Definition 2.1 Let R be a relation between S_M and S_N, then the spread of R at the basis element e_i is defined to be $r_i = sup\{d(\varphi(e_i), \phi(f_j)) \mid f_j \in R(e_i)\}$. The **spread** of R is the supremum of r_i for all i. The spread might be infinite if X is not compact. Given $\delta : X \longrightarrow (0, \infty)$, the spread of R is less than δ if for each i, $d(\varphi(e_i), \phi(f_j)) < \delta(\varphi(e_i))$ for every $f_j \in R(e_i)$.

Let $\alpha : M \longrightarrow N$ be a homomorphism, the *underlying set relation* $\underline{\alpha}$ of α between S_M and S_N is defined to be $\underline{\alpha}(e_i) = \{f_j \mid f_j$ has a nonzero coefficient in $\alpha(e_i) = \sum_j \alpha_{ij} f_j\}$. For a subset E of $S_M = \{e_i\}$, $\underline{\alpha}(E) = \bigcup_{e_i \in E} \underline{\alpha}(e_i)$. A homomorphism $\alpha : M \longrightarrow N$ is a δ-*homomorphism* for $\delta : X \longrightarrow (0, \infty)$ if the spread of $\underline{\alpha}$ is less than δ. If there is also a δ-homomorphism $\beta : N \longrightarrow M$ such that $\alpha\beta = id$, $\beta\alpha = id$, then α is called a δ-*isomorphism*. A *geometric map* from M to N is defined to be the map induced by a bijection of the bases of M and N.

Given two positive functions δ_1, δ_2 on X, the nonabelian composition $\delta_2 \circ \delta_1$ of δ_1 and δ_2 is defined by:

$$\delta_2 \circ \delta_1(x) = sup\{\delta_1(x) + \delta_2(x') \mid x' \in B_{\delta_1(x)}(x)\}.$$

The composition of a δ_1-homomorphism followed by a δ_2-homomorphism is a $\delta_2 \circ \delta_1$-homomorphism.

Suppose K is a subset of X, a submodule $M \mid_K$ of M with the control space K and the control map $\varphi \mid_K$ is defined as follows: $M \mid_K$ is the submodule of M generated by all basis elements e_i satisfying $\varphi(e_i) \in K$ and $\varphi \mid_K$ is the restriction of φ to the basis of $M \mid_K$. The submodule $M \mid_K$ is called the *restriction* of M to K. Given a homomorphism $\alpha : M \longrightarrow N$, then $\alpha \mid_K$ is the map $p_K \cdot \alpha \cdot i_K$, where i_K is the inclusion $M \mid_K \overset{i_K}{\hookrightarrow} M$ and p_K is the projection $N \overset{p_K}{\twoheadrightarrow} N \mid_K$. A δ-homomorphism $\alpha : M \longrightarrow N$ is called a *δ-isomorphism* over K if there is a δ-homomorphism $\beta : N \longrightarrow M$ such that $(\alpha\beta) \mid_K$ and $(\beta\alpha) \mid_K$ are both the identity.

Since M and N both have fixed bases, a homomorphism from M to N is one-one correspondent to an infinite matrix with only finitely many nonzero elements in each row. In our convention, a vector is written as a column. The corresponding matrix of α is denoted as $A(\alpha)$. The spread of α can also be described as follows: let $A(\alpha) = (a_{ij})$, the spread of α at e_i is $sup_{a_{ij} \neq 0}\{d(\varphi(e_i), \phi(f_j))\}$, where i is fixed and j varies through integers. The spread of α is the supremum of its spread at each basis element. In the following, a map is always thought as a matrix with a fixed indexing of the basis in mind.

In the following examples, the control space will be the Euclidean space E^1 with the standard metric, π will be the trivial group. So we have free abelian groups Z^n with the i-th basis element e_i mapped to i of E^1.

Example 2.1 Every $n \times n$ matrix represents a homomorphism of spread $\leq n$. In particular, the spread of the identity matrix is 0.

Example 2.2 Let $A : Z^n \longrightarrow Z^n$ be the following $n \times n$ matrix:

$$
A = \begin{pmatrix}
1 & -1 & & & & \\
 & 1 & -1 & & & \\
 & & 1 & \cdot & & \\
 & & & \cdot & \cdot & \\
 & & & & 1 & -1 \\
 & & & & & 1
\end{pmatrix}
$$

It has spread 1 which does not depend on n. But its inverse

$$
A^{-1} = \begin{pmatrix}
1 & 1 & 1 & \cdot & \cdot & 1 \\
 & 1 & 1 & \cdot & \cdot & 1 \\
 & & 1 & \cdot & \cdot & 1 \\
 & & & \cdot & \cdot & 1 \\
 & & & & \cdot & 1 \\
 & & & & & 1
\end{pmatrix}
$$

has spread n.

Example 2.3 Let $A = (a_{ij}) : Z^{+\infty} \longrightarrow Z^{+\infty}$ be a matrix. If there exists an integer $p > 0$ such that $a_{ij} = 0$ when $\mid i - j \mid > p$, A is called a 2p-band matrix. Its spread is 2p, so A is a 2p-isomorphism. Then the main theorem (Theorem 4.1) says, as a special case, that A can be diagonalized by row operations and stabilizations described below within some larger band about the diagonal.

Now we describe the *elementary operations* that will be used to deform a δ-isomorphism $\alpha : M \longrightarrow N$ to a geometric map. There are two types of them:

Type I. *Stabilizations.* Given a controlled module (M, X, φ), M can be stabilized by adding countably many new basis elements to S_M and specifying an extension of the control map φ. For a δ-homomorphism $\alpha : M \longrightarrow N$, we stabilize it as follows: stabilize M and N by adding new basis elements e to S_M and f to S_N, extend α to $\alpha':Z[\pi]e \oplus M \longrightarrow Z[\pi]f \oplus N$ by sending e to f and the other basis elements the same as α. To extend the control maps, choose any two points p_1, p_2 in X with distance $< \delta(p_1)$, extend φ and ϕ by sending e to p_1 and f to p_2. This procedure is called an *elementary stabilization*. In particular, if $M = N$, f is chosen to be the same as e. This procedure can be repeated as many times as necessary. A *stabilization* of α is a locally finite composition of such elementary stabilizations, where locally finite means only finitely many basis elements can be mapped to any compact subset of X. After a stabilization, α is changed to a map $\alpha \mid_{stabilized}: M \mid_{stabilized} \longrightarrow N \mid_{stabilized}$ which has the same spread of α. If $M = N$, then in matrix notation $A(\alpha)$ is stabilized to $I \oplus A(\alpha)$.

Type II. *Row Operations.* A map $\beta : M \longrightarrow M$ is called an *elementary map* if it satisfies the following: suppose $\beta(e_i) = \sum \beta_{ij} e_j$, then a) $\beta_{ii} \in \pi$ or $-\beta_{ii} \in \pi$ for all i and b) $\beta_{ij} = 0$ for all pairs (i, j) with $i \neq j$ except possibly one pair. Given a homomorphism $\alpha : M \longrightarrow N$, an elementary row operation of α is the composition of α from left with an elementary map. Algebraically, the matrix $A(\alpha)$ is multiplied from left by the following matrix:

$$
\begin{pmatrix}
\beta_{11} & & & & & \\
& \beta_{22} & & & & \\
& & \ddots & a_{ij} & & \\
& & & \ddots & & \\
& & & & \beta_{nn} & \\
& & & & I & \\
& & & & & \ddots
\end{pmatrix}
$$

for some nonnegative integer n, where $\beta_{ij} \in \pi$ or $-\beta_{ij} \in \pi$, and $a_{ij} \in Z[\pi]$. A *row operation* is a locally finite composition of such elementary row operations. Locally finite means that each basis element can be only involved in finitely many elementary row operations.

Definition 2.2 Given a homomorphism $\alpha : (M, X, \varphi) \longrightarrow (N, X, \phi)$. A **deformation** H of α is a locally finite composable string of possibly infinitely many elementary operations. A deformation H can always be written as a composition of a stabilization H_1 followed by a row operation H_2 of the stabilized map $\alpha \mid_{stabilized} = H_1 \cdot \alpha$. So $H_2 = \cdots \beta_2 \cdot \beta_1 : M \mid_{stabilized} \longrightarrow N \mid_{stabilized}$ is a row operation of $\alpha \mid_{stabilized}$, where β_i is an elementary row operation. Given an $\epsilon : X \longrightarrow (0, \infty)$, H is an ϵ-deformation if the following relation R between the basis S'_M of $M \mid_{stabilized}$ and itself has spread less than ϵ: for each $e_i \in S'_M$, $R(e_i) = \{e_j \mid e_j \in \cdots \underline{\beta_2} \cdot \underline{\beta_1}(e_i)$ or $e_j \in \underline{\beta_1^{-1}} \cdot \underline{\beta_2^{-1}} \cdots (e_i)\}$. Here the composition makes sense because of the local finiteness.

If $K \subset X$, then the spread of H restricted to K is the spread of the relation $R \mid_K$, where $R \mid_K$ is the relation between $S \mid_K = \{e_i \mid \varphi(e_i) \in K, e_i \in S'_M\}$ and itself defined by $R \mid_K (e_i) = R(e_i) \bigcap S \mid_K$.

As exercises, if H_1 is a δ_1-deformation and H_2 is a δ_2-deformation, then $H_2 \cdot H_1$ is a $\delta_2 \circ \delta_1$-deformation. If we write H_1, H_2 as matrices A_1 and A_2, then the spread of the composition of H_1 followed by H_2 is $\geq$ the spread of $A_2 \cdot A_1$ in general.

Definition 2.3 Let G be a controlled module over R^m and $h : G \longrightarrow G$ be an isomorphism of G. Let $B_{r_1}(0)$ and $B_{r_2}(0)$ be balls of radii r_1 and r_2 with centers 0, $0 < r_1 < r_2$. An isomorphism $h' : G \longrightarrow G$ is called a **truncation** of h to the ball $B_{r_2}(0)$ if
(i). On $G \mid_{B_{r_1}(0)}$, h' is the same as h up to stabilization.
(ii). On $G \mid_{R^m \backslash B_{r_2}(0)}$, h' is the identity.
After a truncation, h' may be viewed as a map of the finite module G' obtained from G by discarding the basis elements outside the ball $B_{r_2}(0)$.

Whitehead group is well studied in K-theory. The definition is recalled for convenience. Identify each $M \in GL(n, Z[\pi])$ with $\begin{pmatrix} M & 0 \\ 0 & 1 \end{pmatrix} \in GL(n +$ $1, Z[\pi])$, then $GL(1, Z[\pi]) = U(Z[\pi]) \subset GL(2, Z[\pi]) \subset \cdots$. The union $GL(Z[\pi])$ is called the *infinite general linear group.* A matrix is elementary if it is the same as the identity matrix except for one off-diagonal entry. Let $E(Z[\pi])$ be the subgroup generated by elementary matrices, then $E(Z[\pi])$ is a normal subgroup of $GL(Z[\pi])$. Define $K_1(Z[\pi]) = GL(Z[\pi])/E(Z[\pi])$ and $\overline{K_1}(Z[\pi]) = K_1(Z[\pi])/\{\pm I\}$, note that $\pi \subset GL(1, Z[\pi])$, the *Whitehead group* of π is defined to be $\overline{K_1}(Z[\pi])/image(\pi)$. From the definition, it is clear that to deform a δ-isomorphism to the identity is the controlled analogue of the isomorphism being zero in the Whitehead group.

3   Marriage theorem

Given two sets G and B with a relation R between them, say, G is a set of girls and B is a set of boys. We say a girl $g \in G$ knows a boy $b \in B$ if g has the relation R with b. A *marriage* from G to B is an injective map which preserves the relation. So every girl marries a boy that she knew.

Theorem 3.1 *Given a finite set G of girls and a set B of boys with a relation R between them. Let P be a subset of B, boys in P will be called handsome boys. Assume every subset of girls know at least that many boys and every subset of* *handsome boys are known by at least that many girls. Then there is a marriage from G to B such that every handsome boy gets married.*

Proof. The set of girls G is *forced* if there is a subset S of G knowing exactly that many boys or if there is a subset T of P known by exactly that many girls. S and T are called forced subsets of G and P, respectively. Otherwise, G is *unforced*.

Theorem 3.1 will be proved by induction on the number of girls in G. If G has only one girl, it is obvious. Suppose the theorem is true for $|G| < n$, either G is forced or unforced. Then for $|G| = n$, there are two cases:

(i) G is unforced. Pick a handsome boy b in P and a girl g knowing him. Consider $G' = G - \{g\}$ and $B' = B - \{b\}$, then the hypothesis of the theorem is satisfied for G' and B'. By the inductive hypothesis, there is a marriage from G' to B'. Let g marry b, we get a marriage from G to B.

(ii) G is forced. Let S be a forced subset of G with the smallest number of girls, and T be a forced subset of P with the smallest number of boys, and let $k = min\{|S|, |T|\}$. Subcase a): $k < n$. If there is a subset S of girls knowing exactly $k = |S|$ boys, consider $G' = G - S$ and $B' = B - RS$, where RS is the subset of all boys known by at least one girl in S. For any subset Q of G', since $Q \cup S$ know at least $|Q| + |S|$ boys in B and RS have only $|S|$ boys, therefore, Q know at least $|Q|$ boys in B'. For any subset of $P - RS$, they are known by girls that none is in S. By the inductive hypothesis, there is a marriage from G' to B'. It is easy to check that the inductive hypothesis holds for S and RS. Combine the marriage from G' to B' with one from S to RS, we get a marriage from G to B. If it happens there is a subset T of P known by exactly $k = |T|$ girls $R^{-1}T$. Consider $G - R^{-1}T$ and $B - T$. Any subset of $G - R^{-1}T$ know none in T, so they know enough boys, and for any subset of $P - T$, none is known by a girl in $R^{-1}T$. As before, we have a marriage from G to B by combining two marriages. Subcase b): $k = n$, then the n girls know exactly n handsome boys and any proper subset is unforced. The argument is similar to the unforced case.

4 Main theorem

In Section 8 [Q1], Quinn proved the ϵ-δ cancellation theorem. In this section, we prove the following version of the ϵ-δ cancellation theorem which makes explicit the relation between ϵ and δ.

Theorem 4.1 *Given a Riemannian manifold X of dimension n, and a group π. Suppose for each i, $0 \le i \le n$, $Wh(\pi \times Z^i) = 0$. Then there exists an $\bar{\epsilon} : X \longrightarrow (0, \infty)$ such that if $\epsilon : X \longrightarrow (0, \infty)$ and $\delta : X \longrightarrow (0, \infty)$ satisfy $\epsilon < \bar{\epsilon}$ and $\sqrt{n} \cdot 6^{n^2+1} \cdot \delta < \epsilon$, then for any δ-isomorphism α of controlled modules M and N over X, there exists an ϵ-deformation H of α such that $H\alpha : M_{stabilized} \longrightarrow N_{stabilized}$ is a geometric map.*

The geometry of the control space determines the threshold $\bar{\epsilon}$, but for ϵ small, the asymptotic relation between ϵ and δ does not depend on the geometry of X, only on the dimension of X. When ϵ is small enough, the relation between ϵ and δ is linear. The constant for the linear relation between ϵ and δ depends only on the dimension n, and it is $\sqrt{n} \cdot 6^{n^2+1}$. However, if the control space X^n is a subset of Euclidean space E^n, then the constant for the linear relation between ϵ and δ is only $\sqrt{n} \cdot 6^{n+1}$ (see the following Lemma 4.2). A handle decomposition of a Riemannian manifold can be made arbitrarily fine so that each handle is $(1+0.001)$-quasi-isometric to some Euclidean subset, although the shapes of these handles may vary. These handles are subdivided into what we call *small handles* and our modification of α proceeds over these small handles. This two-stage process leads to the n^2 in the exponent.

Some notations:

C^m is the standard m-cube $\{x \mid x \in R^m, \mid x_i \mid \leq 1, i = 1, 2, \cdots, m\}$.

$T^{m-1} \times I$ is the annulus, where T^{m-1} is the $m-1$ *flat* torus and $I = [-1, +1]$ is the 1-cube.

$B_r(x)$ is the ball of radius r and center x.

Suppose X is a metric space, $K \subset X$, $\epsilon > 0$, then $K^{+\epsilon}$ is the ϵ-neighborhood of K, and $K^{-\epsilon} = X - (X - K)^{+\epsilon}$.

Suppose $K \subset R^m$ has a distinguished point x_0, for any $\lambda > 0$, λK denotes the set $\{\lambda x \mid x \in K\}$ when we choose x_0 as the origin of R^m. By a *half k-handle* we mean $C^k \times \frac{1}{2}C^{m-k}$, where the distinguished point for the second factor is the center.

We are constantly deforming maps in the proof of Theorem 4.1 without renaming them to keep the notation simple.

Observation I: If the diameter of the control space X is less than a constant ϵ, then any deformation over X is an ϵ-deformation.

Observation II: Given a family of self homomorphisms α_i of controlled modules M_i, i belongs to an index set I. If the control spaces of M_i are disjoint subsets of a space X, and H_i is a deformation of α_i with spread ϵ_i for each i, then $H = \bigoplus H_i$ is a deformation of $\alpha = \bigoplus \alpha_i : \bigoplus M_i \longrightarrow \bigoplus M_i$, and the spread of H is the supremum of $\{\epsilon_i\}$.

Lemma 4.1 (local cancellation of inverses) *If $A: G \longrightarrow G$ is a δ-isomorphism of a controlled module G of finite rank, then*

$$\begin{pmatrix} A & 0 \\ 0 & A^{-1} \end{pmatrix} : G \bigoplus G \longrightarrow G \bigoplus G$$

is a 3δ-deformation from the identity to $\begin{pmatrix} A & 0 \\ 0 & A^{-1} \end{pmatrix}$ *or a 3δ-deformation from* $\begin{pmatrix} A^{-1} & 0 \\ 0 & A \end{pmatrix}$ *to the identity.*

Proof. If $A : G \longrightarrow G$ is a δ-homomorphism, it follows from the following factorization that $\begin{pmatrix} I & \pm A \\ 0 & I \end{pmatrix}$ is a δ-deformation of any map from $G \oplus G$ to itself:

$$\begin{pmatrix} I & \pm A \\ 0 & I \end{pmatrix} = \prod_{i,j=1}^{d} \{ \begin{pmatrix} I & 0 \\ 0 & I \end{pmatrix} \pm a_{ij} \begin{pmatrix} I & E_{ij} \\ 0 & I \end{pmatrix} \},$$

where E_{ij} is the matrix with all entries 0 except 1 at the (i,j) position. Similarly, $\begin{pmatrix} I & 0 \\ \pm A & I \end{pmatrix}$ is a δ-deformation of $G \oplus G$. We interpret the following matrix identity on $G \oplus G$ as a composition of deformations from the identity to $\begin{pmatrix} A & 0 \\ 0 & A^{-1} \end{pmatrix}$ or from $\begin{pmatrix} A^{-1} & 0 \\ 0 & A \end{pmatrix}$ to the identity. The proof is completed by observing that $\begin{pmatrix} I & 0 \\ \pm I & I \end{pmatrix}$ is a 0-deformation of $G \oplus G$.

$$\begin{pmatrix} A & 0 \\ 0 & A^{-1} \end{pmatrix} = \begin{pmatrix} I & -A \\ 0 & I \end{pmatrix} \begin{pmatrix} I & 0 \\ A^{-1} & I \end{pmatrix} \begin{pmatrix} I & -A \\ 0 & I \end{pmatrix}$$
$$\begin{pmatrix} I & I \\ 0 & I \end{pmatrix} \begin{pmatrix} I & 0 \\ -I & I \end{pmatrix} \begin{pmatrix} I & I \\ 0 & I \end{pmatrix}$$

Addendum. In the following, this lemma will be used several times. But each time, it is a relativized form we use. The modification is easy. For any $K \subset X$, let $P = A \mid_K$, and $Q = A^{-1} \mid_K$, then

$$\begin{pmatrix} I & -P \\ 0 & I \end{pmatrix} \begin{pmatrix} I & 0 \\ Q & I \end{pmatrix} \begin{pmatrix} I & -P \\ 0 & I \end{pmatrix} \begin{pmatrix} I & I \\ 0 & I \end{pmatrix} \begin{pmatrix} I & 0 \\ -I & I \end{pmatrix} \begin{pmatrix} I & I \\ 0 & I \end{pmatrix}$$

is a 3δ-deformation from

$$\begin{pmatrix} (I - PQ)P + P & I - PQ \\ I - QP & Q \end{pmatrix}$$

to the identity. For example, if restricted to the subset $K^{-3\delta} \subset K$, $PQ \mid_{K^{-3\delta}} = $ *identity*. Then, there is a 3δ-deformation from

$$\begin{pmatrix} A \mid_{K^{-3\delta}} & 0 \\ 0 & A^{-1} \mid_{K^{-3\delta}} \end{pmatrix}$$

to the identity.

Lemma 4.2 *(Estimate for the Euclidean case) Suppose Y is a compact subset of E^n, and M is a controlled module over $Y^{+6^n \delta} \subset E^n$, and $\gamma : M \longrightarrow M$ is a δ-homomorphism. If γ is a δ-isomorphism over Y, then there is a $\sqrt{n} \cdot 6^{n+1} \cdot \delta$-deformation H such that $H\gamma : M \mid_{stabilized} \longrightarrow M \mid_{stabilized}$ is the identity map over Y.*

The proof of this lemma is the first induction in the proof of Theorem 4.1.

Proof. (small induction) The standard lattice structure of R^n gives a cubing of R^n. This cubing can be thought as a handle decomposition of R^n whose handles are cubes of the same size (see Fig. 1 for the 2-dimensional case).

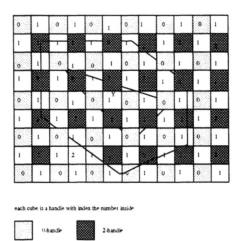

each cube is a handle with index the number inside

□ 0-handle   ■ 2-handle

Figure 1.

By similarity, there is such a handle decomposition of R^n with largest side length of the cubes (handles) so that if a k-handle meets Y, then all adjacent handles of lower indices are inside $Y^{+6"\delta}$. The handles in this decomposition will be called *small handles* or just handles. Over each small handle of index less than n, the argument uses Kirby's torus trick.

Step 1: Work on small 0-handles.

Splitting Procedure. Each small 0-handle is a standard n-cube C^n. Let $j : T^{n-1} \times I \longrightarrow C^n$ be an embedding such that $\frac{1}{2}C^n \subset j(T^{n-1} \times I)$, and the distance between the boundary of $j(T^{n-1} \times I)$ and the boundary of the half 0-handle is large enough(Fig. 2). The precise estimate is given in Step 4.

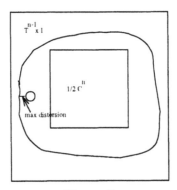

Figure 2.

Pull back γ and γ^{-1} to $T^{n-1} \times I$, and denote them by A and A^{-1} with a fixed indexing of the basis of M. Notice that A^{-1} is not exactly the inverse of A near the boundary of the annulus $T^{n-1} \times I$. Glue the two copies of $T^{n-1} \times I$ with A and A^{-1} over them together along collars of the boundary components of $T^{n-1} \times I$. The two identified collars will be called *flanges*. The cross section looks like Figure 3. Note that we do not reverse the radial orientations of the annuli, and what we get is a torus with two flanges. Suppose each flange has sufficiently large radial length, say 9δ. Again the precise estimate is given in Step 4. Over each flange, the map is $A \oplus A^{-1}$ from $M \oplus M \mid_{flange}$ to itself.

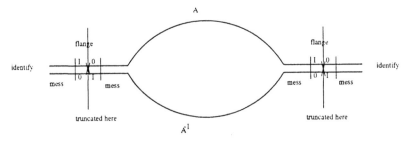

Figure 3.

Apply the 3δ-deformation in Lemma 4.1 to the map over the middle 3δ segment of each flange. After this deformation, $A \oplus A^{-1}$ is deformed to the identity over the middle 3δ segment of each flange. Truncate along the middle, we have a map $U = A \oplus A^{-1} \mid_{truncated}$ of the module over the torus with two half flanges. Since $\gamma \oplus \gamma^{-1}$ is invertible away from the boundary of the torus with two flanges and U is the restriction of $\gamma \oplus \gamma^{-1}$ to a region away from the boundary after a 3δ-deformation, so U is an invertible matrix.

Finally, we project U over the torus with two half flanges to a map over the torus T^n. Since the distances among basis elements decrease, the map over the torus T^n, denoted still by U, is an automorphism of the module $G \mid_{T^n}$ over the torus T^n.

In next step, we will work in a general setting for a map of a module G over the flat torus T^m.

Truncation Procedure. Given a δ-automorphism U of a controlled module G over the torus T^m, let $p : R^m \longrightarrow T^m$ be the universal covering map of T^m. Since δ is very small compared to the injectivity radius of T^m, we can pull back U uniquely to an automorphism $\hat{U} : \hat{G} \longrightarrow \hat{G}$, where the infinitely generated free $Z[\pi]$-module $\hat{G}$ over R^m is the lifting of $G \mid_{T^m}$ by p. The control map extends naturally. It is well known that the covering transformations give $\hat{G}$ a structure of a finitely generated free $Z[\pi \times Z^m]$-module over R^m. So, $\hat{U}$ can be viewed as an automorphism of the free $Z[\pi \times Z^m]$-module $\hat{G}$. As an element in $Wh(\pi \times Z^m)$, $\hat{U}$ is zero. Hence, after a stabilization by direct sum with an identity matrix, $\hat{U}$ is the product of elementary matrices. By factorizing these elementary matrices, it is sufficient to consider the following two different kinds of basic matrices:

$$Type\ I. \begin{pmatrix} 1 & & & & \\ & \ddots & & & \\ & & g & & \\ & & & \ddots & \\ & & & & 1 \end{pmatrix},$$

where $g \in \pi \times Z^m$. It shifts basis elements by g.

$$Type\ II. \begin{pmatrix} 1 & & & & & \\ & \ddots & & & & \\ & & 1 & \cdots & h & \\ & & & \ddots & \vdots & \\ & & & & 1 & \\ & & & & & \ddots \\ & & & & & & 1 \end{pmatrix},$$

where $h \in Z[\pi \times Z^m]$.

Further factorizing type I and II matrices, we may assume that for type I matrices, $g \in Z^m$ (Type I_1) or $\pm g \in \pi$ (Type I_2) and for type II matrices, $h \in Z[\pi]$. Now fix a factorization of $\hat{U}$ as products of such matrices $\hat{U} = E_1 \cdot E_2 \cdots E_t$, where E_i is either a matrix of type I_1, I_2 or type II and is essentially an elementary deformation. But these E_i's are not directly useful because they are maps of a module over R^n, and not δ controlled. So first we will truncate these E_i's to get a map over a large ball. In next step, we will use a retraction to regain δ control.

The spread of $\hat{U}$ over a ball $B_r(0) \subset R^m$ is increasing as r increases. But when r is bigger than some constant $r_0 = r_0(\hat{U})$, the spread of $\hat{U}$ is a constant. The constant r_0 is determined by the number of the type II and type I_2 matrices, and the largest spread of the type I_1 matrices. Pick an $r_1 > r_0$ and fix two concentric balls of radii r_1 and $r_2 > r_1$ with centers 0. If r_1 is large enough, no basis elements of $\hat{G}$ can be mapped out of the ball $B_{r_1}(0)$ by $\hat{U}$. We may assume $B_{r_1}(0)$ contains many copies of the fundamental domains and the region between the two balls consists of many copies of the fundamental domains.

For the type I_2 and type II matrices, the truncation is easy because they leave every fundamental domain invariant. The truncated E_i, denoted by E_i', is defined as follows: inside the ball $B_{r_1}(0)$, it is the same as E_i, outside the larger ball $B_{r_2}(0)$, it is the identity, between the two balls, it is still the same as E_i.

For a type I_1 matrix, assume g is a generator of Z^m. In general, the argument is similar if several fundamental domains together are treated as a "fundamental domain". Then g shifts basis elements from one domain to another. If $m \geq 2$, we use the trick "*ship in water*" to truncate the matrix. As before, choose two fixed concentric balls. We need a map which is a quasi-isometry among the basis elements inside the ball $B_{r_2}(0)$ and is the same as E_i inside the smaller ball $B_{r_1}(0)$. To get such a quasi-isometry, let the basis elements inside the ball $B_{r_1}(0)$ be shifted according to the matrix and the basis elements between the balls follow the water flow as if a ship (the ball $B_{r_1}(0)$) moves forwards in the lake (ball $B_{r_2}(0)$). Figure 4 indicates the directions of the basis elements to go. It is really a permutation outside the ball $B_{r_1}(0)$. If the spread of the truncated map is larger than that of the map before, we add more basis elements between the two balls to get the same spread as before.

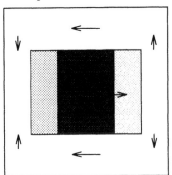

Figure 4. Ship in water.

If m=1, there is the flux problem because there is no extra direction to move basis elements backwards outside $B_{r_1}(0)$. The *flux* of a map is the difference between the number of the basis elements shifted by the map in two opposite directions.

For a type I_1 matrix with g being the generator t of $\pi_1(T^1) = Z$, the element t acts on one basis element e_i by shifting e_i to te_i and on the others by the identity. We index the domain and the basis element in it both by t^l for some integer l. To solve the flux problem, fix $r_1 = r, r_2 = r+1$ and two balls of radii r_1 and r_2 with centers at 0. The obvious solution is to add a new basis element $t^l e_i'$ to the t^l-th domain for each l. Extend the map by sending $t^{r+1}e_i$ to $t^{r+1}e_i'$, $t^l e_i'$ to $t^{l-1}e_i'$, $-r \le l \le r$ and $t^{-r-1}e_i'$ to $t^{-r-1}e_i$. But this map is not in general a δ-map.

To get a δ-map, we introduce more basis elements in each domain. Suppose $\frac{\epsilon}{\delta} \cong k$, we add k new basis elements $e_i^j, j = 1, 2, \cdots, k$ to the fundamental domain and extend the map by:

$$
\begin{cases}
e_i^k & \longrightarrow & e_i^{k-1} \\
\vdots & & \\
e_i^2 & \longrightarrow & e_i^1 \\
e_i^1 & \longrightarrow & t^{-1}e_i^k
\end{cases}
$$

Repeat this procedure in each domain inside the interval except the two end domains. For the left end domain, we define the map as follows:

$$
\begin{cases}
t^{-r}e_i^1 & \longrightarrow \quad t^{-r-1}e_i^k \\
t^{-r-1}e_i^k & \longrightarrow \quad t^{-r-1}e_i^{k-1} \\
\quad \vdots & \\
t^{-r-1}e_i^2 & \longrightarrow \quad t^{-r-1}e_i^1 \\
t^{-r-1}e_i^1 & \longrightarrow \quad t^{-r-1}e_i
\end{cases}
$$

Similarly for the right end domain. The truncated matrix is different from the original one by a δ permutation matrix which is essentially a stabilization after a marriage of basis elements.

After the truncation, we obtain a module $\hat{G}'$ over $B_{r_2}(0)$ and an automorphism of this module $\hat{U}' = E'_t \cdots E'_1$ which can be thought as a deformation from the identity to $\hat{U}'$. The spread of the truncated E'_i is independent of both r_1 and r_2, neither is $\hat{U}' = E'_t \cdots E'_1$. Therefore, we have the freedom to choose r_1 and r_2.

Retraction Procedure. Apply the truncation procedure ($m = n$) to the module $G \mid_{T^n}$ and the map U obtained in the splitting procedure. We get a deformation over $B_{r_2}(0)$ from the identity to $\hat{U}' = E'_1 \cdot E'_2 \cdots E'_t$, where E'_i is the truncated matrix of E_i. As an elementary deformation, E'_i is not δ controlled. But they are all uniformly bounded over $B_{r_2}(0)$. Therefore, we can project them back to a 0-handle C^n to regain δ control. Inside each fundamental domain in $B_{r_1}(0)$, there is a copy of A^{-1} which is the same as γ^{-1} over a neighborhood of $\frac{1}{2}C^n$. Fix such a copy of A^{-1} inside $B_{r_1}(0)$ and identify the control space in $B_{r_1}(0)$ of A^{-1} with $\frac{1}{2}C^n$. In this way, $\frac{1}{2}C^n$ is treated as a subset of $B_{r_1}(0)$. Let $q : R^n \supset B_{r_2}(0) \longrightarrow C^n$ be an embedding which is the identity over $\frac{1}{2}C^n$ and has the following property:

(*) For any constant L, the diameter of the image of the ball $B_L(x)$ under q goes to 0 as x goes to infinity.

It is easy to see there exists such a function. Under q, the truncated module $\hat{G}'$ over $B_{r_2}(0)$ is mapped to a module $q_*\hat{G}'$ over C^n and $\hat{U}'$ is mapped to a deformation $q_*\hat{U}'$. The deformation $q_*\hat{U}'$ can be made into a 3δ-deformation over C^n as follows. The spread of $q_*\hat{U}'$ is bounded since the spread of $\hat{U}'$ over $B_{r_2}(0)$ is a constant and the projection q decreases the spread. But $q_*\hat{U}'$ might not be a 3δ-deformation. For each E'_i, choose r_1 large enough, by the property (*) of q, the spread of $q_*E'_i$ can be made as small as possible outside $q(B_{r_1}(0))$, say for each i, the spread of $q_*E_i{}'$ is $< \frac{\delta}{t}$, then the spread of $q_*\hat{U}'$ is $< t \cdot \frac{\delta}{t} = \delta$. Since $\hat{U}'$ is the same as U over $B_{r_1}(0)$, so $q_*\hat{U}'$ consists of many copies of $\begin{pmatrix} A & 0 \\ 0 & A^{-1} \end{pmatrix}$ over $q_*(B_{r_1}(0))$. By Lemma 4.1, this is a 3δ-deformation. Therefore, if r_1 is large enough, using a tapering retraction we can control the spread of $q_*\hat{U}'$ over the region outside $\frac{1}{2}C^n$ within 3δ. The result is a 3δ-deformation from the identity to a map with A^{-1} over $\frac{1}{2}C^n$. It is important to note that the spread of $q_*\hat{U}'$ does not depend on U and r_1, r_2.

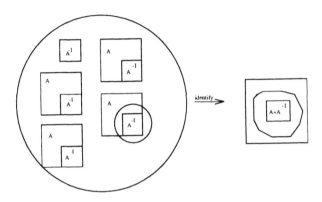

Figure 5.

Cancellation on small 0-handles. Over a small 0-handle, stabilize the module according to $q_*\hat{G}'$ above and run the 3δ-deformation $q_*\hat{U}'$. So we have $\begin{pmatrix} A & 0 \\ 0 & A^{-1} \end{pmatrix}$ over the half 0-handle (Fig. 5). Although A^{-1} is not exactly the inverse of A over the 0-handle, it really is over some neighborhood of the half 0-handle. Applying Lemma 4.1, we run the 3δ-deformation to cancel A and A^{-1} over the half 0-handle. Do this simultaneously over all 0-handles, we deform γ to the identity over all half 0-handles. Notice that the deformation used is a 6δ-deformation as Lemma 4.1 was used twice. It is easy to check that the 3δ-deformation used in the splitting procedure does not contribute to the estimate. Therefore, the resulting map outside all half small 0-handles is a 6δ-isomorphism.

Step 2: Work on general small k-handles. Each small k-handle is a cube $C^k \times C^{n-k}$. Let $j : C^k \times T^{n-k-1} \times I \longrightarrow C^k \times C^{n-k}$ (T^0 is a point) be an embedding whose image contains $C^k \times \frac{1}{2}C^{n-k}$(half k-handle) and is the identity on the factor C^k (Fig. 6).

Assume the distance between the boundary of the image of j and the boundary of the half k-handle is large enough. For the precise estimate, see the following Step 4. Over a k-handle the map is a $6^k \cdot \delta$-isomorphism. Do the analogue of the splitting procedure to get a controlled module over $C^k \times T^{n-k}$. As $Wh(\pi \times Z^{n-k}) = 0$, factorize the matrix analogous to $\hat{U}$, do the truncation procedure for $m = n - k$ with the identity on the factor C^k. The situation is almost exactly the same as the 0-handle case. After the retraction procedure, we get a 3δ-deformation over each k-handle, and this is used to deform a stabilization over each k-handle to a 3δ map with the desired inverse over the half k-handle. After cancellation over the half k-handle using Lemma 4.1, γ has been deformed to the identity over $C^k \times \frac{1}{2}(C^{n-k})$ by a 6δ-deformation. Again we can

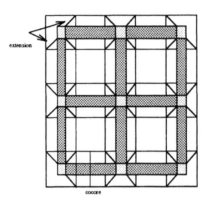

Figure 6.

do this simultaneously over all small k-handles. Continuing inductively when k reaches $n-1$, we find that a $6^n\delta$-deformation has deformed γ to the identity over all half small k-handles ($k \leq n-1$) (Fig. 6). The largest deformation happens over the region near the boundaries of half handles of index $\leq n-1$. Notice that after we deform γ to the identity over half k-handles ($0 \leq k \leq n-2$), it is necessary to extend the (k+1)-handles to the half k-handles left. Figure 6 illustrates the situation. Also note that there exists the flux problem over (n-1)-handles.

Step 3: Cancellation on small n-handles. Now the map is not the identity only over the disjoint n-handles. The n-handles here are the original n-handles enlarged a little. Since $Wh(\pi) = 0$, do a finite algorithm over all n-handles simultaneously, γ is deformed to the identity. Since the side length of each handle (cube) is $6^{n+1} \cdot \delta$ (see following step 4), the diameter of each enlarged n-handle is $(\sqrt{n} \cdot 6^{n+1}) \cdot \delta$. The n-handles are disjoint and the map over them has been already deformed by a $6^n\delta$-deformation, so the deformation used above is a $\sqrt{n} \cdot 6^{n+1} \cdot \delta + 6^n \cdot \delta$-deformation by observations I and II. The key property needed in the second big induction step is although the deformation has spread $\sqrt{n} \cdot 6^{n+1} \cdot \delta$ over $Y^{+6^n\delta}$, the map restricted outside Y has been deformed only by a $6^n\delta$-deformation.

Step 4: Estimate. Let $\bar{\delta}$ denote the side length of small handles (cubes). The side length $\bar{\delta}$ is determined by the radial length of the two flanges in the splitting procedure. We require that the two flanges have radial length 9δ. As over small k-handles γ is a $6^k\delta$-isomorphism, the cocore must be large enough for the torus trick to work. In particular, over a small (n-1)-handle is γ a $6^{n-1}\delta$-isomorphism, so we choose $\bar{\delta}$ to be $4 \cdot 9 \cdot 6^{n-1}\delta = 6^{n+1}\delta$. We will show that if $\bar{\delta} > 6^{n+1}\delta$, then there exists an embedding of $C^k \times T^{n-k-1} \times I$ into $C^k \times C^{n-k}$ so that the flanges have radial length 9δ. In the following, $\bar{\delta}$ will be treated as 1. Let the embedding of $T^{n-k-1} \times I$ into R^{n-k} be as follows:

$$T^{n-k-1} = S^1_{2^{n-k-1}} \times S^1_{2^{n-k-2}} \times \cdots \times S^1_2,$$

where $S^1_{2^{n-k-i}}$ is the circle in the $x_i x_{i+1}$-plane with radius 2^{n-k-i} for $1 \leq i \leq n - k - 1$. Thickening S^1_2 to a two-sided collar of thickness 1, we have an embedding of $T^{n-k-1} \times I$ into R^{n-k}. Choose a cube C^{n-k} with center $x_0 \in S^1_2$, then $\frac{1}{2}C^{n-k}$ is sitting inside $T^{n-k} \times I$. There is a retraction from R^{n-k} into C^{n-k} so that it retracts $T^{n-k} \times I$ to C^{n-k}, and is the identity on $\frac{1}{2}C^{n-k}$, and its maximum distortion is $2^{n-k-1} + 2$. Cross with the identity on the factor C^k, we have a desired embedding of $C^k \times T^{n-k-1} \times I$ into $C^k \times C^{n-k}$. The part with maximum distortion is indicated in Fig. 2. The necessary inequality for $\bar{\delta}$ is $2 \cdot \frac{1}{2^{n-k-1}+2} \cdot \bar{\delta} \geq 18 \cdot 6^k \delta$. This is subsumed by $6^{n+1}\delta < \bar{\delta}$.

Remark. At the end of Step 2, the handle decomposition of X has been extended. This changes the shapes and sizes of the handles. Hence this extension contributes some constant to the final estimate. But note that instead of deforming the map to the identity over only half k-handles at all steps, we can deform the map to the identity over more, say 9/10, of all k-handles. Then the distortion due to the extension of handle decomposition could be neglected.

Proof of the theorem (big induction).

Step I: Fix a fine handle decomposition of X so that each handle is $(1+0.001)$-quasi-isometric to a Euclidean subset of E^n. The handles in this decomposition will be referred to as *big handles*. The mesh of this handle decomposition is the threshold $\bar{\epsilon}$ in the theorem. Given any $\epsilon < \bar{\epsilon}$, if the theorem holds for some $\delta > 0$, then it holds for any $0 < \delta' \leq \delta$. So it is sufficient to determine the largest δ for a given ϵ such that the theorem holds. Let us fix a very small δ now, and the relation of ϵ and δ will be determined in the end of the proof.

Step II: As an application of the marriage theorem, α will be changed to an automorphism γ of M. To do this, we prove there is a bijection between the bases S_M and S_N such that each $f_j \in S_N$ can be identified with an $e_i \in S_M$ and $d(\varphi(e_i), \phi(f_j)) < \delta(\varphi(e_i))$.

Fix a decomposition of $X = \bigcup_{i=1}^{+\infty} K_i$ such that each K_i is compact, $\overline{K_i} \subset intK_{i+1}$ and the distance between K_i and $X - K_{i+1}$ is $> max_{x \in K_i} \delta(x)$. Define a basis element $e_i \in S_M$ knows $f_j \in S_N$ if the coefficient of f_j in $\alpha(e_i)$ is not zero. Let $G_i = \varphi^{-1}(K_i)$ be the set of girls, $B_i = \phi^{-1}(K_i^{+\delta})$ be the set of boys, where $K_i^{+\delta} = \{x \mid d(x, K_i) \leq max_{x \in K_i} \delta(x)\}$ and define the handsome boys P_i of B_i as $\phi^{-1}(K_i^{-\delta})$, where $K_i^{-\delta} = \{x \mid d(x, X - intK_i) \geq mim_{x \in K_i} \delta(x)\}$, then $P_i \subset B_i$. To check that the hypothesis of the marriage theorem holds for G_i and B_i, note that $M \mid_{K_i}$ is the submodule of M generated by G_i, the conditions for G_i and B_i follow from the fact that $\alpha \mid_{K_i}$ is a δ-isomorphism over K_i. The hypothesis for handsome boys follows from the same fact for the inverse of α. By the marriage theorem, there are marriages $\{\rho_i\}$ from G_i to B_i such that $\rho_i(G_k) \subset B_k$ and

$\rho(G_i) \supset B_{i-1}$ for $k \leq i$. Since each B_i is compact (actually finite), so is the product $\times_{i=1}^{+\infty} B_i$. Each marriage ρ_i determines a closed subset F_i of $\times_{i=1}^{+\infty} B_i$ as follows: $F_i = \rho_i(G_1) \times \rho_i(G_2) \times \cdots \times \rho_i(G_i) \times B_{i+1} \times \cdots$. Since F_i has only finite possibilities for each i, there is an $\mathbf{l}$ large enough such that $F_i \supset F_{i+1}$. Pick a subsequence $i_1, i_2 \cdots$ such that $F_{i_1} \supset F_{i_2} \supset \cdots$. Then by the compactness of $\times_{i=1}^{+\infty} B_i$, $F = \cap_{j=1}^{+\infty} F_{i_j} \neq \emptyset$. Since $S_M = \bigcup G_i, S_N = \bigcup B_i = \bigcup P_i$, any element in F is a desired marriage.

Choose a marriage μ between S_N and S_M, if we identify f_j with its image e_i, then $\gamma = \alpha\mu : M \xrightarrow{\alpha} N \xrightarrow{\mu} M$ is a 2δ-automorphism of M. After this marriage, the matrix $A(\alpha)$ is changed by reordering of the columns and the spread of $A(\alpha)$ is changed by 2δ. We will still use $A(\alpha)$ to denote the matrix after the marriage to keep the notation simple.

Step III: Induction on big handles. As each big 0-handle is $(1+0.001)$-quasi-isometric to a Euclidean subset of E^n, so big 0-handles can be safely treated as Euclidean subsets. Applying Lemma 4.2 to each big 0-handle, we deform the map γ over all big 0-handles simultaneously to the identity. As observed in Step 3 of Lemma 4.2, the largest deformation so far, a $(\sqrt{n} \cdot 6^{n+1} + 6^n)\delta$-deformation, was only used to deform the map over all disjoint small n-handles to the identity. The module over these disjoint small n-handles never enters into later modifications. By observation II, they are subsumed by later deformations. The map outside big 0-handles has been deformed only by a $6^n\delta$-deformation. Hence, the resulting map γ is a $6^n\delta$-isomorphism over big 1-handles. Next we proceed to deform γ over all big 1-handles to the identity. As over a big 1-handle γ is a $6^n \cdot \delta$ isomorphism, when Lemma 4.2 is used, the decompositions of big 0-handles and big 1-handles into small handles over a collar of big 0-handles mismatch with different sizes. Now proceed as above, the only difference is that over a big k-handle γ is a $6^{kn} \cdot \delta$-isomorphism. When k reaches $n-1$, we need that the big n-handles are large enough so that the deformation over big k-handles $(k \leq n-1)$ happens over disjoint pieces of the control space. The modification to deform the map over big (n-1)-handles to the identity proceeds over a $6^{n^2} \cdot \delta$ neighborhood of big (n-1)-handles, so the diameter of a big n-handle must be at least $2 \cdot 6^{n^2} \cdot \delta$. Therefore, ϵ and δ should satisfy the inequality: $\epsilon > 2 \cdot 6^{n^2}\delta$. Finally, the map is not the identity only over disjoint big n-handles. Since $Wh(\pi) = 0$, so we can do the usual linear algebra. In the end, the map γ has been deformed to the identity over the whole control space X. The deformation used over big n-handles is a $2 \cdot 6^{n^2} \cdot \delta + 6^{n^2}\delta < 6^{n^2+1}\delta$-deformation by observations I and II. However, the largest deformation happens when Lemma 4.2 is used to deform the map over big (n-1)-handles to the identity. As over a big (n-1)-handle the map is a $6^{(n-1)n}\delta$-isomorphism, so the deformation used is a $\sqrt{n} \cdot 6^{n+1} \cdot 6^{(n-1)n} \cdot \delta = \sqrt{n} \cdot 6^{n^2+1}\delta$-deformation. Composed with the inverse of the marriage, the map α is the same as being deformed to a geometric map. As the marriage used in step II is only a δ marriage, this adds to the spread of the deformation only a negligible 2δ. Thus the proof is completed.

Remarks. (1) If we are not seeking the best constant for the relation between ϵ and δ or consider only compact Riemannian manifolds, the second induction in the proof is not necessary. By the Nash isometric embedding theorem [N], any Riemannian n-manifold can be isometrically embedded into R^{cn^3} (or R^{cn^2} for the compact case) for some universal constant c. So it follows from Lemma 4.2 that the constant for the relation between ϵ and δ is exponential cn^3 (or cn^2 for the compact case)

(2) It follows from Whitney's triangulation method for manifolds [Wh] and the Nash isometric embedding theorem [N] that any compact Riemannian manifold of dimension n can be triangulated with arbitrarily small mesh and every top n-simplex is c_n-quasi-similar to the standard Euclidean n-simplex Δ^n. The constant c_n depends only on n. It seems possible to get a better estimate via this approach, but details have not been worked out.

References

[BFMW] J. Bryant, S. Ferry, W. Mio and S. Weinberger, *Topology of homology manifolds*, Bull. of AMS, **28** (1993), 324-328.

[C1] T.A. Chapman, *A class of homotopy equivalences which are simple homotopy equivalences*, preprint, Univ. of Kentucky, 1977.

[C2] T.A. Chapman, *Controlled Simple Homotopy Theory and Applications*, Springer Lecture Notes, vol. 1009 (1983).

[CH] E.H. Connell, and J. Hollingsworth, *Geometric groups and Whitehead torsion*, Trans. Amer. Math. Soc. **140** (1969), 161-181.

[F1] S. Ferry, *Homotoping ϵ-maps to homeomorphisms*, preprint, Institute for Advanced Study, 1977.

[F2] S. Ferry, *The homeomorphism group of a compact Hilbert cube manifold is an ANR*, Ann. Math. **106** (1977), 101-119.

[FHP] S. Ferry, I. Hambleton and E. Pedersen, *A survey of bounded surgery theory and applications*, Proc. MSRI algebraic topology, 1989-1990.

[FP] S. Ferry, and E. Pedersen, *Controlled algebraic K-theory*, preprint.

[FrP] M. Freedman, and W. Press, *Truncation of wavelet matrices: edge effects and the reduction of topological control*, to appear in J. Linear Algebra, 1995.

[N] J. Nash, *The imbedding problem for Riemannian manifolds*, Ann. Math. **63** (1956), 20-63.

[Q1] F.S. Quinn, *Ends of maps I*, Ann. Math. **110** (1979), 275-331.

[Q2] F.S. Quinn, *Ends of maps II*, Invent. Math. **68** (1982), 353-424.

[RY] A. Ranicki, and M. Yamasaki, *Controlled K-theory*, to appear in Topology and Its Applications.

[W] J.H.C. Whitehead, *On C^1-complexes*, Ann. Math. **41** (1940), 809-824.

[Wh] H. Whitney, "Geometric integration theory," Princeton Univ. Press, 1957.

Non–Linear Similarity Revisited

Ian Hambleton
Erik K. Pedersen

1 Introduction

Let G be a finite group and V, V' finite dimensional real orthogonal representations of G. Then V is said to be *topologically similar* to V' ($V \sim_t V'$) if there exists a homeomorphism $h\colon V \to V'$ which is G-equivariant. If V, V' are topologically similar, but not linearly isomorphic, then such a homeomorphism is called a non-linear similarity.

The topological classification of G-representations was first studied by de Rham [18]. He proved that if a topological similarity $h\colon V \to V'$ of orthogonal representations preserves the unit spheres and restricts to a diffeomorphism between $S(V)$ and $S(V')$, then V and V' are linearly isomorphic. In 1973, Kuiper and Robbin [11] obtained positive results on the general problem and conjectured that topological equivalence implies linear equivalence for all finite groups G. However, in 1981 Cappell and Shaneson [1] constructed the first examples of non-linear similarities. The simplest occurs for $G = \mathbb{Z}/8$, but they also constructed a large class of examples for cyclic groups of the form $G = \mathbb{Z}/4q$. Further results can be found in [2], [3], [4], and [13].

On the other hand, Hsiang and Pardon [10] and Madsen and Rothenberg [12] independently proved the conjecture for all odd–order groups. In addition, the main theorem of [10] ruled out some non-linear similarities for even-order groups G.

The purpose of this paper is to give new restrictions on the existence of non-linear similarities using techniques from bounded topology.

Theorem A *Let V, V' be real orthogonal G-representations, where G is a finite cyclic group. Suppose that $\operatorname{Res}_H V \cong \operatorname{Res}_H V'$ for each proper subgroup $H \subsetneq G$, and that $V^H = V^G$ when $[G : H] = 2$. Then $V \sim_t V'$ if and only if $V \cong V'$.*

This result gives information about topological similarities for non–cyclic groups as well, since linear equivalence of representations is detected by character values. The formulation is also well-adapted to inductive arguments, and we get a new proof for the results of [10], [12].

Corollary B *Let G be a finite group and V, V' be real orthogonal G-representations. Suppose that for each cyclic subgroup $C \subseteq G$ of 2-power order, the*

Ian Hambleton was partially supported by NSERC grant A4000.
Erik Pedersen was partially supported by NSF DMS 9104026.

elements of C act either trivially on W, or freely away from $0 \in W$, for every irreducible C-submodule $W \subset Res_C V$. Then $V \sim_t V'$ if and only if $V \cong V'$.

All previously contructed topological similarities of cyclic groups G contain the non–trivial 1-dimensional representation (i.e. the representation with isotropy group $H \subset G$ of index 2), or are induced from such examples.

For example, suppose that $G = \mathbb{Z}/4q$, $q = 2^{r-2}$, $r \geq 4$. Now let $V_1 = t^i + t^j$ and $V_2 = t^{i+2q} + t^{j+2q}$, with $i \equiv 1 \bmod 4$ and $j \equiv \pm i \bmod 8$. Here t denotes a faithful 2-dimensional representation of G. Let ϵ (resp. δ) denote the 1-dimensional trivial (resp. non–trivial) representation of G. Then $t^i + t^j + \delta + \epsilon \sim_t t^{i+2q} + t^{j+2q} + \delta + \epsilon$ by [4, Thm. 1, Cor. (iii)]. But $Res_H V_1 \cong Res_H V_2$ for every proper subgroup $H \subsetneqq G$, so our result says that $V_1 \oplus W$ is *not* topologically similar to $V_2 \oplus W$ unless δ is a summand of W. More generally,

Corollary C *Suppose that $V_1 \oplus W \sim_t V_2 \oplus W$ is a non–linear similarity for a cyclic group G. Then for some subgroup $H \subseteq G$, $Res_H(V_1 \oplus W) \sim_t Res_H(V_2 \oplus W)$ is also a non–linear similarity and the non–trivial 1-dimensional H-representation is a summand of $Res_H W$.*

Our techniques also give information about the existence and classification of non-linear similarities. We recently noticed that [4, Thm 1(i)] is incorrect as stated. For example for $G = \mathbb{Z}/12$, there are no 6-dimensional non-linear similarities. The problem is that the natural epimorphism $\pi : \mathbb{Z}/4q \to \mathbb{Z}/2^r$ where $q = 2^{r-2}s$, s odd, does not induce an isomorphism $\pi_* : L^p(\mathbb{Z}[\mathbb{Z}/4q]^-) \to L^p(\mathbb{Z}[\mathbb{Z}/2^r]^-)$ as claimed in [4, p. 732 l. -8]. These topics will be discussed elsewhere [9].

Acknowledgement. Both authors would like to thank the Mittag–Leffler Institute for its hospitality in May, 1994 when this work was completed.

2   Preliminaries

Suppose that $V \sim_t V'$ with $Res_H V \cong Res_H V'$ for each proper subgroup $H \subsetneqq G$, and that $V^H = V^G$ whenever $[G : H] = 2$. The result for free representations was proved in [11] or [1, 8.1], and we assume by induction that Theorem A is proved for all groups of smaller order than G. Therefore, we may further assume

(2.1): that V, V' are topologically similar G-representations whose restrictions to every H-fixed set are linearly equivalent as G/H-representations, for each non-trivial subgroup $1 \neq H \subset G$

(2.2): that V, V' are neither free nor quotient G-representations, i.e. for each $g \in G$ there exists $v \in V$ such that $gv \neq v$, and there exists $0 \neq v \in V$ such that $gv = v$ for some $1 \neq g \in G$.

Lemma 2.3 *If $V \sim_t V'$ satisfies the conditions above, then $V = V_1 \oplus W$ and $V' = V_2 \oplus W$, where V_1, V_2 are free G-representations and $W \neq \{0\}$ has no free summands.*

Proof. For each subgroup $K \subseteq G$, let $V(K)$ denote the direct sum of all irreducible subrepresentations of V with kernel K. The G-subspaces $V(K)$ are preserved by G-linear isomorphisms. We let W be the direct sum of all the $V(K)$ for $K \neq 1$. From (2.2) it follows that $V^H \neq \{0\}$ for some subgroup $1 \neq H \subset G$ so $W \neq \{0\}$.

Write $V = V_1 \oplus W$ where V_1 is a free representation. By assumption (2.1), $V^H \cong (V')^H$ as G-representations for each $H \neq 1$. Then for the fixed sets we have

$$V(K)^H = \begin{cases} V(K) & \text{if} \quad H \subseteq K \quad \text{and} \\ 0 & \text{otherwise.} \end{cases}$$

This gives

$$V^H = \bigoplus \{V(K) : H \subseteq K\} \cong (V')^H = \bigoplus \{V'(K) : H \subseteq K\}.$$

Therefore $V(K) \cong V'(K)$ for each subgroup $K \neq 1$ and we can decompose $V' = V_2 \oplus W$ with V_2 a free representation. $\qquad\square$

3 Bounded embedding theorems

For the reader's convenience we will include some material from [9]. First we will need a bounded version of results due to Browder and Wall on smoothing Poincaré embeddings in codimension ≥ 3. Statements for compact smooth or PL manifolds are given in [20, 11.3], and the extension to topological manifolds is sketched on [20, p.245]. The published reference to Rourke and Sanderson's theorem, that the stabilization map

$$F_r / \widetilde{\mathrm{Top}_r} \to F / \mathrm{Top}$$

is a homotopy equivalence for $r \geq 3$ is [19, Thm 2.4]. Here F_r denotes the space of homotopy self-equivalences of the $(r-1)$–sphere, and $F = \varinjlim F_r$.

To state a bounded version, we need to define a *finite bounded Poincaré embedding*. Let X be a metric space on which a finite group G acts by (quasi)isometries. Let $Y \subset X$ be a closed G-invariant subspace, and let $M^m \to Y$, $V^{m+q} \to X$ be finite free bounded G-CW Poincaré complexes [5, Def 2.7]. Then a finite bounded Poincaré embedding of M in V consists of (i) a $(q-1)$-spherical G-fibration ξ, with projection $p \colon E \to M$, (ii) a finite free bounded G-CW Poincaré pair $(C, E) \to (X, Y)$, and (iii) a bounded G-homotopy equivalence $h \colon C \cup M(p) \to V$, bounded over X, where $M(p)$ is the mapping cylinder of p and $C \cap M(p) = E$. Such a Poincaré embedding is "induced" by a locally flat topological embedding if the normal block bundle and complement to the embedding give data which are G-h-cobordant to those of the given Poincaré embedding.

Theorem 3.1 *Suppose given topological manifolds M^m, V^{m+q} with free G-actions, and reference maps $M^m \to Y$, $V^{m+q} \to X$ giving finite free bounded G-CW Poincaré complexes. If $m+q \geq 5$ and $q \geq 3$, then a finite bounded Poincaré embedding of M in V is induced by a locally flat topological G-embedding of $M \to V$.*

Proof. The proof given in [20, §11] generalizes using bounded surgery [6], and there is a relative version as given on [20, p.119] when $m + q \geq 6$.   □

Next we need a "completion" theorem for bounded embeddings (compare [6, §16]. Suppose that Z is an open topological manifold, equipped with a reference map to an open cone $O(K)$, so that $p \colon Z \to O(K)$ is a bounded CW complex [5, Def. 1.5]. The K–completion of Z is the disjoint union $\hat{Z}_K = Z \coprod K$ with a basis for the topology given by (i) open sets in Z, and (ii) sets of the form $p^{-1}(U \times (t, \infty)) \coprod U$, where $t \geq 0$ and $U \subset K$ is open.

We will be interested in comparing the local properties of this completion, when Z is replaced by a manifold Z', bounded homotopy equivalent to Z over $O(K)$. The main ingredient is the following

Lemma 3.2 *Let $X \subset S^n$ be a finite simplicial subcomplex, where $n \geq 5$. Then the bounded structure set $\mathcal{S}_b \begin{pmatrix} S^n - X \\ \downarrow \\ O(X) \end{pmatrix}$ has only one element.*

Proof. Let DX denote the Spanier–Whitehead dual of X in S^n. We use the bounded surgery exact sequence [6]

$$\ldots \to [\Sigma(DX_+), F/\mathrm{Top}] \to L_{n+1}(\mathcal{C}_{O(X)}(\mathbb{Z}))$$

$$\to \mathcal{S}_b \begin{pmatrix} S^n - X \\ \downarrow \\ O(X) \end{pmatrix} \to [DX, F/\mathrm{Top}] \to \ldots$$

If $X = *$ both the normal invariant and the L-group-term are trivial. If $X = S^k$ for some $k < n-2$ then crossing with $\mathbb{R}$ induces an isomorphism on the simply-connected L–groups and at the normal space level from the sequence for S^k to the sequence for S^{k+1}. So starting with $S^{-1} = \emptyset$ and the fact that the structure set of the sphere has only one element we get the result for S^k, $k < n - 2$. For $k = n-2$ it is enough (for our later applications) to assume that the embedding of $X = S^{n-2}$ in S^n is unknotted. Then the effective fundamental group is $\pi_1(S^n - X) = \mathbb{Z}$ so the term behind the structure set is

$$\pi_2(F/\mathrm{Top}) \to L_{n+1}(\mathcal{C}_{\mathbb{R}^{n-1}}(\mathbb{Z}[\mathbb{Z}])) \cong L_2(\mathbb{Z}[\mathbb{Z}]).$$

But since $L_2(\mathbb{Z}[\mathbb{Z}]) = L_2(\mathbb{Z})$ this assembly map is an isomorphism and we are done. Finally for $k = n - 1$ we get 2 copies of R^n, and we use connectedness of F/Top and $L_{n+1}(\mathcal{C}(\mathbb{R}^n)) = 0$. For the general case, we write $X = Y \cup D^k$ and assume that the result is true for Y. We compare the assembly maps for Y to X, where the third term involves "germs away from Y", and reduces to the case of S^k handled above [8, 3.11]. Indeed, on structure sets we have the bijections:

$$
\mathcal{S}_b \left(\begin{array}{c} S^n - X \\ \downarrow \\ O(X) \end{array} \right)_{>O(Y)}
\simeq
\mathcal{S}_b \left(\begin{array}{c} S^n - S^k \\ \downarrow \\ O(S^k) \end{array} \right)_{>O(*)}
\simeq
\mathcal{S}_b \left(\begin{array}{c} S^n - S^k \\ \downarrow \\ O(S^k) \end{array} \right)
$$

where the last step follows by an Eilenberg swindle, showing that $L_i(\mathcal{C}_{O(*)}(\mathbb{Z})) = 0$ for all $i \geq 0$. We now consider the maps of long exact sequences

$$
\begin{array}{ccccc}
[DY, F/\text{Top}] & \longrightarrow & [DX, F/\text{Top}] & \longrightarrow & [DY/DX, F/\text{Top}] \\
\downarrow & & \downarrow & & \downarrow \\
L_n(\mathcal{C}_{O(Y)}(\mathbb{Z})) & \longrightarrow & L_n(\mathcal{C}_{O(X)}(\mathbb{Z})) & \longrightarrow & L_n(\mathcal{C}_{O(X)}^{>O(Y)}(\mathbb{Z}))
\end{array}
$$

The Y–assembly map is an isomorphism by induction, and the assembly map "away from Y" is an isomorphism by the preliminary case above. A similar result holds for the map $[\Sigma(DX), F/\text{Top}] \to L_{n+1}(\mathcal{C}_{O(X)}(\mathbb{Z}))$ and so the structure set is trivial. $\square$

When $(Z, \partial Z)$ is a topological manifold with boundary, we can also consider a relative (K, L) completion in which $p \colon Z \to O(K)$ is a bounded CW complex and $\partial p \colon \partial Z \to O(L)$ is a bounded CW complex with respect to a subcomplex $L \subset K$. If $(F, \partial F) \colon (Z', \partial Z') \to (Z, \partial Z)$ is a bounded homotopy equivalence of pairs over

$$
(p, \partial p) \colon (Z, \partial Z) \to (O(K), O(L))
$$

then $(F, \partial F)$ extends to a homotopy equivalence

$$
(\bar{F}, \partial \bar{F}) \colon (\hat{Z}'_K, \partial \hat{Z}'_L) \to (\hat{Z}_K, \partial \hat{Z}_L)
$$

of pairs by taking the identity on K and L.

Definition 3.3 The completed map $\bar{F} \colon \hat{Z}'_K \to \hat{Z}_K$ is homotopic to a local homeomorphism (relative to L) near K, extending the identity on K, if there exists a neighbourhood $U \subset \hat{Z}_K$ of $x \in K$ such that F restricted to $F^{-1}(U - U \cap K)$ is boundedly homotopic over $O(K)$ to a homeomorphism. When $x \in L$ we further require that ∂F be a local homeomorphism near L, and that ∂F be fixed under the bounded homotopy.

In our applications there is a free G–action on Z, Z', and a G–action on K so that and $p \colon Z \to O(K)$ is G-equivariant. If F is a bounded G–homotopy equivalence so that Z is a finite free bounded G–CW complex then $\bar{F}$ is a G–homotopy equivalence extending the identity on the G–invariant subset K of both domain and range.

Theorem 3.4 *Let $F \colon Z' \rightarrow Z$ be a bounded homotopy equivalence of (open) topological $(m+q)$–manifolds, bounded over the open cone $O(K)$, where K is a finite complex of dimension m. Suppose that the K–completion $\hat{Z}_K$ is a topological $(m+q)$–manifold with $m+q \geq 5$. Then the K–completion $\hat{Z}'_K$ is also a topological $(m+q)$–manifold. Moreover, $\bar{F}$ is homotopy equivalent to a local homeomorphism near K extending the identity on K.*

This result also has a relative version.

Theorem 3.5 *Let $F \colon (Z', \partial Z') \rightarrow (Z, \partial Z)$ be a bounded homotopy equivalence of topological $(m+q)$–manifolds with boundary, bounded over $(O(K), O(L))$ where $L \subset K$ a finite subcomplex with $\dim L < m$. Suppose that the (K, L)–completion $(\hat{Z}_K, \partial \hat{Z}_L)$ is a topological $(m+q)$–manifold with boundary and $m+q \geq 6$. If $\partial \bar{F}$ is a local homeomorphism near L extending the identity on L, then $\bar{F}$ is homotopy equivalent to a local homeomorphism (relative to L) near K, extending the identity on K.*

Proof. We will prove the first result and leave the relative version to the reader. Let $x \in K$ and choose an open disk $D^{m+q} \subset \hat{Z}_K$ around x. Let $L = cls(D^{m+q} \cap K)$ and $X = L/\partial L$. Since

$$\mathcal{S}_b \begin{pmatrix} S^{m+q} - K \\ \downarrow \\ O(K) \end{pmatrix}_{>O(K-L)} \simeq \mathcal{S}_b \begin{pmatrix} S^{m+q} - X \\ \downarrow \\ O(X) \end{pmatrix}_{>O(*)}$$

the result follows from the computation of the local structure sets in Lemma 3.2. $\qquad\qquad\square$

4   Bounded surgery

In this section, the existence of a non-linear similarity $V_1 \oplus W \sim_t V_2 \oplus W$ will be related to the kernel of a bounded transfer map introduced in [8, §3]. For background on bounded surgery we refer to [6].

We begin with an observation from [10, 1.7]: if $V_1 \oplus W \sim_t V_2 \oplus W$, then our inductive assumptions imply that there is a G-homeomorphism

$$h \colon V_1 \oplus W \rightarrow V_2 \oplus W$$

such that

$$h| \bigcup_{H \neq 1} W^H$$

is the identity. One easy consequence (see [1]) is

Lemma 4.1 *There exists a G-homotopy equivalence $S(V_2) \to S(V_1)$.*

Proof. If we 1-point compactify h we obtain a G-homeomorphism

$$h^+ : S(V_1 \oplus W \oplus \mathbb{R}) \to S(V_2 \oplus W \oplus \mathbb{R}).$$

After an isotopy, the image of the free G-sphere $S(V_1)$ may be assumed to lie in the complement $S(V_2 \oplus W \oplus \mathbb{R}) - S(W \oplus \mathbb{R})$ of $S(W \oplus \mathbb{R})$ which is G-homotopy equivalent to $S(V_2)$. $\qquad\square$

Let $f : S(V_2)/G \to S(V_1)/G$ denote the induced homotopy equivalence of the quotient lens spaces. Since for $\dim S(V_i) = 1$ it is clear that G-homotopy equivalence implies $V_1 \cong V_2$, we may assume that $\dim V_i \geq 4$. Another consequence is

Lemma 4.2 *There exists an isovariant G-h-cobordism between $f * 1 : S(V_2 \oplus W) \to S(V_1 \oplus W)$ and the identity on $S(V_1 \oplus W)$, which is a product on all the singular strata $S(W^H)$ for $H \neq 1$.*

Proof. This uses a special case of [4, 1.1]. After radial re-scaling, we may assume that

$$h(D(V_1 \oplus W)) \subset int\, D(V_2 \oplus W),$$

and the region

$$\bar{Z} = D(V_2 \oplus W) - int\, h(D(V_1 \oplus W))$$

is then an isovariant G-h-cobordism from some isovariant G-homotopy equivalence $g : S(V_2 \oplus W) \to S(V_1 \oplus W)$ to the identity. Since h was the identity on all singular strata, it is not hard to check that g and $f * 1$ are isovariantly G-homotopy equivalent, so we may assume that $g = f * 1$. Moreover, the cobordism $\bar{Z}$ is a product $S(W^H) \times I$ along the H-fixed sets for all $H \neq 1$ and a bounded free G-h-cobordism on the complement. $\qquad\square$

Corollary 4.3 *The kernel of the bounded transfer map*

$$\mathrm{trf}_W : \quad \mathcal{S}^h(S(V_1)/G) \to \mathcal{S}_b^h \left(\begin{array}{c} S(V_1) \times_G W \\ \downarrow \\ W/G \end{array} \right) \tag{4.4}$$

contains the element $[f] \in \mathcal{S}^h(S(V_1)/G)$.

Proof. If the whole sphere $S(W)$ is singular (i.e. contains no free orbits), then the vanishing of the bounded transfer $\mathrm{trf}_W([f])$ follows immediately from Lemma 4.2 by removing $S(W) \times I$ from domain and range of the G-h-cobordism. This implies for example that in the present argument we may assume $\dim W \geq 2$, and so $\dim(V_i \oplus W) \geq 6$, since $\dim W = 1$ implies that $W = \mathbb{R}$. Later we will see in Corollary 4.6 that $\mathrm{trf}_{\mathbb{R}}([f]) \neq 0$ and so non-linear similarities do not occur for $\dim(V_i \oplus W) \leq 5$.

In general the problem is that the given G-h-cobordism may not restrict to a G-h-cobordism of $S(W) \times I$. Let

$$\bar{F} \colon (\bar{Z}, \partial_-\bar{Z}, \partial_+\bar{Z}) \to (S(V_1 \oplus W) \times I, S(V_1 \oplus W) \times 0, S(V_1 \oplus W) \times 1)$$

be the G-homotopy equivalence of triads given by (4.2) such that $\bar{F}|_{\partial_-\bar{Z}} = id$ and

$$\bar{F}|_{\partial_+Z} = f * 1 \colon S(V_2 \oplus W) \to S(V_1 \oplus W).$$

In addition, we can assume that for each $1 \neq H \subset G$

$$\bar{F}\big|_{\bar{F}^{-1}(S(W)^H \times I)}$$

is a homeomorphism whose restriction to $S(W)^H \times \partial I$ is the identity. Let

$$X = \bigcup_{1 \neq H \subset G} S(W)^H$$

denote the singular set of $S(V_1 \oplus W)$, and

$$U = \bar{Z} - \bar{F}^{-1}(X \times I)$$

denote the complementary free stratum. The restriction of $\bar{F}$ to this open submanifold gives

$$\begin{aligned} F \colon (U, \partial_-U, \partial_+U) \to &(S(V_1 \oplus W) \times I - X \times I, S(V_1 \oplus W) \\ &\times 0 - X \times 0, S(V_1 \oplus W) \times 1 - X \times 1) \end{aligned}$$

which is a free bounded G-h-cobordism, bounded over the open cone $O(X \times I)$.

Since $\dim V_i \geq 4$ we can regard the bounded G-homotopy equivalence F as a bounded codim ≥ 3 Poincaré embedding of $(S(W) - X) \times I$ in U, relative to the given embedding on $\partial_\pm U$. By Theorem 3.1, there exists a free topological G-embedding inducing the given Poincaré embedding, and extending the embeddings already fixed on $\partial_\pm U$.

By homotopy extension, we can assume that F restricted to this embedding of $(S(W) - X) \times I \subset U$ is a bounded G-h-cobordism, relative to the identity on $\partial_\pm U$, and F is a bounded G-homotopy equivalence over $O(X \times I)$. Now we apply the "completion" construction of Section 2 to adjoin $X \times I$ to both domain and range. By Theorem 3.5 the result is a (new) compact G-h-cobordism

$$\bar{F}' \colon (\bar{Z}', \partial_-\bar{Z}', \partial_+\bar{Z}') \to (S(V_1 \oplus W) \times I, S(V_1 \oplus W) \times 0, S(V_1 \oplus W) \times 1)$$

between $f * 1 \colon S(V_2 \oplus W) \to S(V_1 \oplus W)$ and the identity on $S(V_1 \oplus W)$, with the additional property that the restriction of $\bar{F}'$ to $\bar{F}'^{-1}(S(W) \times I)$ gives a G-h-cobordism with range

$$(S(W) \times I, S(W) \times 0, S(W) \times 1).$$

Now the complement

$$Z' = \bar{Z}' - \bar{F}'^{-1}(S(W) \times I)$$

is a free bounded G-h-cobordism between $f \times 1: S(V_2) \times W \to S(V_1) \times W$ and the identity on $S(V_1) \times W$, bounded with respect to the second factor projection to $W = O(S(W))$. By the definition of the bounded structure set, this means that $\mathrm{trf}_W([f]) = 0$ as required. $\quad\square$

By comparing the ordinary and bounded surgery exact sequences [8, 3.16], and noting that the bounded transfer induces the identity on the normal invariant term, we can assume that f has normal invariant zero. Therefore, under the natural map

$$L_n^h(\mathbb{Z}G) \to \mathcal{S}^h(S(V_1)/G),$$

where $n = \dim V_1$, the element $[f]$ is the image of $\sigma(f) \in L_n^h(\mathbb{Z}G)$, obtained as the surgery obstruction (relative to the boundary) of a normal cobordism from f to the identity. The element $\sigma(f)$ is well-defined in $\tilde{L}_n^h(\mathbb{Z}G) = \mathrm{coker}(L_n^h(\mathbb{Z}) \to L_n^h(\mathbb{Z}G))$.

Lemma 4.5 *For any choice of normal cobordism between f and the identity, the surgery obstruction $\sigma(f)$ is a nonzero element of infinite order in $\tilde{L}_n^h(\mathbb{Z}G)$.*

Proof. Since G is cyclic and V_1 is a free representation, the quotient $X = S(V_1)/G$ is a classical lens space of odd dimension $n - 1$. An element in the surgery obstruction group $\tilde{L}_{2k}^h(\mathbb{Z}G)$ is determined by its discriminant D and multi-signature σ: the odd order case is [20, 13A.5] and the even order case is similar, based on the fact that $\tilde{L}_{2k}^s(\mathbb{Z}G)$ is torsion–free. If X, X' differ by a normal cobordism, then [20, 14E.8] gives the relations $\Delta(X') = D\Delta(X)$ and $\rho(X') = \sigma + \rho(X)$, where $\Delta(X)$ is the Reidemeister torsion $\rho(X)$ is the ρ-invariant [20, 14E.7]. Both Δ and ρ are multiplicative on joins.

If $\sigma(f) \in \tilde{L}_n^h(\mathbb{Z}G)$ were a torsion element, then the relations above would show that a suitable join of copies of $X = S(V_1)/G$ is h-cobordant to the corresponding join of copies of $X' = S(V_2)/G$. But this would imply that $V_1 \oplus \ldots \oplus V_1 \sim_t V_2 \oplus \ldots \oplus V_2$ and since these are free representations, that $V_1 \cong V_2$. $\quad\square$

Corollary 4.6 *Under the natural map $L_n^h(\mathbb{Z}G) \to L_n^p(\mathbb{Z}G)$, the image of $\sigma(f)$ is nonzero.*

Proof. The kernel of the map $L_n^h(\mathbb{Z}G) \to L_n^p(\mathbb{Z}G)$ is the image of $H^n(\mathbb{Z}/2, \tilde{K}_0(\mathbb{Z}G))$ which is a torsion group. $\quad\square$

5   The transfer map

We will now study the transfer map trf_W in (4.3). Since localizing or completing at $p \nmid 2|G|$ gives an injection on the free part (see [20, §13A]):

$$L_n^h(\mathbb{Z}G) \to L_n^h(\mathbb{Z}G) \otimes \mathbb{Z}_{(p)},$$

and we intend to show that $\mathrm{trf}_W(\sigma(f)) \neq 0$ is a p–local injection for G-representations W, with $W^G = W^H$ when $[G : H] = 2$. Thus we will lose no information about elements of infinite order by p-localizing all our L-groups. This will be assumed from now on, without changing notation.

Following [7, §6], [14, §11b], (see also [8, §4] for previous applications in bounded topology) we denote the *top component* of our bounded surgery obstruction group by

$$L_n^h(\mathcal{C}_{W,G}(\mathbb{Z}))(m)$$

where $m = |G|$. The top component of a p-local Mackey functor $(p \nmid m)$ is the intersection of the kernels of all the restriction maps to proper subgroups of G. It turns out to be a natural direct summand of the L-group, associated to an idempotent in the p-local Burnside ring. Moreover the top component has the property that the images of maps induced on L-theory by the inclusion of proper subgroups, project trivially into the top component. In particular, after passing to the top component the indeterminacy in $\sigma(f)$ is zero. Then Theorem A is implied by

Theorem 5.1 *For any G-representation W, with $W^G = W^H$ when $[G : H] = 2$, and any $p \nmid 2|G|$, the transfer*

$$\mathrm{trf}_W \colon L_n^h(\mathbb{Z}G)(m) \to L_{n+k}^h(\mathcal{C}_{W,G}(\mathbb{Z}))(m)$$

is a p-local injection, where $k = \dim W$.

To begin the proof, we will assume that $W = \oplus W_i$ is a direct sum of irreducible 2-dimensional quotient representations. Each W_i has kernel $H_i \neq 1$ which is a proper subgroup of G.

If $W = W_1 \oplus W_2$ there is another inclusion map,

$$c(W_1, W)_* \colon L_n^h(\mathcal{C}_{W_1,G}(\mathbb{Z})) \to L_n^h(\mathcal{C}_{W_1 \oplus W_2,G}(\mathbb{Z}))$$

induced by the subspace inclusion $W_1 \subset W$.

Lemma 5.2 *If $(W_2)^G = \{0\}$ then the subspace inclusion $W_1 \subset W_1 \oplus W_2$ induces an isomorphism*

$$L_n^h(\mathcal{C}_{W_1,G}(\mathbb{Z}))(m) \to L_n^h(\mathcal{C}_{W_1 \oplus W_2,G}(\mathbb{Z}))(m)$$

on the top component for all n.

Proof. The subspace inclusion sits in the exact sequence given in [8, 3.12], and the result is a special case of [8, 4.5].   □

Corollary 5.3 *If $W^G = \{0\}$ then the "cone point" inclusion induces an iso-morphism*

$$c_*: L_n^h(\mathbb{Z}G)(m) \to L_n^h(\mathcal{C}_{W,G}(\mathbb{Z}))(m)$$

on the top component for all n.

Proof. This is just the special case $W_1 = \{0\}$ and [8, 3.10].   □

We can now reduce to the case where W is irreducible.

Lemma 5.4 *Suppose that $W^G = 0$ and $W = W_1 \oplus W_2$ where $\dim W_i = 2l_i$. If trf_{W_i}, for $i = 1, 2$ induces a monomorphism*

$$\mathrm{trf}_{W_i}: L_n^h(\mathbb{Z}G)(m) \to L_{n+2l_i}^h(\mathcal{C}_{W_i,G}(\mathbb{Z}))(m)$$

on the top component for any n, then so does trf_W.

Proof. First note that

$$\mathrm{trf}_{W_1 \oplus W_2}: L_n^h(\mathbb{Z}G) \to L_{n+2l_1+2l_2}^h(\mathcal{C}_{W_1 \oplus W_2,G}(\mathbb{Z}))$$

is the composite of

$$\mathrm{trf}_{W_1}: L_n^h(\mathbb{Z}G) \to L_{n+2l_1}^h(\mathcal{C}_{W_1,G}(\mathbb{Z}))$$

and

$$\mathrm{trf}_{W_2}: L_{n+2l_1}^h(\mathcal{C}_{W_1,G}(\mathbb{Z})) \to L_{n+2l_1+2l_2}^h(\mathcal{C}_{W_1 \oplus W_2,G}(\mathbb{Z})).$$

The first map is a monomorphism on the top component by assumption, and the second can be studied by the commutative diagram

$$
\begin{array}{ccc}
L_{n+2l_1}^h(\mathbb{Z}G) & \xrightarrow{\mathrm{trf}_{W_2}} & L_{n+2l_1+2l_2}^h(\mathcal{C}_{W_2,G}(\mathbb{Z})) \\
\Big\downarrow{\scriptstyle c_*} & & \Big\downarrow{\scriptstyle c(W_1,W)_*} \\
L_{n+2l_1}^h(\mathcal{C}_{W_1,G}(\mathbb{Z})) & \xrightarrow{\mathrm{trf}_{W_2}} & L_{n+2l_1+2l_2}^h(\mathcal{C}_{W_1 \oplus W_2,G}(\mathbb{Z}))
\end{array}
$$

The horizontal maps are transfers trf_{W_2}, and the vertical maps are induced by subspace inclusions. Since the subspace inclusions induce isomorphisms on the top component and the upper horizontal map is a monomorphism by assumption, we are done.   □

We now assume that W is an irreducible 2-dimensional G-representation, with kernel $1 \neq H \neq G$. The G-equivariant projection $S(V_1) \times S(W) \to S(V_1)$ gives a circle bundle with fibre $S(W) = S^1$.

Lemma 5.5 *The group $\Gamma_H = \pi_1(S(V_1) \times_G S(W))$ is isomorphic to $\mathbb{Z} \times H$.*

Proof. By pulling-back the circle bundle to the covering $S(V_1)/H \to S(V_1)/G$ we obtain a commutative diagram

$$
\begin{array}{ccccccccc}
1 & \longrightarrow & \mathbb{Z} & \longrightarrow & \mathbb{Z} \times H & \longrightarrow & H & \longrightarrow & 1 \\
 & & \downarrow & & \downarrow & & \downarrow & & \\
1 & \longrightarrow & \mathbb{Z} & \longrightarrow & \Gamma_H & \longrightarrow & G & \longrightarrow & 1
\end{array}
\tag{5.6}
$$

where the upper row is split exact. The result follows by the classification of extensions since $\mathrm{Ext}^1(G, Z) = H^2(G, \mathbb{Z}) \cong \mathbb{Z}/m$ and the distinct elements are given by the extensions

$$
1 \to \mathbb{Z} \xrightarrow{(k,1)} \mathbb{Z} \times \mathbb{Z}/d \xrightarrow{\binom{1}{-k}} \mathbb{Z}/m \to 1
$$

where $m = dk$. This extension is the unique one which splits when restricted to $\mathbb{Z}/d \subset \mathbb{Z}/m$. $\square$

In order to compute trf_W we will use the following diagram

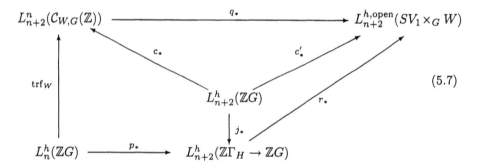

$$\tag{5.7}$$

To construct this diagram we combine maps from three other surgery sequences with those for our bounded surgery groups.

(5.8): The first is just the exact sequence of the pair $\Gamma_H \to G$, corresponding geometrically to the inclusion of the circle bundle defined above into the total space of its disk bundle $S(V_1) \times_G D(W)$:

$$
\dots \to L_{n+2}^h(\mathbb{Z}\Gamma_H) \to L_{n+2}^h(\mathbb{Z}G) \to L_{n+2}^h(\mathbb{Z}\Gamma_H \to \mathbb{Z}G) \to L_{n+1}^h(\mathbb{Z}\Gamma_H) \to \dots
$$

Let

$$
j_*: L_{n+2}^h(\mathbb{Z}G) \to L_{n+2}^h(\mathbb{Z}\Gamma_H \to \mathbb{Z}G)
$$

denote the relativization map.

(5.9): The second sequence arises from the theory of codimension two embeddings in [20, 11.6]:

$$\ldots LS_n(\Phi) \to L_n^h(\mathbb{Z}G) \xrightarrow{p_*} L_{n+2}^h(\mathbb{Z}\Gamma_H \to \mathbb{Z}G) \to LS_{n-1}(\Phi) \to \ldots$$

where the obstruction groups were identified algebraically in [16, 7.8.12] as the L-groups

$$LS_n(\Phi) \cong L_n^h(\mathbb{Z}[\mathbb{Z} \times H], \beta, u)$$

with respect to a "twisted" anti-structure (β, u) for the ring $\mathbb{Z}[\mathbb{Z} \times H]$. In our situation, $\beta(\gamma) = \gamma^{-1}$ for all $\gamma \in \Gamma_H$ and $u = i_*(1)$ where $i_* \colon \mathbb{Z} \to \Gamma_H$ is induced by the inclusion $i \colon S(W) \to S(V_1) \times_G S(W)$ of the fibre. More explicitly, choose generators $a \in G$, $b = a^k \in H$ and $t \in \Gamma_H$ generating the infinite cyclic factor of $\Gamma_H = \mathbb{Z} \times H$ in the exact sequence (5.6). Then $u = t^k b \in \Gamma_H$.

Lemma 5.10 *If $H \subset G$ is a non-trivial subgroup with $[G : H] > 2$, then the image of $LS_n(\Phi) \to L_n^h(\mathbb{Z}G)$ is zero on the top component.*

Proof. After applying the isomorphism $LS_n(\Phi) \cong L_n^h(\mathbb{Z}[\mathbb{Z} \times H], \beta, t^k b)$, we must compute the map

$$L_n^h(\mathbb{Z}[\mathbb{Z} \times H], \beta, t^k b)(m) \to L_n^h(\mathbb{Z}G)(m)$$

where the components for $\mathbb{Z}\Gamma_H$ are given by the preimages of proper subgroups of G under the projection $\Gamma_H \to G$. There are two cases, depending on whether $k = [G : H]$ is even or odd. Note that we only need to discuss the case n even, since the p-localization of the odd L-groups of $\mathbb{Z}G$ is zero.

First suppose that $k = 2l$ is even. Then by "scaling" the anti-structure we can assume that the unit $u = b$ and by [17, §16] there is a natural direct sum splitting

$$L_n^h(\mathbb{Z}[\mathbb{Z} \times H], \beta, b) \cong L_n^h(\mathbb{Z}H, \beta, b) \oplus L_{n-1}^p(\mathbb{Z}H, \beta, b).$$

Since n is even, it is enough to compute the map induced by the inclusion

$$L_n^h(\mathbb{Z}H, \beta, b) \to L_n^h(\mathbb{Z}G, \beta, b) \cong L_n^h(\mathbb{Z}G).$$

However, since $[G : H] = k > 2$ there is a subgroup $H' = \langle c \rangle$ with $b = c^2$, $H \subset H' \subset G$ and $[H' : H] = 2$. Therefore our induction map factors through $L_n^h(\mathbb{Z}H', \beta, c^2) \to L_n^h(\mathbb{Z}G, \beta, c^2)$ which is scale equivalent to the ordinary induction map (whose image has zero projection into the top component).

Next suppose that k is odd. Since H is a proper subgroup of G, with quotient $G/H \cong \mathbb{Z}/k$, the pullback of $H \subset G$ in Γ_H is

$$\Gamma_H' = \mathbb{Z} \times H \xrightarrow{k \times 1} \mathbb{Z} \times H \cong \Gamma_H$$

from the description in Lemma 5.5. By scaling, $(\mathbb{Z}[\mathbb{Z} \times H], \beta, t^k b) \sim (\mathbb{Z}[\mathbb{Z} \times H], \beta, tb)$ so the induced anti-structures are the same on both Γ_H and this proper subgroup Γ'_H. But since k is prime to p, the Mackey double coset formula shows that the induced restriction map on L-groups is a p-local injection. But since the L–groups are isomorphic and finitely generated by (5.9), and the inclusion map integrally has at most 2-primary torsion in its cokernel, we conclude that

$$LS_n(\Phi)(m) \cong L_n^h(\mathbb{Z}[\mathbb{Z} \times H], \beta, t^k b)(m) = 0$$

and p_* is an isomorphism on the top component (compare [20, p.252] for a direct calculation in a special case). □

(5.11): The third sequence comes from the "proper" surgery theory of Maumary-Taylor, resulting in surgery obstruction groups $L_n^{h,\mathrm{open}}(K)$ where K is a locally finite CW complex. We will apply this to $K = S(V_1) \times_G W$ and use the Maumary exact sequence [15, 7.1]

$$\Pi_n^h(K) \xrightarrow{1-s} L_n^h(\mathbb{Z}G) \oplus \Pi_n^h(K) \to L_n^{h,\mathrm{open}}(K) \to \Pi_n^p(K) \to L_n^p(\mathbb{Z}G) \oplus \Pi_{n-1}^p(K)$$

The terms

$$\Pi_n^q(K) = \prod_{i=1}^{\infty} L_n^q(\pi_1(K_i)), \quad q = h, p$$

where $K_1 \supset K_2 \supset K_3 \supset \dots$ is a sequence of neighbourhoods of infinity in K so that each K_i is cocompact and $\cap_{i=1}^{\infty} K_i = \emptyset$. In our case K_i can be taken to be the product $S(V_1) \times_G \{w \in W \mid \|w\| \geq i\}$ so the fundamental groups are all isomorphic to Γ_H.

There are several natural maps relating the groups just introduced. We will need the following ones.

(5.12): The compact relative surgery groups map into the proper groups: a relative problem can be modeled on the disk, sphere bundle pair $S(V_1) \times_G (D(W), S(W))$ and we can complete to the model $S(V_1) \times_G W$ by adding a ray $[1, \infty)$ at each point of $S(W)$. This gives

$$r_*: L_{n+2}^h(\mathbb{Z}\Gamma_H \to \mathbb{Z}G) \longrightarrow L_{n+2}^{h,\mathrm{open}}(S(V_1) \times_G W).$$

(5.13): There is a "cone point" inclusion

$$c'_*: L_{n+2}^h(\mathbb{Z}G) \to L_{n+2}^{h,\mathrm{open}}(S(V_1) \times_G W)$$

induced by the map $S(V_1)/G \times \{0\} \subset S(V_1) \times_G W$. This map appears already in the Maumary exact sequence above.

Lemma 5.14 *The "cone point" inclusion in (5.13) equals the composite*

$$L_{n+2}^h(\mathbb{Z}G) \xrightarrow{j_*} L_{n+2}^h(\mathbb{Z}\Gamma_H \to \mathbb{Z}G) \xrightarrow{r_*} L_{n+2}^{h,\mathrm{open}}(S(V_1) \times_G W).$$

(5.15): There is a "forget some control" map

$$q_*\colon L^h_{n+2}(\mathcal{C}_{W,G}(\mathbb{Z})) \to L^{h,\mathrm{open}}_{n+2}(S(V_1) \times_G W)$$

defined by regarding a bounded surgery problem as a proper surgery problem.

Lemma 5.16 *The "cone point" inclusion in (5.13) equals the composite*

$$L^h_{n+2}(\mathbb{Z}G) \xrightarrow{c_*} L^h_{n+2}(\mathcal{C}_{W,G}(\mathbb{Z})) \xrightarrow{q_*} L^{h,\mathrm{open}}_{n+2}(S(V_1) \times_G W).$$

where c_* is the "cone point" inclusion from (5.3).

Proposition 5.17 *The outer square in diagram (5.7)*

$$
\begin{array}{ccc}
L^h_n(\mathbb{Z}G) & \xrightarrow{\ \mathrm{trf}_W\ } & L^h_{n+2}(\mathcal{C}_{W,G}(\mathbb{Z})) \\[4pt]
{\scriptstyle p_*}\big\downarrow & & \big\downarrow{\scriptstyle q_*} \\[4pt]
L^h_{n+2}(\mathbb{Z}\Gamma_H \to \mathbb{Z}G) & \xrightarrow{\ r_*\ } & L^{h,\mathrm{open}}_{n+2}(S(V_1) \times_G W)
\end{array}
$$

is commutative. The maps r_* and q_* induce an isomorphism on the top component, and the map p_* induces a monomorphism on the top component. Hence trf_W is monic on the top component.

Proof. The commutativity of the diagram is easy to verify from the definitions of the maps given above. We will complete the proof below by checking the isomorphisms for q_*, r_* and the monomorphism for p_*, implying the result for trf_W.

For p_* we apply the top component idempotent to the exact sequence in (5.9). The idempotent comes from the family of proper subgroups of G, which gives by pull-back under the projection $\Gamma_H \to G$, a family of proper subgroups of Γ_H. Now the results of [14, §11b] produce a long exact sequence on the top components. But since k is prime to p and $[G:H] \neq 2$, Lemma 5.10 shows that p_* is monic on the top component.

For q_* we will use the Maumary exact sequence from (5.11). In our situation

$$\Pi^q_n(K) = \prod_{i=1}^{\infty} L^q_n(\mathbb{Z}\Gamma_H), \quad q = h, p$$

and a similar argument shows that the top component of this group is zero. Therefore the map c_* in (5.13) is an isomorphism on the top component. Now by Corollary 5.3 and Lemma 5.16 the map q_* is also an isomorphism on the top component.

For r_* we use Lemma 5.14 to see that $r_* = c_* \circ j_*$ where c_* is the "cone point" inclusion from (5.13), which we have already checked is an isomorphism on the top component. But j_* sits in the exact sequence of (5.8) with the third term being $L^h_n(\mathbb{Z}\Gamma_H)$. Once again, this L-group is zero in the top component and so both j_* and finally r_* induce top component isomorphisms. $\qquad\square$

We now complete the proof of Theorem 5.1 by considering the bounded transfer map for representations which contain trivial subrepresentations (i.e. summands $\mathbb{R}^k$ on which G acts trivially). In this case we again prove that trf induces a p–local monomorphism for $p \nmid 2|G|$.

Lemma 5.18 *Suppose that* $\mathrm{trf}_W \colon L_n^h(\mathbb{Z}G) \to L_{n+\dim(W)}^h(\mathcal{C}_{W,G}(\mathbb{Z}))$ *is a p–local monomorphism on the top component, for all n. Then* $\mathrm{trf}_{W \oplus \mathbb{R}}$ *is also a p–local monomorphism on the top component, for all n.*

Proof. The transfer $\mathrm{trf}_{W \oplus \mathbb{R}}$ can be identified with the natural "change of K-theory" map

$$L_{n+\dim(W)}^h(\mathcal{C}_{W,G}(\mathbb{Z})) \to L_{n+\dim(W)}^p(\mathcal{C}_{W,G}(\mathbb{Z})) \tag{5.19}$$

by means of the isomorphism

$$L_{n+\dim(W)+1}^h(\mathcal{C}_{W \oplus \mathbb{R},G}(\mathbb{Z})) \cong L_{n+\dim(W)}^p(\mathcal{C}_{W,G}(\mathbb{Z}))$$

given in [17, §15]. But the kernel and cokernel of (5.19) are 2-torsion groups, since the change of K-theory map sits in a long exact sequence whose third term is $H^*(\mathbb{Z}/2, \tilde{K}_0(\mathcal{C}_{W,G}(\mathbb{Z})))$.   □

Corollary 5.20 *For any G-representation W, with $W^G = W^H$ when $[G : H] = 2$, the image* $\mathrm{trf}_W(\sigma(f))$ *under the bounded transfer is non-zero in the top component.*

Proof. By Lemma 4.5 the element $\sigma(f)$ has infinite order. We can write any G-representation as a direct sum of irreducible 2-dimensional and trivial subrepresentations. By (5.2) and (5.17) the bounded transfer for the sum of the 2-dimensional factors is a p–local monomorphism on the top component., and the further transfer by a trivial representation $\mathbb{R}^k$ is a p–local monomorphism by (5.18).   □

Remark 5.21 The change of K-theory map $L_n^h(\mathbb{Z}G) \to L_n^p(\mathbb{Z}G)$ is *not* an integral monomorphism in general for G cyclic. If we had only odd index isotropy groups in our irreducible 2-dimensional representations W, then trf_W would induce an isomorphism on the 2–local top components.

Proof of Corollary B. We may assume that G is cyclic. If G has odd order, then the result follows by induction from Theorem A. If G has order $2m$, where m is odd then the odd order theorem determines half the characters of V, V' by restriction to the fixed set of the element of order 2, and the other half by restriction to the subgroup of index 2. If G has order $2^l m$, where m odd and $l \geq 2$ and $V \sim_t V'$ then by induction on $|G|$ we can assume that $Res_H V \cong Res_H V'$ for all proper subgroups $H \subset G$. Since the elements of order $2^l \geq 4$ act trivially or freely in each irreducible subrepresentation of V, the fixed set $V^H = V^G$ for the subgroup $H \subset G$ of index 2. Now Theorem A gives the result.   □

References

[1] S. E. Cappell and J. L. Shaneson, *Non-linear similarity*, Ann. of Math. **113** (1981), 315–355.

[2] _____, *The topological rationality of linear representations*, Publ. Math. I. H. E. S. **56** (1983), 309–336.

[3] S. E. Cappell, J. L. Shaneson, M. Steinberger, S. Weinberger, and J. West, *The classification of non-linear similarities over* $\mathbb{Z}/2^r$, Bull. Amer. Math. Soc. **22** (1990), 51–57.

[4] S. E. Cappell, J. L. Shaneson, M. Steinberger, and J. West, *Non-linear similarity begins in dimension six*, Amer. J. Math. **111** (1989), 717–752.

[5] S. C. Ferry, I. Hambleton, and E. K. Pedersen, *A survey of bounded surgery theory and applications*, Algebraic Topology and Applications, Proceedings of MSRI conference 1990 (New York), Springer-Verlag, 1994, pp. 105–126.

[6] S. C. Ferry and E. K. Pedersen, *Epsilon surgery*, Proceedings of Oberwolfach conference on the Novikov conjecture (S. C. Ferry, J. Rosenberg, and A. A. Ranicki, eds.), 1994, (To appear).

[7] I. Hambleton and I. Madsen, *Actions of finite groups on* R^{n+k} *with fixed set* R^k, Canad. J. Math. **XXXVIII No 4** (1986), 781–860.

[8] I. Hambleton and E. K. Pedersen, *Bounded surgery and dihedral group actions on spheres*, J. Amer. Math. Soc. **4** (1991), 105–126.

[9] _____, *Topological equivalence of linear representations*, To appear.

[10] W-C. Hsiang and W. Pardon, *When are topologically equivalent representations linearly equivalent*, Invent. Math. **68** (1982), 275–316.

[11] N. Kuiper and J. W. Robbin, *Topological classification of linear endomorphisms*, Invent. Math. **19** (1973), 83–106.

[12] I. Madsen and M. Rothenberg, *On the classification of G-spheres I: equivariant transversality*, Acta Math. **160** (1988), 65–104.

[13] W. Mio, *Nonlinearly equivalent representations of quaternionic 2-groups*, Trans. Amer. Math. Soc. **315** (1989), 305–321.

[14] R. Oliver, *Whitehead Groups of Finite Groups*, LMS Lecture Notes Series, vol. 132, Cambridge U. Press, Cambridge, 1988.

[15] E. K. Pedersen and A. A. Ranicki, *Projective surgery theory*, Topology **19** (1980), 239–254.

[16] A. A. Ranicki, *Exact Sequences in the Algebraic Theory of Surgery*, Math. Notes, vol. 26, Princeton University Press, Princeton N. J., 1981.

[17] _____ , *Lower K- and L-theory*, Lond. Math. Soc. Lect. Notes, vol. 178, Cambridge Univ. Press, Cambridge, 1992.

[18] G. de Rham, *Reidemeister's torsion invariant and rotations of S^n*, Diff. Analysis (Oxford), Oxford University Press, 1964, pp. 27–36.

[19] C. Rourke and D. Sanderson, *On topological neighborhoods*, Comp. Math. **22** (1970), 387–424.

[20] C. T. C. Wall, *Surgery on Compact Manifolds*, Academic Press, 1971.

On the Vietoris–Rips Complexes and a Cohomology Theory for Metric Spaces

Jean-Claude Hausmann

1   Introduction

Let X be a pseudo-metric space with pseudo-distance d (i.e. d satisfies the axioms of a distance except that $d(x,y) = 0$ does not imply that $x = y$). Let ε be a positive real number. We denote by X_ε the abstract simplicial complex defined as follows: the vertices of X_ε are the points of X and a q-simplex of X_ε is a subset $\{x_0, \ldots, x_q\}$ of X such that $\mathrm{Diam}\,(\{x_0, \ldots, x_q\}) < \varepsilon$. Recall that if $P \subset X$, the diameter $\mathrm{Diam}\,P$ of P is the least upper bound of $d(x,y)$ for all $x, y \in P$.

Examples.

1. Let $X := \{e^{\frac{2ik\pi}{n}} \mid k = 0, 1, \ldots, n\}$, the set of nth roots of 1 with the geodesic distance on the circle S^1. If $\varepsilon \leq 2\pi/n$, then X_ε is a discrete set with n points. If $2\pi/n < \varepsilon \leq 4\pi/n$ and $n \geq 4$, X_ε is a polygon with n sides. One checks easily that if $4\pi/n < \varepsilon \leq 6\pi/n$ and $n \geq 6$ then X_ε is a combinatorial surface (with boundary if $n \geq 7$). By computing the Euler characteristic and the number of connected components of the boundary, one sees that this surface is

   - the sphere S^2 (an octaedron) if $n = 6$
   - the band $S^1 \times [0,1]$ if $n \geq 8$ is even and
   - the Möbius band if $n \geq 7$ is odd.

   The first case generalizes as follows: if $n = 2m$ and $(m-1)\pi/m < \varepsilon < \pi$ then X_ε is a triangulation of the sphere S^{m-1}.

2. Let X be the set of vertices of a cube of edge 1 in $\mathbb{R}^3$. If $\sqrt{2} < \varepsilon \leq \sqrt{3}$, then X_ε is a triangulation of the sphere S^3.

The complex X_ε was once called the *Vietoris complex* ([Le], p. 271). Since its re-introduction by E. Rips for studying hyperbolic groups, it has been popularized under the name of *Rips complex* ([Gr], § 1.7 and 2.2 or [G–H], p. 68).

   This paper is devoted to the study of X_ε, principally when ε is small (in contrast with the framework of Rips). In § 2, we establish a few general properties of the complexes X_ε, the most interesting being that $|X_\varepsilon|$ is homotopy equivalent to $|\widehat{X}_\varepsilon|$ if $\widehat{X}$ is the metric completion or X. In § 3, we prove theorems concerning the ε-complex of a Riemannian manifold (or of a geodesic space). The main consequence of these results is that for M a compact Riemannian manifold and ε small enough, the geometric realization of M_ε is homotopy equivalent to M.

In § 4, we introduce the *metric cohomology* $\mathcal{H}^*(X;\mathcal{R})$ of a pseudo-metric space X in a coefficient ring $\mathcal{R}$. The group $\mathcal{H}^*(X;\mathcal{R})$ is defined as the inductive limit of the simplicial cohomology $H^*(X_\varepsilon;\mathcal{R})$ when $\varepsilon \to 0$. We show that this produces a cohomology theory for the category of pseudo-metric spaces and uniformly continuous maps (with a "metric version" of the excision axiom and Mayer–Vietoris sequences). In § 5, a short direct proof is given that if X is a compact metric space, then $\mathcal{H}^*(X;\mathcal{R})$ is canonically isomorphic to the Cech cohomology $\check{H}^*(X;\mathcal{R})$. We finish by other properties of the metric cohomology which are quite different than those of previously known cohomology theories: for instance, the metric cohomology of X is isomorphic to that of its metric completion. Also, a contractible metric space might not be acyclic for the metric cohomology.

The corresponding construction for homology ($\mathcal{H}_*(X)$ being the projective limit of $= H_*(X_\varepsilon)$) was the one originally considered by L. Vietoris [Vi] (for compact metric spaces) as one of the first attempt to define homology groups for spaces which were not triangulated. Thus, "Vietoris cohomology" could have been an appropriate denomination for our cohomology. We think however that "metric cohomology" is less confusing, since the definition of Vietoris homology has been subsequently modified in the literature ([Be] and [Do]) in such a way to become the homology counterpart of the Alexander–Spanier cohomology (which is not isomorphic to the metric cohomology; see § 5).

In general, little is known about the ε-complex of a metric space, even for simple spaces like the spheres. Interesting questions seem to arise from the results and techniques of this paper (see, for instance, § 3). We would be happy if these techniques may be of some use in problems involving metrics or pseudo-metrics (approximation by discrete or finite subsets, controlled topology, fractals, image analysis, etc).

I was helped by M. San Millan who worked out some details of § 4 and 5 in her "mémoire de diplôme". I wish also to thank P.Y. Gaillard, J. Roe and H. Munkholm for useful remarks.

2   Some properties of X_ε

(2.1) Lemma. *Let K be a finite pseudo-metric space. Let $\varepsilon > 0$. Then there exists $\delta > 0$ such that id_K induces a simplicial isomorphism from $K_{\varepsilon-2\delta} \xrightarrow{\approx} K_\varepsilon$.*

Proof.   Let

$$m = \max \{d(x,y) \mid x,y \in K \text{ and } d(x,y) < \varepsilon\}.$$

As K is finite, there exist $\delta > 0$ such that $m < \varepsilon - 2\delta$. The simplicial complexes $K_{\varepsilon-2\delta}$ and K_ε have thus the same 1- skeleton. But for any pseudo-metric space X, the complex X_ε is determined by its 1- skeleton ($\{x_0, \dots, x_q\}$ is a q-simplex of X_ε if and only if $\{x_i, x_j\}_{i<j}$ are 1-simplexes of X_ε). $\qquad\square$

Definition. Let X be a pseudo-metric spaces and A a subspace of X. A continuous map $F : X \times [0, 1] \longrightarrow X$ satisfying

a) $F(x, 1) = x$, $F(x, 0) \in A$, $F(a, t) = a$ if $a \in A$

b) $d(F(x, t'), F(y, t')) \leq d(F(x, t), F(y, t))$ whenever $t' \leq t$

is called a *crushing* from X onto A. We call a pseudo-metric space X *crushable* if there is a crushing from X onto one of its point.

Remarks. 1) A crushing is a retraction by deformation and crushable spaces are contractible. The converse is not true, as will be seen in § 3.

2) Let V be a real normed vector space. Then the map $(x, t) \mapsto t \cdot x$ is a crushing of V onto $\{0\}$. Trees (metric or real, see Chapter 2 of [G–H] for definitions) are also crushable.

In the statement below, if P is a simplicial complex, we denote as usual by $|P|$ its geometric realization.

(2.2) Proposition. *Let X be a pseudo-metric space admitting a crushing onto a subspace $A \subset X$. Then, for any $\varepsilon > 0$, the map $|A_\varepsilon| \longrightarrow |X_\varepsilon|$ induced by the inclusion is a homotopy equivalence.*

Proof. Let $g : (|K|, |L|) \longrightarrow (|X_\varepsilon|, |A_\varepsilon|)$ be a continuous map, where K is a finite simplicial complex and L a subcomplex of K. To prove Proposition (2.2), it is enough to show that such a map g is homotopic (rel $|L|$) to a map sending $|K|$ into $|A_\varepsilon|$.

By simplicial approximation we may, after replacing K and L by one of their barycentric subdivisions, suppose that $g = |h|$ for a simplicial map $h : (K, L) \longrightarrow (X_\varepsilon, A_\varepsilon)$. The sets $\overline{K} := h(K^0)$ and $\overline{L} := h(L^0)$ are finite subsets of X and A and h factors through the pair $(\overline{K}_\varepsilon, \overline{L}_\varepsilon)$. By (2.1), there is $\delta > 0$ such that $(\overline{K}_{\varepsilon-2\delta}, \overline{L}_{\varepsilon-2\delta}) = (\overline{K}_\varepsilon, \overline{L}_\varepsilon)$. Therefore, one has a factorization

$$(K, L) \xrightarrow{\ \overline{h}\ } (\overline{K}_{\varepsilon-2\delta}, \overline{L}_{\varepsilon-2\delta}) \xrightarrow{\ j\ } (X_\varepsilon, A_\varepsilon)$$

Let $F : X \times [0, 1] \longrightarrow X$ be a crushing of X onto A. Chose an integer p so that for all $x \in \overline{K} - \overline{L}$ and all $0 \leq k < p$ one has

$$d(F(x, \frac{k}{p}), F(x, \frac{k+1}{p})) < \delta$$

The map $x \mapsto F(x, \frac{k+1}{p})$ induces then a simplicial map

$$f^k : \overline{K}_{\varepsilon-2\delta} \longrightarrow X_\varepsilon$$

such that $f^p|\overline{L}_{\varepsilon-2\delta} = j$ and $f^0(\overline{L}_{\varepsilon-2\delta}) \subset A_\varepsilon$. One has

$$\text{Diam}(\{x_0,\ldots,x_q\}) < \varepsilon - 2\delta$$
$$\Longrightarrow \text{Diam}(\{f^k(x_0),\ldots,f^k(x_q), f^{k+1}(x_0),\ldots,f^{k+1}(x_q)\}) < \varepsilon$$

which shows f^k and f^{k+1} are contiguous (see [Sp], Chapter 3 §5). Therefore, $f^p = j$ and f^0 are in the same contiguity class (rel $\overline{L}_{\varepsilon-2\delta}$). This implies that the induces maps $|j|$ and $|f^0|$ are homotopic (rel $|L_{\varepsilon-2\delta}|$) ([Sp], Lemma 2 p. 130). By composing with $\overline{h}$, this proves that $g = |h|$ is homotopic (rel $|L|$) to $|\overline{h}| \circ |f^0|$ which sends $|K|$ into $|A_\varepsilon|$. $\qquad\square$

(2.3) Corollary. *Let X be a crushable pseudo-metric space. Then, for any $\varepsilon > 0$, the space $|X_\varepsilon|$ is contractible.*

(2.4) Proposition. *Let X be a pseudo-metric space and Y a dense subset of X. Then, for any $\varepsilon > 0$, the map $|Y_\varepsilon| \longrightarrow |X_\varepsilon|$ induced by the inclusion is a homotopy equivalence.*

Proof. As in the proof of (2.2), let us consider a continuous map $g : (|K|,|L|) \longrightarrow (|X_\varepsilon|,|Y_\varepsilon|)$, where K is a finite simplicial complex and L a subcomplex of K and try to prove that g is homotopic (rel $|L|$) to a map sending $|K|$ into $|Y_\varepsilon|$. As in the proof of (2.2), we reduce to the case where g is induced by a simplicial map $h : (K,L) \longrightarrow (X_\varepsilon,Y_\varepsilon)$ with a factorization

$$(K,L) \xrightarrow{\overline{h}} (\overline{K}_{\varepsilon-2\delta}, \overline{L}_{\varepsilon-2\delta}) \xrightarrow{j} (X_\varepsilon,Y_\varepsilon)$$

where $\overline{K} := h(K^0)$ and $\overline{L} := h(L^0)$ are finite subsets of X and Y and $0 < 2\delta < \varepsilon$. For each $u \in \overline{K}$ chose $u' \in Y$ such that $d(u,u') < \delta$ and $u' = u$ if $u \in Y$. The correspondance $u \mapsto u'$ gives a simplicial map $j' : \overline{K}_{\varepsilon-2\delta} \longrightarrow Y_\varepsilon \subset X_\varepsilon$ which is contiguous (rel $\overline{L}_{\varepsilon-2\delta}$) to j. Therefore, $g = |h| = |\overline{j}| \circ |\overline{h}|$ is homotopic (rel $|L|$) to the map $|\overline{j}| \circ |\overline{h}|$ which sends $|K|$ into $|Y|_\varepsilon$. $\qquad\square$

(2.5) Corollary. *Let $j : X \longrightarrow \widehat{X}$ be the inclusion of a metric space into a metric completion $\widehat{X}$ of X. Then, for any $\varepsilon > 0$, the map $|j| : |X_\varepsilon| \longrightarrow |\widehat{X}_\varepsilon|$ is a homotopy equivalence.*

3   The ε-complex of a Riemannian manifold

Let M be a Riemannian manifold. We consider M as equipped with the distance $d(x,y) = \inf\{\text{length}(\gamma)\}$, where γ runs among all smooth curves joining x to y. Define $r(M) \in \mathbb{R}_{\geq 0}$ as the least upper bound of the set of real numbers r satisfying the following three conditions:

   a) for all $x,y \in M$ such that $d(x,y) < 2r$ there is a unique shortest geodesic joining x to y. Its length is $d(x,y)$.

b) let x, y, z, u with $d(x, y) < r$, $d(u, x) < r$, $d(u, y) < r$ and z be a point on the shortest geodesic joining x to y. Then $d(u, z) \le \max\{d(u, x), d(u, y)\}$.

c) if γ and γ' are arc-length parametrized geodesics such that $\gamma(0) = \gamma'(0)$ and if $0 \le s, s' < r$ and $0 \le t \le 1$, then $d(\gamma(ts), \gamma'(ts')) \le d(\gamma(s), \gamma'(s'))$.

Remarks. 1) It is well known that $r(M) > 0$ if M admits a strictly positive injectivity radius and an upper bound on its sectional curvature. In particular, a compact Riemannian manifold M has $r(M) > 0$. For example, the sphere S^n of radius 1 has $r(S^n) = \pi/2$.

2) Conditions a), b) and c) above make sense for a geodesic space (see [G–H], p. 16) and all the theorems of this section which are stated for Riemannian manifolds with $r(M) > 0$ are true for a geodesic space X with $r(X) > 0$.

Chose a total ordering on the points of M. From now on, whenever we describe a finite subset of M by $\{x_0, \dots, x_q\}$, we suppose that $x_0 < x_1 < \dots < x_q$. Suppose that $r(M) > 0$ and let $\varepsilon < r(M)$. we shall associate to each q-simplex $\sigma := \{x_0, \dots, x_q\}$ of M_ε a singular q-simplex $T_\sigma : \Delta^q \longrightarrow M$. Here, Δ^q denotes the standard euclidean q-simplex:

$$\Delta^q := \left\{ \sum_{i=0}^{q} t_i e_i \mid t_i \in [0, 1] \text{ and } \sum t_i = 1 \right\} \tag{3.1}$$

where $e_0, \dots, e_q$ are the standard basis vectors of $\mathbb{R}^{q+1}$. The map T_σ is defined inductively as follows: set $T(e_0) = x_0$. Suppose that $T_\sigma(z)$ is defined for $y = \sum_{i=0}^{p-1} \tau_i e_i$. Let $z := \sum_{i=0}^{p} t_i e_i$. If $t_p = 1$, we pose $T_\sigma(z) = x_p$. Otherwise, let

$$x := T_\sigma \left(\frac{1}{1 - t_p} \sum_{i=0}^{p-1} t_i e_i \right)$$

We define $T_\sigma(z)$ as the point on the shortest geodesic joining x to x_p with $d(x, T_\sigma(z)) = t_p \cdot d(x, x_p)$ (the shortest geodesic exists by Condition b) above). To sum up, T_σ is defined inductively on Δ^p for $p \le q$ as the "geodesic join" of $T_\sigma(\Delta^{p-1})$ with x_p.

If σ' is a face of σ of dimension p we form the euclidean sub-p-simplex Δ' of Δ^q formed by the points $\sum_{i=0}^{q} t_i e_i \in \Delta^q$ with $t_i = 0$ if $x_i \notin \sigma'$. One checks by induction on $\dim \sigma'$ that

$$T_{\sigma'} = T_\sigma | \Delta' \tag{3.2}$$

(3.3) Proposition. *Let M be a Riemannian manifold with $r(M) > 0$. Let $0 < \varepsilon' \le \varepsilon \le r(M)$. Then the inclusion $M_{\varepsilon'} \subset M_\varepsilon$ induces an isomorphism on homology.*

Proof. Let $\sigma = \{x_0, \dots, x_q\}$ be a simplex of M_ε and let $I_\sigma = \text{Image}(T_\sigma)$. If σ' is a face of σ then $I_{\sigma'} \subset I_\sigma$ by (3.2) and thus $(I_{\sigma'})_\delta$ is a subcomplex of

$(I_\sigma)_\delta$ for all $\delta > 0$. On the other hand, $(I_\sigma)_\delta$ is acyclic for all $\delta > 0$. Indeed, I_σ is contained in the geodesic ball $B := B(x_q, r(M))$. By Condition c) of the definition of $r(M)$ the obvious deformation retraction of B on x_q along the geodesics satisfies the hypothesis of (0.179) and I_σ is invariant under this deformation. These considerations show that for $0 < \delta' \leq \delta \leq r(M)$, the correspondence

$$\sigma \mapsto (I_\sigma)_{\delta'} \quad (\sigma \in M_\delta)$$

is an acyclic carrier $\Phi_{\delta'\delta}$ from M_δ to $M_{\delta'}$ (see [Mu], § 13).

We now use the theorem of acyclic carrier ([Mu], Theorem 13.3). It implies that there exists an augmentation preserving chain map $\nu : C_*(M_\varepsilon) \longrightarrow C_*(M_{\varepsilon'})$ which is carried by $\Phi_{\varepsilon'\varepsilon}$. Let μ denote the inclusion from $M_{\varepsilon'}$ into M_ε. Then, $\Phi\varepsilon'\varepsilon'$ is an acyclic carrier for both $\nu \circ \mu_*$ and the identity of $C_*(M\varepsilon')$. By theorem of acyclic carrier again, these two maps are chain homotopic and thus $\nu \circ \mu_*$ induces the identity on $H_*(M_{\varepsilon'})$. The same argument shows that $\mu_* \circ \nu$ induces the identity on $H_*(M_\varepsilon)$ (using the acyclic carrier $\Phi_{\varepsilon\varepsilon}$).   □

We will now compare the simplicial homology of M_ε with the singular homology of M. Formula (3.2) shows that the correspondence $\sigma \mapsto T_\sigma$ gives rise to a chain map

$$T_\sharp : C_*(M_\varepsilon) \longrightarrow SC_*(M)$$

where $SC_*(M)$ denotes the singular chain complex of M.

(3.4) Proposition. *If $0 < \varepsilon \leq r(M)$ the chain map T_* induces an isomorphism on homology.*

Proof. We shall need a few accessory chain complexes. For $\delta > 0$, denote by $SC_*(M; \delta)$ the sub-chain complex of $SC_*(M)$ based on singular simplexes τ such that there exists $x \in M$ with Image (τ) is contained in the ball $B(x, \delta)$. Recall that the inclusion $SC_*(M; \delta) \longrightarrow SC_*(M)$ induces an isomorphism on homology ([Mu], Theorem 31.5).

We shall also use the ordered chain complex $C_*^o(M_\delta)$: the group $C_q^o(M_\delta)$ is free abelian on $(q+1)$-uples $(x_0, \ldots, x_q)$ such that $\{x_0\} \cup \ldots \cup \{x_q\}$ is a simplex of M_δ. On can see $C_*(M_\delta)$ as a sub-chain complex of $C_*^o(M_\delta)$ by associating to a q-simplex $\{x_0, \ldots, x_q\}$ of M_ε (with our convention that $x_0 < x_1 < \ldots < x_q$ for the well ordering on M) the $(q+1)$-uple $(x_0, \ldots, x_q)$. It is also classical that this inclusion is a homology isomorphism ([Mu], Theorem 13.6). Observe that the construction $\sigma \mapsto T_\sigma$ does not required that the vertices of σ are all distincts. One can then define T_σ for a basis element of $C_q^o(M_\delta)$ (if $\delta \leq r(M)$) and $t^\sharp$ thus extends to a chain map $T^\sharp = T^{\sharp, \delta} : C_*^o(M_\delta) \longrightarrow SC_*(M; \delta)$ (the image of $T_{(x_0, \ldots, x_q)}$ being contained in $B(x_q, \delta)$). If $0 < \varepsilon \leq r(M)$, one then has a commutative diagram

$$C_*^o(M_{\varepsilon/2}) \xrightarrow{T^{\sharp,\varepsilon/2}} SC_*(M;\varepsilon/2)$$

$$\downarrow \qquad\qquad\qquad \downarrow$$

$$C_*^o(M_\varepsilon) \xrightarrow{T^{\sharp,\varepsilon}} SC_*(M;\varepsilon)$$

Let $\tau : \Delta^q \longrightarrow M$ be a singular simplex whose image is contained in some ball of radius $\varepsilon/2$. The $(q+1)$-uple $(\tau(e_0),\ldots,\tau(e_q))$ is element of $C_q^o(M_\varepsilon)$. This correspondence gives rise to a chain map

$$R^\varepsilon : SC_*(M;\varepsilon/2) \longrightarrow C_*^o(M_\varepsilon)$$

The composition $R^\varepsilon \circ T^{\sharp,\varepsilon/2}$ is equal to the inclusion $C_*^o(M_{\varepsilon/2}) \subset C_*^o(M_\varepsilon)$ which induces a homology isomorphsim by (3.3). Let us now understand the composition $T^{\sharp,\varepsilon} \circ R^\varepsilon$. Let $\tau : \Delta^q \longrightarrow M$ be a singular simplex such that $\tau(\Delta^q) \subset B(y,\varepsilon/2)$ for some $y \in M$. Therefore, $\tau' := T^{\sharp,\varepsilon} \circ R^\varepsilon(\tau)$ also satisfies $\tau'(\Delta^q) \subset B(y,\varepsilon/2)$ by Condition b). Hence, τ and τ' are cannonically homotopic (following, for each $s \in \Delta^q$, the shortest geodesic joining $\tau(s)$ to $\tau'(s)$). As in the proof of the homotopy axiom for singular homology ([Mu], § 30), these homotopies provide a chain homotopy between $T^{\sharp,\varepsilon} \circ R^\varepsilon$ and the inclusion $SC_*(M;\varepsilon/2) \subset SC_*(M;\varepsilon)$. As said before, this inclusion is known to induce a homology isomorphism. Therefore, $T^{\sharp,\varepsilon} \circ R^\varepsilon$ induces an isomorphism on homology.

We have shown that both $R^\varepsilon \circ T^{\sharp,\varepsilon/2}$ and $T^{\sharp,\varepsilon} \circ R^\varepsilon$ induce homology isomorphisms. Therefore, $T^{\sharp,\varepsilon}$ induces a homology isomorphism. $\qquad\square$

We are now in position to state our final theorem. By (3.2) the correspondence $\sigma \mapsto T_\sigma$ gives rise to a continuous map

$$T : |M_\varepsilon| \longrightarrow M$$

where $|M_\varepsilon|$ denotes the geometric realization of M_ε.

(3.5) Theorem. *Let M be a Riemannian manifold with $r(M) > 0$. Let $0 < \varepsilon \leq r(M)$. Then the map $T : |M_\varepsilon| \longrightarrow M$ is a homotopy equivalence.*

Proof. The proof goes in several steps.

(3.6) We first show that T induces an isomorphism on the fundamental groups. Let $\gamma : [0,1] \longrightarrow M$ represent an element of $\pi_1(M)$. Let $p \in \mathbb{N}$ such that $1/p$ is a Lebesgue number for the covering $\{\gamma^{-1}(B(x,\varepsilon/2))\}_{x \in M}$. The path $\gamma|[k/p,(k+1)/p]$ is then canonically homotopic to a parameterisation of the shortest geodesic joining $\gamma(k/p)$ to $\gamma((k+1)/p)$. This shows that γ is homotopic to a composition γ' of geodesics in balls of radius $\varepsilon/2$. Such a path γ' represents the image by T of an element of $\pi_1(|M_\varepsilon|)$, the later being identifiend with the edge-path group of the simplicial complex M_ε ([Sp], pp. 134–138). This proves that $\pi_1 T : \pi_1(|M_\varepsilon|) \longrightarrow \pi_1(M)$ is surjective. A similar argument using, as usual, a small triangulation of $[0,1] \times [0,1]$ shows the injectivity of $\pi_1 T$.

(3.7) Let $\tilde{M}$ be the universal covering of M endowed with the Riemannian metric so that the covering projection is a local isometry. We leave to the reader to check that $r(\tilde{M}) \geq r(M)$. By Proposition (3.4), the map $\hat{T} : |(\tilde{M})_\varepsilon| \longrightarrow \tilde{M}$ is a homology isomorphism. By (3.6), $\pi_1((\tilde{M})_\varepsilon) \simeq \pi_1(\tilde{M}) \simeq \{1\}$. Therefore, $\hat{T}$ is a homotopy equivalence.

(3.8) Let $p_\varepsilon : \widetilde{|M_\varepsilon|} \longrightarrow M_\varepsilon$ be the universal covering projection over M_ε. By (3.6), the following commutative diagram

$$
\begin{array}{ccc}
\widetilde{|M_\varepsilon|} & \xrightarrow{\tilde{T}} & \tilde{M} \\[2mm]
{\scriptstyle p_\varepsilon}\downarrow & & \downarrow{\scriptstyle p} \\[2mm]
|M_\varepsilon| & \xrightarrow{T} & M
\end{array}
\tag{3.9}
$$

is a pull-back diagram. Therefore, the map $\hat{T} : |(\tilde{M})_\varepsilon| \longrightarrow \tilde{M}$ factors as $\hat{T} = \tilde{T} \circ \check{T}$ where

$$|(\tilde{M})_\varepsilon| \xrightarrow{\check{T}} \widetilde{|M_\varepsilon|} \xrightarrow{\tilde{T}} \tilde{M}$$

(all these maps are unique once choice of base points have been made). We shall see in (3.10) that the map $\check{T}$ is a homeomorphism. Since the map $\hat{T}$ is a homotopy equivalence by (3.7), the maps $\tilde{T}$ (and then T) will be homotopy equivalences and Theorem (3.5) will be proven.

(3.10) Let L be the simplicial complex defined as follows:

– the vertex set $L^{(0)}$ of L is $L^{(0)} := \{p_\varepsilon^{-1}(v) \mid v$ is a vertex of M_ε

– $\bar{\sigma} := \{\bar{v}_0, \ldots, \bar{v}_q\} \subset L^{(0)}$ is a q-simplex of L if there is a q-simplex $\sigma = \{v_0, \ldots, v_q\}$ of M_ε and a lifting $f_{\bar{\sigma}} : |\sigma| \longrightarrow \widetilde{|M_\varepsilon|}$ such that $f_{\bar{\sigma}}(v_i) = \bar{v}_i$. It is known that the collection $\{f_{\bar{\sigma}} \mid \bar{\sigma} \in L\}$ produces a homeomorphism $f : |L| \xrightarrow{\simeq} \widetilde{|M_\varepsilon|}$ ([Sp], pp. 144–145; in other words, (L, f) is a triangulation of $|\widetilde{M_\varepsilon}|$).

Using the pull-back diagram (3.9), one gets a bijection $\tilde{M} \xrightarrow{\simeq} L^{(0)}$. As $(\tilde{M})_\varepsilon^{(0)} = \tilde{M}$ one obtains a bijection $\check{t} : (\tilde{M})_\varepsilon^{(0)} \xrightarrow{\simeq} L^{(0)}$. It is easy to check that $\check{t}$ extends to a simplicial isomorphism $\check{t} : (\tilde{M})_\varepsilon \xrightarrow{\simeq} L$ and $|\check{t}| = \check{T}$. Therefore, $\check{T}$ is a homeomorphism. As said in (0.51), this finishes the proof of Theorem (3.5). $\qquad\square$

Problems.

(3.11) Given a compact Riemannian manifold M, does there exists a finite set $F \subset M$ and an $0 < \varepsilon < r(M)$ such that the restriction of the map T to $|F_\varepsilon|$ gives a homotpopy equivalence $|F_\varepsilon| \xrightarrow{\sim} M$?

Suppose that F is a subset of M which is δ-dense with $2\delta < r(M)$. Then one can choose a (possibly non-continuous) map $h : M \longrightarrow F$ such that $d(x, h(x)) < \delta$ for all $x \in M$. Let $0 < \eta < r(M) - 2\delta$ and let $\varepsilon = 2\delta + \eta$. Then, the composed simplicial map

$$M_\eta \longrightarrow F_\varepsilon \longrightarrow M_\varepsilon$$

is contiguous to the inclusion $M_\eta \subset M_\varepsilon$. By (3.5), this shows that $|F_\varepsilon| \xrightarrow{\sim} M$ is a domination.

(3.12) How does M_ε change when ε grows bigger ? I expected M_ε to become more and more connected till $\varepsilon = \operatorname{Diam} M$ where M_ε is contractible (this is suggested by the examples of the introduction).

4 The metric cohomology

Let $\mathcal{R}$ be a commutative ring. Let X be a pseudo-metric space. One denotes by $H^*(X_\varepsilon; \mathcal{R})$ the simplicial cohomology with coefficient in $\mathcal{R}$ of the ε-complex X_ε of X. One defines

$$\mathcal{H}^*(X; \mathcal{R}) := \lim_{\varepsilon \to 0} H^*(X_\varepsilon; \mathcal{R})$$

Here, one uses the fact that if $\varepsilon' \le \varepsilon$, then $X_{\varepsilon'}$ is a subcomplex of X_ε, giving rise to a homomorphism $H^*(X_\varepsilon; \mathcal{R}) \longrightarrow H^*(X_{\varepsilon'}; \mathcal{R})$. The graded structure of $H^*(X; \mathcal{R})$ passes to the inductive limit and makes $\mathcal{H}^*(X; \mathcal{R})$ a graded ring.

In the same way, let us consider a *pair of pseudo-metric spaces* (X, Y). This means that Y is a subspace of X endowed with the induced pseudo-distance. Then Y_ε is a subcomplex of X_ε for all $\varepsilon > 0$. The simplicial cohomology of the pair $(X_\varepsilon, Y_\varepsilon)$ is then defined and one sets

$$\mathcal{H}^*(X, Y; \mathcal{R}) := \lim_{\varepsilon \to 0} H^*(X_\varepsilon, Y_\varepsilon; \mathcal{R})$$

We shall consider the following category $\mathcal{M}$:

- an object of $\mathcal{M}$ is a pair (X, Y) of pseudo-metric spaces.

- a morphism from f from (X, Y) to (X', Y') is a uniformly continuous map $f : X \longrightarrow X'$ such that $f(Y) \subset Y'$.

- If X is a pseudo-metric space, the space $X \times [0, 1]$ is understood to be endowed with the product pseudo-distance $d((x, t), (x', t')) = d(x, x') + |t - t'|$

The category $\mathcal{M}$ is an admissible category for a homology theory ([Mu], p. 146). This section is devoted to the proof of

(4.1) Theorem.   *The correspondance* $(X,Y) \mapsto \mathcal{H}^*(X,Y;\mathcal{R})$ *satisfies the Eilenberg-Steenrod axioms of an ordinary cohomology theory for the category* $\mathcal{M}$, *with a "metric version" of the excision axiom (see (0,2) below).*

Proof.   Let first establish the functoriality. Let $f : (X,Y) \longrightarrow (X',Y')$ be a uniformly continuous map. For $\varepsilon > 0$, there is $\delta > 0$ such that $d(f(u), f(v)) < \varepsilon$ if $d(u,v) < \delta$. Therfore one has a simplicial map $f_{\delta,\varepsilon} : (X_\delta, Y_\delta) \longrightarrow (X'_\varepsilon, Y'_\varepsilon)$. If $\varepsilon' \le \varepsilon$, one can find $\delta' \le \delta$ so that the following diagram

$$
\begin{array}{ccc}
(X_{\delta'}, Y_{\delta'}) & \xrightarrow{\;f_{\delta',\varepsilon'}\;} & (X'_{\varepsilon'}, Y'_{\varepsilon'}) \\
\downarrow & & \downarrow \\
(X_\delta, Y_\delta) & \xrightarrow{\;f_{\delta,\varepsilon}\;} & (X'_\varepsilon, Y'_\varepsilon)
\end{array}
$$

commutes. The induced homomorphisms $f^*_{\delta,\varepsilon} : H^*(X'_\varepsilon, Y'_\varepsilon; \mathcal{R}) \longrightarrow H^*(X_\delta, Y_\delta; \mathcal{R})$ give, passing to the inductive limit, a well defined homomorphism

$$ f^* : \mathcal{H}^*(X',Y';\mathcal{R}) \longrightarrow \mathcal{H}^*(X,Y;\mathcal{R}) $$

The functorial properties of f^* are easy to check.

Now, for each $\varepsilon > 0$ one has the long exact sequence of the pair $(X_\varepsilon, Y_\varepsilon)$. The exactness being preserved by passing to inductive limits, one gets the long exact sequence

$$ \cdots \longrightarrow \mathcal{H}^{i+1}(X,Y) \xrightarrow{\;\delta\;} \mathcal{H}^i(Y) \longrightarrow \mathcal{H}^i(X) \longrightarrow \mathcal{H}^i(X,Y) \longrightarrow \cdots $$

One easily checks that this exact sequence is functorial.

The dimension axiom is obvious since $\{pt\}_\varepsilon$ is a point for all $\varepsilon > 0$.

For the homotopy axiom, let

$$ F : X \times [0,1] \longrightarrow Z $$

be a uniformly continuous map. Let $\varepsilon > 0$. Choose $\delta > 0$ so that $d(F((x,t)), F((x',t'))) < \varepsilon/3$ when $d((x,t),(x',t')) < \delta$. Chose an integer m such that $1/m < \delta/2$. Define $f_{k/m} : X \longrightarrow Z$ by $f_{k/m}(x) := F(x, k/m)$. With these choices, $f_{k/m}$ induces a simplicial map $f_{k/m} : X_{\delta/2} \longrightarrow Z_\varepsilon$ and if $\{x_0, \ldots, x_q\}$ is a q-simplex of $X_{\delta/2}$ then

$$ \{f_{\frac{k}{m}}(x_0), \ldots, f_{\frac{k}{m}}(x_q), f_{\frac{k+1}{m}}(x_0), \ldots, f_{\frac{k+1}{m}}(x_q)\} $$

span a simplex of Z_ε. The simplicial maps $f_{k/m}$ and $f_{(k+1)/m}$ are thus contiguous and then induce the same homomorphism from $H^*(Z_\varepsilon; \mathcal{R}) \longrightarrow H^*(X\delta/2; \mathcal{R})$. This argument can be made for each ε in a coherent enough way and this proves the equality $f_0^* = f_1^*$ as homomorphisms from $\mathcal{H}^*(Z; \mathcal{R}) \longrightarrow \mathcal{H}^*(X; \mathcal{R})$. The proof for pairs of spaces is similar.

We turn now our attention to the excision axiom or rather its metric version. Its precise statement is as follows

(4.2) Proposition (Metric Excision Axiom). *Let (X, Y) be a pair of pseudo-metric spaces. Let $U \subset Y$ such that $d(U, X - Y) > 0$. Then the homomorphism $\mathcal{H}^*(X, Y; \mathcal{R}) \longrightarrow \mathcal{H}^*(X - U, Y - U; \mathcal{R})$ induced by inclusion is an isomorphism.*

Proof. Let $\mu := d(U, X - Y)$. As soon as $\varepsilon < \mu$, one has

$$X_\varepsilon = (X - U)_\varepsilon \cup Y_\varepsilon \quad , \quad (X - U)_\varepsilon \cap Y_\varepsilon = (Y - U)_\varepsilon$$

and therefore the homomorphism

$$H^*(X_\varepsilon, Y_\varepsilon; \mathcal{R}) \longrightarrow H^*((X - U)\varepsilon, (Y - U)\varepsilon; \mathcal{R})$$

induced by inclusion is an isomorphism. $\qquad\square$

Remark. The usual condition for the excision axiom is $\overline{U} \subset \overset{\circ}{Y}$ which is equivalent to $\overline{U} \cap \overline{X - Y} = \emptyset$. The hypothesis for the metric version is equivalent to $d(\overline{U}, \overline{X - Y}) > 0$. The classical version is false for the theory $\mathcal{H}^*$ as seen by the following example: take $X = \mathbb{R} - \{0\}$ and $U = Y = \mathbb{R}_{<0}$ (an open and closed subset of X). Then $\mathcal{H}^0(X, Y; \mathcal{R}) = 0$ while $\mathcal{H}^0(X - U, Y - U; \mathcal{R}) = \mathcal{H}^0(\mathbb{R}_{<0}, \emptyset; \mathcal{R}) \simeq \mathcal{R}$.

We finish by a Mayer–Vietoris sequence for the metric cohomology. As for the excision property, a slight modification is necessary whit respect to classical statemnts. Let $\mathcal{U}$ be an open covering of the pseudo-metric space X.

(4.3) Proposition. *Let $X = U \cup V$ an open covering of X admitting a Lebesgue number. Then, one has the Mayer–Vietoris exact sequence*

$$\cdots \longrightarrow \mathcal{H}^k(X; \mathcal{R}) \longrightarrow \mathcal{H}^k(U; \mathcal{R}) \oplus \mathcal{H}^k(V; \mathcal{R}) \longrightarrow \mathcal{H}^k(U \cap V; \mathcal{R})$$
$$\longrightarrow \mathcal{H}^{k+1}(X; \mathcal{R}) \longrightarrow \cdots$$

Proof. If the covering $X = U \cup V$ admits a Lebesgue number, then if ε is small enough, one has

$$X_\varepsilon = U_\varepsilon \cup V_\varepsilon \quad , \quad U_\varepsilon \cap V_\varepsilon = (U \cap V)_\varepsilon$$

Thus one has the Mayer–Vietoris for simplicial cohomology which, by passing to the inductive limit, gives the corresponding exact for the metric cohomology. $\qquad\square$

5 Properties of the metric cohomology

Between compact space, uniform continuity is equivalent to continuity and the
metric excision axiom (4.2) is equivalent to the usual excision property. Therefore $\mathcal{H}^*$ can be seen as a ordinary cohomology theory for the admissible category
of pairs of compact metric (or even metrizable) spaces.

(5.1) Theorem. *For the category of compact metric pairs, the metric coho-
mology $\mathcal{H}^*$ and the Cech cohomology $\check{H}^*$ in a ring $\mathcal{R}$ are canonically isomorphic
cohomology theories.*

Proof. Let X be a compact metric space. We denote by $\mathcal{N}_\varepsilon(X)$ the nerve of
$\mathcal{B}_\varepsilon(X) := \{B(x,\varepsilon)\}_{x\in X}$, the covering of X by all balls of radius ε. We label the
vertices of $\mathcal{N}_\varepsilon(X)$ by the points of X, identifying x with $B(x,\varepsilon)$.

If $\sigma = \{x_0,\ldots,x_q\}$ is a q-simplex of X_ε, then $\sigma \subset B(x_0,\varepsilon)\cap\ldots\cap B(x_q,\varepsilon)$
which shows that σ is a q-simplex of $\mathcal{N}_\varepsilon(X)$. This provides us a simplicial map
$g_\varepsilon : X_\varepsilon \longrightarrow \mathcal{N}_\varepsilon(X)$. On the other hand, if $\tau = \{y_0,\ldots,y_q\}$ is a q-simplex
of $\mathcal{N}_\varepsilon(X)$, then $d(x_i,x_j) < 2\varepsilon$ and thus τ is a q-simplex of $X_{2\varepsilon}$. This gives
rise to a simplicial map $h_\varepsilon : \mathcal{N}_\varepsilon(X) \longrightarrow X_{2\varepsilon}$. It is clear that the composition
$h_\varepsilon\circ g_\varepsilon$ is equal to the inclusion $X_\varepsilon \subset X_{2\varepsilon}$ and that $g_{2\varepsilon}\circ h_\varepsilon$ is the canonical
simplicial injection $\mathcal{N}_\varepsilon(X) \longrightarrow \mathcal{N}_{2\varepsilon}(X)$. Passing to the inductive limits, one
gets isomorphisms

$$g^* : \lim_{\varepsilon\to 0} H^*(\mathcal{N}_\varepsilon(X);\mathcal{R}) \xrightarrow{\sim} \mathcal{H}^*(X;\mathcal{R}) \,,$$

$$h^* : \mathcal{H}^*(X;\mathcal{R}) \xrightarrow{\sim} \lim_{\varepsilon\to 0} H^*(\mathcal{N}_\varepsilon(X);\mathcal{R})$$

which are inverses one of another. As X is compact, the coverings $\{B(x,\varepsilon)\}_{x\in X}$
form, by the Lebesgue number, a cofinal system in the projective system of all
the coverings. Therefore, one has an isomorphism

$$\lim_{\varepsilon\to 0} H^*(\mathcal{N}_\varepsilon(X);\mathcal{R}) \xrightarrow{\sim} \check{H}^*(X;\mathcal{R})$$

and thus $\mathcal{H}^*(X;\mathcal{R})$ is isomorphic to $\check{H}^*(X;\mathcal{R})$. The same kind of argument
holds for pairs of compact spaces and one checks easily that one gets this way
an isomorphism od cohomology theories.   □

(5.2) Remark. It is known, but not easy to prove in general, that the Cech
cohomology is isomorphic to the Alexander–Spanier cohomology $\overline{H}^*$. (see, for
instance, [Do], Theorem 2). This shows, using Theorem (5.1) that the metric cohomology $\mathcal{H}^*$ and the Alexander–Spanier cohomology are isomorphic for
compact metric spaces. In addition, explicit isomorphisms (without choices) are
easily obtainable in both directions: let X be a compact metric space. Recall
that one of the definition of $\overline{H}^*(X;\mathcal{R})$ is the inductive limit of $H^*(X(\mathcal{U});\mathcal{R})$
where $\mathcal{U}$ is an open covering of X and $X(\mathcal{U})$ is the abstract simplicial complex

for which $\sigma = \{x_0, \ldots, x_q\} \subset X$ is a q-simplex if and only if there exists $U \in \mathcal{U}$ such that $\sigma \subset U$. As X is compact, the inductive limit can be taken over the coverings $\mathcal{B}_\varepsilon := \{B(x, \varepsilon)\}_{x \in X}$ ([Sp], §6.5). Now, if $\sigma = \{x_0, \ldots, x_q\} \in X_\varepsilon$ then σ is also a q-simplex of $X(\mathcal{B}_\varepsilon)$ since, for instance, $\sigma \subset B(x_0, \varepsilon)$. This gives a simplicial map $\alpha_\varepsilon : X_\varepsilon \longrightarrow X(\mathcal{B}_\varepsilon)$. In the other direction, if τ is a q-simplex of $X(\mathcal{B}_\varepsilon)$, then τ is a q-simplex of $X_{2\varepsilon}$ which provides us a simplicial map $\beta_\varepsilon : X(\mathcal{B}_\varepsilon) \longrightarrow X_{2\varepsilon}$. Passing to the inductive limits, these maps provides, for *compact* metric spaces, isomorphisms between the metric cohomology and the Alexander–Spanier cohomology theories. For general metric spaces, the two cohomologies are not isomorphic. For instance, $\overline{H}^0(\mathbb{R} - \{0\}; \mathcal{R}) \simeq \mathcal{R} \oplus \mathcal{R}$ while $\mathcal{H}^0(\mathbb{R} - \{0\}; \mathcal{R}) \simeq \mathcal{R}$ by (5.5) below.

Even for complete spaces, the metric cohomology might be different from the Cech comomology. For exemple, the space $X := \{(x, y) \in \mathbb{R}^2 \mid (xy)^2 = 1\}$. Then X has the same Cech cohomology as four points while it has the metric cohomology of a circle.

The next results about the metric cohomology are direct consequences of (2.2), (2.3) and (2.5).

(5.3) Proposition. *Let X be a pseudo-metric space admitting a crushing onto a subspace $A \subset X$. Then $\mathcal{H}^*(X, A; \mathcal{R}) = 0$.*

(5.4) Corollary. *Let X be a crushable pseudo-metric space. Then, $\mathcal{H}^*(X; \mathcal{R}) \simeq \mathcal{H}^*(\mathrm{pt}; \mathcal{R})$.*

(5.5) Theorem. *Let $j : X \longrightarrow \widehat{X}$ be the inclusion of a metric space into a metric complection $\widehat{X}$ of X. Then $j_* : \mathcal{H}^*(\widehat{X}; \mathcal{R}) \longrightarrow \mathcal{H}^*(X; \mathcal{R})$ is an isomorphism.*

(5.6) Example. Let $X = S^n - \{\mathrm{pt}\}$. By the above results, one has

$$\mathcal{H}^*(X; \mathbb{Z}) \xrightarrow{\simeq} \mathcal{H}^*(S^n; \mathbb{Z}) \xrightarrow{\simeq} \check{H}^*(S^n; \mathbb{Z})$$

Therefore, though that X is contractible, $\mathcal{H}^*(X; \mathbb{Z}) \not\simeq \mathcal{H}^*(\mathrm{pt}; \mathbb{Z})$. This shows that X is not crushable.

References

[Be] Begle, E.G., The Vietoris mapping theorem for bicompact spaces, *Annals of Math.* **51** (1950), 534–543.

[Do] Dowker, C.H., Homology groups of relations, *Annals of Math.* **56** (1952), 84–95.

[G-H] Ghys. E. – De la Harpe, P., *Sur les groupes hyperboliques d'après M. Gromov*, Birkhäuser, 1990.

[Gr]   Gromov, M., Hyperbolic groups, in *Essays in group theory*, S.M. Gersten (ed.), MSRI Publ. **8** (1987), 75–263.

[Le]   Lefschetz, S., Algebraic Topology, *AMS Coll. Publ.* **27** (1942).

[Mu]   Munkres, J. *Elements of algebraic topology*, Addison–Wesley, 1984.

[Sp]   Spanier, E., *Algebraic Topology*, McGraw–Hill, 1966.

[Vi]   Vietoris, L., Über die höheren Zusammenhang kompakter Räume und eine Klasse von zusammenhangstreuen Abbildungen, *Math. Annalen* **97** (1927), 454–472.

Every Proper Smooth Action of a Lie Group is Equivalent to a Real Analytic Action: A Contribution to Hilbert's Fifth Problem

Sören Illman

In this paper we prove that if a Lie group G acts on a smooth manifold M by a proper and smooth action, then there exists a real analytic structure β on M, compatible with the given smooth structure on M, such that the action of G on M_β is real analytic. In fact we prove a slightly more general result, namely the following.

Theorem A. *Let G be a Lie group and M a Cartan smooth G–manifold. Then there exists a real analytic structure β on M, compatible with the given smooth structure on M, such that the action of G on M_β is real analytic.*

We recall that an action of a Lie group G on a locally compact space X is said to be *proper* if $G_{[A]} = \{g \in G \mid gA \cap A \neq \emptyset\}$ is a compact subset of G for every compact subset A of X, and we say in this case that X is a *proper G–space*. (An equivalent condition is that the map $G \times X \to X \times X$, $(g, x) \mapsto (gx, x)$, is proper, in the sense that the inverse image of any compact set is compact.) This notion of a proper G–space is due to A. Borel. Observe that an action of a compact Lie group is always proper. We say that X is a *Cartan G–space*, and that the action of G on X is *Cartan*, if each point in X has a compact neighborhood A such that $G_{[A]}$ is compact. The term Cartan G–space was introduced by Palais in [29]. There exist smooth actions of Lie groups that are Cartan but not proper, see Section 1. Thus Theorem A actually proves a more general result than the one given in the title of the paper, but since proper actions play a very central role we have chosen to emphasize them. When G is a discrete group the notion of a proper action coincides with the classical notion of a *properly discontinuous* action. (The reader should however note that in the literature the meaning of the term "properly discontinuous" varies. There are also cases where this term is used for what we call a Cartan G–space. On the other hand some authors use the term "properly discontinuous" for a much more restricted class of actions, since they require in addition that a "properly discontinuous" action is a free action.)

We use the notion of a Lie group in the same sense as in Chevalley [6] and Helgason [11], i.e., a Lie group is a group G which is also a real analytic manifold and both the multiplication and the map taking an element to its inverse are real analytic maps. A Lie group G is always paracompact. In general we do not require a manifold M to be second countable, i.e., to have a countable base for its topology, or even to be paracompact. However, we wish to stress that this is not an essential point. The main interest of Theorem A lies of course in the case when the smooth manifold M is a "very ordinary" second countable smooth manifold. See Section 0 for a further discussion of these matters. By smooth we in this paper mean C^∞, but we could as well consider the C^r case, $r \geq 1$, and the main result still holds by essentially the same proof. If G is *compact* (and M is second countable) the result in Theorem A is known before, see Palais [30] and Matumoto–Shiota [23]. However, the technique of proof used in these papers does not apply when G is non–compact.

The importance of Theorem A lies in the fact that real analyticity is a more rigid notion than smooth, and Theorem A thus shows that there is more rigidity present in proper smooth actions of Lie groups than is evident at first hand. The earlier known case of Theorem A, where G is a compact Lie group, is for example used in Hambleton and Lee [10, Section 2]. Here the starting point is a smooth action of a finite group π on a 4–manifold X^4, and the fact that X^4 carries a real analytic structure invariant under the group action playes a crucial role for the ensuing arguments in that paper. Our result opens the way for similar applications in the case of proper, or Cartan, actions of non–compact Lie groups, and thus in particular in the case of properly discontinuous actions of discrete non–finite groups. We also mention here the general result in [14], which says that each proper real analytic paracompact G–manifold, where G is an arbitrary Lie group, has a decomposition into equivariant simplexes, i.e., into pieces that are simple basic objects. Moreover this can be done in such a way that the decomposition is subanalytic. By Theorem A the result in [14] now extends to hold also for proper *smooth* actions of Lie groups. This was in fact our own original motivation for establishing Theorem A.

However, there is also a historic perspective to Theorem A, namely Hilbert's fifth problem [12]. We quote from the discussion in Montgomery–Zippin [26, Section 2.15, p. 70].

> "Let us now consider the following questions, the second and third of which are asked by Hilbert:
>
> If a locally compact group G acts effectively on a manifold M (locally–euclidean space) then
>
> 1) is G necessarily locally euclidean,
>
> 2) if the group G is locally euclidean, is it a Lie group in some appropriate coordinates.
>
> 3) If G is a Lie group, can coordinates be chosen in G and M so that the transforming functions are analytic?"

Although question 1) is not asked by Hilbert, the affirmative answer to 1) is usually known as the Hilbert–Smith conjecture. As far as I know no complete solution of 1) has appeared in print.

Question 2) constitutes the part of Hilbert's fifth problem that deals with the group itself. The answer here is yes, and this answer was given by the contents of two papers, one by Gleason [8] and the other by Montgomery and Zippin [24].

Question 3) is then the part of Hilbert's fifth problem that deals with transformation groups. The question here is whether, given an action of G on M, one can find real analytic structures on G and M such that the action $G \times M \to M$ is real analytic. Since the real analytic structure on a Lie group G is unique, i.e., there exists exactly *one* real analytic structure on G which makes G into a Lie group (see e.g. [11, Theorem II.2.6]), the question here is reduced to finding a real analytic structure on M such that the action $G \times M \to M$ is real analytic. It is to this problem that we address ourselves in this paper. As we shall see below the answer to question 3) is in general *no*, and Theorem A shows when the answer to 3) is *yes*. Concerning 3) Montgomery and Zippin [26, p. 70] write. "The answer to 3) is no. For example a group of reals can act on $\mathbb{R}^2$ by having fixed $x^2 + y^2 \le 1$ and slowly rotating the rest of $\mathbb{R}^2$. This can not be analytic since if it were the existence of an open set of fixed points would imply that all points were fixed. The answer to 3) is no even if G is compact as was first shown by Bing [2] by an example of the cyclic group of order two acting on $\mathbb{R}^3$ in a way which could not be differentiable." In Montgomery–Zippin [25] the example of Bing is modified to give an example of an action of S^1 on $\mathbb{R}^4$, which cannot be equivalent to a smooth action. In both these examples [2] and [25], the action of the group $\mathbb{Z}_2$ and S^1, respectively, is not locally smooth, in the sense of [5, IV.1]. It is interesting to note the remarkable fact that there exists a compact smoothable manifold M^{12} which admits a *locally smooth* effective action of S^1, but which admits no non–trivial smooth action of S^1 in any differentiable structure, see Bredon [5, Corollary VI.9.6]. For the proof of this fact one can use the powerful theorem of Atiyah and Hirzebruch [1], which says that if a compact connected oriented smooth manifold M, of dimension $4m$ and with $w_2(M) = 0$, admits a non–trivial smooth action of S^1, then $\hat{A}(M) = 0$. Thus we see that the answer to 3) is clearly no in the case of topological continuous actions. One may in fact point out that the answer to 3) is no even for the trivial group $G = \{e\}$, since there exist topological manifolds that do not have any smooth, and hence also no real analytic structure, see Kervaire [16].

One should observe that the example of the group of reals $\mathbb{R}$ acting on $\mathbb{R}^2$, which Montgomery and Zippin [26, p. 70] describe, can be chosen such that it is a smooth action, i.e., a C^∞–action. Let for example $\alpha : [0, \infty) \to [0, \infty)$ be a C^∞–function such that $\alpha(x) = 0$, for all $x \in [0, 1]$ and $\alpha(x) = 1$, for all $x \in [2, \infty)$. Then the action $\mathbb{R} \times \mathbb{R}^2 \to \mathbb{R}^2$, $(t, re^{i\varphi}) \mapsto re^{i(\alpha(r)t + \varphi)}$ is a smooth action of the group of reals $\mathbb{R}$ on $\mathbb{R}^2$ that cannot be analytic. Note that this action is not a proper, or even a Cartan, action. Montgomery and Zippin [26, p. 71] conclude their discussion of 3) by, "As mentioned, 3) is true in some cases and it remains to find out when it is true or in general whether the truth resembles the differentiable case in some way". The present paper does exactly this. By Theorem A every Cartan smooth action is equivalent to a real analytic action, and the example above shows that there exist non–Cartan smooth actions which are not equivalent to real analytic actions. Thus we have a complete answer to question 3), and this settles the remaining open part of Hilbert's fifth problem concerning Lie groups of transformations.

We shall now review some earlier results that are relevant to our work. As is well known Whitney proved in [32, Theorem 1] that every second countable, m–dimensional, smooth manifold M is smoothly diffeomorphic to a real analytic manifold, by proving that M is smoothly diffeomorphic to a real analytic submanifold of $\mathbb{R}^{2m+1}$. It is also known that if two second countable real analytic manifolds are smoothly diffeomorphic then they are also real analytically isomorphic. The proof of this fact requires the use of the Grauert–Morrey imbedding theorem [9] and [27], which says that every second countable real analytic manifold has a real analytic imbedding in some euclidean space. (Morrey [27] proved this for compact manifolds, and Grauert [9] in the general case.) In fact it is known that if M and N are second countable real analytic manifolds then $C^\omega(M, N)$ is dense in $C^\infty(M, N)$, see e.g. [13, Section 2.5]. (Here the function spaces $C^\infty(M, N)$ and $C^\omega(M, N)$ have the strong topology, i.e., the Whitney topology.)

The fact that all smooth manifolds, also the ones that are not second countable, can be given a compatible real analytic structure was proved by W. Koch and D. Puppe [19, Theorem 1]. The global part of their proof consists of a maximality argument using Zorn's lemma. (This technique of proof is then also used e.g. in the book [13] by M.W. Hirsch.) We will use a similar maximality argument involving the use of Zorn's lemma in the global part of the proof of Theorem A, but in our case we are led to such a proof for a completely different reason than the one here, see below for a discussion of this matter. A forerunner to the maximality argument introduced by W. Koch and D. Puppe in [19] can be found in H. Kneser [17], where transfinite induction is used to prove that the Alexandroff long line L has a real analytic structure. There exist non–isomorphic real analytic structures on the Alexandroff long line L, which are smoothly diffeomorphic, see H. Kneser and M. Kneser [18].

In the case of smooth manifolds with symmetries, i.e., with group actions, there are earlier known results about raising the differentiability only in the case when the group of symmetries is a compact Lie group. Palais proves in [30, Theorem B] that if K is a compact Lie group, then every compact C^r K–manifold, $1 \leq r < \infty$, is K–equivariantly C^r diffeomorphic to a C^∞ K–manifold. His technique of proof is an equivariant extension of the basic method of Whitney. Matumoto and Shiota [23] use essentially the same method as Palais and their Theorem 1.3 gives the result that if K is a compact Lie group, then every second countable C^r K–manifold, $1 \leq r \leq \infty$, is K–equivariantly C^r diffeomorphic to a real analytic K–manifold. Matumoto and Shiota also prove that if M and N are two second countable real analytic K–manifolds, where K is a compact Lie group, and the number of K–isotropy types occurring in

M and N is finite, then M and N are K–equivariantly and real analytically isomorphic if they are K–equivariantly C^1 diffeomorphic, see [23, Theorem 1.2].

The following question is an open problem. Let G be a Lie group, and suppose that M and N are proper real analytic, second countable, G–manifolds which are smoothly equivalent, i.e., there exists a G–equivariant smooth diffeomorphism $f : M \to N$. Does there then exist a G–equivariant real analytic isomorphism $h : M \to N$? As we noted above the answer is yes if G is compact and one in addition assumes that the number of isotropy types in M is finite [23], but the technique used in [23] cannot be used when G is non–compact, and the question seems to be a very intriguing one for non–compact Lie groups. We prove in [15] that if M and N are proper real analytic second countable G–manifolds, and there exists a G–equivariant smooth diffeomorphism $f : M \to N$, then there exists a G–equivariant *subanalytic isomorphism* $h : M \to N$. Already this uniqueness result has interesting applications. For example, it implies that also the uniquenss part of the equivariant triangulation result in [14] extends to hold in the case of proper *smooth* actions of Lie groups.

Concerning the proof of Theorem A we remark the following. Suppose that M is a Cartan, or let us even say a proper, smooth G–manifold, and also assume that M is second countable. Then one cannot in general imbed M as a G–invariant subset in some finite–dimensional linear representation space for G. This is because the Lie group G need not be a linear group. As is well known every compact Lie group is a linear group. (The first example of a connected Lie group that is not a linear group was given by Birkhoff [3] in 1936, up to then it had been a conjecture that every connected Lie group is isomorphic to a group of matrices.) Thus it is not possible to use an equivariant form of Whitney's method for the proof of Theorem A. We use a maximality argument involving the use of Zorn's lemma, for the global part of the proof of Theorem A. This argument is then analogous to the one used by W. Koch and D. Puppe [19], in a non–equivariant situation. For the actual technical main arguments in proof of Theorem A the following three facts are crucial. First, the result that in a paracompact Cartan smooth G–manifold there exists a smooth slice at each point. This result is due to Palais [29, Proposition 2.2.2], a related weaker result is due to Koszul [20, p. 139], see Section 5. Secondly, a key result on approximations of smooth slices, which we prove in Section 6. Thirdly, the result of T. Matumoto and M. Shiota [23, Theorem 1.2], which says that if K is a compact Lie group and M and N are real analytic K-manifolds, such that the number of K–isotropy types occurring in N is finite, then every K–equivariant smooth map $f : M \to N$ can be approximated arbitrarily well, in the Whitney topology, by a K–equivariant real analytic map $h : M \to N$.

A quick review of the contents of this paper is as follows. Section 0 is concerned with preliminaries, and here we fix our terminology and notation. In Section 1 we consider proper G–spaces and Cartan G–spaces. Here we also prove that in a Cartan G–manifold M, which is not assumed to be paracompact, each point has a paracompact G–invariant open neighborhood. This then shows that locally we are reduced to a situation where, for example, the smooth slice theorem is known to hold. Sections 2–5 give background information and preliminary results needed for the proof of Theorem A. Section 2 is concerned with some results involving the Whitney topology, of these Lemma 2.2 plays an important role in the proof of Lemma 6.1, which is the key result on approximations of smooth slices mentioned before. In Section 3 we consider equivariant real analytic local cross sections in homogenous spaces. In Section 4 we discuss smoothness and real analyticity in twisted products. Section 5 treats slices and near slices and their relation. Here we also define the notion of a K–invariant product neighborhood of a smooth K–slice (K denotes a compact Lie group), which is used in Lemma 6.1. Section 6 is devoted entirely to the proof of Lemma 6.1, and the actual proof of Theorem A is carried out in Section 7.

We announced Theorem A in the preprint "Every smooth properly discontinuous action is equivalent to a real analytic one" [Mathematical Sciences Research Institute, Berkeley, Preprint Nr. 090–92, August 1992], where we also gave a proof in the case of properly discontinuous actions of discrete groups. A preliminary version of the present paper appeared as Preprint, Max–Plank–Institute für Mathematik, Bonn, Nr. 93–3. Some further work on the paper was carried out during my visit to the Forschungsinstitut für Mathematik at ETH in Zürich, in 1993. I wish to thank these research institutes for their kind support.

0   Preliminaries, notation and terminology

By an m–dimensional manifold we mean a Hausdorff topological space M such that each point in M has an open neighborhood which is homeomorphic to an open subset of $\mathbb{R}^m$. We recall here the well–known fact that a manifold M is paracompact if and only if every connected component of M is second countable, i.e., has a countable base for its topology, and this is also equivalent to the condition that every connected component of M is a countable union of compact sets, see Dieudonné [7, Théorème 3 and Corollaire] and Bourbaki [4, Theorem 5 in Chapter I, §9, no. 10].

As we already mentioned in the introduction we use the term *smooth* to mean C^∞ differentiable. A real analytic [smooth] structure on M is, by definition, a maximal real analytic [smooth] atlas on M. A real analytic [smooth] manifold is then a manifold M together with a real analytic [smooth] structure β, i.e., a pair (M, β), for which we also often use the notation M_β. If β' is a real analytic [smooth] atlas on M, we denote by $\langle \beta' \rangle_a$ [$\langle \beta' \rangle_s$] the maximal real analytic [smooth] atlas, i.e., the real analytic [smooth] structure, on M determined by β'. Observe that if β is a real analytic structure on M, then β is in particular a smooth atlas on M and $\alpha = \langle \beta \rangle_s$ is the the smooth structure, on M determined by β. We say that a real analytic atlas β' on M is *compatible* with a smooth structure α on M if $\beta' \subset \alpha$. If this is the case, then also $\beta \subset \alpha$, where $\beta = \langle \beta' \rangle_a$ denotes the real analytic structure on M determined by β'. Note that a real analytic structure β is compatible with a smooth structure α on M if and only if $\langle \beta \rangle_s = \alpha$.

Suppose M is a manifold and let $f : M \to N_\gamma$ be a homeomorphism, where N_γ is a real analytic manifold with real analytic structure $\gamma = \{(V_i, \psi_i)\}_{i \in \Lambda}$. Then $f^* \gamma = \{(U_i, \varphi_i)\}_{i \in \Lambda}$, where $U_i = f^{-1}(V_i)$ and $\varphi_i = \psi_i \circ (f|U_i)$, $i \in \Lambda$, is a real analytic structure on M. We call $\beta = f^* \gamma$ the real analytic structure *induced* from N_γ through f. The map $f : M_\beta \to N_\gamma$ is then a real analytic isomorphism, and the real analytic structure β on M is completely determined by this fact. If M is a smooth manifold and f is a smooth diffeomorphism, then it follows that the induced real analytic structure $\beta = f^* \gamma$ is compatible with the given smooth structure on M. We also say that a smooth manifold $M = M_\alpha$ is *equivalent* to a real analytic manifold if there exists a real analytic manifold $N = N_\gamma$ and a smooth diffeomorphism $f : M \to N$. If this is the case, then $\beta = f^* \gamma$ is a real analytic structure on M and β is compatible with the given smooth structure α on M. On the other hand if β' is a real analytic structure on M, which is compatible with the given smooth structure α on M, then id : $M_\alpha \to M_{\beta'}$ is a smooth diffeomorphism, and hence the smooth manifold $M = M_\alpha$ is equivalent to a real analytic manifold. Thus we see that for a given smooth manifold $M = M_\alpha$ the problem of showing that M is *equivalent* to a real analytic manifold and the problem of finding a real analytic strucutre on M, which is *compatible* with the given smooth structure, are the same problems. The above defined notion of *equivalent* arises in the method of Whitney [32], whereas the notion of a *compatible* real analytic structure is naturally connected with the method of W. Koch and D. Puppe [19].

A Lie group is a topological group G which at the same time is a real analytic manifold such that the multiplication $G \times G \to G$, $(g_1, g_2) \mapsto g_1 g_2$, and the map $\iota : G \to G$, $g \mapsto g^{-1}$, are real analytic, see Helgason [11]. This definition is also equivalent to the one in Chevalley [6]. Some authors define a Lie group by requiring that the multiplication and the map $\iota : G \to G$ are smooth. As is well known this definition is equivalent to the one we have given above, see e.g. Pontryagin [31, Chapter 10]. A Lie group G is always paracompact; a standard argument shows that the connected component G_0 of G containing the identity element $e \in G$ is a second countable manifold, see e.g. Helgason [11, p. 88].

By a G–space X we mean a locally compact space X together with a continuous action of G on X. If X and Y are two G–spaces we say that a map $f : X \to Y$ is G–equivariant if $f(gx) = gf(x)$, for all $g \in G$ and $x \in X$. By a G–isotropy type $[H]$ we mean the conjugacy class in G of a closed subgroup H of G. In this paper we only use this notion in the case when H is compact subgroup of G. We say that a G–isotropy type $[H]$ occurs in X if there exists $x \in X$ such that $[G_x] = [H]$, where G_x denotes the isotropy subgroup of G at x. A smooth [real analytic] G–manifold M is a smooth [real analytic] manifold M together with a smooth [real analytic] action $\varphi : G \times M \to M$, $(g, x) \mapsto gx$, of G on M. We say that a smooth G–manifold M is equivalent to a real analytic G–manifold if there exists a real analytic G–manifold N and a G–equivariant smooth diffeomorphism $f : M \to N$. In complete analogy with the non–equivariant case discussed above we then have that a smooth G–manifold M is equivalent to a real analytic G–manifold if and only if there exists a real analytic structure β on M, compatible with the smooth structure on M, such that the action of G on M_β is real analytic. In this terminology our main result, Theorem A, says that every Cartan smooth G–manifold is equivalent to a real analytic G–manifold.

Let K be a compact Lie group and $\rho : K \to GL(n, \mathbb{R})$ a linear representation of K, i.e., a continuous homomorphism from K into the general linear group $GL(n, \mathbb{R})$. By $\mathbb{R}^n(\rho)$ we denote the corresponding linear representation space for K. That is $\mathbb{R}^n(\rho)$ denotes euclidean space $\mathbb{R}^n$ on which K acts by $kx = \rho(k)x$, all $k \in K$ and $x \in \mathbb{R}^n$. Observe that this action of K on $\mathbb{R}^n(\rho)$ is real analytic. This follows since, by a fundamental theorem for Lie groups (see e.g. [11, Theorem II.2.6]), $\rho : K \to GL(n, \mathbb{R})$ is real analytic and hence the action $K \times \mathbb{R}^n(\rho) \to \mathbb{R}^n(\rho)$, $(k, x) \mapsto \rho(k)x$, is real analytic. Since every linear representation of a compact Lie group K is equivalent to an orthogonal representation, we may as well assume that $\rho : K \to O(n)$. It is a well–known, and easily proved, fact that the number of isotropy types occurring in any linear representation space $\mathbb{R}^n(\rho)$ is finite.

1   Cartan and proper G–spaces

Let G be a Lie group acting on a locally compact space X. Recall that the action is said to be *proper* if the set

$$G_{[A]} = \{g \in A \mid gA \cap A \neq \emptyset\}$$

is a compact subset of G for every compact subset A of X. This is also equivalent to the fact that the map $G \times X \to X \times X$, $(g, x) \mapsto (gx, x)$, is proper. In case G is *discrete* the notion of a proper action is the same thing as the classical notion of a *properly discontinuous* action.

A slightly more general notion than the notion of a proper G–space is the notion of a *Cartan G–space*. We say that a locally compact space X on which G acts is a Cartan G–space if each point x in X has a compact neighborhood A such that $G_{[A]}$ is a compact subset of G. If X is a Cartan G–space it follows in particular that each isotropy subgroup G_x, $x \in X$, is a compact subgroup of G. Furthermore each orbit Gx is a closed subset of X and the map $\alpha : G/G_x \to Gx$, $gG_x \mapsto gx$, is a homeomorphism, see [29, Propositions 1.1.4 and 1.1.5]. Every point x in a Cartan G–space X is contained in a G–invariant open neighborhood W which is a proper G–space. In fact if U is an open neighborhood of x in X such that $A = \overline{U}$ is compact and $G_{[A]}$ is a compact subset of G, then the action of G on $W = GU$ is proper, see [29, Proposition 1.2.4]. The precise relation between proper G–spaces and Cartan G–spaces is given by the following.

Proposition 1.1. *If X is a locally compact G-space then the following are equivalent.*

1. X *is a Cartan G-space and X/G is Hausdorff.*

2. X *is a proper G-space.*

Proof. See Palais [29, Theorem 1.2.9]. □

A good example of a Cartan G-space X, which is not a proper G-space is the following one, cf. Koszul [21, Section I.1].

Example. Define an action of $\mathbb{R}$ on $\mathbb{R}^2 - \{0\}$ by, $(t, (x, y)) \mapsto (e^t x, e^{-t} y)$, for all $t \in \mathbb{R}$ and $(x, y) \in \mathbb{R}^2 - \{0\}$. This action is Cartan but not proper. The orbit space of this action is not Hausdorff, for example the orbits of the points $(1, 0)$ and $(0, 1)$ are two different points in the orbit space which do not have disjoint neighborhoods. If one restricts this action to the integers $\mathbb{Z}$ one obtains an action of $\mathbb{Z}$ on $\mathbb{R}^2 - \{0\}$, which is Cartan but not proper.

Let K be a closed subgroup of G. Then the homogeneous space G/K, with the natural action of G given by $G \times G/K \to G/K$, $(g', gK) \mapsto g'gK$, is a proper G-space if and only K is compact. In fact we have that if $B \subset G/K$, then

$$G_{[B]} = \pi^{-1}(B)(\pi^{-1}(B))^{-1}, \tag{1}$$

where $\pi : G \to G/K$ denotes the natural projection. It is straightforward to verify (1). The isotropy subgroup at eK equals K, so in order for G/K to be a proper G-space, or even a Cartan G-space, it is necessary that K is compact. Now assume that K is compact and let B be a compact subset of G/K. Then $\pi^{-1}(B)$ is compact and hence it follows by (1) that $G_{[B]}$ is compact.

We also note that if $A \subset G$, and we let G act on G by multiplication from the left, then one obtains directly that

$$G_{[A]} = AA^{-1}. \tag{2}$$

One can then also observe that (1) follows from (2) by applying Lemma 1.2 below to the natural projection $\pi : G \to G/K$, which is a G-equivariant map.

Lemma 1.2. *Let X and Y be G-spaces and $f : X \to Y$ a surjective G-map. If $B \subset Y$ we have that*

$$G_{[B]} = G_{[f^{-1}(B)]}.$$

Proof. Direct verification. □

We shall prove in Proposition 1.3 that in a Cartan G-manifold M every point has a paracompact G-invariant open neighborhood, and thus locally M is a paracompact Cartan G-manifold. (In fact M is locally a paracompact proper

G–manifold, see the comment after the proof of Proposition 1.3.) Let us first introduce a notation and make a couple of observations. If A and B are subsets of a G–space X, we denote

$$G_{[A,B]} = \{g \in G \mid gA \cap B \neq \emptyset\}.$$

Thus $G_{[A]} = G_{[A,A]}$. Let J be a subset of G, then we have by direct verification that

$$G_{[A,JB]} = JG_{[A,B]}. \tag{3}$$

If A and B are connected, then the set

$$B \cup G_{[A,B]}A = B \cup \bigcup_{g \in G_{[A,B]}} gA \tag{4}$$

is connected since B is connected and each gA is connected, and $gA \cap B \neq \emptyset$ for all $g \in G_{[A,B]}$.

Proposition 1.3. *Let G be a Lie group and M a Cartan G–manifold, and let $x \in M$. Then there exists a paracompact G–invariant open neighborhood W of x in M.*

Proof. We let U be an open neighborhood of x in M such that U is homeomorphic to $\mathbb{R}^n$ and $A = \overline{U}$ is compact, and $G_{[A]}$ is a compact subset of G. Then U is in particular connected and second countable. We claim that $W = GU$ is paracompact. It is enough to prove that each connected component of W is second countable, cf. Section 0. Let V be the connected component of W which contains U. It is easy to see that each component of W is of the form gV, for some $g \in G$. Hence we only need to prove that V is second countable.

Let us denote $J = G_{[U]}$. We claim that the sets

$$J^n U, \quad n \geq 0 \tag{5}$$

are connected, where $J^0 = \{e\}$. By repeated use of (3) we have that

$$J^{n+1} = G_{[U,J^nU]}, \quad \text{for all } n \geq 0.$$

Assume by induction that $J^n U$ is connected. Since $e \in J$ it follows that $J^n U \subset J^{n+1}U$, and hence

$$J^{n+1}U = J^n U \cup J^{n+1}U = J^n U \cup G_{[U,J^nU]}U,$$

which is connected by (4). This proves that the sets in (5) are connected, and thus

$$V' = \bigcup_{n \geq 0} J^n U$$

is on open and connected subset of W. We claim that V' is closed in $W = GU$. Let $x \in W \cap \overline{V'}$. Then $x = gu$, where $g \in G$ and $u \in U$. Since gU is an open neighborhood of x in W we have that $gU \cap V' \neq \emptyset$. Thus $gU \cap J^n U \neq \emptyset$, for some $n \geq 0$, and hence $g \in G_{[U, J^n U]} = J^{n+1}$. Therefore $x \in gU \subset J^{n+1}U \subset V'$, and we have showed that V' is closed in W. Since V' is an open and closed connected set in W it follows that V' is a connected component of W. Since $U \subset V'$, it follows that $V = V'$.

We can now deduce that V is second countable in the following way. Since $G_{[U]} \subset G_{[A]}$ and $G_{[A]}$ is compact it follows that $J = G_{[U]}$ lies in a subset P_0 of G which is a finite union of components of G. It then follows that $J^* = \bigcup_{n \geq 0} J^n$ lies in a subset P of G which is the union of at most a countable number of components of G, and hence P is second countable. We have that $V = V' = J^*U \subset PU$. The map $G \times U \to GU = W$, $(g, u) \mapsto gu$, is an open map onto W and hence $P \times U \to PU$, $(g, u) \mapsto gu$, is an open map onto PU. Since P and U are second countable spaces it follows that PU is second countable, and since $V \subset PU$ it follows that V is second countable. $\qquad \square$

Since the open neighborhood U of x in M is such that $A = \overline{U}$ is compact and $G_{[A]}$ is a compact subset of G it follows, as we already remarked in the beginning of this section, that the action of G on $W = GU$ is proper.

In connection with Proposition 1.3 we may add the following two remarks.

Remark 1. Suppose that X is a second-countable Cartan G-space. We claim that G can have at most a countable number of connected components, i.e., that G is second-countable. This is seen as follows. Let $x \in X$, then $\alpha : G/G_x \to Gx$, $gG_x \mapsto gx$, is a homeomorphism as we already noted in the beginning of this section. Since X is second-countable it follows that Gx is second-countable and hence G/G_x is second-countable. Thus G/G_x has at most a countable number of connected components, and since G_x is compact it follows from this that G has at most a countable number of connected components.

Remark 2. Suppose that M is a paracompact Cartan G-manifold, and let M_1 be a connected component of M. Then M_1 is a second-countable manifold. Let us denote $G_1 = \{g \in G \mid gM_1 = M_1\}$. Then G_1 is a subgroup of G and $G_0 \subset G_1$, where G_0 denotes the connected component of G containing the identity element $e \in G$, and moreover G_1 is a union of connected components of G. It follows by Remark 1 that G_1 has at most a countable number of connected components. Furthermore the action of G on GM_1 is completely determined by the action of G_1 on M_1, and the set of components of GM_1 is in a natural one-to-one correspondence with G/G_1.

2 Whitney topology

In this section all manifolds, with or without boundary, are assumed to be second countable.

Let M and N be smooth manifolds, with or without boundary. By $C^\infty(M, N)$ we denote the set of all smooth maps from M to N, and we give $C^\infty(M, N)$ the strong topology, i.e., the Whitney topology, see Mather [22, Section 2] and Hirsch [13, Sections 2.1 and 2.4]. The definition of the Whitney topology in [13, Section 2.1] is via charts on M and N, and in [22, Section 2] and [13, Section 2.4] the Whitney topology is defined using the notion of jets, and as noted in [13, Section 2.4] these definitions are equivalent. *In this paper $C^\infty(M, N)$ and other related mapping spaces will always be assumed to carry the Whitney topology.* If M and N are real analytic manifolds, then $C^\omega(M, N)$ denotes the set of all real analytic maps from M to N, and $C^\omega(M, N)$ has the induced topology, i.e., the relative topology from $C^\infty(M, N)$.

Let K be a Lie group, and suppose that M and N are smooth [real analytic] K–manifolds, with or without boundary. We should here point out that we will need the results given in the section only in the case when K is a compact Lie group (a compact subgroup of the Lie group G), but most of the results below hold equally well when K is a Lie group, and we present these results in this generality. We denote by $C_K^\infty(M, N)$ $[C_K^\omega(M, N)]$ the set of all K–equivariant smooth [real analytic] maps from M to N, and we give $C_K^\infty(M, N)$ $[C_K^\omega(M, N)]$ the induced topology, i.e., the relative topology from $C^\infty(M, N)$ $[C^\omega(M, N)]$. We also note here that if $h : N \to P$ is a K–equivariant smooth map, where P is a smooth K–manifold, then the induced map $h_* : C_K^\infty(M, N) \to C_K^\infty(M, P)$, $f \mapsto h \circ f$, is continuous. This follows directly from the fact that the corresponding statement holds in the non–equivariant case, see Mather [22, Proposition 2 in §2].

As we already mentioned in the introduction, it is a well-known and deep result that if M and N are real analytic manifolds then $C^\omega(M, N)$ is dense in $C^\infty(M, N)$, see e.g. [13, Theorem 2.5.1]. We shall use the following equivariant version, due to T. Matumoto and M. Shiota [23, Theorem 1.2], of this result.

Theorem 2.1. *Let K be a compact Lie group and let M and N be real analytic K–manifolds, and assume that the number of K–isotropy types occurring in N is finite. Then $C_K^\omega(M, N)$ is dense in $C_K^\infty(M, N)$.*

In the remaining part of this section, i.e., in the results 2.2.–2.5., K denotes a Lie group, but as we already remarked above we will later in the paper use these results only in the case when K is a compact Lie group.

Lemma 2.2. *Let M and N be smooth K–manifolds and let D be a compact smooth K–manifold possibly with boundary. Then the map*

$$\chi : C_K^\infty(M, N) \to C_K^\infty(D \times M, D \times N), \quad f \mapsto \mathrm{id} \times f,$$

is continuous.

Proof. First of all we note that it is enough to prove our claim in the case when $K = \{e\}$. Furthermore it is known that the natural bijection

$$\iota : C^\infty(D \times M, D \times N) \to C^\infty(D \times M, D) \times C^\infty(D \times M, N), \quad h \mapsto (q_1 \circ h, q_2 \circ h),$$

is a homeomorphism, see Mather [22, Proposition 3 of §2]. (Here $q_1: D \times N \to D$ and $q_2: D \times N \to N$ denote the projections onto the first and second factor, respectively). Therefore we only need to prove that the maps

$$C^\infty(M, N) \to C^\infty(D \times M, D), \quad f \mapsto q_1 \circ (\mathrm{id} \times f) \tag{1}$$

and

$$C^\infty(M, N) \to C^\infty(D \times M, N), \quad f \mapsto q_2 \circ (\mathrm{id} \times f) \tag{2}$$

are continuous. The first one equals the constant map from $C^\infty(M, N)$ onto the element $p_1 \in C^\infty(D \times M, D)$, where $p_1 : D \times M \to D$ denotes the projection, and hence (1) is continuous. Observe that $q_2 \circ (\mathrm{id} \times f) = f \circ p_2$, where $p_2 : D \times M \to M$ denotes the projection. Therefore the map in (2) equals the map $p_2^* : C^\infty(M, N) \to C^\infty(D \times M, N)$, $f \mapsto f \circ p_2$, which is continuous since D is compact, see Mather [22, Proposition 1 of §2].   □

Lemma 2.3. *Let M and N be smooth K-manifolds and let $f : M \to N$ be a K-equivariant smooth map. Let V be a K-invariant open subset of M and let P be a K-invariant smooth submanifold of N such that $f(V) \subset P$. Then there exists an open neighborhood $\mathcal{N}$ of $f|V$ in $C_K^\infty(V, P)$ such that the following holds: If $h \in \mathcal{N}$, and we let*

$$E(h) : M \to N$$

be the extension of $h : V \to P \subset N$ defined by

$$E(h)(x) = \begin{cases} h(x), & x \in V \\ f(x), & x \in M - V, \end{cases}$$

then $E(h)$ is a K-equivariant smooth map. Furthermore the map

$$E : \mathcal{N} \to C_K^\infty(M, N), \quad h \mapsto E(h)$$

is continuous.

Proof. First we note that it is enough to prove the lemma in the case when $K = \{e\}$. This observation follows from the fact that the formula defining $E(h)$ is such that if h is a K-equivariant map then $E(h)$ is also a K-equivariant map. More precisely; let us assume that Lemma 2.3 holds in the non-equivariant case and let $\mathcal{N}^*$ be a neighborhood of $f| : V \to P$ in $C^\infty(V, P)$ such that if $h \in \mathcal{N}^*$ then $E(h)$ is a smooth map and $E : \mathcal{N}^* \to C^\infty(M, N)$ is continuous. Then $\mathcal{N} = \mathcal{N}^* \cap C_K^\infty(V, P)$ is an open neighborhood of $f| : V \to P$ in $C_K^\infty(V, P)$ and if $h \in \mathcal{N}$ then $E(h)$ is a K-equivariant and smooth map, and furthermore $E : \mathcal{N} \to C_K^\infty(M, N)$ is continuous since $C_K^\infty(V, P)$, by definition, has the relative topology from $C^\infty(V, P)$. Now we claim that the non-equivariant version of

Lemma 2.3 expresses a well-known fact. One can for example, by Whitney's imbedding theorem, imbed N smoothly into some euclidean space $\mathbb{R}^q$, and thus we see that it is enough to prove the case where N is a euclidean space $\mathbb{R}^q$ and P is a smooth submanifold of $\mathbb{R}^q$. A proof of the non-equivariant version of Lemma 2.3 in this case is given in [13, Lemma 2.2.8].   □

Theorem 2.4. *Let M and N be smooth K–manifolds possibly with boundary. Then the set $\mathrm{Imb}_K^\infty(M, N)$ of K–equivariant smooth imbeddings of M into N is open in $C_K^\infty(M, N)$.*

Proof. It is well-known that the set $\mathrm{Imb}^\infty(M, N)$ of smooth imbeddings of M into N is open in $C^\infty(M, N)$, see e.g. [13, Theorem 2.1.4]. Hence $\mathrm{Imb}_K^\infty(M, N) = \mathrm{Imb}^\infty(M, N) \cap C_K^\infty(M, N)$ is open in $C_K^\infty(M, N)$.   □

Theorem 2.5. *Let N be a smooth K–manifold, without boundary. Then the set $\mathrm{Diff}_K^\infty(N, N)$ of K–equivariant diffeomorphisms from N to N is open in $C_K^\infty(N, N)$.*

Proof. It is well-known that the set $\mathrm{Diff}^\infty(N, N)$ of diffeomorphisms from N to N is open in $C^\infty(N, N)$, see e.g. [13, Theorem 2.1.6]. Thus $\mathrm{Diff}_K^\infty(N, N) = \mathrm{Diff}^\infty(N, N) \cap C_K^\infty(N, N)$ is open in $C_K^\infty(N, N)$.   □

3   Homogeneous spaces

Let G be a Lie group and K a closed subgroup of G. Then the homogeneous space G/K of left cosets is a real analytic manifold. In fact there is a unique real analytic structure on G/K for which the natural action of G on G/K, i.e., the action $G \times G/K \to G/K$, $(g', gK) \mapsto g'gK$, is real analytic, see [11, Theorem II.4.2]. We will always consider G/K as a real analytic manifold with this real analytic structure. The natural projection $\pi : G \to G/K$ is a real analytic map. Furthermore each point gK in G/K has an open neighborhood U such that there exists a real analytic cross section $\sigma : U \to G$ over U of the projection $\pi : G \to G/K$, see e.g. [6, §V in Chapter IV] or [11, Lemma II.4.1].

Now assume that K is a *compact* subgroup of G. First we note that in this case there exist arbitrarily small K–invariant open neighborhoods of eK in G/K. For if U_0 is an open neighborhood of eK in G/K, then $A = G/K - U_0$ is closed in G/K and since K is compact it follows that KA is closed in G/K, cf. [5, Corollary I.1.3]. Thus $U = (G/K) - KA$ is a K–invariant open neighborhood of eK in G/K, and $U \subset U_0$.

Furthermore the fact that K is a compact subgroup of G implies that we can find a K–invariant open neighborhood U of eK in G/K and a real analytic cross section

$$\sigma : U \to G, \tag{1}$$

which is a K–equivariant map in the sense that

$$\sigma(ku) = k\sigma(u)k^{-1}, \quad \text{for every} \ \ k \in K \ \ \text{and} \ \ u \in U, \tag{2}$$

and moreover $\sigma(eK) = e$. Condition (2) means that σ is a K–equivariant map when the action of K on G is given by conjugation, i.e., by $K \times G \to G$, $(k, g) \mapsto kgk^{-1}$.

It is a well–known fact that such a K–equivariant real analytic cross section σ exists, and the existence of σ can be deduced in the following way. The natural projection $\pi : G \to G/K$ is a real analytic submersion and π is a K–equivariant map, where K acts on G by conjugation and on G/K by the action induced from the natural action of G on G/K. Observe that $e \in G$ and $eK \in G/K$ are fixed points of K. Hence the tangent spaces $T_e(G)$ and $T_{eK}(G/K)$ at $e \in G$ and $eK \in G/K$, respectively, are finite–dimensional linear representation spaces for K and we may assume that they are orthogonal representation spaces since K is compact. The differential $d\pi_e : T_e(G) \to T_{eK}(G/K)$, of π at $e \in G$, is a K–equivariant surjective linear map.

The tangent space $T_e(K)$ to K at $e \in K$ is a K–invariant linear subspace of $T_e(G)$. Let L be the orthogonal complement to $T_e(K)$ in $T_e(G)$. Then L is a K–invariant linear subspace of $T_e(G)$ and $T_e(G) = T_e(K) \oplus L$. Furthermore $\dim L = q = \dim G - \dim K = \dim G/K$, and $d\pi_e| : L \to T_e(G/K)$ is a K–equivariant linear isomorphism. Using the exponential map at $e \in G$ one constructs a q–dimensional K–invariant real analytic submanifold V^* of G, such that $e \in V^*$ and $T_e(V^*) = L$. Then $\pi| : V^* \to G/K$ is a K–equivariant real analytic map and the differential $d(\pi|V^*)_e : T_e(V^*) \to T_e(G/K)$ is a K–equivariant linear isomorphism. It now follows, using the real analytic inverse function theorem (see e.g. [28, 2.2.10]), that there exists a K–invariant open neighborhood V of e in V^* such that $\pi| : V \to \pi(V)$ is a K–equivariant real analytic isomorphism onto a K–invariant open neighborhood $U = \pi(V)$ of eK in G/K. Then

$$\sigma = (\pi|V)^{-1} : U \to V \hookrightarrow G$$

is a K–equivariant real analytic cross section over U of the projection $\pi : G \to G/K$, and $\sigma(eK) = e$.

We also note that by using the exponential map at $eK \in G/K$ one can obtain arbitrarily small K–invariant open neighborhoods of eK in G/K, which are K–equivariantly and real analytically isomorphic to an open disk, of some small radius, in an orthogonal representation space $\mathbb{R}^q(\tau) \cong T_{eK}(G/K)$ for K. Here $\tau : K \to O(q)$ denotes an orthogonal representation of K and $\mathbb{R}^q(\tau)$ denotes the corresponding orthogonal representation space for K. Thus it follows that we can always choose the K–invariant open neighborhood U in (1) such that there exists a K–equivariant real analytic isomorphism

$$h : \mathbb{R}^q(\tau) \to U. \tag{3}$$

In particular we have in this case that the number of K–isotropy types occurring in U is finite.

The following elementary lemma will be used in the proof of Lemma 6.1.

Lemma 3.1. *Let K be a compact subgroup of G and V_1 an open neighborhood of K in G. Then there exists a K–invariant open neighborhood U of eK in G/K such that* $G_{[U]} = \pi^{-1}(U)(\pi^{-1}(U))^{-1} \subset V_1$.

Proof. Consider the map $\nu : G \times G \to G$, $(g_1, g_2) \mapsto g_1 g_2^{-1}$. Then $K \times K \subset \nu^{-1}(V_1)$ and since K is compact it follows that there exists an open neighborhood V_0 of K in G such that $V_0 \times V_0 \subset \nu^{-1}(V_1)$, and hence $V_0 V_0^{-1} \subset V_1$. Now $B = G - V_0$ is closed in G and since K is compact it follows that KBK is closed in G. Clearly KBK is a both left and right K–invariant subset of G and $K \cap KBK = \emptyset$. Hence $V = G - KBK$ is an open neighborhood of K in G such that $KV = V = VK$, and moreover $V \subset V_0$. Thus $U = \pi(V)$ is a K–invariant open neighborhood of eK in G/K, and since $\pi^{-1}(U) = V$ we have that $\pi^{-1}(U)(\pi^{-1}(U))^{-1} \subset V_0 V_0^{-1} \subset V_1$. As we already saw in Section 1 we have that $G_{[U]} = \pi^{-1}(U)(\pi^{-1}(U))^{-1}$. $\qquad\square$

4 Twisted products

We will need the facts discussed in this section only in the case when K is a compact subgroup of the Lie group G. However, some of the facts hold equally well when K is a closed subgroup of G, and we have chosen to present them as such and to invoke the assumption that K is a compact subgroup only when this is essential.

Let G be a Lie group and K a closed subgroup of G. Let N be a smooth [real analytic] K-manifold. We consider the twisted product $G \times_K N$. Recall that $G \times_K N$ is the orbit space of the K–space $G \times N$, where K acts on $G \times N$ by $k(g, x) = (gk^{-1}, kx)$. Let $p : G \times N \to G \times_K N$ be the natural projection, and denote $p(g, x) = [g, x]$. Thus $[gk, x] = [g, kx]$, for all $k \in K$. There is a natural action of G on $G \times_K N$ given by $g'[g, x] = [g'g, x]$.

The map

$$q : G \times_K N \to G/K, \quad [g, x] \mapsto gK, \tag{1}$$

is a well–defined G–equivariant map. We claim that q is a locally trivial projection, with fiber N. This is seen as follows. Suppose $gK \in G/K$. Let U be an open neighborhood of gK in G/K such that there exists a real analytic cross section $\sigma : U \to G$. Then we have a commutative diagram

$$
\begin{array}{ccc}
U \times N & \xrightarrow{\;\;\theta\;\;} & q^{-1}(U) \\
 & {\scriptstyle p_1}\searrow \quad \swarrow {\scriptstyle q|} & \\
 & U &
\end{array}
\tag{2}
$$

where $\theta(u, x) = [\sigma(u), x]$, and p_1 denotes projection onto the first factor. Furthermore θ is a homeomorphism, with inverse given by

$$[g, x] \mapsto (\pi(g), \sigma(\pi(g))^{-1} gx).$$

Observe that $\sigma(\pi(g))^{-1} g \in K$, since $\pi(\sigma(\pi(g))) = \pi(g)$. Here $\pi : G \to G/K$, $g \mapsto gK$, denotes the natural projection.

Now suppose $\sigma' : U' \to G$ is another real analytic cross section over an open subset U' of G/K, and let $\theta' : U' \times N \xrightarrow{\cong} q^{-1}(U')$, $(u', x) \mapsto [\sigma'(u'), x]$, be the corresponding trivialization over U' of the projection q in (1). Assume $U \cap U' \neq \emptyset$. Then the transition function $t_{\theta', \theta} = \theta'^{-1} \circ \theta | (U \cap U') \times N$, from the trivialization θ to the trivialization θ', is given by

$$t_{\theta', \theta} : (U \cap U') \times N \to (U \cap U') \times N, \quad (u, x) \mapsto (u, \sigma'(u)^{-1}\sigma(u)x).$$

Thus we see that the transition functions $t_{\theta', \theta}$ are diffeomorphisms [real analytic isomorphisms]. Hence it follows that there exists a smooth [real analytic] structure on $G \times_K N$, making $G \times_K N$ into a smooth [real analytic] manifold, such that each trivialization map $\theta : U \times N \to q^{-1}(U)$, $(u, x) \mapsto [\sigma(u), x]$, as in (2), is a smooth diffeomorphism [real analytic isomorphism] from $U \times N$ onto the open subset $q^{-1}(U)$ of $G \times_K N$. Furthermore we note that the projection $q : G \times_K N \to G/K$ is a G–equivariant smooth [real analytic] map.

We now claim that the action

$$\Phi : G \times (G \times_K N) \to G \times_K N, \quad (g', [g, x]) \mapsto [g'g, x], \tag{3}$$

of G on $G \times_K N$ is smooth [real analytic]. This can be seen as follows. Let $(g'_0, [g_0, x_0]) \in G \times (G \times_K N)$, then $q([g_0, x_0]) = g_0 K \in G/K$. We recall from Section 3 that the action $G \times G/K \to G/K$, $(g', gK) \mapsto g'gK$, of G on G/K is real analytic. Let U_1 be an open neighborhood of $g'_0 g_0 K$ in G/K such that there exists a real analytic cross section $\sigma_1 : U_1 \to G$. Next we choose an open neighborhood V of g'_0 in G and an open neighborhood U of $g_0 K$ in G/K such that $VU \subset U_1$, and such that there exists a real analytic cross section $\sigma : U \to G$. Now $V \times q^{-1}(U)$ is an open neighborhood of $(g'_0, [g_0, x_0])$ in $G \times (G \times_K N)$ and $\Phi(V \times q^{-1}(U)) \subset q^{-1}(U_1)$. Furthermore we have the commutative diagram

$$
\begin{array}{ccc}
V \times q^{-1}(U) & \xrightarrow{\ \Phi| \ } & q^{-1}(U_1) \\[4pt]
{\scriptstyle id \times \theta} \Big\uparrow {\scriptstyle \cong} & & {\scriptstyle \cong} \Big\uparrow {\scriptstyle \theta_1} \\[4pt]
V \times U \times N & \xrightarrow{\ \Psi \ } & U_1 \times N
\end{array}
\tag{4}
$$

Here the trivializations $\theta : U \times N \to q^{-1}(U)$, $(u, x) \mapsto [\sigma(u), x]$, and $\theta_1 : U_1 \times N \to q^{-1}(U_1)$, $(u, x) \mapsto [\sigma_1(u), x]$, corresponding to the cross sections σ and σ_1, respectively, are smooth diffeomorphisms [real analytic isomorphisms]. The map Ψ is given by

$$\Psi(g, u, x) = (gu, \sigma_1(gu)^{-1}g\sigma(u)x), \quad \text{for all} \quad (g, u, x) \in V \times U \times N,$$

and thus we see that Ψ is a smooth [real analytic] map. Hence it follows that $\Phi|$ in (4) is smooth [real analytic], and thus we have proved that Φ in (3) is a smooth [real analytic] map.

Now suppose that K is a *compact* subgroup of G. As we saw in Section 3, there then exist a K–invariant open neighborhood U of eK in G/K and a real analytic cross section $\sigma : U \to G$, which is K-equivariant in the sense that

$$\sigma(ku) = k\sigma(u)k^{-1}, \quad \text{for all } k \in K, u \in U,$$

and moreover $\sigma(eK) = e$. In this case $q^{-1}(U)$ is a K–invariant open subset of $G \times_K N$ and the trivialization

$$\theta : U \times N \xrightarrow{\cong} q^{-1}(U), \quad (u, x) \mapsto [\sigma(u), x], \tag{5}$$

corresponding to σ, is a K–equivariant smooth diffeomorphism [real analytic isomorphism]. Here K acts diagonally on $U \times N$. The fact that θ is a K–equivariant map is seen by, $\theta(k(u, x)) = \theta(ku, kx) = [\sigma(ku), kx] = [k\sigma(u)k^{-1}, kx] = [k\sigma(u), x] = k[\sigma(u), x] = k\theta(u, x)$.

Lemma 4.1. *Let M be a smooth [real analytic] G–manifold, and let K be a closed subgroup of G and N a smooth [real analytic] K-manifold. Suppose $f : N \to M$ is a K-equivariant smooth [real analytic] map. Then*

$$\mu(f) : G \times_K N \to M, \quad [g, x] \mapsto gf(x),$$

is a G-equivariant smooth [real analytic] map.

Proof. Let $[g_0, x_0] \in G \times_K N$. We choose an open neighborhood U of $q([g_0, x_0]) = g_0 K$ in G/K such that there is a real analytic cross section $\sigma : U \to G$. Then $q^{-1}(U)$ is an open neighborhood of $[g_0, x_0]$ in $G \times_K N$ and $\theta : U \times N \xrightarrow{\cong} q^{-1}(U)$, $(u, x) \mapsto [\sigma(u), x]$, is a diffeomorphism [real analytic isomorphism]. The composite map

$$(\mu(f)|) \circ \theta : U \times N \to M$$

is given by $(u, x) \mapsto \sigma(u)f(x)$, and hence it is smooth [real analytic]. Since θ is a diffeomorphism [real analytic isomorphism] it follows that $\mu(f)| : q^{-1}(U) \to M$ is a smooth [real analytic] map, and this proves our claim. $\square$

5   Slices

Definition 5.1. Let M be a smooth G–manifold and K a closed subgroup of G. We say that a K–invariant smooth submanifold S of M is a smooth K–slice in M if GS is open in M and the map

$$\mu : G \times_K S \to GS, \quad [g, x] \to gx,$$

is a G–equivariant diffeomorphism.

Let S be a smooth K–slice in M, and let $x \in S$. Then GS is a G–invariant open neighborhood of the orbit Gx in M, and GS is called a *tube* about Gx.

By a *presentation* of S we mean a smooth K–manifold N and a K–equivariant smooth imbedding

$$j : N \to M$$

such that $j(N) = S$. Given such a presentation j of S we obtain a G–equivariant diffeomorphism

$$\mu(j) : G \times_K N \xrightarrow{\cong} GS, \quad [g, x] \mapsto gj(x),$$

which we call a presentation of the tube GS. Conversely, if $j : N \to M$ is a K–equivariant smooth imbedding such that $Gj(N)$ is open in M and $\mu(j) : G \times_K N \to Gj(N)$, $[g, x] \mapsto gj(x)$, is a G–equivariant diffeomorphism, then $S = j(N)$ is a smooth K–slice in M and $j : N \to M$ is a presentation of S.

By a *smooth slice at a point x in M* we mean a smooth G_x–slice S such that $x \in S$. A smooth slice S at $x \in M$ is said to be *linear* if S has a presentation of the form

$$j : \mathbb{R}^n(\rho) \xrightarrow{\cong} S \subset M,$$

where $\mathbb{R}^n(\rho)$ denotes a linear representation space for K and $j(0) = x$.

The notion of a slice is defined in a slightly different way in Palais [29], where a subset S of a G–space X is said to be a K–slice in X if GS is open in X and there exists a G–equivariant map $r : GS \to G/K$ such that $r^{-1}(eK) = S$, see [29, Definition 2.1.1]. In the smooth case the obvious smooth version of the above definition is employed in [29]. It follows by Lemma 5.2. below that Definition 5.1 is equivalent to the one used in [29]. In [29] smooth manifolds are assumed to be paracompact, but this distinction does not matter here in Lemma 5.2. However, in order to obtain the "smooth slice theorem" also in the cases when smooth manifolds are not assumed to be paracompact we will have to invoke Proposition 1.3.

Lemma 5.2. *Let S be a smooth submanifold of M. Then the following are equivalent:*

1. *S is a smooth K–slice in M.*

2. *GS is open in M and there exists a G–equivariant smooth map $r : GS \to G/K$ such that $r^{-1}(eK) = S$.*

Proof. Suppose S is a smooth K–slice in M. Then S is a K–invariant smooth submanifold of M, and GS is open in M and $\mu : G \times_K S \to GS$, $[g, x] \mapsto gx$, is a G–equivariant diffeomorphism. Thus

$$r = q \circ \mu^{-1} : GS \to G/K, \quad gx \mapsto gK,$$

is a G–equivariant smooth map. Here $q : G \times_K S \to G/K$, $[g, x] \mapsto gK$, denotes the natural projection, which as we noted in Section 4 is a G–equivariant smooth map. Clearly $r^{-1}(eK) = S$.

Now suppose that S is a smooth submanifold of M such that GS is open in M and there exists a G–equivariant smooth map $r : GS \rightarrow G/K$, with $r^{-1}(eK) = S$. Then $r(gx) = gK$, for all $g \in G$ and $x \in S$, and S is K–invariant and $gS \cap S = \emptyset$ for every $g \in G - K$. Therefore

$$\mu : G \times_K S \rightarrow GS, \quad [g, x] \mapsto gx, \tag{1}$$

is a well–defined G–equivariant bijective map, and μ is smooth by Lemma 4.1. The fact that the inverse $\mu^{-1} : GS \rightarrow G \times_K S$ is a smooth map is seen as follows. Let $y \in GS$ and denote $r(y) = gK \in G/K$. Choose an open neighborhood U of gK in G/K such that there exists a smooth cross section $\sigma : U \rightarrow G$. Then $W = r^{-1}(U)$ is an open neighborhood of y in GS. Furthermore

$$\mu^{-1}| : W \rightarrow G \times_K S$$

is given by $z \mapsto [\sigma(r(z)), \sigma(r(z))^{-1}z]$, for all $z \in W$, and hence $\mu^{-1}|W$ is smooth. Thus μ^{-1} is smooth and μ in (1) is a G–equivariant diffeomorphism. This proves that S is a smooth K–slice in M.   □

Let M be a Cartan smooth G–manifold and let $x \in M$, and denote $G_x = K$. Then there exists a smooth slice S at x, and in addition S may be chosen to be linear. This was proved by Palais [29, Proposition 2.2.2], in the case when M is paracompact. (The weaker statement that there is a smooth near K–slice at x was proved by Koszul [20, p. 139].) It follows by Proposition 1.3 that the existence of smooth slices in the paracompact case implies the existence of smooth slices in arbitrary Cartan smooth G–manifolds. Hence we have.

Theorem 5.3 (The smooth slice theorem). *Let M be a Cartan smooth G–manifold and let $x \in M$. Then there exists a smooth slice S at x, and we may choose S be to a linear slice.*

Let S be a smooth K–slice in M. Suppose $\sigma : U \rightarrow G$ is a smooth cross section over an open subset U of G/K. Then $W = \sigma(U)S$ is open in M and

$$\gamma : U \times S \rightarrow W, \quad (u, x) \mapsto \sigma(u)x, \tag{2}$$

is a diffeomorphism. This is seen as follows. As we saw in Section 4 the cross section σ gives rise to a diffeomorphism $\theta : U \times S \xrightarrow{\cong} q^{-1}(U), (u, x) \mapsto [\sigma(u), x]$, onto the open subset $q^{-1}(U)$ of $G \times_K S$. Since the map γ in (2) is obtained by composing θ with $\mu|q^{-1}(U)$, where μ is as in Definition 5.1, it now follows that γ is diffeomorphism onto the open subset $\sigma(U)S = W$. If U is an open neighborhood of eK in G/K, then $W = \sigma(U)S$ is an open neighborhood of S in M. We may in this case choose σ such that $\sigma(eK) = e$, and then γ has the additional property that $\sigma(eK, x) = x$, for all $x \in S$.

If K is a *compact* subgroup of G we can choose U to be a K–invariant open neighborhood of eK in G/K and $\sigma : U \rightarrow G$ to be a K–equivariant smooth cross section, with $\sigma(eK) = e$, see Section 3. Then W is a K–invariant open neighborhood of S in M, and γ in (2) is a K–equivariant diffeomorphism, where K acts diagonally on $U \times S$. In this case we call W a K-*invariant product neighborhood* of S in M, and γ a *presentation* of W. If $j : N \xrightarrow{\cong} S \subset M$ is a presentation of the smooth K–slice S we obtain that

$$\gamma_\# : U \times N \to W, \quad (u, x) \mapsto \sigma(u)j(x),$$

is a K–equivariant diffeomorphism, and we say that $\gamma_\#$ is a presentation of W corresponding to the presentation j of S.

Now suppose S is a linear slice and let $j : \mathbb{R}^n(\rho) \to S$ be a presentation of S. As we noted in (3) of Section 3 we may choose the K–invariant open neighborhood U of eK such that there is a K–equivariant diffeomorphism (in fact a real analytic isomorphism) $h : \mathbb{R}^q(\tau) \to U$, where $\mathbb{R}^q(\tau)$ denotes a linear representation space for K. Then

$$\gamma_* : \mathbb{R}^q(\tau) \times \mathbb{R}^n(\rho) \to W, \quad (a, b) \mapsto \sigma(h(a))j(b), \tag{3}$$

is a K–equivariant diffeomorphism. Since $\mathbb{R}^q(\tau) \times \mathbb{R}^n(\rho) = \mathbb{R}^{q+n}(\tau \oplus \rho)$, and the number of K–isotropy types occurring in a linear representation space for K is finite, it follows in particular that the number of K–isotropy types occurring in W is finite.

Definition 5.4. A K–invariant smooth submanifold S of M is said to be a *smooth near K–slice* in M, if there exists an open neighborhood U of eK in G/K and a smooth cross section $\sigma : U \to G$ such that $\sigma(U)S$ is open in M and the map

$$\gamma : U \times S \to \sigma(U)S, \quad (u, x) \mapsto \sigma(u)x,$$

is a diffeomorphism.

Suppose that S is a smooth near K–slice in M, and let $g_0 K \in G/K$. Let U and $\sigma : U \to G$ be as in Definition 5.4. Then $U_0 = g_0 U$ is an open neighborhood of $g_0 K$ in G/K and $\sigma_0 : U_0 \to G$, $u' \mapsto g_0 \sigma(g_0^{-1} u')$ is a smooth cross section over U_0 and $\sigma_0(U_0)S = g_0 \sigma(U)S$ is open in M and $\gamma_0 : U_0 \times S \to \sigma_0(U_0)S$, $(u', x) \mapsto g_0 \sigma(g_0^{-1} u')x$, is a diffeomorphism.

Furthermore we note that if $\sigma : U \to G$ and $\sigma' : U \to G$ are two smooth cross sections over an open subset U of G/K, then $\kappa(u) = \sigma(u)^{-1}\sigma'(u) \in K$, for all $u \in U$, and $\kappa : U \to K$ is a smooth map. Then $\sigma(u)S = \sigma'(u)S$, and we see that $\gamma : U \times S \to \sigma(u)S$, $(u, x) \mapsto \sigma(u)s$, is a diffeomorphism if and only if $\gamma' : U \times S \to \sigma'(u)S$, $(u, x) \mapsto \sigma'(u)s$, is a diffeomorphism. Thus it follows that the notion of a smooth near K–slice, as defined in Definition 5.4, is in fact independent of the choice of the smooth cross section $\sigma : U \to G$.

Lemma 5.5. *Let M be a smooth G–manifold and K a closed subgroup of G. Let S be a K–invariant smooth submanifold of M. Then the following are equivalent:*

1. S is a smooth K–slice in M.

2. *S is a smooth near K–slice in M and $gS \cap S = \emptyset$, for all $g \in G - K$.*

Proof. The discussion before Definition 5.4 shows that a smooth K–slice in M is always a smooth near K–slice in M. Moreover, if S is a smooth K–slice in M, then $gS \cap S = \emptyset$ for all $g \in G - K$, and thus we see that 1 implies 2.

Now assume that 2. holds. We consider the map

$$\mu : G \times_K S \to M, \quad [g, x] \mapsto gx.$$

The map μ is clearly a G–equivariant map onto GS. The fact that $gS \cap S = \emptyset$, for all $g \in G - K$, implies that μ is injective. We shall prove that μ is a diffeomorphism. Let $[g_0, x_0] \in G \times_K S$, then $q([g_0, x_0]) = g_0 K \in G/K$. By the discussion before Lemma 5.5 there is an open neighborhood U_0 of $g_0 K$ in G/K and a smooth cross section $\sigma_0 : U_0 \to G$ such that $\sigma_0(U_0)S$ is open in M and $\gamma_0 : U_0 \times S \to \sigma_0(U_0)S$, $(u, x) \mapsto \sigma_0(u)x$, is diffeomorphism. Now $q^{-1}(U_0)$ is an open neighborhood of $[g_0, x_0]$ in $G \times_K S$ and $\theta : U_0 \times S \xrightarrow{\cong} q^{-1}(U_0)$, $(u, x) \mapsto [\sigma_0(u), x]$, is a diffeomorphism, see Section 4. The composite map $(\mu|) \circ \theta : U_0 \times S \to M$, is given by $(u, x) \mapsto \sigma_0(u)x$, and hence equals $\gamma_0 : U_0 \times S \xrightarrow{\cong} \sigma_0(U_0)S \subset M$, which is a diffeomorphism onto the open subset $\sigma_0(U_0)S$. Since θ is a diffeomorphism we have that $\mu| : q^{-1}(U_0) \to \sigma_0(U_0)S$ is a diffeomorphism onto $\sigma_0(U_0)S$. It now follows that $GS = G(\sigma_0(U_0)S)$ is open in M, and that $\mu : G \times_K S \to GS$ is a diffeomorphism. This proves that S is a smooth K–slice in M. $\square$

6   Approximations of smooth slices

Lemma 6.1. *Let M be a smooth G–manifold, and let K be a compact subgroup of G and N a smooth K–manifold. Suppose $i : N \to M$ is a K–equivariant smooth imbedding such that $i(N) = S$ is a smooth K–slice in M. Then there exist a K–invariant product neighborhood W of S in M and an open neighborhood $\mathcal{W}$ of $i : N \to W$ in $C_K^\infty(N, W)$ such that the following holds. If $j \in \mathcal{W}$, then $j : N \to W$ is a K–equivariant smooth imbedding such that $j(N) = S'$ is a smooth K–slice in M and $GS' = GS$.*

Proof. Since S is a smooth K–slice in M and $i : N \to S$ is a presentation of S, we have that GS is open in M and

$$\mu(i) : G \times_K N \to GS, \quad [g, x] \mapsto gi(x),$$

is a G–equivariant diffeomorphism. As we saw in Section 5 we can find a K–invariant product neighborhood W^* of S in M, and we let

$$\gamma : U^* \times N \xrightarrow{\cong} W^*, \quad (u, x) \mapsto \sigma(u)i(x), \tag{1}$$

be a presentation of W^* corresponding to the presentation $i : N \xrightarrow{\cong} S \subset M$ of the smooth K–slice S. Here U^* denotes a K–invariant open neighborhood of eK in G/K and $\sigma : U^* \to G$ is a K–equivariant real analytic cross section, with $\sigma(eK) = e$, as in Section 3. (It is not essential here in this proof that σ is real analytic, smooth would be enough.) The map γ is a K–equivariant diffeomorphism onto W^*, where K acts diagonally on $U^* \times N$.

Next we choose a K–invariant open neighborhood U_1 of eK in G/K such that

$$D = \overline{U}_1 \subset U^*,$$

and such that D is a K–invariant compact real analytic submanifold with boundary of G/K. We recall from Section 3 that we can always find such a neighborhood U_1; in fact we may choose U_1 to be such that its closure $\overline{U}_1 = D$ is K–equivariantly and real analytically isomorphic to a unit representation disk $D^q(\tau)$ in an orthogonal representation space $\mathbb{R}^q(\tau)$ for K. Finally we have by Lemma 3.1 that there exists a K–invariant open neighborhood U of eK in G/K such that

$$G_{[U]} = \pi^{-1}(U)(\pi^{-1}(U))^{-1} \subset \pi^{-1}(U_1), \tag{2}$$

where $\pi : G \to G/K$ denotes the natural projection. We now have that

$$U \subset U_1 \subset D \subset U^*,$$

and we define

$$W = \sigma(U)S.$$

Then W is a K–invariant product neighborhood of S in M, and the restriction of γ in (1) to $U \times N$ gives us a K–equivariant diffeomorphism

$$\gamma : U \times N \xrightarrow{\cong} W, \quad (u, x) \mapsto \sigma(u)i(x). \tag{3}$$

We also have that

$$S \subset W \subset W^* \subset GS.$$

Let $r : GS \to G/K$, where $r^{-1}(eK) = S$, be the G–equivariant smooth map onto G/K determined by the smooth K–slice S, see Lemma 5.2. Then we have that $W = r^{-1}(U)$ and hence it follows by Lemma 1.2 that $G_{[W]} = G_{[U]}$, and thus we obtain from (2) that

$$G_{[W]} \subset \pi^{-1}(U_1). \tag{4}$$

Since D is compact we have by Lemma 2.2 that the map

$$\chi : C_K^\infty(N, W) \to C_K^\infty(D \times N, D \times W), \quad f \mapsto \mathrm{id} \times f,$$

is continuous. Furthermore the K–equivariant smooth map $\eta : D \times W \to GS$, $(u, y) \mapsto \sigma(u)y$, induces a continuous map

$$\eta_* : C_K^\infty(D \times N, D \times W) \to C_K^\infty(D \times N, GS).$$

Therefore the map

$$\Gamma = \eta_* \circ \chi : C_K^\infty(N, W) \to C_K^\infty(D \times N, GS) \qquad (5)$$

is continuous. Observe that for the given presentation $i : N \xrightarrow{\cong} S \subset W$ of the smooth K–slice S we have that

$$\Gamma(i) = \gamma : D \times N \to GS, \quad (u, x) \mapsto \sigma(u)i(x),$$

and $\gamma : D \times N \to GS$ is a K–equivariant smooth imbedding (in fact a closed imbedding), since it is obtained by restricting the K–equivariant diffeomorphism γ in (1) to $D \times N$.

Let $\mathcal{M}$ be an open neighborhood of $\Gamma(i) = \gamma$ in $C_K^\infty(D \times N, GS)$ such that each element of $\mathcal{M}$ is a K–equivariant smooth imbedding of $D \times N$ into GS, see Theorem 2.4. We set

$$\mathcal{W}' = \Gamma^{-1}(\mathcal{M}) \subset C_K^\infty(N, W).$$

Since the map Γ in (5) is continuous it follows that $\mathcal{W}'$ is an open neighborhood of $i : N \to W$ in $C_K^\infty(N, W)$.

We now claim that if $j \in \mathcal{W}'$, then $j : N \to W$ is a K–equivariant smooth imbedding and

$$j(N) = S' \qquad (6)$$

is a smooth K–slice in M. This is seen as follows. Since $\Gamma(j) \in \mathcal{M}$ we have that $\Gamma(j) : D \times N \to GS$ is a K–equivariant smooth imbedding, and therefore the restriction

$$\Gamma(j) : U_1 \times N \to GS, \quad (u, x) \mapsto \sigma(u)j(x), \qquad (7)$$

is also a K–equivariant smooth imbedding. Also $\Gamma(j)|(eK \times N) = j$ and hence $j : N \to W$ is a K–equivariant smooth imbedding. Furthermore $\Gamma(j)(U_1 \times N) = \sigma(U_1)j(N) = \sigma(U_1)S'$ is open in GS, and hence also in M. This follows, for example by invariance of domain, since $\dim (U_1 \times N) = \dim GS$. Since $\sigma(U_1)S'$ is open in M and $\Gamma(j) : U_1 \times N \xrightarrow{\cong} \sigma(U_1)S'$ is a K–equivariant diffeomorphism it follows that S' is a smooth near K–slice in M, see Definition 5.4.

In order to prove that S' is a smooth K–slice in M it is by Lemma 5.5 enough to prove that

$$gS' \cap S' = \emptyset, \quad \text{for all} \ g \in G - K. \qquad (8)$$

Suppose that $g \in G$ is such that $gS' \cap S' \neq \emptyset$. Since $S' = j(N) \subset W$ it follows that $gW \cap W \neq \emptyset$, and thus $g \in G_{[W]}$. Hence we have by (4) that $g \in \pi^{-1}(U_1)$. Therefore $\pi(g) = u \in U_1$, and we have that $g = \sigma(u)k$, for some $k \in K$. Hence

$$gS' = \sigma(u)kS' = \sigma(u)S' = \Gamma(j)(u \times N).$$

Since $\Gamma(j) : U_1 \times N \to GS$ in (7) is injective it follows that $\Gamma(j)(u \times N)$ is disjoint from $S' = \Gamma(j)(eK \times N)$, for all $u \in U_1$, except $u = eK$. Thus the assumption $gS' \cap S' \neq \emptyset$ implies that $u = eK$. Since $\pi(g) = u = eK$ it follows that $g \in K$, and we have shown that (8) holds. We have now proved our claim in (6) that if $j \in W'$ then $j : N \to W$ is a K-equivariant smooth imbedding such that $j(N) = S'$ is a smooth K-slice in M.

We shall complete the proof of Lemma 6.1 by showing that we can choose a possibly smaller neighborhood W, i.e., $W \subset W'$, of $i : N \to W$ in $C_K^\infty(N, W)$ such that we in addition obtain that the condition $GS' = GS$ holds.

Since the map $\gamma : U \times N \to W$, $(u, x) \mapsto \sigma(u)i(x)$, in (3), is a K-equivariant diffeomorphism it follows that each element of W can be expressed uniquely in the form $\sigma(u)i(x)$, where $u \in U$ and $x \in N$, and that

$$p = p_2 \circ \gamma^{-1} : W \to N, \quad p(\sigma(u)i(x)) = x, \quad u \in U, \ x \in N, \tag{9}$$

is a K-equivariant smooth map. (Here $p_2 : U \times N \to N$ denotes the projection onto the second factor.) Then p induces a continuous map

$$p_* : C_K^\infty(N, W) \to C_K^\infty(N, N).$$

Observe that $p_*(i) = \mathrm{id}_N$, where $i : N \xrightarrow{\cong} S \subset W$ denotes the given presentation of the slice S. Let $\mathcal{N}$ be an open neighborhood of id_N in $C_K^\infty(N, N)$ such that each element of $\mathcal{N}$ is a K-equivariant diffeomorphism from N onto N, see Theorem 2.5. We set

$$\mathcal{W}'' = p_*^{-1}(\mathcal{N}) \subset C_K^\infty(N, W),$$

and we claim that if $h \in \mathcal{W}''$ then

$$Gh(N) = GS. \tag{10}$$

Since $h(N) \subset W$ we have that $Gh(N) \subset GW = G(\sigma(U)S) = GS$. We shall prove the converse inclusion. If $h \in \mathcal{W}''$ then $p_*(h) \in \mathcal{N}$ and thus $p_*(h) = p \circ h : N \to N$ is a K-equivariant diffeomorphism, and hence in particular a surjective map onto N. It follows that the map

$$p| : h(N) \to N \tag{11}$$

is surjective. Now suppose that $gz_0 \in GS$, where $g \in G$ and $z_0 \in S$, and let $z_0 = i(x_0)$, where $x_0 \in N$. Since the map $p|$ in (11) is surjective there exists $y_0 \in h(N)$ such that $p(y_0) = x_0$. Since the map $p : W \to N$ in (9) is given by $p(\sigma(u)i(x)) = x$, and $p(y_0) = x_0$, it follows that $y_0 = \sigma(u)i(x_0)$, for some $u \in U$. Thus $i(x_0) = \sigma(u)^{-1}y_0$ and hence

$$gz_0 = gi(x_0) = g\sigma(u)^{-1}y_0 \in Gh(N).$$

This proves that $GS \subset Gh(N)$, and we have shown that (10) holds.

Now define

$$\mathcal{W} = \mathcal{W}' \cap \mathcal{W}'' \subset C_K^\infty(N, W).$$

Then W is an open neighborhood of $i : N \to W$, and by what we have proved above we have that if $j \in W$ then $j : N \to W$ is a K–equivariant smooth imbedding such that $S' = j(N)$ is a smooth K–slice in M and $GS' = GS$. $\square$

Addendum 6.2. If the smooth K–slice S in Lemma 6.1 is a linear slice we may add to the conclusion of Lemma 6.1 that the K–invariant product neighborhood W of S is such that the number of K–isotropy types occurring in W is finite.

Proof. In this case we choose the K–invariant product neighborhood W^* of S, given in (1) in the proof of Lemma 6.1, to be such that the number of K–isotropy types occurring in W^* is finite. This is always possible as we saw in (3) in Section 5. Since the K–invariant product neighborhood W, constructed in the proof of Lemma 6.1, is such that $W \subset W^*$ the desired conclusion follows.

$\square$

7 Proof of the main result

Theorem 7.1. Let M be a Cartan smooth G–manifold. Then there exists a real analytic structure β on M, compatible with the given smooth structure on M, such that the action of G on M_β is real analytic.

Proof. We define $\mathcal{B}$ to be the family of all pairs (B, β), where B is a non–empty G–invariant open subset of M and β is a real analytic structure on B, compatible with the smooth structure on B, such that the action of G on $B_\beta = (B, \beta)$ is real analytic.

Let us first note that the family $\mathcal{B}$ is non–empty. This is seen as follows. Let $x_0 \in M$ and denote $G_{x_0} = K_0$. Then K_0 is a compact subgroup of G. By the smooth slice theorem, see Theorem 5.3, there exists a G–equivariant diffeomorphism

$$\mu_0 : G \times_{K_0} \mathbb{R}^{n_0}(\rho_0) \to GS_0,$$

where $B_0 = GS_0$ is a G–invariant open neighborhood of x_0 in M, and $\mathbb{R}^{n_0}(\rho_0)$ denotes a linear representation space for K_0. As we noted in Section 4, the twisted product $G \times_{K_0} \mathbb{R}^{n_0}(\rho_0)$ is a real analytic G–manifold, and we give B_0 the real analytic structure β_0 induced from $G \times_{K_0} \mathbb{R}^{n_0}(\rho_0)$ through μ_0^{-1}. Since μ_0 is a diffeomorphism it follows that β_0 is compatible with the smooth structure on B_0. Since the action of G on $G \times_{K_0} \mathbb{R}^{n_0}(\rho_0)$ is real analytic and since μ_0 is G–equivariant, it follows that the action of G on $(B_0)_{\beta_0}$ is real analytic. Thus $(B_0, \beta_0) \in \mathcal{B}$, and we have shown that $\mathcal{B}$ is non–empty. We define an order in $\mathcal{B}$ by setting

$$(B_1, \beta_1) \leq (B_2, \beta_2)$$

if and only if:

   (i) $B_1 \subset B_2$.

(ii) The real analytic structure β_1 on B_1 is the one induced from the real analytic structure β_2 on B_2.

Now suppose C is a chain in B, i.e., if (B_1, β_1) and (B_2, β_2) belong to C then either $(B_1, \beta_1) \leq (B_2, \beta_2)$ or $(B_2, \beta_2) \leq (B_1, \beta_1)$. Let C_1 denote the family of all B occurring as the first coordinate of a pair in C, and let C_2 be the family of all β occurring as the second coordinate of a pair in C. Using this notation we define

$$B^* = \bigcup_{B \in C_1} B, \quad \text{and} \quad \beta' = \bigcup_{\beta \in C_2} \beta.$$

Then B^* is a non–empty G–invariant open subset of M, and β' is a real analytic atlas on B^*, compatible with the smooth structure on B^*. Let $\beta^* = \langle \beta' \rangle$ be the real analytic structure, i.e., the maximal real analytic atlas, on B^* generated by β'. Then the real analytic structure β^* on B^* has the property that for each $(B, \beta) \in C$ the real analytic structure that β^* induces on B equals β. It can now easily be seen that the action of G on $B_{\beta^*}^*$ is real analytic. For if $x \in B^*$, then there exists an open neighborhood U of x in B^* with $\overline{U}$ is compact and $\overline{U} \subset B^*$. Since $\overline{U}$ is compact it follows that there exists $(B, \beta) \in C$, such that $U \subset \overline{U} \subset B$. Now $GU \subset B$ and the real analytic structure on GU induced from B_β equals the one induced from $B_{\beta^*}^*$. Since the action of G on B_β is real analytic it now follows that the action of G on $(GU)_{\beta^*}$ is real analytic, and hence we have shown that the action of G on $B_{\beta^*}^*$ is real analytic. Thus $(B^*, \beta^*) \in B$, and furthermore we have that

$$(B, \beta) \leq (B^*, \beta^*), \quad \text{for all} \quad (B, \beta) \in C.$$

This shows that (B^*, β^*) is an upper bound for C in B. Hence we obtain by Zorn's lemma that there exists a maximal element (B, β) in B. We claim that $B = M$.

Suppose the contrary and assume that $B \subsetneq M$. If B is closed in M, then $M - B$ is a non–empty G–invariant open subset of M. In this case we could, as in the beginning of the proof, find a non–empty G–invariant open subset B_0, where $B_0 \subset M - B$, such that B_0 has a real analytic structure β_0, compatible with the smooth structure on B_0, and the action of G on $(B_0)_{\beta_0}$ is real analytic. Now $\beta \dot{\cup} \beta_0$ is a real analytic atlas on $B \dot{\cup} B_0$, which is compatible with the smooth structure on $B \dot{\cup} B_0$, and we let $\langle \beta \dot{\cup} \beta_0 \rangle$ denote the real analytic structure determined by $\beta \dot{\cup} \beta_0$. Then $\langle \beta \dot{\cup} \beta_0 \rangle$ is compatible with the smooth structure on $B \dot{\cup} B_0$, and it is also clear that the action of G on $(B \dot{\cup} B_0, \langle \beta \dot{\cup} \beta_0 \rangle)$ is real analytic. Thus $(B \dot{\cup} B_0, \langle \beta \dot{\cup} \beta_0 \rangle) \in B$ and $(B, \beta) < (B \dot{\cup} B_0, \langle \beta \dot{\cup} \beta_0 \rangle)$, which contradicts the fact that (B, β) is a maximal element in B. Thus B is not closed in M, and hence $\overline{B} - B \neq \emptyset$.

Let $x \in \overline{B} - B$, and denote $G_x = K$. By the smooth slice theorem, see Theorem 5.3, there exists a linear smooth slice S at x in M, and we let

$$i : \mathbb{R}^n(\rho) \xrightarrow{\cong} S \subset M$$

be a K–equivariant smooth imbedding into M such that $i(\mathbb{R}^n(\rho)) = S$, and $i(0) = x$. Here $\mathbb{R}^n(\rho)$ denotes a linear representation space for K. Then GS is open in M and

$$\mu(i) : G \times_K \mathbb{R}^n(\rho) \xrightarrow{\cong} GS, \quad [g, x] \mapsto gi(x), \tag{1}$$

is a G–equivariant diffeomorphism.

We now choose a K–invariant product neighborhood W of S in M and an open neighborhood $\mathcal{W}$ of

$$i : \mathbb{R}^n(\rho) \to W \tag{2}$$

in $C_K^\infty(\mathbb{R}^n(\rho), W)$, such that Lemma 6.1 holds for W and $\mathcal{W}$. Furthermore we may assume that the number of K–isotropy types occurring in W is finite, see Addendum 6.2.

Since $x \in \overline{B} - B$ and GS is an open neighborhood of x in M it follows that

$$GS \cap B \neq \emptyset, \quad \text{and} \quad GS \cap (M - B) \neq \emptyset. \tag{3}$$

Furthermore we have that $G(S \cap B) = GS \cap B$, since B in a G–invariant set, and hence $S \cap B \neq \emptyset$. Thus we see that $S \cap B$ is a non–empty K–invariant open subset of S, and therefore

$$V = i^{-1}(S \cap B)$$

is a non–empty K–invariant open subset of $\mathbb{R}^n(\rho)$, and $i(V) = S \cap B \subset W \cap B$.

The fact that V is a K–invariant open subset of $\mathbb{R}^n(\rho)$ implies in particular that V is a real analytic K–manifold. Also $W \cap B_\beta$ is a real analytic K–manifold, since $W \cap B_\beta$ is a K–invariant open subset of the real analytic G–manifold B_β. We now consider the K–equivariant smooth imbedding

$$i|V : V \to W \cap B_\beta.$$

By Lemma 2.3 there exists an open neighborhood $\mathcal{N}$ of $i|V$ in $C_K^\infty(V, W \cap B_\beta)$ such that we obtain a continuous map

$$E : \mathcal{N} \to C_K^\infty(\mathbb{R}^n(\rho), W)$$

by defining for each $h \in \mathcal{N}$,

$$E(h)(x) = \begin{cases} h(x), & x \in V \\ i(x), & x \in \mathbb{R}^n(\rho) - V. \end{cases} \tag{4}$$

Observe that $E(i|V) = i$. Since E is continuous, and $\mathcal{W}$ is an open neighborhood of i in $C_K^\infty(\mathbb{R}^n(\rho), W)$, there exists an open neighborhood $\mathcal{N}_1$ of $i|V$ in $\mathcal{N}$ such that

$$E(\mathcal{N}_1) \subset \mathcal{W}. \tag{5}$$

The number of K–isotropy types occurring in W is finite and hence the same holds for $W \cap B_\beta$. By Theorem 2.1 the map $i|V : V \to W \cap B_\beta$ can be approximated arbitrarily well, in the Whitney topology, by a K–equivariant real analytic map. Thus there exists a K–equivariant real analytic map $h_1 : V \to W \cap B_\beta$ such that $h_1 \in \mathcal{N}_1$, i.e., we have that

$$h_1 \in \mathcal{N}_1 \cap C_K^\omega(V, B_\beta \cap W). \tag{6}$$

We now define

$$j = E(h_1) : \mathbb{R}^n(\rho) \to W. \tag{7}$$

Then we have by (5) and (6) that $j \in \mathcal{W}$. By the choice of $\mathcal{W}$, Lemma 6.1 holds for $\mathcal{W}$ and hence j is a K–equivariant smooth imbedding such that

$$j(\mathbb{R}^n(\rho)) = S'$$

is a smooth K–slice in M and $GS' = GS$. (In fact $j(0) = x$, so S' is a smooth slice at x, but this is not essential here.) Hence

$$\mu(j) : G \times_K \mathbb{R}^n(\rho) \xrightarrow{\cong} GS' = GS, \quad [g, x] \mapsto gj(x), \tag{8}$$

is a G–equivariant diffeomorphism.

We claim that the restriction

$$\mu(j)| : G \times_K V \to GS \cap B_\beta \tag{9}$$

is a G–equivariant real analytic isomorphism. First of all we claim that $\mu(j)(G \times_K V) = GS \cap B$. It follows directly from the definition (4) that the maps $j = E(h_1)$ in (7) and i in (2) agree on the set $\mathbb{R}^n(\rho) - V$. Hence the maps $\mu(j)$ in (8) and $\mu(i)$ in (1) agree on $G \times_K (\mathbb{R}^n(\rho) - V)$. Since $\operatorname{im}(\mu(j)) = GS' = GS = \operatorname{im}(\mu(i))$, and both $\mu(j)$ and $\mu(i)$ are bijective maps it now follows that $\mu(j)(G \times_K V) = \mu(i)(G \times_K V) = Gi(V) = G(S \cap B) = GS \cap B$. Thus we see that $\mu(j)|$ in (9) is a G–equivariant diffeomorphism onto $GS \cap B_\beta$. Since $j|V = h_1 : V \to W \cap B_\beta \subset B_\beta$ is a K–equivariant real analytic map it follows by Lemma 4.1 that $\mu(j)|$ in (9) is a real analytic map. It now follows by the real analytic inverse function theorem (see e.g. [28, 2.2.10]) that $(\mu(j)|)^{-1}$ is real analytic. This proves that $\mu(j)|$ in (9) is a G–equivariant real analytic isomorphism.

Let us now denote $GS = B_1$. Then B_1 is a G–invariant open subset of M. We give B_1 the real analytic structure β_1 induced from $G \times_K \mathbb{R}^n(\rho)$ through $\mu(j)^{-1}$, i.e., the real analytic structure β_1 for which $\mu(j) : G \times_K \mathbb{R}^n(\rho) \to (B_1)_{\beta_1}$ is a real analytic isomorphism. Since $\mu(j)$ in (8) is a diffeomorphism it follows that the real analytic structure β_1 is compatible with the smooth structure on B_1. Since $\mu(j)$ is G–equivarint and the action of G on $G \times_K \mathbb{R}^n(\rho)$ is real analytic it follows that the action of G on $(B_1)_{\beta_1}$ is real analytic. The fact that $\mu(j)|$ in (9) is a real analytic isomorphism onto the open subset $B_1 \cap B_\beta$ of B_β implies that the real analytic structure on $B_1 \cap B$ induced from $(B_1)_{\beta_1}$ is the same as the one induced from B_β. Hence $\beta \cup \beta_1$ is a real analytic atlas on $B \cup B_1$, which is compatible with the smooth structure on $B \cup B_1$, and we let $\langle \beta \cup \beta_1 \rangle$ denote the real analytic structure determined by $\beta \cup \beta_1$. It is clear that the action of G on $(B \cup B_1, \langle \beta \cup \beta_1 \rangle)$ is a real analytic, and thus $(B \cup B_1, \langle \beta \cup \beta_1 \rangle) \in \mathcal{B}$. Furthermore we have by the latter part of (3) that $B \subsetneq B \cup B_1$, and hence $(B, \beta) < (B \cup B_1, \langle \beta \cup \beta_1 \rangle)$, but this contradicts the maximality of (B, β). Thus $B = M$. $\qquad\square$

References

[1] M.F. Atiyah and F. Hirzebruch, Spin–manifolds and group actions, in, Essays on Topology and Related Topics, Memoires dédiés à Georges de Rham, Springer–Verlag, Berlin and New York, 1970, 18–28.

[2] R. Bing, A homeomorphism between the 3–sphere and the sum of two solid horned spheres, Ann. of Math. 56 (1952), 354–362.

[3] G. Birkhoff, Lie groups simply isomorphic with no linear group, Bull. Amer. Math. Soc. 42 (1936), 883–888.

[4] N. Bourbaki, General Topology, Part 1, Hermann, Paris and Addison–Wesley, Reading, Mass., 1966.

[5] G. Bredon, Introduction to compact transformation groups, Academic Press, New York and London, 1972.

[6] C. Chevalley, Theory of Lie Groups, Vol. 1, Princeton University Press, Princeton, N.J., 1946.

[7] J. Dieudonné, Une généralisation des espaces compacts, J. Math. Pures Appl. 23 (1944), 65–76.

[8] A. Gleason, Groups without small subgroups, Ann. of Math. 56 (1952), 193–212.

[9] H. Grauert, On Levi's problem and the imbedding of real analytic manifolds, Ann. of Math. 68 (1958), 460–472.

[10] I. Hambleton and R. Lee, Perturbation of equivariant moduli spaces, Math. Ann. 293 (1992), 17–37.

[11] S. Helgason, Differential geometry and symmetric spaces. Academic Press, 1962.

[12] D. Hilbert, Mathematische Probleme, Nachr. Akad. Wiss. Göttingen 1900, 253–297.

[13] M.W. Hirsch, Differential Topology, Springer–Verlag, New York, Heidelberg, Berlin, 1976.

[14] S. Illman, Subanalytic equivariant triangulation of real analytic proper G–manifolds for G a Lie group, Preprint, Princeton University, June 1992. (To be published.)

[15] S. Illman, Smoothly equivalent proper real analytic G–manifolds are sub-analytically equivalent. (In preparation.)

[16] M.A. Kervaire, A manifold which does not admit any differentiable structure, Comment. Math. Helv. 34 (1960), 257–270.

[17] H. Kneser, Analytische Struktur und Abzählbarkeit, Ann. Acad. Sci. Fennicae A I, 251/5, 1958.

[18] H. Kneser and M. Kneser, Reell–analytische Strukturen der Alexandroff–Halbgeraden und der Alexandoff–Geraden, Arch. Math. 11 (1960), 104–106.

[19] W. Koch and D. Puppe, Differenzierbare Strukturen auf Mannigfaltigkeiten ohne abzählbare Basis, Arch. Math. 19 (1968), 95–102.

[20] J.L. Koszul, Sur certains groupes des transformations de Lie, Colloque de Géométrie Différentiable, Strasbourg, 1953, 137–141.

[21] J.L. Koszul, Lectures on Groups of Transformations, Tata Institute of Fundamental Researach, Bombay, 1965.

[22] J. Mather, Stability of C^∞–mappings II. Infinitesimal stability implies stability, Ann. Math. 89 (1969), 254–291.

[23] T. Matumoto and M. Shiota, Unique triangulation of the orbit space of a differentiable transformation group and its applications, Homotopy Theory and Related Topics, Advanced Studies in Pure Math., vol. 9, Kinokuniya, Tokyo, 1987, 41–55.

[24] D. Montgomery and L. Zippin, Small subgroups of finite–dimensional groups, Ann. of Math. 56 (1952), 213–241.

[25] D. Montgomery and L. Zippin, Examples of transformation groups, Proc. Amer. Math. Soc. 5 (1954), 460–465.

[26] D. Montgomery and L. Zippin, Topological Transformation Groups, Interscience Publishers, New York and London, 1955.

[27] C.B. Morrey, Jr., The analytic embedding of abstract real–analytic manifolds, Ann. of Math. 68 (1958), 159–201.

[28] R. Narasimhan, Analysis on real and complex manifolds, North–Holland, 1968.

[29] R.S. Palais, On the existence of slices for actions of non–compact Lie groups, Ann. of Math. 73 (1961), 295–323.

[30] R.S. Palais, C^1 actions of compact Lie groups on compact manifolds are C^1–equivalent to C^∞ actions, Amer. J. Math. 92 (1970), 748–760.

[31] L.S. Pontryagin, Topological Groups, Second Edition, Gordon and Breach, New York, 1966.

[32] H. Whitney, Differentiable manifolds, Ann. of Math. 37 (1936), 645–680.

Formal Deformations of Equivariant Genera, Fixed Point Formula and Universal Symmetry Blocks

Gabriel Katz

Introduction

This paper is the first in a series. The series originated as our attempt to understand some of the recent work on elliptic genera by Bott, Hirzebruch, Landweber, Liu, Ochanine, Segal, Stong, Taubes, Witten and others (cf. [BT], [H], [L], [LS], [Li], [01], [02], [Se], [W]).

We started by building a framework that would be very much adjusted to [BT], but soon found ourselves crossing the borders of the Elliptic kingdom to study other equivariant genera of K-theoretical nature. Despite similarity with [BT], we feel that the philosophy of our approach is much closer to some old works of the Russian school on equivariant bordism, especially to those of Kričever [Kr1], [Kr2]. For instance, it relies on [K1], where for equivariant bundles we have developed (in a quite geometrical manner) the notion of bordism-valued characteristic classes.

Thus, the present paper bridges the results of [K1] on equivariant bordism with more familiar material on equivariant K-theory (developed by Atiyah, Singer and Segal in the early 1960s). We also go beyond this classical treatment in a number of directions: 1) we obtain fixed-point formulae for computing formal deformations of classical G-equivariant indices of several important elliptic operators; 2) the answer is expressed through a *new type of* ("non-equivariant") *invariants* which makes the study of the possible *range* of a G-index more transparent; 3) we treat a *special class* of G-actions which is larger than the classical class of cyclic or cyclically dense ones (although classical fixed point formulae, coupled with the cyclic induction, implicitly tell the whole story, an explicit answer in terms of the G-fixed point set is probably a new wrinkle).

Let G be a compact Lie group acting on a smooth closed manifold M in such a way that the normal bundle $\nu(M^G, M)$ admits an equivariant complex or quaternionic structure. We denote by ψ a typical complex (quaternionic) G-representation on a fiber of $\nu(M^G, M)$.

Given a finite list $\mathfrak{R} = \{\rho\}$ of irreducible complex (quaternionic) G-representations ρ, we shall restrict our considerations to a class of G-actions which we call $\mathfrak{R}$-*linearly modeled*. They satisfy the following hypotheses:

(1) the set $\bigcap_{\rho \in \mathfrak{R}} \{g \in G \mid \mathrm{Det}(1 - \rho(g)) \neq 0\}$ is non-empty; in particular, for a connected Lie group G, this means that, for any $\rho \in \mathfrak{R}$, all the weights of ρ are non-trivial.

(2) *any normal representation ψ splits into a sum of the irreducible ρ's from the list $\mathfrak{R}$.*

(3) *For any $x \in M \backslash M^G$, its isotropy group G_x is contained in an isotropy group of a non-zero vector from the space of some G-representation $\chi = \bigoplus_{\rho \in \mathfrak{R}} n_\rho^\chi \cdot \rho$ (the non-negative integers n_ρ^χ denoting the multiplicity of ρ in χ).*

Intuitively, one might think about linearly modeled actions as having a non-empty open set of elements $g \in G$, so that $M^g = M^G$, as well as having the isotropy type combinatorics of a linear G-representation.

Note that about "half" of the connected Lie groups G fail to satisfy (1) and, as a result, have no linearly modeled actions. On the other hand, for an appropriate choice of $\mathfrak{R}$, any circle action or any $\mathbf{SU}(2)$-action, with the isotropy groups all being finite or $\mathbf{SU}(2)$ itself, is linearly modeled.

Although in [K2] we deal mostly with connected Lie groups G, the results of this paper are valid for (linearly modeled) actions of generic compact Lie groups, in particular, one can think of G as being finite.

Let τM stand for the tangent bundle of M, $\dim M = 2n$.

Let $\mathcal{G}$ be a compact Lie group acting transitively on the unit sphere S^{2n-1} (such $\mathcal{G}$ have been classified). Thus, S^{2n-1} is orientably isomorphic to a homogeneous space $\mathcal{G}/\mathcal{H}$. Assume that τM admits a G-equivariant $\mathcal{G}$-structure, i.e. the structure group of τM equivariantly reduces to $\mathcal{G}$ (see Sect. 2.1 for more details). Then, starting with a virtual $\mathcal{G}$-representation σ, whose restriction to $\mathcal{H} \subset \mathcal{G}$ is trivial, one can construct an element (a symbol) σ_M, living in the equivariant complex K-theory (with compact support) $K_G(\tau M)$ of τM. It can be thought of as the symbol of a pseudo-differential elliptic operator D_σ on M, intimately related to the $\mathcal{G}$-structure. Model examples are provided by the equivariant almost complex, Riemannian and Spin-structures on M, together with a special choices of σ. They give rise correspondingly to the Euler-Todd (i.e., $\bar{\partial}$), Signature and Dirac operators.

Let $\pi_!^M : K_G(\tau M) \longrightarrow K_G(pt) \approx R(G)$ be the Gysin transfer homomorphism, $R(G)$ denoting the ring of virtual finite dimensional G-representations.

For G cyclic, the Atiyah-Segal-Singer Index Theorem [ASi] makes it possible to compute the equivariant index of the operator D_σ,

$$\text{ind}(D_\sigma, M) = \pi_!^M(\sigma_M) \in R(G)$$

in local terms of $\nu(M^G, M)$. Although, for generic G-actions, there is no immediate connection between $\text{ind}(D_\sigma, M)$ and the structure of $\nu(M^G, M)$, [1] for linearly modeled ones, such a strong connection can be established. The goal of this paper is, for the category of linearly modeled G-actions, to obtain local formulae for computing a formal t-deformation $\text{ind}_t(D_\sigma, M)$ of the invariant

---

[1]Instead, one computes the value $\text{ind}(D_\sigma, M)(g), g \in G$, in terms of $\nu(M^g, M)$, i.e. takes advantage of cyclic induction.

$\mathrm{ind}(D_\sigma, M)$ (cf. Theorem 2.3). This deformation is produced by means of a stable exponential operation

$$\phi : \ K_G(M) \longrightarrow \mathcal{K}_G(M)[[t]],$$

where $\mathcal{K}_G(M) := K_G(M) \otimes \mathbb{C}$ and $\mathcal{R}(G) := R(G) \otimes \mathbb{C}$, as follows. Put

$$\mathrm{ind}_t(D_\sigma, M) = \pi_!^M(\sigma_M \otimes \phi(\tau_{\mathbb{C}})) \in \mathcal{R}(G)[[t]],$$

$\tau_{\mathbb{C}}$ denoting the complexification of τM. One should think of $\mathrm{ind}_t(D_\sigma, M)$ as a formal sum

$$\sum_{j \geq 0} \pi_!^M(\sigma_M \otimes Q_j(\tau_{\mathbb{C}})) \, t^j,$$

where $Q_j(\tau_{\mathbb{C}})$ are certain universal polynomials in the exterior powers $\Lambda^1 \tau_{\mathbb{C}}$, $\Lambda^2 \tau_{\mathbb{C}}, \ldots, \Lambda^k \tau_{\mathbb{C}}, \Lambda^{-k} \tau_{\mathbb{C}}$ and $k = \dim M$. In other words, $\mathrm{ind}_t(D_\sigma, M)$ is a formal sum $\sum_{j \geq 0} \chi_j t^j$, where each χ_j is a central algebraic function on G and can be viewed as an index of a basic operator D_σ, twisted with the appropriate (virtual) tensorial fields on M. The operation ϕ determines explicitly the type of the twist.

Our local Lefschetz-type formula (2.6.1) in Theorem 2.3 provides a convenient setting for studying the Witten's rigidity phenomenon [W] for $\chi_j = \pi_!^M(\sigma_M \otimes Q_j(\tau_{\mathbb{C}}))$, $j = 0, 1, 2, \ldots$, in the spirit of the beautiful paper [BT]. Here by "rigidity" we mean that, under certain combinatorial conditions on the normal representation $\{\psi\}$ (usually implied by the existence of an equivariant Spin-structure), each function $\chi_j : G \longrightarrow \mathbb{C}$ is, in fact, a *constant*. In [K2] we are going to employ (2.6.1) to derive some of these rigidity results.

In fact, the stable exponential operations $\phi : \ K_G(X) \longrightarrow \mathcal{K}_G(X)[[t]]$ are in one-to-one correspondence with the elements

$$\varphi_t(u) = \sum_{j \geq 0} \varphi_j(u) \, t^j \in \mathbb{C}[u, u^{-1}][[t]], \quad (\varphi_0 \in \mathbb{C}^\times).$$

For instance, picking

$$\varphi_t(u) \ = \ \prod_{n \geq 1} \left(\frac{1 + t^n u}{1 - t^n u} \right) \qquad \text{and } D_\sigma = \text{ Signature operator}$$

or

$$\varphi_t(u) \ = \ \prod_{n \geq 1} \left(\frac{1 + t^{2n-1} u}{1 - t^{2n} u} \right) \quad \text{and } D_\sigma = \text{ Dirac operator,}$$

one gets the invariants $\mathrm{ind}_t(D_\sigma, M)$ whose rigidity has been conjectured by Witten [W] and proven by Taubes and Bott-Taubes [T], [BT] (see also [LS] for an important special case).

Our basic observation is that, for linearly modeled actions and the operators D_σ as above, one has the G-fixed point formula:

$$\mathrm{ind}_t(D_\sigma, M)(g) = \sum_{\psi, \omega} \Phi_{\omega, t}(\nu_\psi) \cdot \Theta_{\omega, t}^\psi(g), \qquad g \in G. \tag{0.1}$$

Here ω is a multi-index (a tableau), $\nu_\psi = \nu(M_\psi^G, M)$, where M_ψ^G stands for the component of M^G having the normal slice-type ψ, $\Phi_{\omega,t}(\nu_\psi) \in \mathbb{C}[[t]]$ is a characteristic element (a characteristic "number") of the normal G-bundle ν_ψ and $\Theta_{\omega,t}^\psi : G \longrightarrow \mathbb{C}[[t]]$ is a universal central function, well-defined on the complement to the divisor $\mathcal{D}_\psi = \{g \in G \mid \mathrm{Det}(1 - \psi(g)) = 0\}$ in G.

A distinctive feature of (0.1) is the effective *separation* of *the topology* of the action, encoded through the "non-equivariant" characteristic elements ("numbers") $\{\Phi_{\omega,t}(\nu_\psi)\}$ of ν_ψ and the *G-symmetry*, manifesting itself through universal (i.e., M-independent) "equivariant" block-functions $\{\Theta_{\omega,t}^\psi\}$. This separation also makes it possible to understand the range of $\mathrm{ind}_t(D_\sigma, \sim)$ and of $\mathrm{ind}(D_\sigma, \sim)$. In fact, $\Phi_{\omega,t}(\nu_\psi) = \mathrm{ind}_t(D_\sigma, \mathcal{S}_\omega(\nu_\psi))$ for a characteristic manifold $\mathcal{S}_\omega(\nu_\psi)$ carring a trivial G-action. Letting $t = 0$, it follows that $\mathrm{ind}(D_\sigma, M)$ is an *integral* combination of the universal functions $\{\Theta_\omega^\psi := \Theta_{\omega,0}^\psi\}$ on G. In a sense, (0.1) makes it possible to treat any linearly modeled action as one having only *isolated* fixed points.

Formula (0.1) (see also (2.6.1) and (2.6.2)) is in the same line with the local formula in [BeK] (the modified Atiyah-Singer G-signature formula). It is a descendent of a universal local formula from [K1] living at the upper floor — in the bordism of equivariant bundles. Note that even the description of $\mathrm{ind}(D_\sigma, M)$—the 0-term of $\mathrm{ind}_t(D_\sigma, M)$—is a new result (cf. Theorem 2.4).

It turns out, that when the basic formal power series $\varphi_t(u) \in \mathbb{C}[u, u^{-1}][[t]]$ is actually the Taylor t-expansion of an analytic (holomorphic, meromorphic) function in the variables (t, u), then, roughly speaking, the invariant $\mathrm{ind}_t(D_\sigma, M)(g)$ (as a function of $t \in \mathbb{C}$ and $g \in G^\mathbb{C}$) possesses similar analytic properties (here $G^\mathbb{C}$ denotes the complexification of the Lie group G). Thus, one can regard t as a "small" complex number, not a formal variable. Moreover, when the auxiliary function $H_t(u)$ (cf. (2.4.5)), closely related to $\varphi_t(u)$, has automorphic properties (is modular) and some combinatorial conditions on $\{\psi\}$ are satisfied, the Witten's rigidity follows. These implications of Theorem 2.3 are the subject of the next paper [K2].

It is my pleasant duty to thank G. Schwarz for very friendly and stimulating discussions.

1 Local formula in the G-bordism

We start with a brief review of some results from [K1] which will be used in the following.

Let G be a compact Lie group and M a compact smooth G-manifold. Assume that the G-bundle $\nu = \nu(M^G, M)$, normal to the G-fixed point set M^G, admits an equivariant complex or quaternionic structure. Let ψ stand correspondingly, for a typical complex or quaternionic G-representation on a fiber of ν. It can be decomposed into a direct sum $\bigoplus_{\rho \in \mathcal{I}rr} n_\rho^\psi \cdot \rho$ of non-trivial irreducible G-

representations $\{\rho\}$, $n_\rho^\psi \in \mathbb{Z}$ denoting the multiplicity of ρ in ψ. Let M_ψ^G be the set of points in M^G of the normal slice-type ψ and let $\nu_\psi = \nu(M_\psi^G, M)$.

Given a finite list $\mathfrak{R}$ of non-trivial irreducible G-representations ρ, consider a family $\mathcal{F} = \mathcal{F}_\mathfrak{R}$ of compact subgroups in G which arise as the isotropy groups of vectors in the spaces of various representations of the form $\bigoplus_{\rho \in \mathfrak{R}} n_\rho \cdot \rho$. Put $\mathcal{F}' = \mathcal{F} \setminus \{G\}$.

One can introduce a G-bordism ring $\Omega_*^G(\mathcal{F})$ based on the G-oriented (cf. [K1]), or depending on the category, stably complex (symplectic) G-manifolds with the isotropy types in $\mathcal{F}$ and with the normal G-representations $\{\psi\}$ as above. The relative version $\Omega_*^G(\mathcal{F}, \mathcal{F}')$ of such G-bordism is available: it employs the M's, so that the isotropy types of their boundaries ∂M's belong to the subfamily $\mathcal{F}'$ (i.e. ∂M^G is empty).

There exists an obvious ring homomorphism $j_* : \Omega_*^G(\mathcal{F}) \longrightarrow \Omega_*^G(\mathcal{F}, \mathcal{F}')$. Our goal is to study "local" invariants of M, depending solely on its j_*-image. To achieve this goal, we shall describe $j_*(M) \in \Omega_*^G(\mathcal{F}, \mathcal{F}')$ in terms of certain characteristic elements of $\nu(M^G, M)$. Prior to it, we introduce a few combinatorial notations.

Denote by $\mathfrak{G}_\psi$ the set $\prod_{\rho \in \mathfrak{R}} S^{n_\rho^\psi} \mathbb{Z}_+$, where $S^n \mathbb{Z}_+$ stands for the n-th symmetric power of the set of non-negative integers $\mathbb{Z}_+$. The elements $\omega \in \mathfrak{G}_\psi$ can be interpreted as tables, doubly-indexed by $\rho \in \mathfrak{R}$ and $\ell \in \mathbb{N}$, $1 \leq \ell \leq n_\rho^\psi$, and filled with integers $\omega(\rho, \ell) \geq 0$. By definition, such two fillings are equivalent, if they differ at most by permutations of $\{\omega(\rho, \ell)\}_{1 \leq \ell \leq n_\rho^\psi}$, i.e. by the $\prod_{\rho \in \mathfrak{R}} S(n_\rho^\psi)$-action.

In [K1], for each table $\omega \in \mathfrak{G}_\psi$, a generalized flag manifold Flag_ω has been constructed. It is a homogeneous space for a product of classical unitary (symplectic) Lie groups. The space Flag_ω carries certain "tautological" vector bundle ζ^ω. One can associate with the G-bundle ν_ψ a (non-equivariant) fibration $\pi : \mathrm{Flag}_\omega(\nu_\psi) \longrightarrow M_\psi^G$ with a fiber Flag_ω. The bundle ζ^ω gives rise to a vector bundle $\tilde{\zeta}^\omega$ over $\mathrm{Flag}_\omega(\nu_\psi)$. Denote by τ_π the bundle, formed by vectors tangent to the fibers of π.

Given an oriented (complex, quaternionic) vector bundle ξ over a closed oriented (stably complex, stably symplectic) manifold base L, let $\chi(\xi)$ denote an oriented (stably complex, stably symplectic) submanifold of L obtained as an intersection of a generic section of ξ with its zero-section. The bordism class of $\chi(\xi)$—the Euler class of ξ—does not depend on a particular choice of the section.

Now, following [K1], for each $\omega \in \mathfrak{G}_\psi$, put:

$$\mathcal{S}_\omega(\nu_\psi) = \chi(\tilde{\zeta}^\omega \oplus \tau_\pi) \in \Omega_*(pt). \tag{1.1}$$

Remarkably, the class $j_*(M) \in \Omega_*^G(\mathcal{F}, \mathcal{F}')$ is detected by finitely many "non-equivariant" characteristic elements ("numbers") $\{\mathcal{S}_\omega(\nu_\psi)\}_{\psi, \omega}$. Thus, any G-bordism invariant of M, depending on $\nu(M^G, M)$ only (equivalently, which factors through j_*), is computable in terms of the appropriate non-equivariant invariants of the characteristic manifolds $\{S_\omega(\nu_\psi)\}$.

Let P_r denote the complex (quaternionic) r-dimensional projective space and η_r—the normal bundle of the standard imbedding $P_r \subset P_{r+1}$. The G-bundles $\rho \otimes \eta_r$ —the normal G-bundles of the imbeddings $P_r \subset P(\rho \oplus \underline{r+1})$ —are the main building blocks of the local formula below (depending on the context, the tensor product is taken over the complex numbers $\mathbb{C}$ or over the quaternions $\mathbb{H}$). Note that, for $\rho \in \mathfrak{R}$, they can be viewed as elements of the ring $\Omega_*^G(\mathcal{F}, \mathcal{F}')$, $\mathcal{F} = \mathcal{F}_\mathfrak{R}$.

We emphasize that in formulae (1.2), (1.3) the "+" operation stands for the *disjoint union* of G-bundles and manifolds and products are *cartesian products* of the G-spaces with the diagonal G-action.

Theorem 1.1 [K1, Prop. 4.4]. *Under the assumptions and the notations above, the following formula is valid in the ring $\Omega_*^G(\mathcal{F}, \mathcal{F}')$:*

$$j_*(M) = \coprod_\psi \nu(M_\psi^G, M) = \sum_{\psi, \omega \in \mathcal{B}_\psi} T_{\psi, \omega}(\{\rho \otimes \eta_r\}, \{P_r\}) \cdot S_\omega(\nu_\psi), \qquad (1.2)$$

where the $T_{\psi, \omega}$'s are certain universal integral polynomials in the variables $\{\rho \otimes \eta_r\}$, $\{P_r\}$, $r = 0, 1, 2, \ldots$. They can be computed using the generating function (1.3) in the formal variables $\{x_{\rho, \ell}\}$:

$$\prod_{\rho \in \mathfrak{R}, 1 \le \ell \le n_\rho^\psi} \left(\frac{\rho + (\rho \otimes \eta_1) x_{\rho, \ell} + (\rho \otimes \eta_2) x_{p, \ell}^2 + \cdots}{1 + P_1 x_{\rho, \ell} + P_2 x_{\rho, \ell}^2 + \cdots} \right)$$

$$= \sum_{\omega \in \mathcal{B}_\psi} T_{\psi, \omega}(\{\rho \otimes \eta_r\}, \{P_r\}) \cdot S_\omega(\{x_{\rho, \ell}\}). \qquad (1.3)$$

Here $S_\omega(\{x_{\rho, \ell}\})$ is the "minimal" $\prod_{\rho \in \mathfrak{R}} S(n_\rho^\psi)$-symmetric polynomial containing the monomial $\prod_{\rho \in \mathfrak{R}, 1 \le \ell \le n_\rho^\psi} x_{\rho, \ell}^{\omega(\rho, \ell)}$. Identity (1.3) takes place in the ring $\mathbb{Z}[\{\rho \otimes \eta_r\}, \{P_r\}][[\{x_{\rho, \ell}\}]]$.

Example 1.2. Let $G = T$ be a 2-dimensional torus, viewed as a subgroup $S^1 \times S^1 \subset \mathbb{C}^\times \times \mathbb{C}^\times$ ($\mathbb{C}^\times$ denotes the multiplicative group of complex numbers $\mathbb{C}$ and $S^1 = \{u \in \mathbb{C} : |u| = 1\}$).

Denote by ρ_1 the complex irreducible T-representation $(\tau_1, \tau_2) \mapsto \tau_1^2 \tau_2$, $(\tau_1, \tau_2) \in T$, and by ρ_2 —the representation $(\tau_1, \tau_2) \mapsto \tau_1 \tau_2^{-1}$. Let T act projectively on the complex projective 3-space $P_3 = P(\mathbb{C}^4)$ via the T-representation $1 \oplus (\rho_1^* \otimes \rho_2) \oplus \rho_1^* \oplus \rho_1^*$, "$*$" standing for the $\mathbb{C}$-dual representation. The T-fixed point set of this action consists of the two points $P(1) = [1 : 0 : 0 : 0]$ and $P(\rho_1^* \otimes \rho_2) = [0 : 1 : 0 : 0]$ together with the line $P(\rho_1^* \oplus \rho_1^*) = [0 : 0 : z_2 : z_3]$. The normal T-representations of this action are: $\psi_1 = (\rho_1^* \otimes \rho_2) \oplus 2\rho_1^*$ at $P(1)$, $\psi_2 = (\rho_1 \otimes \rho_2^*) \oplus 2\rho_2^*$ at $P(\rho_1^* \otimes \rho_2)$ and $\psi = \rho_1 \oplus \rho_2$ along $P(2\rho_1^*)$. Hence, the list $\mathfrak{R}$ is $\{\rho_1, \rho_1^*, \rho_2, \rho_2^*, \rho_1 \otimes \rho_2^*, \rho_1^* \otimes \rho_2\}$. The family $\mathcal{F}' = \mathcal{F}_\mathfrak{R}'$ of the isotropy groups associated with $\mathfrak{R}$ comprises three 1-dimensional subtori of T :

$$T_0 = \{\tau_1 \tau_2^{-1} = 1\}, \quad T_1 = \{\tau_1^2 \tau_2 = 1\}, \quad T_2 = \{\tau_1 \tau_2^2 = 1\},$$

their pairwise intersections—the cyclic 3-group $C_3 = \{(1,1), (\lambda, \lambda), (\lambda^2, \lambda^2)\}$, $\lambda = \exp(2\pi i/3)$, and the trivial group $e = (1,1)$.

Now we turn to an arbitrary 6-dimensional oriented T-manifold M with the isotropy types in $\mathcal{F} = \mathcal{F}_{\mathfrak{R}}$ and with the normal slice-types ψ_1, ψ_2, ψ. Under these hypotheses, formulae (1.2), (1.3) collapse to:

$$j_*(M) = \# M^T_{\psi_1} \cdot \psi_1 + \# M^T_{\psi_2} \cdot \psi_2 + \nu(M^T_\psi, M)$$

(we use the same notation for a T-representation and for the trivial T-bundle it defines). The complex bundle $\nu_\psi = \nu(M^T_\psi, M)$ admits a canonical equivariant splitting $(\rho_1 \otimes \xi_1) \oplus (\rho_2 \otimes \xi_2)$, ξ_1, ξ_2 being some complex line bundles over the surface M^T_ψ. By (1.3), in the ring $\Omega^T_*(\mathcal{F}, \mathcal{F}')$, we have:

$$\nu_\psi = [\rho_1 \times \rho_2] \times \mathcal{S}_{0,0}(\nu_\psi) \quad + \quad [\rho_2 \times (\rho_1 \otimes \eta_1) - \rho_1 \times \rho_2 \times P_1] \times \mathcal{S}_{1,0}(\nu_\psi) +$$
$$+ \quad [\rho_1 \times (\rho_2 \otimes \eta_1) - \rho_1 \times \rho_2 \times P_1] \times \mathcal{S}_{0,1}(\nu_\psi) \quad (1.4)$$

where $\mathcal{S}_{0,0}(\nu_\psi) = M^T_\psi$, $\mathcal{S}_{1,0}(\nu_\psi) = \chi(\xi_1) = M^T_\psi \underset{\xi_1}{\pitchfork} M^T_\psi$, $\mathcal{S}_{0,1}(\nu_\psi) = \chi(\xi_2) = M^T_\psi \underset{\xi_2}{\pitchfork} M^T_\psi$. The oriented transversal self-intersections of the surface M^T_ψ are taken, correspondingly, in the spaces of the line bundles ξ_1, ξ_2 and "$\times$" stands for the cartesian product of bundles and manifolds (so that "$\rho_1 \times \rho_2$" actually means $\rho_1 \oplus \rho_2$). If one deals with the oriented bordisms, rather than with the stably complex ones, $\mathcal{S}_{0,0}(\nu_\psi) = M^T_\psi$ and P_1 vanish in (1.4).

Example 1.3. This example is provided by the natural $\mathbf{SU}(3)$-action on the Cayley projective plane $\mathcal{C}P_2$ of real dimension 16 (cf. [D]). The fixed point set of this action is $\mathbb{C}P_2$, the normal complex representation ψ is a sum of two copies of the standard $\mathbf{SU}(3)$-representation ρ in $\mathbb{C}^3$, the isotropy types of the action coincide with the isotropy types of ρ. Therefore, $\mathfrak{R} = \{\rho\}$ and the family $\mathcal{F} = \mathcal{F}_\rho$ consists of three copies of $\mathbf{SU}(2) \subset \mathbf{SU}(3)$ (permuted by the S_3-action), the trivial group, and $\mathbf{SU}(3)$ itself.

Now let M be a 16-dimensional $\mathbf{SU}(3)$-manifold, modeled after $\mathcal{C}P_2$. We denote by ξ^2 a complex 2-bundle over $M^{\mathbf{SU}(3)}$, so that $\nu_\psi \approx (\rho \oplus \rho) \otimes \xi^2$. Let $\pi : P(\xi^2) \xrightarrow{P_1} M^{\mathbf{SU}(3)}$ be the complex projectivization of ξ^2. Denote by $\tilde\zeta$ the tautological line bundle over $P(\xi^2)$ and by τ_π—the line bundle tangent to the fibers of π (so that $\tau_\pi \oplus 1_{\mathbb{C}} \approx 2\tilde\zeta^*$). We shall consider a real oriented surface

$$\mathcal{S}_{1,0} = \chi(\tilde\zeta \oplus \tau_\pi) = P(\xi^2) \underset{\tilde\zeta \oplus \tau_\pi}{\pitchfork} P(\xi^2)$$

—the transversal self-intersection of $P(\xi^2)$ in the space of $\tilde\zeta \oplus \tau_\pi$, as well as an oriented 0-dimensional manifold $\mathcal{S}_{2,0} = \chi(2\tilde\zeta \oplus \tau_\pi)$.

According to Theorem 1.1, the bordism class of the $\mathbf{SU}(3)$-bundle ν_ψ is determined by the bordism classes of the manifolds:

$$S_{0,0} = M^{\mathbf{SU}(3)}, \ S_{1,0} = \chi(\tilde\zeta \oplus \tau_\pi), \ S_{1,1} = \chi(\xi^2), \ S_{2,0} = \chi(2\tilde\zeta \oplus \tau_\pi)$$

of real dimensions $4, 2, 0, 0$ correspondingly. The exact expression of ν_ψ through the S_ω's can be calculated from (1.3). Hence, in stably-complex bordism $\Omega_*^{\mathbf{SU}(3)}$ $(\mathcal{F}, \mathcal{F}')$, one has an identity:

$$
\begin{aligned}
\nu_{\rho\oplus\rho} \ = \ & [\rho \times \rho] \times S_{0,0} + [\rho \times (\rho \otimes \eta_1) - \rho \times \rho \times P_1] \\
& \times S_{1,0} + [(\rho \otimes \eta_1) - \rho \times P_1]^2 \times S_{1,1} \\
& + [\rho \times (\rho \otimes \eta_2) - \rho \times (\rho \otimes \eta_1) \times P_1 + \rho \times \rho \times (P_1^2 - P_2)] \times S_{2,0},
\end{aligned}
$$

where $S_\omega = S_\omega(\nu_{\rho\oplus\rho})$. In oriented bordism, $P_1 = 0, S_{1,0} = 0$, so the formula above collapses to:

$$
\begin{aligned}
\nu_{\rho\oplus\rho} \ = \ & [\rho \times \rho] \times S_{0,0} + [(\rho \otimes \eta_1) \times (\rho \otimes \eta_1)] \times S_{1,1} + \\
& + \ [\rho \times (\rho \otimes \eta_2) - \rho \times \rho \times P_2] \times S_{2,0}.
\end{aligned}
\tag{1.5}
$$

2 Local formula for formal deformations of equivariant genera, arising from exponential operations in K_G-theory

In this section we shall derive a fixed-point formula for the invariant $\Phi_t(M) := \mathrm{ind}_t(D_\sigma, M) \in \mathcal{R}(\mathcal{G})[[\sqcup]]$ in terms of the invariants

$$\{\Phi_{\omega,t}(\nu_\Psi) := \mathrm{ind}_t(D_\sigma, S_\omega(\nu_\ominus)) \in \mathbb{C}[[\sqcup]]\}.$$

Due to Theorem 1.1, this derivation amounts to calculations of the invariants, $\Phi_t(\sim)$ for the projective spaces P_r's and the G-bundles $\rho \otimes \eta_r$'s. In principle,[2] while calculating $\Phi_t(M)$ there is no need to rely on the G-index formulae of [ASi]. In a sense, our computation can be viewed as an alternative proof of the classical equivariant index formulae (for the special D_σ's under consideration). Prior to the computation, we would like to recall a basic connection between the representation theory of compact Lie groups and the symbols of elliptic operators. A similar treatment can be found, for example, in [Ba].

2.1 $\mathcal{G}$-structure and the universal symbol

Let G be a compact Lie group and X a locally compact G-space. We denote by $K_G(X)$ the equivariant complex K-theory with compact support. Put $\mathcal{K}_G(X) =$

---

[2] Nevertheless we do take a short-cut employing [ASi].

$K_G(X) \otimes \mathbb{C}$. The ring of complex virtual finite dimensional G-representations will be denoted by $R(G)$ and $R(G) \otimes \mathbb{C}$—by $\mathcal{R}(G)$.

Let M be a $2n$-dimensional oriented G-manifold. Denote by τM its tangent bundle and by DM, SM—the associated disk and spherical bundles. Put $\tau_{\mathbb{C}} M := \tau M \otimes \mathbb{C}$. Let $\mathcal{G} = \mathcal{G}(n)$ be one of the following Lie groups: $\mathbf{U}(n)$, $\mathbf{SO}(2n)$, $\mathbf{Spin}(2n)$. Denote by $\mathbf{T}(n) \subset \mathbf{U}(n) \subset \mathbf{SO}(2n)$ the maximal torus.

Let ξ be an oriented $2n$-dimensional real bundle over M. We shall say that ξ admits a $\mathcal{G}(n)$-*structure*, if there exists a principal $\mathcal{G}(n)$-fibration $\mathcal{P}_{\mathcal{G}}\xi \longrightarrow M$ over M, so that ξ is orientably isomorphic to the bundle $\mathcal{P}_{\mathcal{G}}\xi \times_{\mathcal{G}} \mathbb{R}^{2n} \longrightarrow M$, the $\mathcal{G}$-structure in $\mathbb{R}^{2n}$ being the obvious one. In particular, we shall say that M admits a $\mathcal{G}$-structure if τM does. In this case we shall abbreviate $\mathcal{P}_{\mathcal{G}}(\tau M)$ as $\mathcal{P}_{\mathcal{G}} M$.

Starting with an element $\sigma \in K_{\mathcal{G}(n)}(D^{2n}, S^{2n-1})$, in a quite functorial way, one can construct an element σ_ξ of $K(D\xi, S\xi)$ [ASi]. Let us recall this construction. First, consider the fragment of the long exact sequence in the $K_{\mathcal{G}}$-theory:

$$\longrightarrow K_{\mathcal{G}}(D^{2n}, S^{2n-1}) \xrightarrow{j^*} K_{\mathcal{G}}(D^{2n}) \xrightarrow{i^*} K_{\mathcal{G}}(S^{2n-1}).$$

One can identify the ring $K_{\mathcal{G}}(D^{2n}) \approx K_{\mathcal{G}}(pt)$ with $R(\mathcal{G})$ and $K_{\mathcal{G}}(S^{2n-1})$ with $K_{\mathcal{H}}(pt) \approx R(\mathcal{H})$, where $\mathcal{H} = \mathbf{U}(n-1)$, $\mathbf{SO}(2n-1)$, $\mathbf{Spin}(2n-1)$ are the isotropy groups of a vector from the homogeneous space $\mathcal{G}/\mathcal{H} \approx S^{2n-1} \subset \mathbb{R}^{2n}$. The homomorphism i^* from the exact sequence above can be identified with the restriction homomorphism $\mathrm{res}_{\mathcal{H}} : R(\mathcal{G}) \longrightarrow R(\mathcal{H})$. Thus, $\mathrm{Im} j^* = \mathrm{Ker}\, i^*$ consists of the characters which vanish under $\mathrm{res}_{\mathcal{H}}$. The map $\mathrm{res}_{\mathcal{H}}$ is classically well-understood (cf. [Hu], Chap13, §10): in fact, $\chi \in \mathrm{Ker}(\mathrm{res}_{\mathcal{H}})$ if and only if it is divisible in $R(\mathcal{G})$ by the character

$$
\sigma_n^\dagger =
\begin{cases}
\displaystyle\prod_{i=1}^{n}(1 - \alpha_i) & \text{for } \mathcal{G} = \mathbf{U}(n), \\[2.5ex]
\displaystyle\prod_{i=1}^{n}(\alpha_i^{-1} - \alpha_i) & \text{for } \mathcal{G} = \mathbf{SO}(2n), \\[2.5ex]
\displaystyle\prod_{i=1}^{n}(\alpha_i^{-\frac{1}{2}} - \alpha_i^{\frac{1}{2}}) & \text{for } \mathcal{G} = \mathbf{Spin}(2n).
\end{cases}
$$

Here the α_i's are the basic characters from $R(\mathbf{T}(n))$ and $\alpha_i^{\frac{1}{2}}$'s are interpreted as basic characters on the maximal torus $\widetilde{\mathbf{T}}(n) \subset \mathbf{Spin}(2n)$. Recall, that $\mathrm{res}_{\mathcal{H}}$ is identical on $\alpha_1, \dots, \alpha_{n-1}$ and takes α_n to 1. As a result, $\mathrm{Im} j^*$ coincides with the ideal $\sigma_n^\dagger \cdot R(\mathcal{G}) \lhd R(\mathcal{G})$. In fact, one can show that $K_{\mathcal{G}}(D^{2n}, S^{2n-1})$ can be canonically identified with the ideal $\sigma_n^\dagger \cdot R(\mathcal{G})$.

Let $q : D\xi \longrightarrow M$ denote the natural projection. Given an element $\sigma \in \sigma_n^\dagger \cdot R(\mathcal{G})$, represented as a virtual difference of $\mathcal{G}$-representations σ^+, σ^-, so that $\mathrm{res}_{\mathcal{H}} \sigma^+ \approx \mathrm{res}_{\mathcal{H}} \sigma^-$, form vector bundles $\xi_\sigma^\pm := \mathcal{P}_{\mathcal{G}}\xi \times_{\mathcal{G}} (V_\sigma^\pm)$ over M, where $V_\sigma^\pm$ denote the spaces of the $\mathcal{G}$-representations $\sigma^\pm$. Define an isomorphism $\gamma_\sigma = \gamma(\sigma^+, \sigma^-): \ q^*(\xi_\sigma^+)\big|_{S\xi} \longrightarrow q^*(\xi_\sigma^-)\big|_{S\xi}$ by the formula

$$\gamma_\sigma(p \times [w, v]) = p \times [w, \quad \sigma_-(g_w^{-1}) \cdot \sigma_+(g_w)v \ ].$$

Here $p \in \mathcal{P}_\mathcal{G}\xi$, $w \in S^{2n-1}$, $v \in V_\sigma^+$ and $g_w \in \mathcal{G}$ is an element whose class in $\mathcal{G}/\mathcal{H}$ is w. The triple $(q^*\xi_\sigma^+, q^*\xi_\sigma^-, \gamma_\sigma)$ defines an element $\sigma_\xi \in K(D\xi, S\xi)$. We shall refer to it as obtained by the *universal construction* from $\sigma \in \sigma_n^\dagger \cdot R(\mathcal{G}(n))$.

Suppose a Lie group G acts as a group of bundle transformations of $\xi \longrightarrow M$, and the G-action on M lifts to a G-action on the principal $\mathcal{G}$-fibration $\mathcal{P}_\mathcal{G}\xi \longrightarrow M$ in such a way that: (i) it commutes with the natural right $\mathcal{G}$-action on $\mathcal{P}_\mathcal{G}\xi$, and (ii) the G-bundles $\mathcal{P}_\mathcal{G}\xi \times_\mathcal{G} V \longrightarrow M$ and $\xi \longrightarrow M$ are G-isomorphic. Then, the universal construction is G-equivariant and σ gives rise to an element $\sigma_\xi \in K_G(D\xi, S\xi)$. We shall view σ_ξ as a symbol of a pseudodifferential operator D_σ related to the $\mathcal{G}$-structure on ξ and mapping sections of the bundle ξ_σ^+ to the ones of ξ_σ^-.

From this moment on we shall restrict our discussion to special "*totally splittable*" σ's from $\sigma_n^\dagger \cdot R(\mathcal{G})$ and to the corresponding D_σ's. To take advantage of the Borel-Hirzebruch type formalism, we shall consider only $\sigma = \sigma_n$'s whose characters are of the form:

$$\sigma_n = \prod_{i=1}^n [\sigma^\dagger(\alpha_i) \cdot Q(\alpha_i)], \tag{2.1.1}$$

where

$$\sigma^\dagger(u) = \begin{cases} 1 - u & \text{for } \mathcal{G} = \mathbf{U}(n), \\ u^{-1} - u & \text{for } \mathcal{G} = \mathbf{SO}(2n), \\ u^{-\frac{1}{2}} - u^{\frac{1}{2}} & \text{for } \mathcal{G} = \mathbf{Spin}\ (2n), \end{cases}$$

$Q(u) \in \mathbb{Z}[u, u^{-1}], Q(1) \neq 0$, and, for $\mathcal{G} = \mathbf{SO}(2n)$, $\mathbf{Spin}(2n)$, $Q(u^{-1}) = Q(u)$. Evidently, $\sigma_n \in \sigma_n^\dagger \cdot R(\mathcal{G})$ and therefore, it gives rise to a symbol and an operator. If $Q(u) = 1$, the appropriate operators are [ASi]:

(i) the "Euler-Todd operator" $d + \delta : \ C^\infty(\Lambda^{\mathrm{ev}}(\xi)) \longrightarrow C^\infty(\Lambda^{\mathrm{odd}}(\xi))$ for $\mathcal{G} = \mathbf{U}(n)$,

(ii) the Signature operator $d + \delta : \ C^\infty(\Lambda_+(\xi_\mathbb{C})) \longrightarrow C^\infty(\Lambda_-(\xi_\mathbb{C}))$ for $\mathcal{G} = \mathbf{SO}(2n)$,

(iii) the Dirac operator $D : \ C^\infty(\Delta_+(\xi)) \longrightarrow C^\infty(\Delta_-(\xi))$ for $\mathcal{G} = \mathbf{Spin}(2n)$.

They are produced, correspondingly, by the complex $\mathcal{G}$-representations:

(i) $\sigma_n^+ = \Lambda^{\mathrm{ev}}(\mathbb{C}^n), \sigma_n^- = \Lambda^{\mathrm{odd}}(\mathbb{C}^n)$,

(ii) $\sigma_n^\pm = \Lambda_\pm(\mathbb{C}^{2n})$,

(iii) $\sigma_n^\pm = \Delta_\pm(\mathbb{R}^{2n})$.

Here $\Lambda^{\mathrm{ev}}(\mathbb{C}^n)$, $\Lambda^{\mathrm{odd}}(\mathbb{C}^n)$ stand for the $\mathcal{G}$-modules of the even- and odd- dimensional exterior $\mathbb{C}$-forms on $\mathbb{C}^n$, $\Lambda_{\pm}(\mathbb{C}^{2n})$—are the (± 1)-eigenspaces of an involution τ acting of $\Lambda^*(\mathbb{C}^{2n})$ and related to the Hodge operator $*$, and $\Delta_{\pm}(\mathbb{R}^{2n})$ are the $(\pm)$-eigenspaces of an involution ω (the Cliford multiplication by the volume element ω) acting on the space of the standard $\mathbf{Spin}(2n)$-representation $\Delta(\mathbb{R}^{2n})$. We shall abbreviate $\sigma_\tau M$ as $\sigma_M \in K_G(DM, SM)$.

2.2 Formal deformations of equivariant genera

Now, let us consider the Gysin homomorphism in K_G-theory—the transfer $\pi_!^M$: $K_G(\tau M) \longrightarrow K_G(pt) \approx R(G)$ (cf. [ASi]). By the Atiyah-Singer Index Theorem [ASi], the $\pi_!^M$-image of the symbol $\sigma_M \in K_G(DM, SM) = K_G(\tau M)$ is the G-equivariant index $\mathrm{ind}(D_\sigma, M)$ of the operator D_σ.

Our goal is to obtain a local formula for computing $\mathrm{ind}(D_\sigma, M) = \pi_!^M(\sigma_M)$ in terms of the G-equivariant normal bundle $\nu = \nu(M^G, M)$ (Theorem 2.4). Moreover, we shall derive such formulae also for formal deformations of $\mathrm{ind}(D_\sigma, M)$ (Theorem 2.3). Such a deformation is produced employing a stable exponential operation

$$\phi: K_G(X) \longrightarrow \mathcal{K}_G(X)[[t]].$$

In fact, any formal sum

$$\varphi(t, u) = \sum_{j \geq 0} \varphi_j(u) \, t^j, \qquad \varphi_j(u) \in \mathbb{C}[u, u^{-1}], \qquad (2.2.1)$$

(with $\varphi_0 \in \mathbb{C}^\times$) gives rise to an exponential operation ϕ of this sort. First note that, for a complex line bundle η over X, $\phi(\eta) := \varphi(t, \eta)$ is a well-defined element of $\mathcal{K}_G(X)[[t]]$. Here $\eta^{-1} = \eta^*$ is the $\mathbb{C}$-dual of η. If a complex n-dimensional bundle ξ splits into a Whitney sum $\bigoplus_{1 \leq \ell \leq n} \eta_\ell$ of line bundles $\{\eta_\ell\}$, then put $\phi(\xi) = \prod_{1 \leq \ell \leq n} \phi(\eta_\ell)$. This product can be developed into a sum:

$$\phi(\xi) = \sum_{j \geq 0} Q_{n,j}^\varphi(\sigma_1, \sigma_2, \dots, \sigma_n, \sigma_n^{-1}) \, t^j,$$

where $\{Q_{n,j}^\varphi\}$ are certain "universal" polynomials and σ_i stands for the i-th elementary symmetric polynomial in the variables $\{\eta_\ell\}$. Since $\sigma_i(\eta_1, \dots, \eta_n)$ can be identified with the i-th exterior power $\Lambda^i \xi$, one has

$$\phi(\xi) = \sum_{j \geq 0} Q_{n,j}^\varphi(\Lambda^1 \xi, \Lambda^2 \xi, \dots, \Lambda^n \xi, \Lambda^n \xi^*) \, t^j, \qquad (2.2.2)$$

$\Lambda^n \xi^*$ being the line bundle $\mathbb{C}$-dual of $\Lambda^n \xi$. Evidently, (2.2.2) makes perfect sense as an element of $\mathcal{K}_G(X)[[t]]$ for *any* n-dimensional G-bundle ξ (and not only for the splittable ones). In particular, ϕ is well-defined for complex n-dimensional G-representations. It can be verified that ϕ in (2.2.2) has the exponential property:

$$\phi(\xi' \oplus \xi'') = \phi(\xi') \cdot \phi(\xi'').$$

This defines an operation $\phi : K_G(X) \longrightarrow \mathcal{K}_G(X)[[t]]$. The transfer $\pi_!^M :$ $K_G(\tau M) \longrightarrow R(G)$ extends in an obvious way to an $R(G)$-module homomorphism

$$\pi_!^M : K_G(\tau M)[[t]] \longrightarrow R(G)[[t]]. \qquad (2.2.3)$$

For a $2n$-dimensional M admitting a G-equivariant $\mathcal{G}$-structure, put

$$\Phi(M) = \mathrm{ind}_t(D_\sigma, M) = \begin{cases} \pi_!^M(\sigma_M \cdot \phi(\tau M)) & \text{for } \mathcal{G} = \mathbf{U}(n) \\ \pi_!^M(\sigma_M \cdot \phi(\tau_\mathbb{C} M)) & \text{for } \mathcal{G} = \mathbf{SO}(2n), \ \mathbf{Spin}(2n), \end{cases} \qquad (2.2.4)$$

where the symbol $\sigma_M \in K_G(\tau M)$ is produced by the universal construction from a virtual $\mathcal{G}$-representation σ_n described in (2.1.1).

One can think of $\Phi(M) \in \mathcal{R}(G)[[t]]$ as a formal sum $\sum_{j \geq 0} \chi_j \cdot t^j$, where the "characters" $\chi_j \in \mathcal{R}(G)$ are the indices of the operator $\tilde{D}_\sigma$, twisted with the complex virtual tensorial bundles

$$Q_{2n,j}^\varphi(\Lambda^1(\tau_\mathbb{C} M), \dots, \Lambda^{2n}(\tau_\mathbb{C} M), \Lambda^{2n}(\tau_\mathbb{C}^* M)), \qquad j = 0, 1, 2, \dots,$$

(cf. (2.2.2)).

By a Stokes' Theorem-type argument in K_G-theory (cf. [Se]) applied to the Thom-Pontrjagin construction, $\Phi(M)$ is an invariant of the G-bordism class of M in the stably complex bordism for $\mathcal{G} = \mathbf{U}(n)$, oriented bordism for $\mathcal{G} = \mathbf{SO}(2n)$ and $\mathbf{Spin}$-bordism for $\mathcal{G} = \mathbf{Spin}(2n)$. The appropriate bordism rings are denoted by $_G\Omega_*^G$ or just by Ω_*^G. It is easy to check that $\Phi(M_1 \times M_2) = \Phi(M_1) \cdot \Phi(M_2)$. Therefore, any character σ as in (2.2.1) and an element $\varphi(t, u) \in \mathbb{C}[u, u^{-1}][[t]]$ with $\varphi_0(u) \in \mathbb{C}^\times$ give rise to a ring homomorphism

$$\Phi : \ _G\Omega_*^G \longrightarrow \mathcal{R}(G)[[t]].$$

2.3 Localization and local formulae in equivariant K-theory

Let the normal bundle $\nu(M^G, M)$ admit a G-equivariant *complex* structure when $\mathcal{G} = \mathbf{U}(n)$ or $\mathbf{SO}(2n)$ and a *quaternionic* one when $\mathcal{G} = \mathbf{Spin}(2n)$. We shall generally follow the Atiyah-Segal treatment [ASe] while discussing localization in K_G-theory. Let the normal bundle $\nu = \nu(M^G, M)$, the list $\mathfrak{R}$ of irreducible representations $\{\rho\}$ and the families $\mathcal{F} = \mathcal{F}_\mathfrak{R}, \mathcal{F}' = \mathcal{F}'_\mathfrak{R}$ be as in Section 1.

With each $\rho \in \mathfrak{R}$ we associate a virtual G-representation

$$\lambda_{-1}(\rho) = \sum_{i \geq 0} (-1)^i \Lambda^i \rho.$$

Denote the character of $\lambda_{-1}(\rho)$ by χ_ρ. Let $\Lambda_\mathfrak{R}$ be a multiplicative system in $R(G)$, generated by $\{\lambda_{-1}(\rho_\mathbb{C})\}$. Our intention is to invert the elements of $\Lambda_\mathfrak{R}$. Hence, we consider $K_G(\sim)_{\Lambda_\mathfrak{R}}$—equivariant complex K-theory, localized at $\Lambda_\mathfrak{R}$. It is a module over the ring $R(G)_{\Lambda_\mathfrak{R}}$.

Lemma 2.1. *Let $\mathfrak{R}$ be a finite family of complex (quaternionic) irreducible G-representations ρ. Assume that a G-action on M is such, that any normal G-representation ψ in $\nu = \nu(M^G, M)$ is of the type $\bigoplus_{\rho \in \mathfrak{R}} n_\rho^\psi \cdot \rho$.*
Then, the Thom-Bott element $\lambda_{-1}(\nu_\mathbb{C})$ is invertible in $K_G(M^G)_{\Lambda_\mathfrak{R}}$.

Proof. By an argument similar to one in [ASe, Lemma 2.2], an element of $K_G(M^G)_{\Lambda_\mathfrak{R}}$ is invertible, if and only if its restriction at any point of M^G is invertible in $R(G)_{\Lambda_\mathfrak{R}}$. The character χ of $\lambda_{-1}(\psi)$ is given by the formula: $\chi(g) = \mathrm{Det}(I - \psi(g))$, $g \in G$. Note that $\chi(\sim) \neq 0$ exactly at the set $G \setminus \bigcup_{x \in S_\psi} G_x$, where S_ψ stands for the unit sphere in the space of ψ and G_x is the isotropy group of x. Since $\psi = \bigoplus_{\rho \in \mathfrak{R}} n_\rho^\psi \cdot \rho$, χ is a monomial in $\{\chi_\rho, \chi_{\rho^*}\}$. Thus, $\chi \in \Lambda_\mathfrak{R}$ and becomes invertible in $R(G)_{\Lambda_\mathfrak{R}}$. $\qquad\square$

By the formula $\chi(g) = \mathrm{Det}(I - \psi(g))$, the character $\chi_\rho \in \Lambda_\mathfrak{R}$ being restricted to G_x, $x \in S_\rho$, vanishes. Therefore, by recalling the definition of the family $\mathcal{F}_\mathfrak{R}$ and by a Mayer-Vietoris spectral sequence argument as in [ASe], the inclusion $i : (\tau M)^G \hookrightarrow \tau M$ induces an isomorphism $i_* : K_G(\tau M^G)_{\Lambda_\mathfrak{R}} \longrightarrow K_G(\tau M)_{\Lambda_\mathfrak{R}}$. This (again, as in [ASe]), combined with the functoriality $\pi_!^M \circ i_! = \pi_!^{M^G}$ of the transfer map, leads to the formula in $R(G)_{\Lambda_\mathfrak{R}}$:

$$\pi_!^M(\xi) = \pi_!^{M^G}\left(\frac{i^*\xi}{\lambda_{-1}(\nu_\mathbb{C})}\right), \tag{2.3.1}$$

where $\xi \in K_G(\tau M)$, $\nu = \nu(M^G, M)$ and $K_G(\tau M)$ is viewed as a $K_G(M)$-module.

Applying (2.3.1) to the invariant (2.2.4), we get the following equality in $\mathcal{R}(G)[[t]]_{\Lambda_\mathfrak{R}}$:

$$\Phi(M) = \mathrm{ind}_t(D_\sigma, M) = \pi_!^{M^G}\left[\frac{i^*\sigma_M}{\lambda_{-1}(\nu_\mathbb{C})} \cdot \phi(i^*\tau_\mathbb{C} M)\right], \tag{2.3.2}$$

which is valid for $\mathcal{G} = \mathbf{SO}(2n)$, $\mathbf{Spin}(2n)$. When $\mathcal{G} = \mathbf{U}(n)$, one replaces $\tau_\mathbb{C} M$ by τM in (2.3.2). Note, that $i^*\tau M \approx \tau M^G \oplus \nu$. By the exponentiality of ϕ, $\phi(i^*\tau_\mathbb{C} M) = \phi(\tau_\mathbb{C} M^G) \cdot \phi(\nu_\mathbb{C})$. According to our basic assumption about ν, in a tubular neighborhood $D\nu_\psi$ of M_ψ^G in M, the $\mathcal{G}$-structure reduces to: $\mathbf{SO}(2k) \times \mathbf{U}(n-k)$ when $\mathcal{G} = \mathbf{SO}(2n)$, to $\mathbf{Spin}(2k) \times \mathbf{Sp}(\frac{n-k}{2})$ when $\mathcal{G} = \mathbf{Spin}(2n)$ and to $\mathbf{U}(k) \times \mathbf{U}(n-k)$ when $\mathcal{G} = \mathbf{U}(n)$ ($2k = \dim M_\psi^G$). This makes it possible to split the tangent bundle τ of $D\nu_\psi$ into a sum of its normal component τ_ν (formed by vectors along the fibers of $D\nu_\psi \longrightarrow M_\psi^G$) and its orthogonal complement $\tau_\nu^\perp$. Because of the special "splitting" property of the character σ in (2.1.1) and by the universal construction, the symbol σ_M is the exterior tensor product $\sigma_{\tau_\nu} \widehat{\otimes} \sigma_{\tau_\nu^\perp}$ of the symbols $\sigma_{\tau_\nu} \in K_G(D\tau_\nu, S\tau_\nu)$ and $\sigma_{\tau_\nu^\perp} \in K_G(D\tau_\nu^\perp, S\tau_\nu^\perp)$. The exterior pairing

$$K_G(D\tau_\nu, S\tau_\nu) \widehat{\otimes} K_G(D\tau_\nu^\perp, S\tau_\nu^\perp) \longrightarrow K_G(D\tau, S\tau)$$

is given by the formula $\xi \widehat{\otimes} \eta = p_\nu^*(\xi) \otimes p_{\nu^\perp}^*(\eta)$, where $\xi \in K_G(D\tau_\nu, S\tau_\nu)$, $\eta \in K_G(D\tau_\nu^\perp, S\tau_\nu^\perp)$ and $p_\nu : D\tau \longrightarrow D\tau_\nu$, $p_{\nu^\perp} : D\tau \longrightarrow D\tau_{\nu^\perp}$ are the obvious projections. Therefore, $i^* \sigma_M = i^*(\sigma_{\tau_\nu} \widehat{\otimes} \sigma_{\tau^\perp}) = i^*_\nu(\sigma_{\tau_\nu}) \otimes \sigma_{M^G}$, $i_\nu : M^G \longrightarrow$ $D(\tau_\nu)$ being the natural imbedding (the zero section). Note that $i^*_\nu(\sigma_{\tau_\nu}) = \nu^+_\sigma - \nu^-_\sigma$. Thus, (2.3.2) can be written as

$$\Phi(M) = \pi_!^{M^G}\left\{ \left[\frac{(\nu^+_\sigma - \nu^-_\sigma)}{\lambda_{-1}(\nu_{\mathbb{C}})} \cdot \phi(\nu_{\mathbb{C}})\right] \cdot \left[\sigma_{M^G} \cdot \phi(\tau_{\mathbb{C}} M^G)\right]\right\}, \qquad (2.3.3)$$

where $\nu^\pm_\sigma \in K_G(M^G), \sigma_{M^G} \in K_G(\tau M^G)$.

Again, by a Stokes' Theorem-type argument, it is possible to verify that, for any closed oriented (almost complex or **Spin**-) manifold L and a complex (correspondingly, complex or quaternionic) G-bundle $\xi \approx \bigoplus_{\rho \in \mathfrak{R}} \rho \otimes \xi_\rho$, the element

$$\Phi(L, \xi) := \pi_!^L\left\{ \left[\frac{(\xi^+_\sigma - \xi^-_\sigma)}{\lambda_{-1}(\xi_{\mathbb{C}})} \cdot \phi(\xi_{\mathbb{C}})\right] \times \left[\sigma_L \cdot \phi(\tau_{\mathbb{C}} L)\right]\right\} \qquad (2.3.4)$$

in $\mathcal{R}(G)[[t]]_{\Lambda_\mathfrak{R}}$ is an invariant of the oriented (correspondingly, stably complex or **Spin**-) bordism class of L equipped with the G-bundle data ξ. It is straightforward to check that $\Phi(L' \amalg L'', \xi' \amalg \xi'') = \Phi(L', \xi') + \Phi(L'', \xi'')$ and $\Phi(L' \times L'', \xi' \times \xi'') = \Phi(L', \xi') \cdot \Phi(L'', \xi'')$.

As (2.3.4) demonstrates, it is possible to calculate $\Phi(M) \in \mathcal{R}(G)[[t]]_{\Lambda_\mathfrak{R}}$ just in terms of the class $[M] \in \Omega^G_*(\mathcal{F}, \mathcal{F}')$, $\mathcal{F} = \mathcal{F}_\mathfrak{R}$, or equivalently, in terms of the G-bordism class of the pair (M^G, ν).

2.4   Computing $\mathrm{ind}_t(D_\sigma, \sim)$ for the generators

The next basic observation is that, for a complex ν_ψ and an oriented or stably complex M^G_ψ, by (1.1.1), all the characteristic manifolds $\{S_w(\nu_\psi)\}$ are oriented or correspondingly, stably complex. Note, that all the G-bundles $\rho \otimes \eta_r$'s over the P_r's are complex G-bundles over complex varieties. Similarly, if ν_ψ is quaternionic and M^G_ψ admits a **Spin**-structure, the $S_w(\nu_\psi)$'s are **Spin**-manifolds. Also note that $\rho \otimes_\mathbb{H} \eta_r$'s are symplectic (and, hence, **Spin**-) bundles over the **Spin**-manifolds P_r's, Therefore, under the assumptions above, all the ingredients of (1.2), (1.3), as well as the corresponding cobordisms, lie in the same category, and Theorem 1.1 is applicable to the computation of $\Phi(M)$.

We start with the computation of $\Phi(P_r, \rho \otimes \eta_r)$.

Let ℓ denote the trivial bundle $\mathrm{Hom}_\mathbb{C}(\eta^*_r, \eta^*_r) \approx P_r \times \mathbb{C}$ when P_r is a complex projective space and the **Spin**(4)-bundle $\mathrm{Hom}_\mathbb{H}(\eta^*_r, \eta^*_r)$ when P_r is a quaternionic projective space. Since $\tau P_r \oplus \ell \approx (r+1)\eta_r$, $\phi(\tau_{\mathbb{C}} P_r) = \phi((\eta_r)_{\mathbb{C}})^{r+1}/\phi(\ell_{\mathbb{C}})$. Note, that $\varphi_0(u) \in \mathbb{C}^\times$ implies that $\phi(\ell_{\mathbb{C}})$ is invertible in $\mathcal{K}_G(P_r)[[t]]$.

Therefore,

$$\Phi(P_r, \rho \otimes \eta_r) = \pi_!^{P_r}\left\{ \left[\frac{(\rho \otimes \eta_r)^+_\sigma - (\rho \otimes \eta_r)^-_\sigma}{\lambda_{-1}((\rho \otimes \eta_r)_{\mathbb{C}})} \cdot \phi((\rho \otimes \eta_r)_{\mathbb{C}})\right] \times \left[\sigma_{P_r} \cdot \frac{\phi((\eta_r)_{\mathbb{C}})^{r+1}}{\phi(\ell_{\mathbb{C}})}\right]\right\}$$
$$(2.4.1)$$

where $\sigma_{P_r} \in K(\tau P_r)$ ($\mathcal{G} = \mathbf{SO}(2n)$, $\mathbf{Spin}(2n)$). For $\mathcal{G} = \mathbf{U}(n)$ one replaces $\phi((\rho \otimes \eta_r)_{\mathbb{C}})$ by $\phi(\rho \otimes \eta_r)$, $\phi((\eta_r)_{\mathbb{C}})$ by $\phi(\eta_r)$ and $\phi(\ell_{\mathbb{C}})$ by $\phi(\ell)$ in (2.4.1).

We would like to elaborate upon the computation of $(\rho \otimes \eta_r)_\sigma^+ - (\rho \otimes \eta_r)_\sigma^- \in K_G(P_r)$, where $\sigma = \sigma_k \in \sigma_k^\dagger \cdot R(\mathcal{G})$ ($k = \dim_{\mathbb{C}} \rho$) has been introduced in (2.1.1).

Let $\beta : \mathcal{G}(k) \longrightarrow \mathbf{SO}(2k)$ be the standard homomorphism (monic for $\mathcal{G}(k) = \mathbf{U}(k), \mathbf{SO}(2k)$). The corresponding $\mathcal{G}(k)$-module $\mathbb{R}^{2k}$ will be denoted by V_β.

One can represent σ as $\sigma^+ - \sigma^-$, where $\sigma^\pm$ are some integral polynomials (depending on k) with *positive* coefficients in the variables

(i) $\qquad \{\Lambda^i(V_\beta)\}_{1 \le i \le k}, \quad \Lambda^k(V_\beta^*) \qquad$ when $\mathcal{G} = \mathbf{U}(k),$]

(ii) $\qquad \{\Lambda^i(V_{\beta,\mathbb{C}})\}_{1 \le i \le k-1}, \quad \Lambda_\pm^*(V_{\beta,\mathbb{C}}) \quad$ when $\mathcal{G} = \mathbf{SO}(2k)$,

(iii) $\qquad \{\Lambda^i(V_{\beta,\mathbb{C}})\}_{1 \le i \le k-2}, \quad \Delta_\pm(V_{\beta,\mathbb{C}}) \quad$ when $\mathcal{G} = \mathbf{Spin}(2k)$

(although $\sigma^\pm$ are not uniquely defined, $\sigma = \sigma^+ - \sigma^-$ is).

If, in addition, V_β is a G-module via a representation $\psi : G \longrightarrow \mathbf{SO}(V_\beta)$ and the $\mathcal{G}$-structure is G-equivariant, then, evidently, (i)–(iii) are complex G-representations. Hence, so are the $\sigma^\pm$. We shall denote them $\sigma_\psi^\pm$. Note, that when the centralizer $\mathcal{Z}\psi$ of ψ in $\mathbf{SO}(V_\beta)$ contains the operator $i, i^2 = -id$, V_β admits a G-equivariant $\mathbf{U}(k)$-structure. When $\mathcal{Z}\psi$ contains the operators I, J, K (related to a quaternionic structure in V_β), then V_β admits an equivariant $\mathbf{Sp}(k/2)$—and, hence, an equivariant $\mathbf{Spin}(2k)$-structure as well. Therefore, under the hypotheses on $\mathcal{Z}\psi$ above, $\sigma_\psi^\pm$ are complex G-modules. Let $\chi_\psi^\pm$ stand for the characters of the G-representations $\sigma_\psi^\pm$. If, in the first place, V_β is a complex G-module, $\psi(g)$, viewed as an unitary transformation, is diagonalizable and, choosing the maximal torus in $\mathcal{G}$ in a way compatible with the diagonalization, by (2.1.1), we get:

$$\chi_\psi(g) := \chi_\psi^+(g) - \chi_\psi^-(g) = \prod_{\lambda \in \mathrm{Spec}\,\psi(g)} \sigma(\lambda).$$

Here Spec A stands for the spectrum of a $k \times k$-matrix A, the eigenvalues of A being taken with the appropriate multiplicities (in other words, Spec A is viewed as an effective divisor in $\mathbb{C}$) and $\sigma(\lambda) = \sigma_1^\dagger(\lambda) \cdot Q(\lambda)$ (cf. (2.1.1)).

For a complex projective space P_r, one can "spread" the argument above "along the P_r" via a balanced-product construction which identifies $(\rho \otimes \eta_r)_\sigma^\pm$ with $S^{2r+1} \times_{\mathbf{U}(1)} [\sigma_\rho^\pm] \longrightarrow P_r$:

$$[(\rho \otimes \eta_r)_\sigma^+ - (\rho \otimes \eta_r)_\sigma^-](g) = \prod_{\lambda \in \mathrm{Spec}\,\rho(g)} \sigma(\lambda \cdot \eta_r), \quad g \in G. \qquad (2.4.2)$$

Also,

$$\lambda_{-1}((\rho \otimes \eta_r)_{\mathbb{C}})(g) = \prod_{\lambda \in \mathrm{Spec}\,\rho(g)} (1 - \lambda\eta_r)(1 - \lambda^{-1}\eta_r^{-1})$$

and

$$\phi((\rho \otimes \eta_r)c)(g) = \prod_{\lambda \in \operatorname{Spec}\rho(g)} \varphi_t(\lambda \eta_r)\, \varphi_t(\lambda^{-1}\eta_r^{-1}).$$

Therefore, in the localized ring $\{\mathcal{R}(G) \otimes K(P_r)\}[[t]]_{\Lambda_{\mathfrak{R}}}$, employing (2.4.2), the fragment of (2.4.1) in the square brackets (preceding the multiplication sign "$\times$") is equal to

$$\prod_{\lambda \in \operatorname{Spec}\rho(g)} \frac{\sigma(\lambda \eta_r)}{(1 - \lambda \eta_r)(1 - \lambda^{-1}\eta_r^{-1})}\, \tilde{\varphi}_t(\lambda \cdot \eta_r), \qquad (2.4.3)$$

where

$$\sigma(u) = \begin{cases} (1 - u)Q(u) \\ (u^{-1} - u)Q(u^{-1} + u) \end{cases} \text{ and } \tilde{\varphi}_t(u) = \begin{cases} \varphi_t(u) & \text{for } \mathcal{G} = \mathbf{U} \\ \varphi_t(u) \cdot \varphi_t(u^{-1}) & \text{for } \mathcal{G} = \mathbf{SO}. \end{cases} \qquad (2.4.4)$$

At this point, we shall introduce an important expression $H_t(u)$ (an element of the ring $\{\mathbb{C}[u, u^{-1}][[t]]\}_{[1-u]}$ when $\mathcal{G} = \mathbf{U}, \mathbf{SO}$ and of the ring $\{\mathbb{C}[u^{\frac{1}{2}}, u^{-\frac{1}{2}}][[t]]\}_{[1-u]}$ when $\mathcal{G} = \mathbf{Spin}$) in terms of the auxiliary expressions $S^{\dagger}(u), \tilde{Q}(u), \tilde{\varphi}_t(u)$ described in the table below:

$\mathcal{G}$	$S^{\dagger}(u) =$	$\tilde{Q}(u) =$	$\tilde{\varphi}_t(u) =$
$\mathbf{U}$	$u/(u-1)$	$Q(u) \in \mathbb{Z}[u, u^{-1}]$	$\varphi_t(u) \in \mathbb{C}[u, u^{-1}][[t]]$
$\mathbf{SO}$	$(u+1)/(u-1)$	$Q(u) \in \mathbb{Z}[u + u^{-1}]$	$\varphi_t(u) \cdot \varphi_t(u^{-1})$
$\mathbf{Spin}$	$\sqrt{u}/(u-1)$	$Q(u) \in \mathbb{Z}[u + u^{-1}]$	$\varphi_t(u) \cdot \varphi_t(u^{-1})$

Let

$$H_t(u) := S(u) \cdot \tilde{\varphi}_t(u) := S^{\dagger}(u) \cdot \tilde{Q}(u) \cdot \tilde{\varphi}_t(u). \qquad (2.4.5)$$

Note that, for $\mathcal{G} = \mathbf{SO}, \mathbf{Spin}$, $H_t(u^{-1}) = -H_t(u)$. In the new notations, (2.4.3) transforms into

$$\prod_{\lambda \in \operatorname{Spec}\rho(g)} H_t(\lambda \eta_r). \qquad (2.4.6)$$

Note that (2.4.6) is well-defined for $\lambda \neq 1$ and, hence, can be regarded as partially-defined central functions $G \longmapsto \mathcal{K}(P_r)[[t]]$.

For a trivial G-space X, let $\operatorname{ch} : K_G(X) \longrightarrow H^{**}(X; \mathbb{Q}) \otimes R(G)$ be the equivariant Chern character defined by the formula $\operatorname{ch}(\bigoplus_{\rho}(\rho \otimes \xi_r)) = \sum_{\rho} \rho \otimes \operatorname{ch}(\xi_{\rho})$. Denote by $\operatorname{Td}_{\mathbb{C}}(\xi)$ the Todd class of the complexification $\xi_{\mathbb{C}}$. For any $\xi \in K_G(\tau L), L$ carrying a trivial G-action, by a topological Index Formula [ASi],

$$\begin{aligned} \pi_!^L(\xi) &= (-1)^d \langle \operatorname{ch}(\xi) \cdot \operatorname{Td}_{\mathbb{C}}(\tau L), [\tau L] \rangle = \\ &= (-1)^{\frac{1}{2}d(d+1)} \langle \chi(\tau L)^{-1} \cdot \operatorname{ch}(\xi|_L) \cdot \operatorname{Td}_{\mathbb{C}}(\tau L), [L] \rangle, \qquad (2.4.7) \end{aligned}$$

where $d = \dim L$ and $\chi(\tau L)$ stands for the Euler class of τL.

In the following, v denotes both the canonical generator of $H^2(P_r; \mathbb{Z})$ and a formal variable. Thus, $\mathrm{ch}(\eta_r) = e^v$. In view of (2.4.2), (2.4.6), to apply the index formula (2.4.7) to (2.4.1), one needs to analyze the contribution of the block $\sigma_{P_r} \cdot \phi((\eta_r)_\mathbb{C})^{r+1}/\phi(\mathbf{1}_\mathbb{C})$ from (2.4.1). We start by computing

$$\chi(\tau P_r)^{-1} \cdot \mathrm{ch}(\sigma_{P_r}\Big|_{P_r}) \cdot \mathrm{Td}_\mathbb{C}(\tau P_r) = (-1)^r \left[\frac{v \cdot \sigma(e^v)}{(1 - e^v)(1 - e^{-v})}\right]^{r+1}/\epsilon$$

$$= (-1)^r \cdot [v \cdot S(e^v)]^{r+1}/\epsilon,$$

where $S(u)$ has been defined in (2.4.5) and $\epsilon = \lim_{v \to 0}(v \cdot S(e^v))$. It equals $Q(1)$ for $\mathcal{G} = \mathbf{U}(n)$, and $2Q(1)$ for $\mathcal{G} = \mathbf{SO}(2n)$. Hence, with $\epsilon_t = \epsilon \cdot \tilde{\varphi}_t(1) \in \mathbb{C}[[t]]$, and the definition (2.4.5) in mind, we get:

$$\Phi(P_r, \rho \otimes \eta_r) = \epsilon_t^{-1} \cdot \left\{z^{r+1} \prod_{\lambda \in \mathrm{Spec}\, \rho(\sim)} H_t(\lambda e^z) \cdot (H_t(e^z))^{r+1}\right\}_{\langle r \rangle}. \qquad (2.4.8)$$

Here $\{\ \}_{\langle r \rangle}$ denotes the r-th coefficient (with the value in $\mathcal{R}(G)[[t]]_{\Lambda_\mathfrak{R}}$) of the z-expansion of the expression in the figure brackets.

Put $h_t(z) = H_t(e^z)$. By (2.4.8), the Φ-image of a typical numerator of the local formula (1.3) in Theorem 1.1 is:

$$\sum_{r=0}^{\infty} \Phi(P_r, \rho \otimes \eta_r)\, x^r =$$

$$= \epsilon_t^{-1} \cdot \sum_{r=0}^{\infty} \mathrm{Res}_{z=0} \left\{\prod_{\lambda \in \mathrm{Spec}\, \rho(\sim)} h_t(z + \ln \lambda) \cdot h_t(z)^{r+1}\right\} x^r,$$

where $\mathrm{Res}_{z=0}\{\ \}$ denotes the residue of the corresponding Laurent z-expansion. The residue takes its value in the ring $\mathcal{R}(G)[[t]]_{\Lambda_\mathfrak{R}}$.

A similar computation for $\Phi(P_r) = \pi_!^{P_r}(\sigma_{P_r} \cdot \phi(\tau_\mathbb{C} P_r))$ leads to the following formula for the Φ-image of a typical denominator in (1.3):

$$\sum_{r=0}^{\infty} \Phi(P_r)\, x^r = \epsilon_t^{-1} \cdot \sum_{r=0}^{\infty} \mathrm{Res}_{z=0}\{h_t(z)^{r+1}\} x^r. \qquad (2.4.10)$$

The statement below is a version of the Lagrange's Inversion Formula [cf. C]:

Lemma 2.2. *For any $f_t(z) \in \mathbb{C}[[t, z]]$ and $h_t(z) = g_t(z)/z$, where $g_t(z) \in \mathbb{C}[[t, z]]$, $g_0(0) \neq 0$, the following identity is valid in $\mathbb{C}[[t, x]]$:*

$$\frac{\displaystyle\sum_{r=0}^{\infty} \mathrm{Res}_{z=0}\{f_t(z) \cdot h_t(z)^{r+1}\}\, x^r}{\displaystyle\sum_{r=0}^{\infty} \mathrm{Res}_{z=0}\{h_t(z)^{r+1}\}\, x^r} = f_t((1/h_t)^{-1}(x)).$$

Here $(1/h_t)^{-1}$ denotes the formal functional inverse of the reciprocal of $h_t(\sim)$.

In fact, if in the left-hand side of the identity above one multiplies each term in the figure brackets by the same factor, say $\alpha_t(z) \in \mathbb{C}[[t, z]]$, the ratio will not change.

Note, that $h_t(z) = H_t(e^z)$ from (2.4.10) has a simple t-independent pole at $z = 0$, implying that the equation $x = 1/h_t(z)$ has a formal solution $z = z(t, x) \in \mathbb{C}[[t, x]]$. Examining (2.4.5) and the preceding table and taking into consideration that $\tilde{Q}(1) \neq 0$, $\tilde{\varphi}_0(u) \in \mathbb{C}^\times$, one concludes that $1/H_t(u) \in (u-1) \cdot \mathbb{C}[[t, u-1]]$. Therefore, $H_t^{-1}(1/x) \in \mathbb{C}[[t, x]]$.

Applying Lemma 2.2 to formulae (2.4.9), (2.4.10), we get that their ratio $\Theta^{\langle \rho \rangle}$ (viewed as a central $\mathbb{C}[[t, x]]$-valued function on G), is given by a nice formula

$$\Theta^{\langle \rho \rangle}(g) \qquad = \qquad \prod_{\lambda \in \mathrm{Spec}\, \rho(g)} h_t(\ell n\, \lambda + h_t^{-1}(1/x))$$

$$= \prod_{\lambda \in \mathrm{Spec}\, \rho(g)} H_t(\lambda \cdot H_t^{-1}(1/x)), \qquad\qquad (2.4.11)$$

suggesting a formal group law.

We emphasize that the functions $\Theta^{\langle \rho \rangle} : G \longrightarrow \mathbb{C}[[t, x]]$ in (2.4.11) are well-defined on the complement in G of the set $\mathcal{D}_\rho = \{g \in G | 1 \in \mathrm{Spec}\, \rho(g)\}$: indeed, examining formula (2.4.5) and the table preceding it, we see that the only potential for trouble while computing $H_t(\lambda \cdot H_t^{-1}(1/x))$, is the procedure of dividing by $1 - \lambda H_t^{-1}(1/x)$. Since $H_t^{-1}(1/x) = 1 + \sum_{j \geq 1} F_j x^j$, $F_j \in \mathbb{C}[[t]]$, this division is possible in $\mathbb{C}[[t, x]]$, provided $\lambda \neq 1$.

Now, for any finite-dimensional G-representation ψ, we introduce a sequence of partially defined central functions $\Theta_j^{\langle \psi \rangle} : G \longrightarrow \mathbb{C}[[t]]$ (equivalently, of elements from $\mathcal{R}(G)[[t]]_{\Lambda_\mathfrak{R}}$), $j = 0, 1, 2, \ldots$, by the generating formulae below:

$$\prod_{\lambda \in \mathrm{Spec}\, \psi(g)} H_t(\lambda \cdot H_t^{-1}(1/x)) = \sum_{j \geq 0} \Theta_j^{\langle \psi \rangle}(g)\, x^j \qquad (2.4.12)$$

(ψ being a complex representation).

The elements $\Theta_j^{\langle \psi \rangle}(\sim)$ are the key players in the local Lefschetz-type formula in Theorem 2.3 to follow.

Finally, in view of (2.4.12), combining (2.4.9), (2.4.10) with Theorem 1.1, and taking advantage of the multiplicative and additive properties of the invariant $\Phi(L, \xi)$ (with respect to the cartesian products and disjoint unions of the pairs (L, ξ)) (cf. (2.3.4)), we shall obtain our main result—Theorem 2.3. Let us outline how the argument goes. According to the calculations above,

$$\frac{\Phi(\rho) + \Phi(\rho \otimes \eta_1)x_{\rho,\ell} + \Phi(\rho \otimes \eta_2)x_{\rho,\ell}^2 + \cdots}{1 + \Phi(P_1)x_{\rho,\ell} + \Phi(P_2)x_{\rho,\ell}^2 + \cdots} = \sum_{j \geq 0} \Theta_j^{\langle \rho \rangle} x_{\rho,\ell}^j,$$

where $\Theta_j^{\langle\rho\rangle}$ has been defined in (2.4.12). Thus,

$$\prod_{\rho\in\mathfrak{R},\,1\le\ell\le n_\rho^\psi}\left(\sum_{j\ge0}\Theta_j^{\langle\rho\rangle}x_{\rho,\ell}^j\right)=$$

$$=\sum_{\omega\in\mathfrak{G}_\psi}\left[\prod_{\rho\in\mathfrak{R},\,1\le\ell\le n_\rho^\psi}\Theta_{\omega(\rho,\ell)}^{\langle\rho\rangle}\right]S_\omega(\{x_{\rho,\ell}\}),$$

and, in view of the formal character of the identity (1.3), is equal to

$$\sum_{\omega\in\mathfrak{G}_\psi}T_{\psi,\omega}\Big(\{\Phi(\rho\otimes\eta_r)\},\{\Phi(P_r)\}\Big)S_\omega(\{x_{\rho,\ell}\}).$$

Therefore, by (1.2),

$$\Phi(\nu_\psi)=\sum_{\omega\in\mathfrak{G}_\psi}\left[\prod_{\rho\in\mathfrak{R},\,1\le\ell\le n_\rho^\psi}\Theta_{\omega(\rho,\ell)}^{\langle\rho\rangle}\right]\Phi(S_\omega(\nu_\psi)). \qquad (2.4.13)$$

2.5 Quaternionic case

Since the calculations for the case $\mathcal{G}=\mathbf{Spin}$ and $\nu(M^G,M)$ being quaternionic are nastier, we choose to encase them in a separate subsection.

Let P_r be a quaternionic projective space and η_r- the normal bundle of P_r in P_{r+1}. We regard the $\mathbf{Sp}(1)$-bundle η_r as an $\mathbf{SU}(2)$-bundle under the isomorphism $\mathbf{Sp}(1)\approx\mathbf{SU}(2)$. Since ρ is a quaternionic representation, for any $g\in G$, $\rho(g)\in\mathbf{U}(2k)$ commutes with the diagonal $\mathbf{SU}(2)$-action on $\mathbb{C}^{2k}$. If $v\in\mathbb{C}^{2k}$ is an eigenvector for $\psi(g)$, then so is Jv, $J=\begin{pmatrix}0&1\\-1&0\end{pmatrix}$, 1 standing for the unit $k\times k$-matrix. Thus, one can choose a basis $(e_1,f_1,e_2,f_2,\ldots,e_k,f_k)$ in $\mathbb{C}^{2k}$, so that $V_i=\mathrm{Span}\{e_i,f_i\}$, $1\le i\le k$, is $\mathbf{SU}(2)$-invariant and $\psi(g)e_i=\lambda e_i$, $\psi(g)f_i=\lambda f_i$ ($\lambda\in\mathbb{C},|\lambda|=1$). In other words, for a cyclic g, one can view V_i as an irreducible 2-dimensional $\{g\}\times\mathbf{SU}(2)$-module via the epimorphism $\{g\}\times$ $\mathbf{SU}(2)\longrightarrow\mathbf{U}(2)$ which takes g to $\begin{pmatrix}\lambda&0\\0&\lambda\end{pmatrix}$ and is the standard monomorphism $\mathbf{SU}(2)\subset\mathbf{U}(2)$ on $\mathbf{SU}(2)$.

Regarding $R(\mathbf{Spin}(4k))\approx\mathbb{Z}[\Lambda_1,\Lambda_2,\ldots,\Lambda_{2k-2},\Delta_+,\Delta_-]$ as the Weyl-invariant part of $\mathbb{Z}[\alpha_1,\alpha_1^{-1},\ldots,\alpha_{2k},\alpha_{2k}^{-1},(\alpha_1\cdots\cdots\alpha_{2k})^{\frac12}]$, the symbol-character $\sigma_{2k}=\prod_{i=1}^{2k}(\alpha_i^{-\frac12}-\alpha_i^{\frac12})Q(\alpha_i)$ gives rise to the polynomials $\sigma_{2k}^\pm$ in the variables $\{\Lambda_i\}_{1\le i\le 2k-2}$, $\Delta_\pm$ with positive integral coefficients.

This argument makes it possible to identify $(\rho\underset{\mathbb{H}}{\otimes}\eta_r)_{\sigma_{2k}}^\pm$ with the G-bundles $S^{4r+3}\times_{\mathbf{SU}(2)}\sigma_\rho^\pm\longrightarrow P_r$, where the $G\times\mathbf{SU}(2)$-representations $\sigma_\rho^\pm$ have been produced with the help of the $\sigma_{2k}^\pm$ as follows. Let $\chi_\rho^\pm$ stand for the characters of

$\sigma_\rho^\pm$. Substituting $\alpha_i = \beta \cdot \lambda_i$, $1 \leq i \leq 2k$, ($\lambda_i \in \operatorname{Spec} \psi(g)$, $\beta + \beta^{-1}$ being the character of the standard 2-dimensional $\mathbf{SU}(2)$-representation) into the expression for σ_{2k}, one gets:

$$\chi_\rho^+(g) - \chi_\rho^-(g) =$$

$$= \prod_{i=1}^{k} [(\beta \lambda_i)^{-\frac{1}{2}} - (\beta \lambda_i)^{\frac{1}{2}}][(\beta^{-1} \lambda_i)^{-\frac{1}{2}} - (\beta^{-1} \lambda_i)^{\frac{1}{2}}] Q(\beta \lambda_i) \cdot Q(\beta^{-1} \lambda_i)$$

$$= \prod_{\lambda \in \frac{1}{2} \operatorname{Spec} \rho(g)} [(\lambda^{-1} + \lambda) - (\beta^{-1} + \beta)] \, \widehat{Q}(\lambda^{-1} + \lambda, \beta^{-1} + \beta). \qquad (2.5.1)$$

Here $\frac{1}{2} \operatorname{Spec} \rho(g)$ denotes the divisor in $\mathbb{C}$ whose double is $\operatorname{Spec} \rho(g)$ and the polynomial $\widehat{Q}(x, y)$ is defined by:

$$\widehat{Q}(\lambda^{-1} + \lambda, \beta^{-1} + \beta) = Q(\beta \lambda) \cdot Q(\beta^{-1} \lambda), \qquad Q(u) \in \mathbb{Z}[u + u^{-1}].$$

"Spreading" (2.5.1) along the P_r via the balanced-product construction (over $\mathbf{SU}(2)$), in $K_G(P_r)$ we get:

$$[(\rho \underset{\mathbb{H}}{\otimes} \eta_r)_{\sigma_{2k}}^+ - (\rho \underset{\mathbb{H}}{\otimes} \eta_r)_{\sigma_{2k}}^-](g) =$$

$$= \prod_{\lambda \in \frac{1}{2} \operatorname{Spec} \rho(g)} [(\lambda^{-1} + \lambda) - \eta_r] \, \widehat{Q}(\lambda^{-1} + \lambda, \eta_r). \qquad (2.5.2)$$

We introduce a formal variable ζ_r (a "ghost" of a complex line bundle), so that $\eta_r = \zeta_r + \zeta_r^{-1}$ (η_r, being lifted to the projectivization $P_{\mathbb{C}}(\eta_r)$ of η_r, splits as $\zeta_r \oplus \zeta_r^{-1}$). This enables us to verify that

$$\lambda_{-1}((\rho \underset{\mathbb{H}}{\otimes} \eta_r)_{\mathbb{C}})(g) = \prod_{\lambda \in \operatorname{Spec} \rho(g)} [\lambda^{-1} + \lambda - \eta_r]$$

and

$$\phi((\rho \underset{\mathbb{H}}{\otimes} \eta_r)_{\mathbb{C}}) = \prod_{\lambda \in \frac{1}{2} \operatorname{Spec} \rho(g)} \widehat{\varphi}_t(\lambda^{-1} + \lambda, \eta_r).$$

Here $\widehat{\varphi}_t \in \mathbb{C}[\lambda^{-1} + \lambda, u^{-1} + u][[t]]$ is defined by

$$\widehat{\varphi}_t(\lambda^{-1} + \lambda, \zeta^{-1} + \zeta) = \varphi_t(\lambda \zeta) \cdot \varphi_t(\lambda \zeta^{-1}) \cdot \varphi_t(\lambda^{-1} \zeta^{-1}) \cdot \varphi_t(\lambda^{-1} \zeta). \qquad (2.5.3)$$

In the quaternionic case, the fragment of (2.4.1) in the square brackets preceding the multiplication sign "$\times$" is equal (in the localized ring $\mathcal{K}_G(P_r)[[t]]_{\Lambda_{\mathfrak{R}}}$) to

$$\prod_{\lambda \in \frac{1}{2} \operatorname{Spec} \rho(g)} \frac{\widehat{Q}(\lambda^{-1} + \lambda, \eta_r)}{(\lambda^{-1} + \lambda) - \eta_r} \cdot \widehat{\varphi}_t(\lambda^{-1} + \lambda, \eta_r). \qquad (2.5.4)$$

Note, that $\eta_r - 2$ is nilpotent in $K(P_r)$. Hence, $[(\lambda^{-1} + \lambda - 2) + (2 - \eta_r)]^{-1} = [(\lambda^{-1} + \lambda) - \eta_r]^{-1} \in \mathbb{Z}[\lambda^{-1} + \lambda, \eta_r]_{[\lambda^{-1} + \lambda - 2]}$.

By (2.4.5) and the preceeding table, one can represent (2.5.4) as

$$\prod_{\lambda \in \frac{1}{2}\operatorname{Spec}\rho(g)} H_t(\lambda\zeta_r) \cdot H_t(\lambda\zeta_r^{-1}), \qquad (2.5.5)$$

$\eta_r = \zeta_r + \zeta_r^{-1}$. In fact, since $\eta_r - 2$ is nilpotent in $K(P_r)$, (2.5.5) defines an element of $\mathbb{C}[\lambda^{-1} + \lambda, \eta_r][[t]]_{[1-\lambda]}$.

Let z be the canonical generator of $H^4(P_r; \mathbb{Z})$. Recall, that the total Pontrjagin class of η_r is $(1 + z)^2$ and that of $\ell = \operatorname{Hom}_{\mathbb{H}}(\eta_r^*, \eta_r^*)$ is $1 + 4z$; moreover, the Euler class of ℓ is trivial [Sz]. We shall introduce a formal variable v, so that $v^2 = z$ (one might like to think of v as being the first Chern class of the ghost line bundle ζ_r). With these notations,

$$\chi(\tau P_r)^{-1} \operatorname{ch}(\sigma_{P_r|P_r}) \cdot \operatorname{Td}_\mathbb{C}(\tau P_r) = \left[\frac{-v^2 \sigma(e^v) \cdot \sigma(e^{-v})}{(1-e^v)^2(1-e^{-v})^2}\right]^{r+1} \Big/ \epsilon$$

$$= \left[\left(\frac{v}{1-e^{-v}}\right)\left(\frac{-v}{1-e^v}\right)Q(e^v)^2\right]^{r+1} \Big/ \epsilon,$$

where $\sigma(u) = (u^{-\frac{1}{2}} - u^{\frac{1}{2}})Q(u), Q(u^{-1}) = Q(u)$, and $\epsilon = \operatorname{ch}(\sigma_\ell|P_r) \cdot \tilde{\operatorname{Td}}_\mathbb{C}(\ell) = \left(\frac{2ve^v}{e^{2v}-1}\right)Q(e^{2v})$. The modified "class" $\tilde{\operatorname{Td}}_\mathbb{C}(\sim)$ is determined by the generating function $x/(1 - e^x)(1 - e^{-x})$.

This formula and formula (2.5.5) imply that

$$\Phi(P_r, \rho \underset{\mathbb{H}}{\otimes} \eta_r) = \left\{\frac{1}{2}v^{2r+1} \prod_{\lambda \in \frac{1}{2}\operatorname{Spec}\rho(\sim)} H_t(\lambda e^v) \cdot H_t(\lambda e^{-v}) \times \right.$$

$$\left. \times [H_t(e^v) \cdot H_t(e^{-v})]^{r+1} / H_t(e^{2v}) \right\}_{(2r)}, \qquad (2.5.6)$$

where $H_t(u) = \frac{\sqrt{u}}{u-1} Q(u)\varphi_t(u)\varphi_t(u^{-1})$. (Note that $H_t(u^{-1}) = -H_t(u)$.) Similarly,

$$\Phi(P_r) = \left\{\frac{1}{2}v^{2r+1}[H_t(e^v)H_t(e^{-v})]^{r+1}/H_t(e^{2v})\right\}_{(2r)}. \qquad (2.5.7)$$

Thus,

$$\sum_{r=0}^{\infty} \Phi(P_r, \rho \underset{\mathbb{H}}{\otimes} \eta_r)\, x^r =$$

$$= \frac{1}{2}\sum_{r=0}^{\infty} \operatorname{Res}_{v=0}\left\{\left[\prod_{\lambda \in \frac{1}{2}\operatorname{Spec}\rho(\sim)} H_t(\lambda e^v)H_t(\lambda e^{-v})\right] \Big/ H_t(e^{2v})\right.$$

$$\left. \times \left[H_t(e^v)H_t(e^{-v})\right]^{r+1}\right\}x^r \qquad (2.5.8)$$

and

$$\sum_{r=0}^{\infty} \Phi(P_r)\, x^r =$$

$$= \frac{1}{2} \sum_{r=0}^{\infty} \operatorname{Res}_{v=0} \left\{ (1/H_t(e^{2v})) \times [H_t(e^v) H_t(e^{-v})]^{r+1} \right\} x^r. \quad (2.5.9)$$

Note that $\tilde{h}_t(z) := H_t(e^v) \cdot H_t(e^{-v}), z = v^2$, from (2.5.9) has a simple t-independent pole at $z = 0$ (equivalently, a double pole at $v = 0$), implying the existence of formal solutions $z = z(t, x) \in \mathbb{C}[[t, x]]$, $v = v(t, x) \in \mathbb{C}[[t, \sqrt{x}\,]]$ for the equations $x = 1/h_t(z)$, $x = 1/\tilde{h}_t(v^2) = -1/H_t(e^v)^2$.

Applying Lemma (2.2) (with $\alpha_t(v) = 1/H_t(e^{2v})$) to formulae (2.5.8), (2.5.9), their ratio, $\Theta^{\langle \rho \rangle}$, is

$$\Theta^{\langle \rho \rangle}(g) = \prod_{\lambda \in \frac{1}{2}\operatorname{Spec}\rho(g)} H_t(\lambda \cdot H_t^{-1}(i/\sqrt{x})) \cdot H_t(\lambda \cdot H_t^{-1}(-i/\sqrt{x})), \quad (2.5.10)$$

where $i^2 = -1$. This suggests a bi-valued formal group law. To derive (2.5.10) one uses the symmetry $H_t(u^{-1}) = -H_t(u)$ to conclude that $1/H_t^{-1}(i/\sqrt{x}) = H_t^{-1}(-i/\sqrt{x})$. The substitution $\sqrt{x} \longrightarrow -\sqrt{x}$ does not change (2.5.10), therefore, despite its appearance, $\Theta^{\langle \rho \rangle}$ expands as a power series in x.

For a quaternionic ψ, let us introduce central functions $\Theta_j^{\langle \psi \rangle} : G \setminus \mathcal{D}_\psi \to \mathbb{C}[[\sqcup]]$, $j = 0, 1, 2, \ldots$, via the formula

$$\prod_{\lambda \in \frac{1}{2}\operatorname{Spec}\psi(g)} H_t(\lambda \cdot H_t^{-1}(i/\sqrt{x})) \cdot H_t(\lambda \cdot H_t^{-1}(-i/\sqrt{x}))$$

$$= \sum_{j \geq 0} \Theta_j^{\langle \psi \rangle}(g)\, x^j. \quad (2.5.11)$$

With these notations, repeating the argument of Section 2.4 leading to (2.4.13), one concludes that (2.4.13) is valid in the quaternionic case as well.

2.6   Formulations of the main results

Prior to the formulation of Theorem 2.3, we shall provide our reader with a short-cut entrance to the paper. Let us recall the relevant general setting and the hypothesis.

We start with a finite set $\mathfrak{R}$ of complex (quaternionic) irreducible G-representations ρ. Assume that the set $G_{\mathfrak{R}}^0 := \bigcap_{\rho \in \mathfrak{R}}\{g \in G \mid \operatorname{Det}(1 - \rho(g)) \neq 0\}$ is non-empty (for a connected Lie group G, this means that all the weights of each $\rho \in \mathfrak{R}$ are non-trivial).

Let M be a closed G-oriented G-manifold with the isotropy groups belonging to the family $\mathcal{F}_{\mathfrak{R}}$. It is formed by the isotropy groups of vectors from the spaces of various G-representations of the form $\bigoplus_{\rho \in \mathfrak{R}} n_\rho^\psi \cdot \rho$. Assume that M admits a G-equivariant $\mathcal{G}$-structure (cf. Section 2.1), $\mathcal{G} = \mathbf{U}(n), \mathbf{SO}(2n), \mathbf{Spin}(2n)$. In addition, one requires that, for $\mathcal{G} = \mathbf{U}(n), \mathbf{SO}(2n)$, the normal G-bundle $\nu = \nu(M^G, M)$ admits a G-equivariant complex and, for $\mathcal{G} = \mathbf{Spin}(2n)$, an equivariant quaternionic structure. Moreover, we assume that all the normal G-representations $\{\psi\}$ in the fibers of ν are sums of the irreducible ρ's from the list $\mathfrak{R}$.

Starting with a virtual $\mathcal{G}$-representation $\sigma_n \in \sigma_n^\dagger \cdot R(\mathcal{G})$ as in (2.1.1), there exists an equivariant pseudo-differential elliptic operator D_σ on M, intimately related to its $\mathcal{G}$-structure.

Formula (2.4.13) implies the following statement.

Theorem 2.3. *Under the assumptions and notations above, the following fixed point formula for the formal t-deformation* [3] *$\Phi(M) = \mathrm{ind}_t(D_\sigma, M)$ of $\mathrm{ind}(D_\sigma, M)$ (cf. (2.2.4)) is valid in the ring $\mathcal{R}(G)[[t]]_{\Lambda_\mathfrak{R}}$:*

$$\Phi(M) = \sum_{\psi, \omega \in \mathfrak{B}_\psi} \left[\sum_{\substack{\rho \in \mathfrak{R} \\ 1 \le \ell \le n_\rho^\psi}} \Theta_{\omega(\rho, \ell)}^{\langle \rho \rangle} \right] \cdot \Phi_\omega(\nu_\omega). \tag{2.6.1}$$

Here the universal central functions

$$\Theta_j^{\langle \rho \rangle} : G_\mathfrak{R}^0 \longrightarrow \mathbb{C}[[t]], \qquad j = 0, 1, 2, \ldots,$$

for a complex ν, have been introduced in (2.4.11), (2.4.12) and, for a quaternionic ν, in (2.5.10), (2.5.11) via the generating function $H_t(u)$. This basic element $H_t(u)$ from the ring $\{\mathbb{C}[u, u^{-1}][[t]]\}_{[1-u]}$ or the ring $\{\mathbb{C}[u^{\frac{1}{2}}, u^{-\frac{1}{2}}][[t]]\}_{[1-u]}$, has been defined in (2.4.5). The invariants $\{\Phi_\omega(\nu_\psi) \in \mathbb{C}[[t]]\}$ are the (non-equivariant) Φ-invariants of the characteristic manifolds $\{S_\omega(\nu_\psi)\}$ described in (1.1).

In fact, $\Phi(M)$ is an invariant of the G-bordism class $[M] \in {}_G \Omega_*^G(\mathcal{F}_\mathfrak{R})$.

Remark. As a function of $g \in G$, the right-hand side of (2.6.1) is well-defined on the complement to the "divisor" $\mathcal{D}_\mathfrak{R} = \{g \in G | \ 1 \in \mathrm{Spec}\,\rho(g)$ for some $\rho \in \mathfrak{R}\}$. By definition (2.2.4), the left-hand side of (2.6.1)—the invariant $\Phi(M) \in \mathcal{R}(G)[[t]]$—gives rise to a central function $\Phi(M) : G \longrightarrow \mathbb{C}[[t]]$ well-defined everywhere. Theorem 2.3 implies that the "singularities" of the right-hand side cancel along $\mathcal{D}_\mathfrak{R}$.

Recall, that any finite dimensional complex G-representation ψ uniquely extends to a finite dimensional representation $\tilde{\psi}$ of the complexification $G^\mathbb{C}$ of the Lie group G.

The theorem below is "the 0-approximation" of Theorem 2.3.

---

[3] The t-deformation $\Phi(M)$ is produced by employing an exponential operation $\phi : K_G(\sim) \longrightarrow K_G(\sim)[[t]]$, generated by an element $\varphi_t(u) = \sum_{j \le 0} \varphi_j(u)\, t^j$, where $\varphi_j(u) \in \mathbb{C}[u, u^{-1}]$ and $\varphi_0(u) \in \mathbb{C}^\times$.

Theorem 2.4. (i) *Under the hypotheses preceding Theorem 2.3, the following formula is valid in the ring of central rational functions on the closure $G_{\mathfrak{R}}$ of $G_{\mathfrak{R}}^0$ in G:*

$$\operatorname{ind}(D_\sigma, M) = \sum_{\psi, \omega \in \mathcal{B}_\psi} \left[\prod_{\substack{\rho \in \mathfrak{R} \\ 1 \le \ell \le n_\rho^\psi}} \vartheta_{\omega(\rho,\ell)}^{\langle \rho \rangle} \right] \operatorname{ind}(D_\sigma, S_\omega(\nu_\psi)). \qquad (2.6.2)$$

Here the central rational functions $\{\vartheta_j^{\langle \rho \rangle} : G_{\mathfrak{R}} \longrightarrow \mathbb{C}\}$ are defined by the formula

$$\sum_{j \ge 0} \vartheta_j^{\langle \rho \rangle}(g)\, x^j = \begin{cases} \displaystyle\prod_{\lambda \in \operatorname{Spec} \rho(g)} H(\lambda \cdot H^{-1}(1/x)) \\ \qquad \text{for } \mathcal{G} = \mathbf{U}, \mathbf{SO} \\ \displaystyle\prod_{\lambda \in \frac{1}{2}\operatorname{Spec}\rho(g)} H(\lambda \cdot H^{-1}(i/\sqrt{x})) \cdot H(\lambda \cdot H^{-1}(-i/\sqrt{x})) \\ \qquad \text{for } \mathcal{G} = \mathbf{Spin}, \end{cases}$$

$$(2.6.3)$$

where

$$H(u) = \begin{cases} \dfrac{u}{u-1} \cdot Q(u), & Q(u) \in \mathbb{Z}[u, u^{-1}], & \text{for } \mathcal{G} = \mathbf{U} \\[2mm] \dfrac{u+1}{u-1} \cdot Q(u), & Q(u) \in \mathbb{Z}[u + u^{-1}], & \text{for } \mathcal{G} = \mathbf{SO} \\[2mm] \dfrac{\sqrt{u}}{u-1} \cdot Q(u), & Q(u) \in \mathbb{Z}[u + u^{-1}], & \text{for } \mathcal{G} = \mathbf{Spin} \end{cases}$$

and $Q(1) \ne 0$. The invariants $\operatorname{ind}(D_\sigma, S_\omega(\nu_\psi))$ are, evidently, integers. The operator D_σ, correspondingly, is the Q-twisted Euler-Todd, Signature and Dirac operator, whose symbol is defined (in terms of virtual $\mathcal{G}$-representations) by (2.1.1).

(ii) *The character $\operatorname{ind}(D_\sigma, M) \in R(G)$ uniquely extends to a character $\widetilde{\operatorname{ind}}(D_\sigma, M) \in R(G^{\mathbb{C}})$. At the same time, replacing ρ in (2.6.3) by its analytic continuation—the $G^{\mathbb{C}}$-representation $\tilde{\rho}$, one extends each $\vartheta_j^{\langle \rho \rangle}$ to a central meromorphic function $\tilde{\vartheta}_j^{\langle \rho \rangle} : G_{\mathfrak{R}}^{\mathbb{C}} \longrightarrow \mathbb{C}$, holomorphic on*

$$\overset{\circ}{G}_{\mathfrak{R}}^{\mathbb{C}} = \bigcap_{\rho \in \mathfrak{R}} \{g \in G^{\mathbb{C}} \mid \operatorname{Det}(1 - \tilde{\rho}(g)) \ne 0\}.$$

In fact, formula (2.6.2) holds for the extended functions on $G_{\mathfrak{R}}^{\mathbb{C}}$ as well, implying that the poles of $\{\tilde{\vartheta}_j^{\langle \rho \rangle}\}$ cancel in the right-hand side of the extended version of (2.6.2).

Proof. Theorem 2.4 basically follows from Theorem 2.3 by letting $t = 0$ in the arguments of Sect.II. The only statements that may need a commentary are the ones describing the analyticity properties of $\operatorname{ind}(D_\sigma, M)$ and of the $\tilde{\vartheta}_j^{\langle \rho \rangle}$'s. Since $\operatorname{Spec} \rho(g) \subset \mathbb{C}^\times$ and $H(\lambda) = H_0(\lambda)$ is meromorphic in $\mathbb{C}^\times$ and holomorphic in $\mathbb{C}^{\times}\backslash\{1\}$, by (2.6.3), the functions $\{\tilde{\vartheta}_j^{\langle\rho\rangle}\}$ are meromorphic on $G_{\mathfrak{R}}^{\mathbb{C}}$ and holomorphic on $\overset{\circ\;\mathbb{C}}{G_{\mathfrak{R}}}$ (indeed, the right-hand side of (2.6.3), with the ρ being replaced by $\tilde{\rho}$, is an analytic function of $g \in G^{\mathbb{C}}$ and $x \in \mathbb{C}$, small enough). Note that, in the $\mathcal{G} = \mathbf{Spin}$-case, despite its appearance, (2.6.3) is a *single-valued* function both in g (in λ) and in x (which, of course, is consistent with the nature of $\widetilde{\mathrm{ind}}(D_\sigma, M)$). Since $G_{\mathfrak{R}} \subset G_{\mathfrak{R}}^{\mathbb{C}}$ is a totally real subvariety and each component of $G^{\mathbb{C}}$ contains a unique component of G and since, by Theorem 2.3, the formula (2.6.2) is valid on $G_{\mathfrak{R}}$, its continuation is valid on $G_{\mathfrak{R}}^{\mathbb{C}}$ as well. $\qquad\square$

There is one corollary of Theorem 2.4 that uses very little of the specifics of (2.6.2), besides the fact that the invariant $\mathrm{ind}(D_\sigma, M)$ is a linear combination (over the field of rational functions on $G_{\mathfrak{R}}$) of the invariants $\{\mathrm{ind}(D_\sigma, S_\omega(\nu_\psi))\}_{\psi,\omega}$. Note that, if ν_ψ admits an equivariant quaternionic structure, then, for any ω, $\dim(M) \equiv \dim S_\omega(\nu_\psi) \bmod 4$ Thus, if $\dim M \equiv 2 \bmod 4$, and ν is quaternionic, $\dim S_\omega(\nu_\Psi) \equiv 2 \bmod 4$. A manifold of dimension $4k+2$ represents an element of order 2 in $\Omega_{4k+2}^{\mathrm{SO}}(pt)$ and also of order 2 in $\Omega_{4k+2}^{\mathrm{Spin}}(pt)$. Thus, $\mathrm{ind}(D_\sigma, S_\omega(\nu_\psi)) = 0$, implying

Corollary 2.5. *Assume that G acts on $M = M^{4k+2}$ in a way described prior to Theorem 2.3 (i.e. the action in linearly modeled). Let M^{4k+2} admit a G-equivariant $\mathbf{SO}$- (or $\mathbf{Spin}$-) structure, while $\nu(M^G, M)$ admits an equivariant quaternionic one, subordinate to the ambient $\mathbf{SO}$-(or $\mathbf{Spin}$-)structure. Then the G-index of any twisted Signature (correspondingly, twisted Dirac) operator is zero.*

Corollary 2.6. *For any $\mathfrak{R}$-linearly modeled G-action on a $2d$-dimensional manifold with a fixed list of normal G-representations $\{\psi\}$, the invariant $\mathrm{ind}(D_\sigma, \sim) \in R(G_{\mathfrak{R}})$ is a $\mathbb{Z}$-linear combination of at most $\Sigma_\psi A_\psi(d - \dim_{\mathbb{C}} \psi)$ universal rational functions*

$$\Theta_\omega^\psi := \prod_{\rho\in\mathfrak{R},\,1\leq\ell\leq n_\rho^\psi} \vartheta_{\omega(\rho,\ell)}^{\langle\rho\rangle} : G_{\mathfrak{R}} \longrightarrow \mathbb{C},$$

where $A_\psi(k)$ stands for the number of elements $\omega = \{\omega_{\rho,\ell}\}$ in $\prod_{\rho\in\mathfrak{R}} S^{n_\rho^\psi}\mathbb{Z}_+$ whose norm $\|\omega\| = \Sigma_{\rho,\ell}\,\omega(\rho, \ell)$ does not exceed k.

There is one aspect of formula (2.6.2) that is worth mentioning, but hard to formulate explicitly without cumbersome combinatorial notations. Therefore, we shall state

Meta-Theorem 2.7. *One can view (2.6.2) as a Laurent expansion-type formula: the higher the norm $\|\omega\|$ of a multi index ω, the higher is the pole order (along the appropriate component of the divisor $\mathcal{D}_{\mathfrak{R}}^{\mathbb{C}}$ in $G_{\mathfrak{R}}^{\mathbb{C}}$) of the local contribution to $\mathrm{ind}(D_\sigma, M)$ in (2.6.2), indexed by ω. Therefore, in the pole cancellation* *game, first, one makes choices for the invariants* $\{\mathrm{ind}(D_\sigma, \mathcal{S}_\omega(\nu_\psi))\}$ *with large* $\|\omega\|$ *'s.*

In particular, the highest position in this hierarchy is occupied by the 0-dimensional $\mathcal{S}_\omega(\nu_\psi)$'s. For them $\mathrm{ind}(D_\sigma, \mathcal{S}_\omega(\nu_\psi))$ coincides with the virtual cardinality $\#\mathcal{S}_\omega(\nu_\omega)$ of $\mathcal{S}_\omega(\nu_\omega)$. This gives a number of universal relations among $\{\#\mathcal{S}_\omega(\nu_\psi)\}$, $\|\omega\| = \dim M/2$ (or $\dim M/4$ in the quaternionic case).

Proof. Since $H(\lambda)$ has a simple pole at $\lambda = 1$, the function $\vartheta_j(\lambda) = \frac{1}{j!} \frac{\partial^j}{\partial x^j} \{H(\lambda \cdot H^{-1}(1/x))\}_{x=0}$ has a pole of order $j + 1$ at $\lambda = 1$. Similarly, the function $\vartheta_j(\lambda) = \frac{1}{j!} \frac{\partial^j}{\partial x^j} \{H(\lambda \cdot H^{-1}(i/\sqrt{x})) \cdot H(\lambda \cdot H^{-1}(-i/\sqrt{x}))\}_{x=0}$ has a pole of order $j + 2$ at $\lambda = 1$. $\qquad\square$

Here is an illustration of how the principles of the Meta-Theorem work.

Example 2.8. We continue to discuss the setting of Example 1.2, where the 2-torus T acts on an oriented manifold M^6 with the normal T-representation ψ_1, ψ_2, ψ. The divisor $\mathcal{D}_\mathfrak{R}$ consists of three irreducible curves in T: the hyperbola T_0 and the two cubics T_1, T_2. The multiplicative system $\Lambda_\mathfrak{R}$ is generated by $1 - \rho_1$, $1 - \rho_2$, $1 - \rho_1 \rho_2^{-1}$.

With the help of (2.6.1), formula (1.4) for oriented bordism transforms into an identity in the ring $\{\mathbb{C}[\rho_1, \rho_1^{-1}, \rho_2, \rho_2^{-1}][[t]]\}_{\Lambda_\mathfrak{R}}$:

$$
\begin{aligned}
\Phi(M) &= \Theta_0^{\langle \rho_1^{-1}\rho_2 \rangle}(\Theta_0^{\langle \rho_1^{-1} \rangle})^2 \cdot \#M_{\psi_1}^T + \Theta_0^{\langle \rho_1 \rho_2^{-1} \rangle}(\Theta_0^{\langle \rho_2^{-1} \rangle})^2 \cdot \#M_{\psi_2}^T \\
&+ \Theta_1^{\langle \rho_1 \rangle}\Theta_0^{\langle \rho_2 \rangle} \cdot \chi(\xi_1) + \Theta_0^{\langle \rho_1 \rangle}\Theta_1^{\langle \rho_2 \rangle} \cdot \chi(\xi_2), \qquad (2.6.3)
\end{aligned}
$$

where

$$
\Theta_0^{\langle \psi \rangle}(g) = \prod_{\lambda \in \mathrm{Spec}\, \psi(g)} H_t(\lambda)
$$

$$
\Theta_1^{\langle \psi \rangle}(g) = \frac{\partial}{\partial x} \left\{ \prod_{\lambda \in \mathrm{Spec}\, \psi(g)} H_t(\lambda \cdot H_t^{-1}(1/x)) \right\}_{x=0}
$$

and, for the twisted Signature operator,

$$
H_t(u) = \left(\frac{u+1}{u-1} \right) Q(u) \varphi_t(u) \varphi_t(u^{-1}),
$$

with $Q(u) \in \mathbb{Z}[u + u^{-1}]$, $\varphi_t(u) \in \mathbb{C}[u, u^{-1}][[t]]$ $(Q(1) \neq 0, \varphi_0(u) \in \mathbb{C}^\times)$.

In (2.6.3) $\chi(\sim)$ stands for the Euler number of a bundle. In this example all the relevant characteristic manifolds $\{\mathcal{S}_\omega(\nu_\psi)\}_{\psi, \omega}$ are zero-dimensional. Since $\Phi(pt) = 1$ for any Φ, remarkably, the four integers $\{\#M_{\psi_i}^T, \chi(\xi_i)\}_{i=1,2}$ do not depend on the choice of $H_t(u)$. By (2.6.3), they determine the twisted T-signature of M.

A priori, the right-hand side of (2.6.3) is $\mathbb{C}[[t]]$-valued function, well-defined at $T \backslash \mathcal{D}_\mathfrak{R}$ In fact, by Theorem 2.3, it extends over T. Now let $t = 0$ in (2.6.3) and let $H(u) = H_0(u)$. We shall think of $g \in T^\mathbb{C}$ as of a pair (τ_1, τ_2), $\tau_1, \tau_2 \in \mathbb{C}^\times$.

Consider the following characters on $T^{\mathbb{C}}$: $\mu_1(g) = \tau_1^2 \tau_2$, $\mu_1^*(g) = \tau_1^{-2} \tau_2^{-1}$, $\mu_2(g) = \tau_1 \tau_2^{-1}, \mu_2^*(g) = \tau_1^{-1} \tau_2$, $\mu_3(g) = \tau_1 \tau_2^2, \mu_3^* = \tau_1^{-1} \tau_2^{-2}$ (so that μ_1 is the character of ρ_1, μ_2 of ρ_2 and μ_3 of $\rho_1 \otimes \rho_2^{-1}$). The curves $\mu_i = 1$ and $\mu_i^* = 1$, $1 \leq i \leq 3$, define the same element in the local ring of $1 = (1,1) \in T^{\mathbb{C}}$ and distinct $\mu_i = 1, 1 \leq i \leq 3$, give rise to the distinct elements. For each of the divisors $\mathcal{D}_i = \{g \in T \,|\, \mu_i(g) = 1\}$, we shall pick the terms of a modified formula (2.6.3), i.e. of the formula

$$\Phi(M)(g) = H(\mu_3^*) \cdot H(\mu_1^*)^2 \cdot \#M^T_{\psi_1} + H(\mu_3)H(\mu_2^*)^2 \cdot \#M^T_{\Psi_2}$$
$$+ \Theta_1(\mu_1)H(\mu_2) \cdot \chi(\xi_1) + H(\mu_1)\Theta_1(\mu_2) \cdot \chi(\xi_2),$$

having poles of the maximal order along a specific $\mathcal{D}_i$. Put $c_j = \lim_{u \to 1}(u - 1)^{j+1}\Theta_j(u)$ $(c_j \neq 0)$. Since Φ is holomorphic, multiplying it by an appropriate power of $\mu_i - 1$ and letting $g \longrightarrow \mathcal{D}_i$, we get the following three relations:

$$-c_0 H(\mu_1^*) \cdot \#M^T_{\psi_1} + c_0 H(\mu_2^*)\#M^T_{\psi_2}\big|_{\mathcal{D}_3} = 0.$$

The functions μ_1^* and μ_2^* coincide on $\mathcal{D}_3$ and aren't identically zero. Thus,

$$\#M^T_{\psi_1} = \#M^T_{\psi_2}.$$

Multiplying $\Phi(M)$ by $(\mu_1 - 1)^2$ and letting $g \longrightarrow \mathcal{D}_1$,

$$c_0^2 \cdot H(\mu_3^*) \cdot \#M^T_{\psi_1} + c_1 \cdot H(\mu_2) \cdot \chi(\xi_1)\big|_{\mathcal{D}_1} = 0.$$

Since $\mu_3^* = \mu_2$ on $\mathcal{D}_1$ and since $c_1 = -c_0^2$, we get

$$\#M^T_{\psi_1} = \chi(\xi_1).$$

An analogous treatment of $\mathcal{D}_2$ leads to

$$\#M^T_{\psi_2} = \chi(\xi_2).$$

Therefore, for any T^2-action on M^6 as in Example 1.3, we have

$$\#M^T_{\psi_1} = \#M^T_{\psi_2} = \chi(\xi_1) = \chi(\xi_2).$$

As a result, these invariants are *proportional* to the similar invariants of the model linear T^2-action on $\mathbb{C}P_3$.

Example 2.9. We continue with Example 1.3. In this setting (in the category of oriented $\mathbf{SU}(3)$-manifolds) formulae (1.5) and (2.6.1) produce

$$\Phi(M) = (\Theta_0^{\langle\rho\rangle})^2 \cdot \Phi_{0,0}(\nu) + (\Theta_1^{\langle\rho\rangle})^2 \cdot \Phi_{1,1}(\nu) + \Theta_0^{\langle\rho\rangle}\Theta_2^{\langle\rho\rangle} \cdot \Phi_{2,0}(\nu), \qquad (2.6.4)$$

where, for $g \in \mathbf{SU}(3)$,

$$\Theta_0^{\langle\rho\rangle}(g) + \Theta_1^{\langle\rho\rangle}(g)x + \Theta_2^{\langle\rho\rangle}(g)x^2 + \cdots = \prod_{\lambda \in \mathrm{Spec}\,\rho(g)} H_t(\lambda \cdot H_t^{-1}(1/x)),$$

$H_t(u)$ being as in Example 2.8, and

$$\Phi_{0,0}(\nu) = \Phi(M^{\mathbf{SU}(3)}),\, \Phi_{1,1}(\nu) = \chi(\xi^2),\, \Phi_{2,0}(\nu) = \chi(2\zeta \oplus \tau_\pi)$$

(see Example 1.3 for the notations). A priori, the right-hand side of (2.6.4) is well-defined on the complement in $\mathbf{SU}(3)$ of the bouquet of three copies of $\mathbf{SU}(2)$. By Theorem 2.4, it extends over $\mathbf{SU}(3)$ and even over $\mathbf{SL}_{\mathbb{C}}(3)$.

Now let $t = 0$. Let $\mu_i : T \longrightarrow \mathbb{C}^\times$, $1 \leq i \leq 3$ denote the weights of the standard 3-dimensional $\mathbf{SU}(3)$-representation ρ. Restricting (2.6.4) to the maximal torus $T \subset \mathbf{SU}(3)$, we get

$$\Phi(M) = \Phi(M^{\mathbf{SU}(3)}) \cdot \prod_{i=1}^{3} H(\mu_i)^2$$
$$+\chi(\xi^2)[\Theta_1(\mu_1)H(\mu_2)H(\mu_3) + \cdots]^2$$
$$+\chi(2\zeta \oplus \tau_\pi)[\Theta_2(\mu_1)H(\mu_1)H(\mu_2)^2 H(\mu_3)^2 + \cdots],$$

where "$+\cdots$" stands for the missing part of the S_3-symmetric expression.

The divisors $\mathcal{D}_i = \{g \in T | \mu_i(g) = 1\}$, $1 \leq i \leq 3$, through $\mathbf{1} \in T$ are distinct. It follows from the formula above that $\Phi(M)|_T$ has a pole of order at most 4 along each of the $\mathcal{D}_i$'s. Multiplying $\Phi(M)$ by $(\mu_1 - 1)^4$ and letting $g \longrightarrow \mathcal{D}_1$, one gets

$$c_1^2[H(\mu_2)H(\mu_3)]^2 \cdot \chi(\xi^2) + c_2 c_0 [H(\mu_2)H(\mu_3)]^2 \cdot \chi(2\zeta \oplus \tau_\pi)\big|_{\mathcal{D}_1} = 0.$$

Since $H(\mu_2)H(\mu_3)\big|_{\mathcal{D}_1}$ is not identically zero and since $c_1^2 = c_2 c_0$,

$$\chi(\xi^2) = -\chi(2\zeta \oplus \tau_\pi).$$

Analysis of $\mathcal{D}_2, \mathcal{D}_3$ doesn't give a new information. In other words, $\mathcal{S}_{1,1}(\nu) = -\mathcal{S}_{2,0}(\nu)$, $\nu = \nu(M^{\mathbf{SU}(3)}, M)$, for any $\mathbf{SU}(3)$-action as in Example 1.3, in particular, they are proportional to the similar invariants of the model $\mathbf{SU}(3)$-action on the Cayley projective plane.

References

[ASe] Atiyah, M.F., Segal, G.B., The Index of Elliptic Operators II, *Ann. of Math.* **87** (1968), 531–545.

[ASi] Atiyah, M.F., Singer, 1.M., The Index of Elliptic Operators III, *Ann. of Math.* **87** (1968), 546–604.

[Ba] Bär, C., *Elliptic Operators and Representation Theory of Compact Groups*, preprint (1993).

[BeK] Berend D., Katz, G., Separating Topology and Number Theory in the Atiyah-Singer *G*-signature Formula, *Duke Math. J.* **61** (1990), 939–971.

[BT] Bott, R., Taubes, C., On the Rigidity Theorems of Witten, *Journal of AMS* **2** (1989), 137–186.

[C] Copson, E.T., *Theory of Functions of a Complex Variable*, Oxford University Press, 1935.

[D] Davis, M., Multiaxial Actions on Manifolds, *Lect. Notes in Math.* **643**, Springer-Verlag (1978).

[H] Hirzebruch, F., Elliptic Genera of Level *N* for Complex Manifolds, in: *Differential Geometrical Methods in Theoretical Physics*, Kluwer Acad. Publishers, Dordrecht, (1988), 37–63.

[Hu] Husemoller, D., *Fibre Bundles*, McGraw-Hill Book Co, New York-St.Louis-San Francisco-Toronto-London-Sydney, 1966. [now Springer GTM 20, 3rd ed.]

[K1] Katz, G., Local Formulae in Equivariant Bordism, *Topology* **31:4** (1992), 713–733.

[K2] ——, Analytical Deformations of Equivariant Genera and the Witten's Rigidity à la Bott-Taubes, manuscript (1993).

[Kr1] Kričever, I.M., Formal Groups and the Atiyah-Hirzebruch Formula, *Math. USSR Izvestija* **8:6** (1974), 1271–1285.

[Kr2] ——, Obstructions to the Existence of S^1-Actions, *Math. USSR Izvestija* **10:4** (1976), 783–797.

[LS] Landweber, P.S., Stong, R.E., Circle Actions on Spin Manifolds and Characteristic Numbers, *Topology* **27** (1988), 145–162.

[L] Landweber, P.S., *Elliptic Curves and Modular Forms in Algebraic Topology*, Proceedings (Princeton 1986), *Lect. Notes in Math.* **1326**, Springer-Verlag (1988).

[Li] Liu, K., On Elliptic Genera and Theta-Functions, preprint (1992).

[O1] Ochanine, S., Sur les Genres Multiplicatifs Définis par les Intégrales Elliptiques, *Topology* **26** (1987), 143–151.

[O2] ——, Genres Elliptique Equivariants, in: *Elliptic Curves and Modular Forms in Algebraic Topology*, Proceedings (Princeton, 1986), *Lect. Notes in Math.* **1326**, Springer-Verlag (1988), 107–122.

[Se] Segal, G., Elliptic Cohomology, *Séminaire Bourbaki* **695** (1988), 1987–88.

[Sz]   Szczarba, R.H., On Tangent Bundles of Fibre Spaces and Quotient
        Spaces, *Amer. J. of Math.* **86** (1964), 685–697.

[T]   Taubes, C., S^1-Actions and Elliptic Genera, *Comm. in Math. Physics*
        122 (1989), 455–526.

[W]   Witten, E., The Index of the Dirac Operator in Loop Space , *Elliptic
        Curves and Modular Forms in Algebraic Topology*, Proceedings, (Prince-
        ton 1986) *Lect. Notes in Math.* **1326**, Springer-Verlag, (1988), 161–186.

Stable Prime Decompositions of Four-Manifolds

Matthias Kreck

Wolfgang Lück

Peter Teichner

Abstract

The main result of this paper is a four-dimensional stable version of Kneser's conjecture on the splitting of three-manifolds as connected sums. Namely, let M be a topological respectively smooth compact connected four-manifold (with orientation or $Spin$-structure). Suppose that $\pi_1(M)$ splits as $*_{i=1}^n \Gamma_i$ such that the image of $\pi_1(C)$ in $\pi_1(M)$ is subconjugated to some Γ_i for each component C of ∂M. Then M is stably homeomorphic respectively diffeomorphic (preserving the orientation or $Spin$-structure) to a connected sum $\natural_{i=1}^n M_i$ with $\Gamma_i = \pi_1(M_i)$. Stably means that one allows additional connected sums with some copies of $S^2 \times S^2$ on both sides. We also prove a uniqueness statement. As a consequence we obtain the existence and uniqueness of the stable prime decomposition of compact connected four-manifolds (with orientation or $Spin$-structure). The main technical ingredients are the bordism approach to the stable classification of manifolds due to the first author and the Kurosh Subgroup Theorem.

Introduction

A compact connected orientable smooth three-manifold M has a so called prime decomposition. Namely, M is oriented diffeomorphic to a connected sum $\natural_{i=1}^n M_i$ of oriented manifolds M_i which are prime, i.e. if M_i is diffeomorphic to $M_i' \natural M_i''$, then M_i' or M_i'' is oriented diffeomorphic to S^3. The manifolds M_i are unique up to order and oriented diffeomorphism.

The corresponding result cannot hold for four-manifolds. For example $(S^2 \times S^2)\natural\mathbb{C}P^2$ is diffeomorphic to $\mathbb{C}P^2\natural\overline{\mathbb{C}P^2}\natural\mathbb{C}P^2$. Or, for a simply connected four-dimensional $Spin$-manifold M with non-trivial signature $M\natural M^-$ is homeomorphic and often diffeomorphic to a connected sum of $(S^2 \times S^2)$'s. The problem here is, that the value of the signature for different pieces is not determined by the large manifold. A natural way to overcome these difficulties is to allow connected sum with an arbitrary *simply connected* closed four-manifold. Up to connected sum with simply connected closed manifolds, we prove the corresponding result in dimension four.

A *stable oriented diffeomorphism* from a four-manifold M to N is an orientation preserving diffeomorphism from $M \natural k(S^2 \times S^2)$ to $N \natural \overline{k}(S^2 \times S^2)$ for some non-negative integers k and $\overline{k}$. If the manifolds are equipped with a *Spin*-structure, we can in addition require that these structures are preserved. We call a connected compact orientable smooth four-manifold M *stably prime* if M_i stably oriented diffeomorphic to $M_i' \sharp M_i''$ implies that M_i' or M_i'' is simply connected and closed.

Theorem 0.1 (Stable Prime Decomposition) *Let M be a connected compact oriented smooth four-manifold. Then:*

1. *There are stably prime oriented four-manifolds $M_1, M_2, \ldots, M_n$ and a stable oriented diffeomorphism*

$$f : M \longrightarrow \sharp_{i=1}^n M_i.$$

2. *Let $f' : M \longrightarrow \sharp_{i=1}^{n'} M_i'$ be another stable oriented diffeomorphism for stably prime oriented four-manifolds $M_1', M_2', \ldots, M_{n'}'$. Suppose that none of the M_i's and M_i''s is simply connected and closed. Then $n = n'$ and $M_i \sharp S_i$ and $M_{\sigma(i)}' \sharp S_i'$ are stably oriented diffeomorphic for $i \in \{1, 2, \ldots, n\}$, appropriate simply connected closed oriented four-manifolds S_i and S_i' and a permutation σ.*

Closely related to prime decompositions is Kneser's conjecture. Let M be a compact connected three-manifold with incompressible boundary whose fundamental group admits a splitting $\alpha : \pi_1(M) \longrightarrow \Gamma_1 * \Gamma_2$. Kneser's conjecture whose proof can be found in [6, chapter 7] says that there are manifolds M_1 and M_2 with Γ_1 and Γ_2 as fundamental groups and a homeomorphism $M \longrightarrow M_1 \sharp M_2$ inducing α on the fundamental groups. Kneser's conjecture fails even in the closed case in dimensions ≥ 5 by results of Cappell [2],[3]. Counterexamples of closed orientable four-manifolds which even do not split up to homotopy and examples of closed orientable four-manifolds which split topologically but not smoothly are constructed by the authors of this article in [9]. But again it holds stably. We restrict ourselves to oriented manifolds. For simplicity we state in the introduction only an easy to formulate special case of our more general results whose precise statements are given in section 1. A group π is *indecomposable* if π is non-trivial and $\pi \cong \Gamma_1 * \Gamma_2$ implies that Γ_1 or Γ_2 is trivial.

Theorem 0.2 (Stable Kneser Decomposition) *If M is a closed connected smooth oriented four-manifold with non-trivial fundamental group, then there are oriented smooth four-manifolds $M_1, M_2, \ldots, M_n$ with indecomposable $\pi_1(M_i)$ for $i \in \{1, 2, \ldots, n\}$, such that M and $\sharp_{i=1}^n M_i$ are stably oriented diffeomorphic.*

If we have two splittings $\sharp_{i=1}^n M_i$ and $\sharp_{i=1}^{n'} M_i'$ of M as above, then $n = n'$ and $M_i \sharp S_i$ and $M_{\sigma(i)}' \sharp S_i'$ are oriented diffeomorphic for $i \in \{1, 2, \ldots, n\}$, appropriate simply connected closed smooth manifolds S_i, S_i' and a permutation $\sigma \in \Sigma_n$. If M is a Spin-manifold and we equip M_i and M_i' with the Spin-structures induced from a stable diffeomorphism as in Theorem 0.1, we can take S_i and S_i' as Spin-manifolds and the diffeomorphisms Spin-structure preserving.

Since the stable diffeomorphism type of a simply connected closed smooth four-manifold is determined by the type of the intersection form (I = odd = non-*Spin* or II = even = *Spin*) and the signature, we can take for the manifolds S either $r(\mathbb{C}P^2) \,\natural\, p(S^2 \times S^2)$ in the case I or $rK \,\natural\, p(S^2 \times S^2)$ in the case II, where $\mathbb{C}P^2$ is the complex projective space of complex dimension two and K is the Kummer surface (note that by Rohlin's theorem the signature is divisible by $16 = -\mathrm{sign}\,(K)$).

Both results have topological versions. All manifolds are topological, "diffeomorphic" must be substituted by "homeomorphic" and in the last paragraph $r(\mathbb{C}P^2) \,\natural\, p(S^2 \times S^2)$ respectively $rK \,\natural\, p(S^2 \times S^2)$ must be substituted by $r(\mathbb{C}P^2) \,\natural\, s(E_8) \,\natural\, p(S^2 \times S^2)$ respectively $r(E_8) \,\natural\, p(S^2 \times S^2)$ where E_8 is the simply connected closed topological four-manifold with E_8 as intersection form whose existence is proved by Freedman [5, Theorem 1.7].

We mention that this article is motivated by a paper of Hillman [7] which shows the existence of a stable splitting for a closed connected four-manifold M with fundamental group $\pi_1(M) = \Gamma_1 * \Gamma_2$. (Actually Hillman only proves that after adding $\mathbb{C}P^2$ or $\overline{\mathbb{C}}P^2$ one gets a splitting but his argument can be modified to give the original statement).

The paper is organized as follows :

The precise statements of our results, also for compact connected four-manifolds with boundary are given in section 1 and from this we deduce Theorems 0.1 and 0.2 using Kurosh's Subgroup Theorem. Section 2 summarizes the bordism approach to the stable classification due to the first author in a setup which is adequate for the purposes of this article and contains some preliminary results. One may skip section 2 and turn directly to the proofs of the main theorems in the following sections and get back to section 2 when necessary.

1 Kneser splittings for manifolds with boundary

All manifolds are assumed to be compact. We will formulate and prove our results for smooth manifolds. With the same modifications as explained in the introduction the analogous results hold for topological manifolds. The proofs are also identically the same replacing everywhere the smooth objects by the corresponding topological ones.

We will use the following convention on fundamental groups. Let $f : X \longrightarrow Y$ be a map of path-connected spaces. If we write $\pi_1(X)$, we mean $\pi_1(X, x)$ after some choice of base point $x \in X$. The homomorphism $\pi_1(f) : \pi_1(X) \longrightarrow \pi_1(Y)$ is the composition of $\pi_1(f, x) : \pi_1(X, x) \longrightarrow \pi_1(Y, f(x))$ and the isomorphism $c(w) : \pi_1(Y, f(x)) \longrightarrow \pi_1(Y, y)$ given by conjugation with a path w joining $f(x)$ and y. Notice that $\pi_1(f)$ is only well-defined up to inner automorphisms of $\pi_1(Y)$. Given a connected sum $\natural_{i=1}^n M_i$ we will use the following composition of isomorphisms as identification

$$*_{i=1}^n \pi_1(M_i) \xrightarrow{j^{-1}} *_{i=1}^n \pi_1(M_i - \mathrm{int}(D^4)) \xrightarrow{k} \pi_1(\natural_{i=1}^n M_i)$$

where j and k are induced by the inclusions in the obvious way. Notice again that this identification is only well-defined up to inner automorphisms. We call a component C of ∂M π_1-*null* if the inclusion induces the trivial map $\pi_1(C) \longrightarrow \pi_1(M)$.

Theorem 1.3 (Existence of Stable Splitting) *Let M be an oriented connected four-manifold with non-trivial fundamental group. Let*

$$\alpha : \pi_1(M) \longrightarrow *_{i=1}^n \Gamma_i$$

be a group isomorphism such that each Γ_i is non-trivial. Suppose for any component C of ∂M that the image of the composition $\alpha \circ \pi_1(j) : \pi_1(C) \longrightarrow *_{i=1}^n \Gamma_i$ is subconjugated to one of the Γ_i's for $j : C \longrightarrow M$ the inclusion.

Then there are oriented connected four-manifolds $M_1, M_2, \ldots, M_n$ with identifications $\pi_1(M_i) \longrightarrow \Gamma_i$ and oriented simply connected four-manifolds $N_1, N_2, \ldots, N_p$ and a stable oriented diffeomorphism

$$f : M \longrightarrow \natural_{i=1}^n M_i \natural \natural_{j=1}^p N_j$$

such that the composition

$$\pi_1(M) \xrightarrow{\pi_1(f)} \pi_1(\natural_{i=1}^n M_i \natural \natural_{j=1}^p N_j) \longrightarrow *_{i=1}^n \pi_1(M_i) \longrightarrow *_{i=1}^n \Gamma_i$$

agrees with α up to inner automorphisms, no boundary component of the M_i's is π_1-null and each N_i has a connected non-empty boundary.

Theorem 1.4 (Uniqueness of Stable Splitting) *Let $M_1, M_2, \ldots, M_n$ and $M_1', M_2', \ldots, M_n'$ be oriented connected four-manifolds with non-trivial fundamental groups $\Gamma_i = \pi_1(M_i)$ and $\Gamma_i' = \pi_1(M_i')$ such that no boundary component of them is π_1-null. Let $N_1, N_2, \ldots, N_p$ and $N_1', N_2', \ldots, N_q'$ be oriented simply connected four-manifolds whose boundaries are connected and non-empty. Let*

$$f : \natural_{i=1}^n M_i \natural \natural_{j=1}^p N_j \longrightarrow \natural_{i=1}^n M_i' \natural \natural_{j=1}^q N_j'$$

be a stable diffeomorphism, which is either oriented or Spin-structure preserving, if the underlying manifolds are Spin. Denote the homomorphism induced by f on the fundamental groups by

$$f_* : *_{i=1}^n \Gamma_i \longrightarrow *_{i=1}^n \Gamma_i'.$$

Suppose for $i \in \{1, 2, \ldots, n\}$ that $\text{pr}_i' \circ f_* \circ j_i$ is an isomorphism where $\text{pr}_i' : *_{i=1}^n \Gamma_i' \longrightarrow \Gamma_i'$ is the canonical projection and $j_i : \Gamma_i \longrightarrow *_{i=1}^n \Gamma_i$ is the canonical inclusion. Then:

For $i \in \{1, 2, \ldots, n\}$ we have $f(\partial M_i) = \partial M_i'$ and there are simply connected oriented closed four-manifolds S_i and S_i' and oriented diffeomorphisms

$$f_i : M_i \sharp S_i \longrightarrow M_i' \sharp S_i'$$

which extend $f|_{\partial M_i} : \partial M_i \longrightarrow \partial M_i'$ and induce up to inner automorphism $\text{pr}_i' \circ f_* \circ j_i$ on the fundamental groups. Moreover, we have $p = q$ and there is an appropriate permutation σ such that $f(\partial N_j) = \partial N_{\sigma(j)}'$ and there are oriented simply connected closed four-manifolds T_j and T_j' and oriented diffeomorphisms

$$g_j : N_j \sharp T_j \longrightarrow N_{\sigma(j)}' \sharp T_j'$$

which extend $f|_{\partial N_j} : \partial N_j \longrightarrow \partial N_j'$. If the manifolds are Spin-manifolds we can choose the manifolds S_i, S_i', T_j and T_j' as Spin-manifolds and f_i and g_i Spin-structure preserving.

We finish this section by deriving Theorems 0.2 and 0.1 from these results and the following version of Kurosh's Subgroup Theorem (see [4, Theorem 8, chapter 7 on page 175]).

Theorem 1.5 (Kurosh Subgroup Theorem) *Let H be a subgroup of the free product $G = *_{i \in I} G_i$. There is a suitable chosen set of representatives $g \in \overline{g}$ for the double cosets $\overline{g} \in H \backslash G / G_i$ for each $i \in I$ and a free subgroup $F \subset G$ satisfying*

$$H = F * *_{i \in I} \left(*_{\overline{g} \in H \backslash G / G_i} gG_i g^{-1} \cap H \right).$$

Let π be a non-trivial finitely generated group. A *Kurosh splitting* is an isomorphism

$$\alpha : \pi \longrightarrow *_{i=0}^n \Gamma_i$$

such that Γ_0 is free and Γ_j is indecomposabel and not infinite cyclic for $j = 1, 2, \ldots, n$. Recall that Γ_i is called indecomposable if Γ_i is non-trivial and $\Gamma_i \cong \Gamma_i' * \Gamma_i''$ implies that Γ_i' or Γ_i'' is trivial. The existence and the following uniqueness statement for a second Kurosh splitting $\alpha' : \pi \longrightarrow *_{i=0}^{n'} \Gamma_i'$ follow from Kurosh Subgroup Theorem 1.5. If $j_i : \Gamma_i \longrightarrow *_{i=0}^n \Gamma_i$ and $\text{pr}_i' : *_{i=0}^{n'} \Gamma_i' \longrightarrow \Gamma_i'$ are the inclusion and projection, then $n = n'$ and there is a permutation σ such that $\text{pr}_{\sigma(i)}' \circ \alpha' \circ \alpha^{-1} \circ j_i$ is an isomorphism for $i \in \{0, 1, \ldots, n\}$. Theorem 0.2 now follows from the above Theorems.

Before we prove Theorem 0.1, we characterize the property "stably prime" in terms of the fundamental group data.

Lemma 1.6 *A connected compact orientable four-manifold M is stably prime if and only if it satisfies the following conditions.*

1. *There is no isomorphism $\alpha : \pi_1(M) \longrightarrow \Gamma_1 * \Gamma_2$ for non-trivial groups Γ_1 and Γ_2 such that for each component C of the boundary the composition of α and the map induced by the inclusion $\pi_1(C) \longrightarrow \pi_1(M)$ has an image which is subconjugated to Γ_1 or Γ_2.*

2. *If M has a π_1-null boundary component, then M is simply connected and ∂M is non-empty and connected.*

Proof. If one of the conditions above is violated, Theorem 1.3 gives a splitting of M into $M_1 \sharp M_2$ such that neither M_1 nor M_2 is simply connected and closed. Conversely, given such a splitting, one sees immediately, that at least one of the conditions above is not fullfilled.   □

In particular a connected closed orientable four-manifold is stably prime if and only if $\pi_1(M)$ is trivial or indecomposable. Now, we prove Theorem 0.1.

Proof. 1) The existence of f follows from the following inductive process. If M is stably prime, the process stops. If M is not stably prime, choose a stable oriented diffeomorphism $M \longrightarrow M_1 \sharp M_2$ such that neither M_1 nor M_2 is simply connected and closed. Now apply this process to both M_1 and M_2. It remains to show that this process stops after a finite number of steps. This follows from Lemma 1.6, the Grushko-Neumann Theorem [10, Theorem 1.8 and Corollary 1.9 on page 178] which implies that the rank of a group, i.e. the minimal number of generators, is additive under free products and the simple fact that M has only finitely many π_1-null boundary components.

2) Consider a stable oriented diffeomorphism

$$f : (\sharp_{i=1}^l L_i) \sharp (\sharp_{i=1}^n M_i) \sharp (\sharp_{i=1}^p N_i) \longrightarrow (\sharp_{i=1}^{l'} L_i') \sharp (\sharp_{i=1}^{n'} M_i') \sharp (\sharp_{i=1}^{p'} N_i')$$

such that each L_i, L_i', M_i, M_i', N_i and N_i' is stably prime, each L_i is closed and has infinite cyclic fundamental group, none of the M_i's and M_i''s is simply connected or has both infinite cyclic fundamental group and empty boundary, and each N_i and N_i' is simply connected and has a non-empty boundary. Notice that any finite connected sum of stably prime connected four-manifolds can be written in this way if none of the summands is simply connected and closed. We conclude from Lemma 1.6 that none of the M_i's and M_i''s has a π_1-null boundary component and that the boundary of each N_i and N_i' is non-empty and connected. We abbreviate in the sequel $\Gamma_i = \pi_1(M_i)$ and $\Gamma_i' = \pi_1(M_i')$ and introduce the finitely generated free groups $\Gamma_0 = *_{i=1}^l \pi_1(L_i)$ and $\Gamma_0' = *_{i=1}^{l'} \pi_1(L_i')$. The map induced on the fundamental groups by f is denoted by

$$f_* : *_{i=0}^n \Gamma_i \longrightarrow *_{j=0}^{n'} \Gamma_j'.$$

Fix an index $i \in \{1, 2, \ldots, n\}$. We apply Kurosh Subgroup Theorem 1.5 to $f_*(\Gamma_i) \subset *_{j=0}^n \Gamma'_j$ and obtain

$$f_*(\Gamma_i) = F * *_{j=0}^{n'} \left(*_{\overline{g} \in f_*(\Gamma_i) \backslash *_{j=0}^{n'} \Gamma'_j / \Gamma'_j} \, g\Gamma'_j g^{-1} \cap f_*(\Gamma_i) \right).$$

Let C be a boundary component of M_i. There is an index $j \in \{1, 2, \ldots, n'\}$ such that $f(C) \subset M'_j$. If $j_* : \pi_1(C) \longrightarrow \Gamma_i$ is the map induced by the inclusion, then $f_*(j_*(\pi_1(C)))$ is subconjugated to Γ'_j. Hence there is $g_0 \in *_{i=0}^{n'} \Gamma'_i$ such that $f_*(j_*(\pi_1(C))) \subset g_0\Gamma'_j g_0^{-1} \cap f_*(\Gamma_i)$ holds. We conclude from Kurosh Subgroup Theorem 1.5 that $f_*(j_*(\pi_1(C)))$ is subconjugated to $g\Gamma'_j g^{-1} \cap f_*(\Gamma_i)$ for appropriate $\overline{g} \in f_*(\Gamma_i) \backslash *_{j=0}^{n'} \Gamma'_j / \Gamma'_j$. Recall that M_i is stably prime, has no π_1-null boundary component and it is not true that M_i has both infinite cyclic fundamental group and empty boundary. We derive from Lemma 1.6 applied to the isomorphism induced by f_*

$$\pi_1(M_i) = \Gamma_i \longrightarrow f_*(\Gamma_i) = F * *_{j=0}^{n'} \left(*_{\overline{g} \in f_*(\Gamma_i) \backslash *_{j=0}^{n'} \Gamma'_j / \Gamma'_j} \, g\Gamma'_j g^{-1} \cap f_*(\Gamma_i) \right).$$

that there is a unique index $\sigma(i) \in \{1, 2, \ldots, n\}$ and $\overline{g} \in f_*(\Gamma_i) \backslash *_{j=0}^{n'} \Gamma'_j / \Gamma'_j$ satisfying

$$f_*(\Gamma_i) = g\Gamma'_{\sigma(i)} g^{-1}.$$

We get a map $\sigma : \{1, 2, \ldots, n\} \longrightarrow \{1, 2, \ldots, n'\}$. Completely analogously one defines a map $\sigma' : \{1, 2, \ldots, n'\} \longrightarrow \{1, 2, \ldots, n\}$ such that for each $j \in \{1, 2, \ldots, n'\}$ there is $g \in *_{i=0}^n \Gamma_i$ satisfying

$$f_*^{-1}(\Gamma'_j) = g\Gamma_{\sigma'(j)} g^{-1}.$$

Let $\mathrm{pr}'_j : *_{j=0}^n \Gamma'_i \longrightarrow \Gamma'_j$ be the canonical projection and $j_i : \Gamma_i \longrightarrow *_{i=0}^n \Gamma_i$ be the canonical inclusion. We conclude for each $i \in \{1, 2, \ldots, n\}$ and $j \in \{0, 2, \ldots, n'\}$ that the composition $\mathrm{pr}'_j \circ f_* \circ j_i$ is an isomorphism if $j = \sigma(i)$ and trivial otherwise. Hence $\sigma' \circ \sigma = \mathrm{id}$ and following diagram commutes for appropriate $\overline{f_*}$

$$
\begin{array}{ccc}
*_{i=0}^n \Gamma_i & \xrightarrow{\;f_*\;} & *_{i=0}^n \Gamma'_i \\
\downarrow {\scriptstyle \mathrm{pr}_0} & & \downarrow {\scriptstyle \mathrm{pr}'_0} \\
\Gamma_0 & \xrightarrow[\;\overline{f_*}\;]{} & \Gamma'_0
\end{array}
$$

The same argument applied to f_*^{-1} shows that $\sigma \circ \sigma' = \mathrm{id}$ and that $\overline{f_*}$ has an inverse. Hence σ and σ' are inverse to one another, $n = n'$ and the composition $\mathrm{pr}'_{\sigma(i)} \circ f_* \circ j_i$ is an isomorphism for $i \in \{0, 1, 2, \ldots, n\}$ if we put $\sigma(0) = 0$.

From Theorem 1.4 we conclude that $M_i \sharp S_i$ and $M_{\sigma(i)} \sharp S'_i$ are stably oriented diffeomorphic for each $i \in \{1, 2, \ldots, n\}$, appropriate simply connected closed oriented four-manifolds S_i and S'_i and that $p = p'$ and $N_i \natural T_i$ and $N'_{\tau(i)} \natural T'_i$ are stably oriented diffeomorphic for each $i \in \{1, 2, \ldots, p\}$, appropriate simply connected closed oriented four-manifolds T_i and T'_i and permutation τ. Since Γ_0 and Γ'_0 are isomorphic, we get $l = l'$. Each L_i and L'_i is stably isomorphic to $S^1 \times S^3$ after adding simply connected closed oriented four-manifolds for $i \in \{1, 2, \ldots, l\}$ by Theorem 2.1 and Lemma 2.3 since L_i and L'_i are closed and have infinite cyclic fundamental groups. This finishes the proof of Theorem 0.1.

$\square$

2   Stable classification and bordism theory

In this section we explain the necessary details of the bordism approach to the stable classification of manifolds due to the first author and prove some preliminary lemmas. Recall that all manifolds are assumed to be compact and we restrict ourselves to smooth manifolds.

We begin with organizing the bookkeeping of the fundamental group data. We consider pairs (π, w_2) which consist of a finitely presented group π and an element w_2 in $H^2(\pi; \mathbb{Z}/2) \coprod \{\infty\}$. We call two such pairs (π, w_2) and (π', w'_2) *equivalent* if there is an isomorphism $f : \pi \longrightarrow \pi'$ with the properties that either $w_2 = \infty$ and $w'_2 = \infty$ or $w_2 \in H^2(\pi; \mathbb{Z}/2)$, $w'_2 \in H^2(\pi'; \mathbb{Z}/2)$ and $f^*(w'_2) = w_2$ holds. A *type* T is an equivalence class $[\pi, w_2]$ of such pairs.

An oriented manifold determines a type $T(M)$, called the *normal 1-type*, for which a representative is given as follows. Put $\pi = \pi_1(M)$. Let $g : M \longrightarrow K(\pi, 1)$ be a classifying map of the universal covering and denote by $w_k(M) \in H^k(M; \mathbb{Z}/2)$ the k-th Stiefel-Whitney class of the normal bundle of M. If $w_2(\widetilde{M}) \neq 0$ holds for the universal covering $\widetilde{M}$, then put $w_2 = \infty$. Otherwise let w_2 be the unique element satisfying $g^*(w_2) = w_2(M)$. The unique existence follows from the exact sequence coming from the Serre spectral sequence of the fibration $\widetilde{M} \longrightarrow M \longrightarrow K(\pi, 1)$

$$0 \longrightarrow H^2(K(\pi, 1); \mathbb{Z}/2) \xrightarrow{g^*} H^2(M; \mathbb{Z}/2) \longrightarrow H^2(\widetilde{M}; \mathbb{Z}/2).$$

Two homotopy equivalent manifolds have the same normal 1-type.

Before we introduce the relevant bordism groups, we recall how to convert a continous map $u : X \longrightarrow K$ into a fibration $u' : P(u) \longrightarrow K$. Define

$$P(u) = \{(x, w) \mid w(0) = u(x)\} \subset X \times \mathrm{map}\,(I, K)$$

and $u'(x, w) = w(1)$. Define the map $u'' : P(u) \longrightarrow X$ by sending (x, w) to x and define the homotopy $\psi : u \circ u'' \simeq u'$ by sending $((x, w), t)$ to $w(t)$. The triple $(P(u), u'', \psi)$ has the universal property that for any space Z together with maps $f' : Z \to K$ and $f'' : Z \longrightarrow X$ and homotopy $\phi : u \circ f'' \simeq f'$ there is precisely one map $g : Z \longrightarrow P(u)$ such that

$$f'' = u'' \circ g, \quad f' = u' \circ g \quad \text{and} \quad \phi = \psi \circ (g \times \mathrm{id}).$$

Namely, define $g(z) = (f''(z), \psi_z)$ for ψ_z the path sending t to $\psi(z, t)$. There is a map $i : X \longrightarrow P(u)$ sending x to $(x, c_{u(x)})$ where $c_{u(x)}$ is the constant path in K at $u(x)$. It is a homotopy inverse of u'' and its composition with u' is u.

A type T determines a fibration $\mathcal{B}(T)$ over BSO or over $BSpin$, if $w_2 = 0$, as follows. Let $[\pi, w_2]$ be a representative of T. If $w_2 = \infty$ define it as the trivial fibration

$$\mathcal{B}(T) = BSO \times K(\pi, 1) \to BSO$$

over BSO. If $w_2 = 0$ we define it as the trivial fibration

$$\mathcal{B}(T) = BSpin \times K(\pi, 1) \to BSpin$$

over $BSpin$. If $w_2 \neq 0, \infty$ represent w_2 by a map $u : K(\pi, 1) \to K(\mathbb{Z}/2, 2)$ with corresponding fibration $P(u)$ over $K(\mathbb{Z}/2, 2)$. Represent the universal Stiefel Whitney class by a map $q : BSO \to K(\mathbb{Z}/2, 2)$. Then define our fibration by the pullback

$$\mathcal{B}(T) = q^*(P(u)) \to BSO.$$

These fibrations are up to fibre homotopy equivalence uniquely determined by T. In all three cases there are projection maps to $K(\pi, 1)$ denoted by $p_{K(\pi,1)}$.

Suppose that M has normal 1-type $T(M)$ and let $g : M \to K(\pi, 1)$ be a map satisfying $g^* w_2 = w_2(\nu(M))$ if $w_2(M) = w_2(\nu(M)) \neq \infty$. Then the normal Gauss map $\nu : M \to BSO$ or $\nu : M \to BSpin$, if M is has a *Spin*-structure, admits a lift ρ over $\mathcal{B}(T)$ as follows. If $w_2 = \infty$ it is given by the normal Gauss map together with g. If $w_2 \neq \infty$ it is given by the normal Gauss map together with g and with a homotopy between the composition of the two maps to $K(\mathbb{Z}/2, 2)$. We call such a lift ρ a *normal structure* of M in $\mathcal{B}(T(M))$ compatible with g and the orientation resp. *Spin*-structure. If a normal structure ρ is a 2-equivalence, it is called a *normal 1-smoothing*. Notice that a normal structure ρ is a normal 1-smoothing if and only if the underlying map g induces an isomorphism on π_1.

Given a fibration $\mathcal{B} \to BSO$ or $\mathcal{B} \to BSpin$ we denote the bordism group of n-dimensional closed oriented or *Spin*-manifolds together with a lift of the normal Gauss map over $\mathcal{B}$ by

$$\Omega_n(\mathcal{B}).$$

If ρ is a normal 1-smoothing of M in $\mathcal{B}(T(M))$, then the pair (M, ρ) determines an element in $\Omega_4(\mathcal{B}(T(M)))$.

Now we can formulate the main result of the bordism approach to the stable classification of connected four-manifolds due to the first author [8].

Theorem 2.1 (Stable Classification of Four-Manifolds by Bordism Theory)
Let M_1 and M_2 be connected four-manifolds with orientation respectively Spin-structure and $\partial f : \partial M_1 \longrightarrow \partial M_2$ be a diffeomorphism which preserves the induced orientation respectively Spin-structure. Suppose that the normal 1-type

of M_1 and M_2 is equal to T and denote by $\mathcal{B}(T)$ any representation of the associated fibration. Let $g_i : M_i \to K(\pi, 1)$ be classifying maps of the universal covering respecting w_2 such that $g_2|_{\partial M_2} \circ \partial f = g_1|_{\partial M_1}$.

1. There exists a stable oriented (Spin-structure preserving, if M_i are Spin) diffeomorphism

$$f : M_1 \longrightarrow M_2$$

   extending ∂f such that the maps $g_2 \circ f$ and g_1 to $K(\pi, 1)$ are homotopic if and only if there are normal 1-smoothings ρ_i of M_i compatible with g_i and the orientations resp. Spin-structures, such that $\rho_2|_{\partial M_2} \circ \partial f$ and $\rho_1|_{\partial M_1}$ agree and

$$[M_1^- \cup_{\partial f} M_2, \rho_1^- \cup_{\partial f} \rho_2] = 0 \qquad \in \Omega_4(\mathcal{B}(T))$$

2. Given a manifold with boundary together with a lift of the normal Gauss-map to $\mathcal{B}(T)$, it is bordant relative boundary to a normal 1-smoothing.

The strategy for proving the main Theorems is to analyse how the bordism group decomposes if the fundamental group splits as a free product. For this the following categorial considerations are useful.

Denote $K = K(\mathbb{Z}/2, 2)$ or $K = *$. Define a category $\mathcal{C}$ as follows. An object (X, u) is a map $u : X \longrightarrow K$ and a morphism $(f, \phi) : (X, u) \longrightarrow (Y, v)$ consists of a map $f : X \longrightarrow Y$ together with a homotopy $\phi : v \circ f \simeq u$. The composition $(g, \psi) \circ (f, \phi)$ is defined by $(g \circ f, (\psi \circ (f \times \mathrm{id})) * \phi)$ where $*$ denotes the composition of homotopies. If the homotopy ψ is the constant homotopy, we abbreviate (f, ψ) by f. Two morphisms (f_0, ψ_0) and (f_1, ψ_1) from (X, u) to (Y, v) are called homotopic if they can be connected by a continuous one parameter family of morphisms (f_t, ϕ_t). The following elementary facts will frequently be used in the sequel.

Lemma 2.2 Let $(f, \phi) : (X, u) \longrightarrow (Y, v)$ be a morphism. If $g : X \longrightarrow Y$ is a map homotopic to f, then there is a homotopy $\psi : v \circ g \simeq u$ such that the morphisms (f, ϕ) and (g, ψ) are homotopic. If $f : X \longrightarrow Y$ is a homotopy equivalence, then there is a morphism $(g, \psi) : (Y, v) \longrightarrow (X, u)$ such that both compositions of (f, ϕ) and (g, ψ) are homotopic to the identity morphism.

Let $q : B \to K$ be a fixed map. Given an object (X, u), we have the pullback

$$\begin{array}{ccc} \mathcal{B}(u) & \xrightarrow{\ \bar{q}\ } & P(u) \\ {\scriptstyle \bar{u}}\downarrow & & \downarrow{\scriptstyle u'} \\ B & \xrightarrow{\ q\ } & K \end{array}$$

For a manifold M and appropriate choices of q and u, the fibration $\mathcal{B}(u)$ over B corresponds to a normal 1-type as described above. More precisely, if

$w_2 = \infty$, let $B = BSO$, $K = *$ and $X = K(\pi_1(M), 1)$. For $w_2 = 0$ choose $B = BSpin$ instead of BSO. For $w_2 \neq 0, \infty$ choose $B = BSO$, $K = K(\mathbb{Z}/2, 2)$, $X = K(\pi_1(M), 1)$ and q and u maps representing the second Stiefel Whitney classes. Then in all three cases $\mathcal{B}(T(M)) = \mathcal{B}(u)$.

For the special purpose of this paper the formulation of $\mathcal{B}(u)$ has the advantage that it separates the categorial input, namely the fundamental group data encoded in $K(\pi, 1)$ from the other data like orientation and *Spin*-structure.

Define

$$\Omega_n(X, u) = \Omega_n(\mathcal{B}(u))$$

A morphism $(f, \phi) : (X, u) \longrightarrow (Y, v)$ defines by the universal property of the construction $P(-)$ a fiber map $P(f, \phi) : P(u) \longrightarrow P(v)$ where fiber map means $P(f, \phi) \circ v' = u'$. If we apply Ω_n to it, we obtain a homomorphism denoted by

$$\Omega_n(f, \phi) : \Omega_n(X, u) \longrightarrow \Omega_n(Y, v).$$

Clearly this is a functor on $\mathcal{C}$. Moreover, it is a generalized homology theory in the sense that it has the following properties. It is homotopy invariant, i.e., homotopic morphisms induce the same homomorphism. There is a Mayer-Vietoris sequence in the following sense. Consider the following pushout

$$
\begin{array}{ccc}
X_0 & \xrightarrow{\ i_1\ } & X_1 \\
{\scriptstyle i_2}\big\downarrow & & \big\downarrow{\scriptstyle j_1} \\
X_2 & \xrightarrow[\ j_2\ ]{} & X
\end{array}
$$

with i_1 a cofibration. Put $j_0 = j_2 \circ i_2 = j_1 \circ i_1$. Let (X, u) be an object. We obtain objects (X_k, u_k) by $u_k = u \circ j_k$ and morphisms $j_k : (X_k, u_k) \longrightarrow (X, u)$ for $k = 0, 1, 2$. Recall that we omit constant homotopies in our notation for morphisms. Now there is a long exact Mayer-Vietoris sequence

$$\cdots \xrightarrow{\ \delta\ } \Omega_n(X_0, u_0) \xrightarrow{(\Omega_n(i_1), \Omega_n(i_2))} \Omega_n(X_1, u_1) \oplus \Omega_n(X_2, u_1)$$

$$\xrightarrow{\Omega_n(j_1) - \Omega_n(j_2)} \Omega_n(X, u) \xrightarrow{\ \delta\ } \Omega_{n-1}(X_0, u_0) \longrightarrow \cdots$$

Namely, we obtain a pushout with a cofibration as horizontal upper arrow

$$
\begin{array}{ccc}
\overline{P(u_0)} & \xrightarrow{\ \overline{P(i_1)}\ } & \overline{P(u_1)} \\
{\scriptstyle \overline{P(i_2)}}\big\downarrow & & \big\downarrow{\scriptstyle \overline{P(j_1)}} \\
\overline{P(u_2)} & \xrightarrow[\ \overline{P(j_2)}\ ]{} & \overline{P(u_0)}
\end{array}
$$

Notice that the bordism group $\Omega_n(\mathcal{B})$ can be identified with the bordism group of the stable vector bundle over $\mathcal{B}$ which is the pullback of the universal bundle over

BSO respectively $BSpin$ and thus the existence of the Mayer-Vietoris sequence follows by standard arguments, namely the Pontrjagin-Thom construction and the fact that stable homotopy is a generalized homology theory [1, Kapitel II].

Let (X, u) be an object with path-connected X. Denote by $*$ the space consisting of one point. Consider an object $(*, v)$ and a morphism $(j, \mu) : (*, v) \longrightarrow (X, u)$. Define

$$\widetilde{\Omega}_n(X, u) := \mathrm{cok}\,(\Omega_n(j, \mu) : \Omega_n(*, v) \longrightarrow \Omega_n(X, u))$$

We want to show that the definition of $\widetilde{\Omega}_n(X, u)$ is independent of the choice of v and (j, μ) and that a morphism $(f, \phi) : (X, u) \longrightarrow (Y, v)$ induces a homomorphism making the following diagram commute for pr the canonical projection.

$$
\begin{array}{ccc}
\Omega_n(X, u) & \xrightarrow{\;\mathrm{pr}\;} & \widetilde{\Omega}_n(X, u) \\
\Big\downarrow{\scriptstyle \Omega_n(f,\phi)} & & \Big\downarrow{\scriptstyle \widetilde{\Omega}_n(f,\phi)} \\
\Omega_n(Y, v) & \xrightarrow[\;\mathrm{pr}\;]{} & \widetilde{\Omega}_n(Y, v)
\end{array}
$$

This follows from the following fact and Lemma 2.2. If $(j', \mu') : (*, v') \longrightarrow (Y, v)$ is a morphism and (j, μ) and (f, ϕ) are as above, then there is a morphism $(\mathrm{id}, \psi) : (*, v) \longrightarrow (*, v')$ such that $(f, \phi) \circ (j, \mu)$ and $(j', \mu') \circ (\mathrm{id}, \psi)$ are homotopic morphisms.

The reduced group $\widetilde{\Omega}_n(X, u)$ is relevant for our Uniqueness Theorem 1.4 since there we classify stably up to connected sum with a simply connected oriented resp. $Spin$-manifold and by Theorem 2.1 this is decided in the reduced bordism group.

Next we make some computations for this generalized homology theory. Recall that sign denotes the signature. The group $\Omega_n(*, v)$ is either equal to Ω_n^{SO} if $B = BSO$, or to Ω_n^{Spin} if $B = BSpin$.

Lemma 2.3   1. *The following table gives generators and explicit isomorphisms for the various bordism groups:*

$$
\begin{array}{ll}
\Omega_3^{SO} = 0 & \\
\Omega_3^{Spin} = 0 & \\
\mathrm{sign} : \Omega_4^{SO} \xrightarrow{\;\cong\;} \mathbb{Z} & \;\Big|\; \mathbb{C}P^2 \\
\mathrm{sign} : \Omega_4^{Spin} \xrightarrow{\;\cong\;} 16 \cdot \mathbb{Z} & \;\Big|\; K
\end{array}
$$

2. $\widetilde{\Omega}_4(K(F, 1), u)) = 0$ for F a finitely generated free group and both cases $B = BSO$ or $B = BSpin$.

Proof. 1.) is standard. 2.) follows from the Mayer-Vietoris sequence applied to a wedge of S^1's and to the pushout which decribes S^1 as the identification of the two end points of $[0, 1]$ to one point.   □

3 Proof of the existence of a stable Kneser splitting

In this section we prove Theorem 1.3. We recall that the normal 1-type of a manifold M determines a fibration $\mathcal{B}(T(M))$. With the notation of the last section, if M has $w_2 = \infty$, let $B = BSO$, $K = *$ and $X = K(\pi_1(M), 1)$. For $w_2 = 0$ choose $B = BSpin$ instead of BSO. For $w_2 \neq 0, \infty$ choose $B = BSO$, $K = K(\mathbb{Z}/2, 2)$, $X = K(\pi_1(M), 1)$ and q and u maps representing the second Stiefel Whitney classes. Then in all three cases $\mathcal{B}(T(M)) = \mathcal{B}(u)$.

Firstly we show that we can assume without loss of generality that no boundary component C of M is π_1-null, i.e. the inclusion induces the trivial map $\pi_1(C) \longrightarrow \pi_1(M)$. Let $C_1, C_2, \ldots, C_m$ be the π_1-null boundary components of M. Since $\Omega_3(*, v)$ is trivial by Lemma 2.3, there is a nullbordism N_i for each C_i with respect to $(*, v)$. By 0- and 1-dimensional surgery on the interior of N_i we can achieve that N_i is simply connected. Define

$$\widehat{M} = M \cup_{C_1} N_1^- \cup_{C_2} \ldots \cup_{C_m} N_m^-$$

By Theorem 2.1 there is a stable oriented diffeomorphism $f : M \longrightarrow \widehat{M} \,\natural\, N_1 \,\natural\, \ldots$ $\natural\, N_m$ which induces on the fundamental groups the isomorphism induced by the inclusion of M in $\widehat{M}$. No boundary component of $\widehat{M}$ is π_1-null. Obviously it suffices to prove the claim for $\widehat{M}$.

If C is a component of ∂M, there is by assumption an index $i \in \{1, 2, \ldots, n\}$ such that the image of $\alpha \circ \pi_1(j)$ for $j : C \longrightarrow M$ the inclusion is subconjugated to Γ_i. Since we also assume that this image is non-trivial, this index is unique. For $i \in \{1, 2, \ldots, n\}$ let $\partial_i M$ be the union of those components C of ∂M for which this index is i. Since the inclusion $\partial M \longrightarrow M$ is a cofibration, we can construct maps $g : M \longrightarrow \vee_{i=1}^n K(\Gamma_i, 1)$ and $\partial_i g : \partial_i M \longrightarrow K(\Gamma_i, 1)$ for $i \in \{1, 2, \ldots, n\}$ such that the restriction of g to $\partial_i M$ is the composition of g_i with the canonical inclusion $j_i : K(\Gamma_i, 1) \longrightarrow \vee_{i=1}^n K(\Gamma_i, 1)$ and g induces α on the fundamental groups. Choose pointed maps $u_i : K(\Gamma_i, 1) \longrightarrow K$ such that the composition $u = (\vee_{i=1}^n u_i) \circ g : M \longrightarrow K$ corresponds to the Stiefel-Whitney classes of M in the case where K is not a point, but $K(\mathbb{Z}/2, 2)$. Let ρ be normal 1-smoothings of M in $\mathcal{B}(\sqcap)$ compatible with g and the orientation resp. $Spin$-structure. Denote the restriction of ρ to $\partial_i M$ by $\partial_i \rho$. By construction the homomorphism

$$\oplus_{i=1}^n \Omega_3(j_i) : \oplus_{i=1}^n \Omega_3(K(\Gamma_i, 1), u_i) \longrightarrow \Omega_3(\vee_{i=1}^n K(\Gamma_i, 1), \vee_{i=1}^n u_i)$$

sends $([\partial_i M, \partial_i \rho] \mid i = 1, 2, \ldots, n)$ to the element $[\partial M, \rho|_{\partial M}]$ which is zero since (M, ρ) is a nullbordism for its representative. This homomorphism is injective by a Mayer-Vietoris argument and Lemma 2.3. Hence we can find nullbordisms (V_i, σ_i) for $(\partial_i M, \partial_i \rho)$ with respect to $(K(\Gamma_i, 1), u_i)$ for $i \in \{1, 2, \ldots, n\}$. By the same argument as above the homomorphism

$$\oplus_{i=1}^n \Omega_4(j_i) : \oplus_{i=1}^n \Omega_4(K(\Gamma_i, 1), u_i) \longrightarrow \Omega_4(\vee_{i=1}^n K(\Gamma_i, 1), \vee_{i=1}^n u_i)$$

is surjective. Let $([W_i, \tau_i] \mid i = 1, 2, \ldots, n)$ be a preimage of $-[M^- \cup_{\partial M} \coprod_{i=1}^n V_i,$
$\rho^- \cup \coprod_{i=1}^n \sigma_i]$ and then we get

$$\left[M^- \cup_{\partial M} \coprod_{i=1}^n (V_i \coprod W_i), \rho^- \cup \coprod_{i=1}^n (\sigma_i \coprod \tau_i) \right] = 0 \qquad \in \Omega_4(\vee_{i=1}^n K(\Gamma_i, 1), \vee_{i=1}^n u_i).$$

By Theorem 2.1 $(V_i \coprod W_i), (\sigma_i \coprod \tau_i)$ is bordant relative boundary to a normal
1-smoothing (M_i, ρ_i). We have

$$[M \cup_{\partial M} \natural_{i=1}^n M_i, \rho \cup \natural_{i=1}^n (j_i \circ \rho_i)] = 0 \qquad \in \Omega_4(\vee_{i=1}^n K(\Gamma_i, 1), \vee_{i=1}^n u_i).$$

By Theorem 2.1 there is a stable oriented diffeomorphism

$$f : M \longrightarrow \natural_{i=1}^n M_i$$

such that the composition $\natural_{i=1}^n (j_i \circ g_i) \circ f$ is homotopic to $g : M \to \vee_{i=1}^n K(\Gamma_i, 1)$.
This finishes the proof of Theorem 1.3.

4   Proof of the uniqueness result

This section is devoted to the proof of Theorem 1.4. We firstly show that we
can assume without loss of generality that none of the manifolds N_j respectively
N_j' are present. By counting the π_1-null components we conclude $p = q$. After
possibly renumbering the N_j''s, we can assume that f maps ∂N_j to $\partial N_j'$ for all
$j \in \{1, 2, \ldots, q\}$. Since each N_j and N_j' is simply connected, the desired oriented diffeomorphism $g_j : N_j \natural T_j \longrightarrow N_j' \natural T_j'$ exists by Theorem 2.1. Again by
Theorem 2.1 there is a stable orientation respectively $Spin$-structure preserving
diffeomorphism

$$\natural_{i=1}^n M_i \longrightarrow \natural_{i=1}^n M_i \natural \natural_{j=1}^p N_j \cup_{\partial N_j} N_j^-$$

inducing on the fundamental group the obvious isomorphism and the claim
follows.

Suppose for a moment that $\natural_{i=1}^n M_i$ and $\natural_{i=1}^n M_i'$ are $Spin$. We want to show
that M_i and M_i' are diffeomorphic modulo connected sum with appropriate
simply connected four-manifolds with $Spin$-structure. Notice that for all M_i
and M_i' the normal 1-type has $w_2 = 0$. Thus we have to show that M_i and
M_i' have same normal 1-type and admit normal 1-smoothings in $\mathcal{B}_i$ which induce the right map on π_1 as stated in Theorem 1.4 and are compatible with
$f|_{\partial M_i}$ such that M_i and M_i' are bordant rel. boundary (identified via $f|_{\partial M_i}$)
in the reduced bordism group corresponding to the normal 1-type, which here
is $\widetilde{\Omega}_4^{Spin}(K(\Gamma_i', 1))$. If $\natural_{i=1}^n M_i$ and $\natural_{i=1}^n M_i'$ are just oriented manifolds we are
allowed to modify M_i and M_i' by connected sum with any oriented simply con-
nected four-manifold and after adding copies of $\mathbb{C}P^2$ we can assume that the
normal 1-type for all M_i and M_i' has $w_2 = \infty$. Then we have to show that M_i

and M_i' have same normal 1-type and admit normal 1-smoothings in $\mathcal{B}_i$ which induce the right map on π_1 as stated in Theorem 1.4 and are compatible with $f|_{\partial M_i}$ such that M_i and M_i' are bordant relative boundary (identified via $f|_{\partial M_i}$) in the reduced bordism group corresponding to the normal 1-type, which here is $\widetilde{\Omega}_4(K(\Gamma_i', 1)) = \widetilde{\Omega}_4^{SO}(K(\Gamma_i', 1))$. In the following proof the argument is identically the same in the *Spin*-case and in the oriented case and thus we restrict ourselves for simplicity to the oriented case.

We first show $f(\partial M_i) = f(\partial M_i')$. Let C be a component of ∂M_i for $i \in \{0, 1, \ldots, n\}$. Since $\mathrm{pr}_i' \circ f_* \circ j_i$ is an isomorphism and C is not π_1-null in M_i, the image of the composition of pr_i' and the homomorphism $\pi_1(f(C)) \longrightarrow \pi_1(\#_{i=1}^n M_i')$ $= *_{i=1}^n \Gamma_i'$ induced by the inclusion is non-trivial. This implies $f(C) \subset M_i'$.

Let W' be obtained from $\coprod_{i=1}^n M_i' \times [0, 1]$ by attaching 1-handles to $\coprod_{i=1}^n M_i' \times \{1\}$ such that

$$\partial W' = \coprod_{i=1}^n (M_i')^- \cup_{\coprod_{i=1}^n \partial M_i'} \#_{i=1}^n M_i'$$

where we identify M_i' with $M_i' \cup_{\coprod_{i=1}^n \partial M_i' \times \{0\}} \coprod_{i=1}^n \partial M_i' \times [0, 1]$. Define analogously W for the M_i's. Let $V = W^- \cup_f W'$ be obtained by glueing W and W' together along f. Choose a map $h_i' : M_i' \longrightarrow K(\Gamma_i', 1)$ inducing the identity on the fundamental groups and mapping the embedded disk where the 1-handles are attached to the base point. Let $h : W' \longrightarrow \vee_{i=1}^n K(\Gamma_i', 1)$ be the map which is on $M_i \times [0, 1]$ the composition of the projection $M_i \times [0, 1] \longrightarrow M_i$, h_i and the canonical inclusion of $K(\Gamma_i', 1)$ into $\vee_{i=1}^n K(\Gamma_i, 1)$ and on the one-handles the constant map. Since the inclusion of W into V is 3-connected we can extend this map to a map $h : V \longrightarrow \vee_{i=1}^n K(\Gamma_i, 1)$. Notice for the sequel that the restriction of this map to M_i' composed with the projection $\mathrm{pr}_k' : \vee_{i=1}^n K(\Gamma_i', 1) \longrightarrow K(\Gamma_k, 1)$ is the constant map for $k \neq i$. Restricting a normal structure of V compatible with h to $M_i \cup_{f|_{\partial M_i}} M_i'$ yields a normal structure ρ_i for $M_i \cup_{f|_{\partial M_i}} M_i'$ with respect to $\vee_{i=1}^n K(\Gamma_i', 1)$. In the sequel we abbreviate $M_i^- \cup_{f|_{\partial M_i}} M_i'$ by $M_i \cup M_i'$. We conclude

Lemma 4.1 *We have*

$$\sum_{i=1}^n [M_i \cup M_i', \rho_i] = 0 \qquad \in \widetilde{\Omega}_4(\vee_{i=1}^n K(\Gamma_i', 1)).$$

The projection $\mathrm{pr}_j : \vee_{i=1}^n K(\Gamma_i', 1) \longrightarrow K(\Gamma_j, 1)$ induces a homomorphisms

$$\widetilde{\Omega}_4(\mathrm{pr}_j) : \widetilde{\Omega}_4(\vee_{i=1}^n K(\Gamma_i', 1)) \longrightarrow \widetilde{\Omega}_4(K(\Gamma_j', 1)).$$

Notice that

$$\oplus_{i=1}^n \widetilde{\Omega}_4(j_i) : \oplus_{i=1}^n \widetilde{\Omega}_4(K(\Gamma_i', 1)) \longrightarrow \widetilde{\Omega}_4(\vee_{i=1}^n K(\Gamma_i', 1))$$

and

$$\oplus_{i=1}^n \widetilde{\Omega}_4(\mathrm{pr}_i) : \widetilde{\Omega}_4(\vee_{i=1}^n K(\Gamma_i', 1)) \longrightarrow \oplus_{i=1}^n \widetilde{\Omega}_4(K(\Gamma_i', 1))$$

are isomorphisms, inverse to one another, by a Mayer-Vietoris argument and Lemma 2.3.

Lemma 4.2 *For $i, k \in \{1, 2, \ldots, n\}$ with $i \neq k$ we get*

$$\widetilde{\Omega}_4(\mathrm{pr}\,'_k)\,[M_i \cup M_i', \rho_i] = 0 \qquad \in \widetilde{\Omega}_4(K(\Gamma_k', 1)).$$

Notice that Theorem 1.4 follows from Lemma 4.1 and Lemma 4.2 because they imply together with the pair of inverse isomorphisms above

$$[M_i \cup M_i', \mathrm{pr}\,'_i \circ \rho_i] = 0 \qquad \in \widetilde{\Omega}_4(K(\Gamma_i', 1)).$$

for $i \in \{1, 2, \ldots, n\}$ and then one can apply Theorem 2.1. So it remains to prove Lemma 4.2.

Let $C_1, C_2, \ldots, C_m$ be the components of ∂M_i. Let $k_j : \pi_1(C_j) \longrightarrow \Gamma_i$ be the homomorphism induced by the inclusion. Similarly, define $k_j' : \pi_1(f(C_j)) \longrightarrow \Gamma_i'$. If $g : G \longrightarrow H$ is a group homomorphism, denote by $H /\!\!/ g$ the pushout of groups of $* \longleftarrow G \xrightarrow{g} H$. This is the same as the quotient of H by the normal subgroup generated by the image of g.

Lemma 4.3 *Suppose that ∂M_i is non-empty. Then there is an isomorphism*

$$\alpha : \left(\Gamma_i /\!\!/ *_{j=1}^m k_j\right) * F \longrightarrow \pi_1(M_i/\partial M_i)$$

for F a finitely generated free group of rank $m - 1$ and a map

$$\beta : \Gamma_i /\!\!/ *_{j=1}^m k_j \longrightarrow \left(\Gamma_i' /\!\!/ *_{j=1}^m k_j'\right) * *_{1 \leq l \leq n, l \neq i} \Gamma_l'$$

such that the composition of β with the projection

$$\overline{\mathrm{pr}}\,'_i : \left(\Gamma_i' /\!\!/ *_{j=1}^m k_j'\right) * *_{1 \leq l \leq n, l \neq i} \Gamma_l' \longrightarrow \Gamma_i' /\!\!/ *_{j=1}^m k_j'$$

is an isomorphism and the following diagram commutes up to inner automorphisms of Γ_k' for $j : \Gamma_i /\!\!/ *_{j=1}^m k_j \longrightarrow \Gamma_i /\!\!/ *_{j=1}^m k_j * F$ the canonical inclusion and $\overline{g}_i$ induced by $\mathrm{pr}\,'_k \circ g_i$

$$
\begin{array}{ccc}
\Gamma_i /\!\!/ *_{j=1}^m k_j & \xrightarrow{\;\alpha \circ j\;} & \pi_1(M_i/\partial M_i) \\
\beta \downarrow & & \downarrow \pi_1(\overline{g}_i) \\
\left(\Gamma_i' /\!\!/ *_{j=1}^m k_j'\right) * *_{1 \leq l \leq n, l \neq i}\Gamma_l' & \xrightarrow[\;\mathrm{pr}\,'_k\;]{} & \Gamma_k'
\end{array}
$$

Before we prove Lemma 4.3, we explain how Lemma 4.2 and hence Theorem 1.4 follow from it. We only treat the more difficult case where ∂M_i is non-empty, the other case is similiar and does not use Lemma 4.3.

The underlying map of the normal structure ρ_i is

$$g_i \cup g_i' : M_i \cup M_i' \longrightarrow \vee_{i=1}^n K(\Gamma_i', 1).$$

The composition $\mathrm{pr}'_k \circ g_i'$ is the constant map to the base point. Therefore, we obtain a factorization of $\mathrm{pr}'_k \circ (g_i \cup g_i') : M_i \cup M_i' \longrightarrow K(\Gamma_k', 1)$ as the composition of the projection $q : M_i \cup M_i' \longrightarrow M_i/\partial M_i$ and the map $\overline{g}_i : M_i/\partial M_i \longrightarrow K(\Gamma_k', 1)$ induced by $\mathrm{pr}'_k \circ g_i$. The element $\widetilde{\Omega}_4(\mathrm{pr}'_k) [M_i \cup M_i', \rho_i]$ in $\widetilde{\Omega}_4(K(\Gamma_k', 1))$ lies in the image of the composition $\widetilde{\Omega}_4(\overline{g}_i) \circ \widetilde{\Omega}_4(q)$. The map $\overline{g}_i$ induces a map $K(\overline{g}_i, 1) : K(\pi_1(M_i/\partial M_i), 1) \longrightarrow K(\Gamma_k', 1)$ which induces a homomorphism

$$\widetilde{\Omega}_4(K(\overline{g}_i, 1)) : \widetilde{\Omega}_4(K(\pi_1(M_i/\partial M_i), 1)) \longrightarrow \widetilde{\Omega}_4(K(\Gamma_k', 1))$$

As $\overline{g}_i$ factorizes over $K(\overline{g}_i, 1)$, it suffices to show that $\widetilde{\Omega}_4(K(\overline{g}_i, 1))$ is trivial. If $j : G \longrightarrow G * F$ is the inclusion for F a finitely generated free group, then the homomorphism

$$\widetilde{\Omega}_4(K(j, 1)) : \widetilde{\Omega}_4(K(G, 1)) \longrightarrow \widetilde{\Omega}_4(K(G * F, 1))$$

is an isomorphism by a Mayer-Vietoris argument and Lemma 2.3. Hence it suffices to show because of Lemma 4.3 that the homomorphism

$$\widetilde{\Omega}_4(K(\mathrm{pr}'_k \circ \beta, 1)) : \widetilde{\Omega}_4(K(\Gamma_i /\!/ *_{j=1}^m k_j, 1)) \longrightarrow \widetilde{\Omega}_4(K(\Gamma_k', 1))$$

is trivial. From the Kurosh Subgroup Theorem 1.5 there are subgroups A_p and a free subgroup F' of $(\Gamma_i' /\!/ *_{j=1}^m k_j') * *_{1 \leq l \leq n, l \neq i} \Gamma_l'$ such that each A_p is subconjugated to $\Gamma_i' /\!/ *_{j=1}^m k_j'$ or some of the Γ_l's and the image of β is given by

$$\mathrm{im}(\beta) = F' * *_{p=1}^q A_p.$$

Since the composition of β with

$$\overline{\mathrm{pr}'_i} : (\Gamma_i' /\!/ *_{j=1}^m k_j') * *_{1 \leq l \leq n, l \neq i} \Gamma_l' \longrightarrow \Gamma_i' /\!/ *_{j=1}^m k_j'$$

is injective by Lemma 4.3 each A_p is subconjugated to $\Gamma_i' /\!/ *_{j=1}^m k_j'$. The inclusion $\iota : *_{p=1}^q A_p \longrightarrow \mathrm{im}(\beta)$ induces by the argument above an isomorphism

$$\widetilde{\Omega}_4(K(\iota, 1)) : \widetilde{\Omega}_4(\vee_{p=1}^q K(A_p, 1)) \longrightarrow \widetilde{\Omega}_4(\mathrm{im}(\beta))$$

Since $\widetilde{\Omega}_4(K(\mathrm{pr}'_k \circ \beta, 1))$ factorizes through $\widetilde{\Omega}_4(\mathrm{im}(\beta))$ and the composition of ι with the projection

$$\mathrm{pr}'_k : (\Gamma_i' /\!/ *_{j=1}^m k_j') * *_{1 \leq l \leq n, l \neq i} \Gamma_l' \longrightarrow \Gamma_k'$$

is trivial, $\widetilde{\Omega}_4(K(\mathrm{pr} \circ \beta, 1))$ is trivial. This finishes the proof that Lemma 4.3 implies Lemma 4.2 and hence Theorem 1.4.

It remains to prove Lemma 4.3. The map

$$\alpha : \left(\Gamma_i /\!/ *_{j=1}^m k_j \right) * F \longrightarrow \pi_1(M_i/\partial M_i)$$

is given on $\Gamma_i /\!/ *_{j=1}^m k_j$ by the map which is induced by the projection $M_i \longrightarrow M_i/\partial M_i$. The free group F has $m-1$ generators and α sends the i-th generator to the class in $\pi_1(M_i/\partial M_i)$ represented by some path in M_i joining C_1 and C_{j+1}. Notice that $M_i/\partial M_i$ is up to homotopy the same as attaching to each boundary component C_j the cone over C_j and then attaching $m-1$ one-cells such that the j-th one-cell joins the top of the cone of C_1 and C_{j+1} for $1 \leq j \leq m-1$. One easily checks using Seifert-van Kampen Theorem that α is an isomorphism.

For appropriate choices of elements w_j in $*_{i=1}^n \Gamma_i'$ the following diagram commutes

$$
\begin{array}{ccc}
*_{j=1}^m \pi_1(C_j) & \xrightarrow{\;*_{j=1}^m \pi_1(f|_{C_j})\;} & *_{j=1}^m \pi_1(f(C_j)) \\
{\scriptstyle *_{j=1}^m k_j} \downarrow & & \downarrow {\scriptstyle *_{j=1}^m c(w_j) \circ k_j'} \\
\Gamma_i & \xrightarrow[\;f_* \circ j_i\;]{} & *_{i=1}^n \Gamma_i'
\end{array}
$$

where $c(w_j)$ denotes conjugation with w_j. The map β is the homomorphism making the following diagram commute

$$
\begin{array}{ccc}
\Gamma_i & \xrightarrow{\;f_* \circ j_i\;} & *_{i=1}^n \Gamma_i' \\
{\scriptstyle \mathrm{pr}} \downarrow & & \downarrow {\scriptstyle \mathrm{pr}} \\
\Gamma_i /\!/ *_{j=1}^m k_j & \xrightarrow[\;\beta\;]{} & \left(\Gamma_i' /\!/ *_{j=1}^m k_j' \right) * *_{1 \leq l \leq n, l \neq i} \Gamma_l'
\end{array}
$$

Since $*_{j=1}^m \pi_1(f|_{C_j})$ is an isomorphism and $\mathrm{pr}_i' \circ f_* \circ j_i$ is an isomorphism by assumption, the map $\mathrm{pr}_i' \circ f_* \circ j_i$ induces an isomorphism β_1 making the following diagram commute

$$
\begin{array}{ccc}
\Gamma_i & \xrightarrow{\;\mathrm{pr}_i' \circ f_* \circ j_i\;} & \Gamma_i' \\
{\scriptstyle \mathrm{pr}} \downarrow & & \downarrow {\scriptstyle \mathrm{pr}} \\
\Gamma_i /\!/ *_{j=1}^m k_j & \xrightarrow[\;\beta_1\;]{} & \Gamma_i' /\!/ *_{j=1}^m c(\mathrm{pr}_i'(w_j)) \circ k_j'
\end{array}
$$

The identity on Γ_i' induces an isomorphism

$$\beta_2 : \Gamma_i' /\!/ *_{j=1}^m c(\mathrm{pr}_i'(w_j)) \circ k_j' \longrightarrow \Gamma_i' /\!/ *_{j=1}^m k_j'$$

The composition of β with the projection

$$\overline{\mathrm{pr}}_i' : \left(\Gamma_i' /\!/ *_{j=1}^m k_j' \right) * *_{1 \leq l \leq n, l \neq i} \Gamma_l' \longrightarrow \Gamma_i' /\!/ *_{j=1}^m k_j'$$

is the composition of the isomorphisms β_1 and β_2 and hence bijective. The diagram commutes up to inner automorphisms of Γ_k' by construction. This finishes the proof of Lemma 4.3 and hence of Theorem 1.4. $\square$

References

[1] Bröcker, T. and tom Dieck, T. : *"Kobordismentheorie"*, Lecture Notes in Mathematics 178 (1970)

[2] Cappell, S.E.: *"On connected sums of manifolds"*, Topology 13, 395 - 400 (1974)

[3] Cappell, S.E.: *"A spitting theorem for manifolds"*, Inventiones Math. 33, 69 - 170 (1976)

[4] Cohen, M.M.: *"Combinatorial group theory: a topological approach"*, LMS student texts 14 (1989)

[5] Freedman, M.H.: *"The topology of four-dimensional manifolds"*, J. of Differential Geometry 17, 357 - 453 (1982)

[6] Hempel, J.: *"3-manifolds"*, Annals of Mathematics Studies 86, Princeton University Press (1976)

[7] Hillman, J.A.: *"Free products and 4-dimensional connected sums"*, preprint (1993)

[8] Kreck, M.: *"Surgery and duality"*, to appear, Vieweg (1994)

[9] Kreck, M., Lück, W. and Teichner, P.: *"Counterexamples to the Kneser conjecture in dimension four"*, preprint, Mainz (1994)

[10] Lyndon, R.C. and Schupp, P.E.: *" Combinatorial Group Theory"*, Ergebnisse der Mathematik und ihrer Grenzgebiete 89, Springer (1977)

Smooth Correspondences

Jack Morava

Abstract

This is a brief account of a category of smooth complex-oriented correspondences, which is related to the category of motives roughly as complex
cobordism is related to the Chow ring.

Introduction

The theory of correspondences is a old tool with a history going back to the
early days of algebraic topology, but in recent years it has been used primarily by
algebraic geometers in their attempts to construct a theory of motives; this note
is an account of a similar formalism, expressed in terms of smooth manifolds
and cobordism rather than in terms of algebraic cycles on algebraic varieties.
I first became interested in this material in the early 1970's, while trying to
understand work of Quillen; in fact Proposition 2.2 below was the result of
some conversations in those days with Bill Browder. Recently, interest in such
techniques seems to have reawakened; they play an ideological role, for example,
in recent work of Cohen, Jones, and Segal on Floer homotopy type (cf. [2,§5]).

In very vague terms, a category of correspondences is a construction which
enriches a category where some objects [e.g. Poincaré spaces] have good duality
properties, into a category in which that duality is built directly into its foun-
dations. Smooth manifolds provide a particularly clear example, and I have
tried to keep this sketch as simple as possible. The first section sets up basic
properties; the second invokes the theory of operations in complex cobordism to
show that this geometrically defined category can be enriched to a category in
which the hom-objects become representations of something like what algebraic
geometers call a 'motivic group'. There are then two sections of applications:
the first, which is suggested by some of the work of Cohen, Jones, and Segal, is
a construction which makes a suitable moduli space of Riemann surfaces into
a kind of monoid in the category of correspondences; the point is that glueing surfaces together may make better sense as a correspondence than it does
as a map. The second section shows that some, at least, of Thom's theory of
singularities of maps continues to make sense for correspondences.

It is said that we calculate with one side of our brain, and that we think
about geometry with the other; whether this is true or not, in Browder's papers
these aspects of mathematical thought seem to be beautifully balanced. The
material in this note is not very deep, but I hope it reflects a concern for this
duality, which runs through so much of his work.

1 The category

1.0. The graded additive category (**Corr***) will have compact closed even-dimensional complex-oriented manifolds (P,Q etc.) as its objects; the graded group

$$\mathrm{mor}^*(P,Q) := MU^{*+2q}(P \times Q)[\tfrac{1}{2}],$$

(where p and q are the complex dimensions of P and Q respectively, and 2 is inverted for technical reasons) is to be the set of morphisms from P to Q. Composition is defined by the diagram

$$\mathrm{mor}^*(P,Q) \times \mathrm{mor}^*(Q,R) \xrightarrow{\quad *\quad} \mathrm{mor}^*(P,R)$$

$$\downarrow \qquad\qquad\qquad\qquad\qquad\qquad \downarrow$$

$$MU^{*+2q}(P \times Q)[\tfrac{1}{2}] \times MU^{*+2r}(Q \times R)[\tfrac{1}{2}]$$

$$\downarrow$$

$$MU^{*+2q}(P \times Q) \otimes_{MU^*(Q)} MU^{*+2r}(Q \times R)[\tfrac{1}{2}]$$

$$\downarrow$$

$$MU^{*+2(q+r)}(P \times Q \times R)[\tfrac{1}{2}] \xrightarrow{\quad pr_{Q!}\quad} MU^{*+2r}(P \times R)[\tfrac{1}{2}]$$

where $pr_{Q!}$ is the Gysin transformation defined by the projection

$$pr_Q : P \times Q \times R \to P \times R.$$

An object P of this category has an identity map $[1_P]$ in $\mathrm{mor}^0(P,P)$, defined by its diagonal map Δ_P. More generally, if $f : P \to Q$ is a smooth map of compact complex-oriented manifolds, then its graph

$$P \to P \times Q$$

is a complex-oriented map of complex codimension q which defines a correspondence $[f]$ of grade zero, from P to Q.

Proposition 1.1. *If*

$$P \xrightarrow{f} Q \xrightarrow{g} R$$

is a composition of smooth maps, then

$$[g \circ f] = [g] \star [f].$$

Corollary. *The operation which sends a compact complex-oriented manifold to itself, and which sends the smooth map f to the correspondence $[f]$, defines a functor from the usual category of such manifolds, to the category of smooth correspondences.*

Remarks. These definitions admits many variants. The motives of algebraic geometry are constructed from algebraic varieties, with morphism-sets defined by groups of algebraic cycles; different notions of equivalence of cycles define different categories of motive-like objects [6]. Similarly, in algebraic topology any multiplicative cohomology theory E with Gysin transformations [e.g. rational cohomology, or PL-cobordism] defines a category of E-correspondences.

There are other kinds of variants as well. Below we will use a category (**Corr**) constructed as above, but in which the hom-sets are given their even-odd grading. Similarly, these categories have equivariant versions: if G is a compact Lie group, P and Q are G-spaces, and $f : P \to Q$ is a G-equivariant map, then the graph of f defines a G-equivariant map of P to $P \times Q$. Any multiplicative G-cohomology theory thus leads to a category of equivariant correspondences, cf. [4]. It may be helpful to note that related constructions are well-established in the theory of transformation groups. In for example [13, Ch. IV §8] the subgroups of G are made into a category $\Omega(G)$, and with suitably defined induction and restriction functors, the category of correspondences equivariant with respect to subgroups of G is fibered over $\Omega(G)$. An adequate discussion of this structure would require too great a digression; but note that if H is a subgroup of G, and P is a manifold with H-action, then $G \times_H P$ is a manifold with G-action; and if

$$P \longrightarrow Q \times R$$

is an H-equivariant correspondence, then

$$G \times_H P \longrightarrow (G \times G) \times_{(H \times H)} (Q \times R)$$

is a G-equivariant correspondence from $G \times_H Q$ to $G \times_H R$.

It is occasionally confusing that the objects of these categories of correspondences are called manifolds: it is the morphisms that are the correspondences. Perhaps the objects of such categories could be called pastiches, in homage to the motivic terminology of algebraic geometry.

1.2. The category of smooth correspondences possesses a monoidal structure, defined by the cartesian product of manifolds. More precisely, if P, Q, R, S are objects of the category, then the usual twist map defines a product, by the diagram

$$
\begin{array}{ccc}
\mathrm{mor}^*(P,Q) \times \mathrm{mor}^*(R,S) & \longrightarrow & \mathrm{mor}^*(P \times R, Q \times S) \\
\downarrow & & \\
MU^{*+2q}(P \times Q)[\tfrac{1}{2}] \times MU^{*+2s}(R \times S)[\tfrac{1}{2}] & & \Big\downarrow \\
\downarrow & & \\
MU^{*+2(q+s)}(P \times Q \times R \times S)[\tfrac{1}{2}] & \xrightarrow[1 \times t_{Q,R} \times 1]{} & MU^{*+2(q+s)}(P \times R \times Q \times S)[\tfrac{1}{2}];
\end{array}
$$

moreover the 'forgetful' functor (from the category of smooth compact complex-oriented manifolds to the category of smooth correspondences) respects this product.

There is also a contravariant endofunctor

$$\text{op} : (\mathbf{Corr}) \to (\mathbf{Corr})$$

which we will call the adjoint; it is the identity on objects, but it sends the correspondence C in $\text{mor}^*(P, Q)$ to C^{op} in $\text{mor}^{*+2(q-p)}(Q, P)$, by exchanging domain and range.

Exercise.

i) op is in fact a contravariant (monoidal) functor.

ii) if C is the graph of an isomorphism of P with Q then C^{op} is the graph of its inverse.

iii) C is an isomorphism iff both $C \star C^{\text{op}}$ and $C^{\text{op}} \star C$ are isomorphisms.

1.3. It is reasonable (though we will not use this construction in the rest of this note) to enlarge the set of objects of $(\mathbf{Corr}^*)$ to include symbols $P[n]$, where n is an integer and P is a compact complex-oriented manifold as before, by setting

$$\text{mor}^*(P[n], Q[m]) := \text{mor}^{*+2(m-n)}(P, Q).$$

This allows us to define an object

$$Q^P := (P \times Q)[-p]$$

with some of the properties of an object of morphisms from P to Q. In particular, the diagonal of $P \times Q$ defines a canonical 'evaluation' correspondence

$$P \times Q^P \to Q;$$

similarly $(Q^P)^R$ is canonically isomorphic to $Q^{P \times R}$. Moreover, if P and Q are complex manifolds, and $\text{Hol}(P, Q)$ is the space of holomorphic maps from P to Q, then the graph of the evaluation map

$$P \times \text{Hol}(P, Q) \to Q$$

defines a kind of correspondence from $\text{Hol}(P, Q)$ to Q^P. [Some qualification is necessary, because the space $\text{Hol}(P, Q)$, though finite-dimensional, will not necessarily be compact.]

2 (Some of) its symmetries

2.0. In this section we construct a group of invertible additive monoidal functors

$$(\mathbf{Corr}^*) \to (\mathbf{Corr}^*)$$

which equal the identity on objects; in a sense (cf. [7]) the existence of this group expresses the fact that $(\mathbf{Corr}^*)$ is rationally a kind of deformation of the category of homology correspondences.

Theorem. *The Landweber-Novikov group* $\mathbf{G}_0$ *(of formal power series*

$$T \mapsto t(T) = \sum_{n \geq 0} t_n T^{n+1}$$

with $t_0 = 1$, *under composition) acts by such automorphisms of (**Corr**).*

To see this we need some familiar machinery from [12]. First of all, the composition-law of $\mathbf{G}_0$ defines the structure of a Hopf algebra on the polynomial ring

$$S_* = \mathbf{Z}[t_n | n > 0];$$

if M_* is an S_*-comodule, I will write

$$s_t : M_* \to M_* \otimes S_*$$

for its coaction. If P is a complex-oriented manifold, then the formula

$$c_t(P) = \phi^{-1} s_t \phi(1),$$

where

$$\phi : MU^*(P) \to MU^{*+2p}(T_P^+)$$

denotes the Thom isomorphism for the tangent bundle of P, defines the total Conner–Floyd–Chern class of P; this is a polynomial in the elements t_i with coefficients from $MU^*(P)$, having leading term 1. It is standard that

$$s_u c_t(P) = c_u(P)^{-1} c_{tou}(P),$$

where $(t \circ u)(T) = t(u(T))$ is the composition of the power series t and u.

Now if C is an element of $MU^*(P \times Q)[\frac{1}{2}]$, let

$$\tilde{s}_t(C) = c_t(P \times Q)^{-\frac{1}{2}} s_t(C);$$

the square root exists because two is a unit of the coefficient ring.

Proposition 2.1.

i) $\tilde{s}_t(C \star D) = \tilde{s}_t(C) \star \tilde{s}_t(D)$

ii) $\tilde{s}_t(\tilde{s}_u(C)) = \tilde{s}_{tou}(C)$

iii) $\tilde{s}_t(\Delta) = \Delta$;

where D *is a correspondence of* Q *with* R, *and* Δ *is the diagonal class in* $MU^{2p}(P \times P)[\frac{1}{2}]$.

The first assertion follows because

$$\tilde{s}_t(C \star D) = c_t(P \times R)^{-\frac{1}{2}} pr_{Q!}(s_t(C \otimes D)c_t(Q)^{-1})$$

by the Riemann–Roch theorem for cobordism. This expression can be rewritten as

$$c_t(P \times R)^{-\frac{1}{2}} pr_{Q!}(\tilde{s}_t(C)c_t(P \times Q)^{\frac{1}{2}} \otimes \tilde{s}_t(D)c_t(Q \times R)^{\frac{1}{2}} c_t(Q)^{-1})$$
$$= \tilde{s}_t(C) \star \tilde{s}_t(D).$$

The second assertion follows similarly, because

$$\tilde{s}_t(s_u(C)c_u(P \times Q)^{-\frac{1}{2}})$$
$$= s_{tou}(C)s_t(c_u(P \times Q)^{-\frac{1}{2}})c_t(P \times Q)^{-\frac{1}{2}}$$
$$= s_{tou}(C)c_{tou}(P \times Q)^{-\frac{1}{2}} = \tilde{s}_{tou}(C).$$

Finally, note that the correspondence Δ is just $\Delta_!(1)$, so

$$\tilde{s}_t(\Delta) = s_t(\Delta)c_t(P \times P)^{-\frac{1}{2}};$$

but

$$s_t(\Delta) = \Delta_! c_t(\nu_\Delta),$$

and therefore

$$
\begin{aligned}
\tilde{s}_t(\Delta) &= \Delta_!(c_t(P)(\Delta^* c_t(P \times P)^{-\frac{1}{2}})) \\
&= \Delta_!(1) = \Delta.
\end{aligned}
$$

Proposition 2.2. *Suppose a smooth map $f : P \to Q$ becomes invertible in* (**Corr**); *then (away from the prime two) some suspension*

$$f \times 1_{\mathbf{C}^n} : P \times \mathbf{C}^n \to Q \times \mathbf{C}^n$$

is properly homotopic to a diffeomorphism.

Indeed, since the map f^* induced in complex cobordism by f commutes with the operators s_t, its inverse must do the same; by the Riemann–Roch theorem, the total Chern class of the normal bundle to f must then be trivial. But by a theorem of Conner and Floyd, the vanishing of these classes entails the stable triviality of this bundle; hence f is a tangential homotopy equivalence, and so by a theorem of Mazur is stably homotopic to a diffeomorphism.

Remark. This might be paraphrased by saying that at odd primes, a smooth complex - oriented manifold can be described, up to stable diffeomorphism, as a Poincaré complex with normal data encoded by an MU-orientation. This is suggested by Sullivan's theorem, which identifies a PL-manifold, at odd primes, as a Poincaré complex with a $K\otimes$-orientation.

2.3. Modulo some formalities, it is now a consequence of §2.1 that the Landweber–Novikov group acts as claimed. To be precise, if A is a commutative ring in which two is invertible, let the category $(\mathbf{Corr})_A$ have the same objects as $(\mathbf{Corr})$, but with (the $\mathbf{Z}/2\mathbf{Z}$-graded group of) morphisms

$$\mathrm{mor}^*(P,Q) \otimes A$$

between two objects; $(\mathbf{Corr})_A$ is thus an A-linear category. I claim that the group

$$\mathbf{G}_0(A) := \mathrm{Hom}_{\mathrm{rings}}(S, A)$$

of A-valued points of the Landweber–Novikov group acts by A-linear monoidal automorphisms of $(\mathbf{Corr})_A$: given the series $t(T)$ with coefficients t_i in A, let the functor

$$I_t : (\mathbf{Corr})_A \to (\mathbf{Corr})_A$$

be the identity on objects; then if C is a correspondence from P to Q,

$$I_t(C) = \tilde{s}_t(C)$$

defines an A-linear map

$$\mathrm{mor}^*(P,Q) \otimes A \to \mathrm{mor}^*(P,Q) \otimes A.$$

This construction preserves identity correspondences, by iii), and it preserves the composition of correspondences, by i); so it is indeed a functor. Part ii) of the proposition implies that

$$I_t \circ I_u = I_{t \circ u}.$$

It remains to see that the functor I_t is monoidal: but it is classical that the automorphisms s_t are multiplicative, and that $c_t(P \times Q) = c_t(P) \otimes c_t(Q)$, from which the assertion is immediate. It follows similarly that the functors I_t are pivotal: in other words the operations $\tilde{s}_t$ are self-adjoint, in the sense that

$$(\tilde{s}_t(C))^{op} = \tilde{s}_t(C^{op}).$$

2.4. The argument in the preceding paragraph used an enrichment of the category of correspondences: in it the morphism-sets of $(\mathbf{Corr})$ were shown to be linear representations of a certain groupscheme. This kind of construction can be pushed a bit further, by recalling that a graded module M_* can be interpreted as a linear representation of the multiplicative algebraic group $\mathbf{G}_m$. [With the diagonal which sends t_0 to $t_0 \otimes t_0$, the Hopf algebra $\mathbf{Z}[t_0, t_0^{-1}]$ of $\mathbf{G}_m$ represents the functor which sends a commutative ring to its group of units; the ungraded module M underlying M_* becomes a comodule over this Hopf algebra, with a coaction sending an element x homogeneous of degree i to xt_0^i.] However, there are some subtleties.

The functor $\mathbf{G}$, which assigns to a commutative ring A, the group of all invertible power series with coefficients from A, fits in an exact sequence

$$1 \longrightarrow \mathbf{G}_0 \longrightarrow \mathbf{G} \longrightarrow \mathbf{G}_m \longrightarrow 1 \; ;$$

it is represented by a Hopf algebra

$$\mathbf{Z}[t_i | i \geq 0][t_0^{-1}].$$

There is also an exact sequence

$$1 \longrightarrow \mu_2 \longrightarrow \mathbf{G}_m \longrightarrow \mathbf{G}_m \longrightarrow 1$$

of commutative groupschemes, defined by the squaring operation. Let $\mathbf{G}(2)$ denote the pullback of $\mathbf{G}$ along this map; it is thus a double cover of $\mathbf{G}$, represented by a Hopf algebra

$$\mathbf{Z}[t_i | i \geq 0][t_0^{-\frac{1}{2}}].$$

Over $\mathbf{Z}[\frac{1}{2}]$, $\mathbf{G}(2)$ is just the groupscheme of power series substitutions

$$T^{\frac{1}{2}} \mapsto \tau(T^{\frac{1}{2}}) = \sum_{n \geq 0} \tau_n T^{n+\frac{1}{2}} = (t_0 T)^{\frac{1}{2}} \left(1 + \sum_{n \geq 1} t_0^{-1} t_n T^n\right)^{\frac{1}{2}},$$

cf. [1]. I claim that the action of $\mathbf{G}_0$ extends to an action of $\mathbf{G}(2)$ as automorphisms of the graded category of correspondences; that the action of $\mathbf{G}_0$ on $MU^*(-)$ extends to an action of $\mathbf{G}$ is quite familiar, but this action of the double cover is something of a surprise. If t is an element of $\mathbf{G}$, then we have

$$c_t(P) = t_0^p + \text{ terms of higher order.}$$

The adjunction to the algebra of functions on $\mathbf{G}$ of the element $t_0^{\frac{1}{2}}$ is therefore exactly the information required to make sense of the expression $c_t(P)^{\frac{1}{2}}$.

Corollary. *The group $\mathbf{G}(2)$ acts by monoidal additive automorphisms of the graded category of correspondences.*

This action, however, is not quite pivotal: for if C is a correspondence mapping P to Q, then

$$\tilde{s}_t(C^{op}) = t_0^{q-p}(\tilde{s}_t(C))^{op}.$$

In other words, the action has a kind of projective anomaly.

2.5. We can now make an elementary observation about 'motivic cohomology':

$$P \mapsto \text{mor}^*(P, pt) := MU^*(P)[\tfrac{1}{2}]$$

defines a contravariant functor from $(\mathbf{Corr}^*)$ to graded abelian groups. In fact, the induced homomorphism

$$\text{mor}^*(P, Q) \longrightarrow \text{Hom}^{*+2q}(MU^*(Q), MU^*(P))[\tfrac{1}{2}]$$

is a homomorphism of S_*-comodules, provided that we use the appropriate definition for the action on the right.

The point is that the natural S_*-coaction on $\mathrm{mor}^*(P, pt)$ is defined by the operators $\tilde{s}_t$, while the usual coaction on $MU^*(P)[\frac{1}{2}]$ is defined by s_t. Thus the cobordism groups of a complex-oriented manifold, regarded as an object of the category (**Corr***), agree with its cobordism groups as usually defined; but the S_*-comodule structures on these groups will in general be distinct. This 'motivic cohomology' functor is not quite monoidal, because cobordism has a Kunneth spectral sequence rather than a Kunneth formula. However, this distinction is a matter of torsion, which disappears after tensoring with the rationals. In that case there are other simplifications; in particular, the S_* - coaction can be reconstructed from the induced action of the Lie algebra of $\mathbf{G}(2)$ [or $\mathbf{G}$]. If we write $e(n)$ for the power series $T + eT^{n+1}$ then the derivatives v_n of $s_{e(n)}$ at $e = 0$ satisfy the relations

$$[v_n, v_m] = (n - m)v_{n+m}$$

on the groups

$$MU_{\mathbf{Q}}^*(X) := MU^*(X) \otimes \mathbf{Q}$$

associated to a topological space X; but on the 'motivic' cohomology of an object of the category of correspondences we have an action by natural transformations

$$\tilde{v}_n = v_n - \tfrac{1}{2}c_n$$

which satisfy the same Lie bracket relations.

2.6. The constructions of this section show only that $\mathbf{G}(2)$ is contained in the group of automorphisms of the category of correspondences, but it seems reasonable to conjecture that $\mathbf{G}(2)$ is in fact the entire group of automorphisms of the category. It follows from general theory [5, §6] that the category of smooth manifolds and rational cohomology correspondences has only trivial automorphisms, but (for example) in the category of ordinary cohomology correspondences modulo two, I do not even know how to carry out the analogue of the construction of this section.

3   Monoids of surfaces

3.0. This section sketches an application of the preceding machinery. The problem concerns the construction of an analogue of the connected sum operation of differential topology, but in the context of Riemann surfaces: we start with a compact Riemann surface endowed with a base point, as well as a holomorphically embedded (closed) disk, centered at the base point. Given two such decorated surfaces, say of genus g and h respectively, there is a well-defined surface of genus $g + h$ obtained by first excising the interiors of the disks from the two original surfaces, followed by glueing together the resulting surfaces with boundary; the embeddings define a canonical holomorphic structure in a neighborhood of the suture. This suggests that we might try to think of the moduli spaces of such Riemann surfaces as some kind of monoid.

More specifically, there is a moduli space M_g of compact (connected) Riemann surfaces of genus g, as well as a space $M_g(1)$ parameterizing Riemann surfaces marked with a point; indeed $M_g(1)$ can be regarded as a kind of bundle over M_g, with the surface as fiber. There is also a space $\tilde{M}_g(1)$ which parameterizes surfaces together with embedded holomorphic disks, and it is known that this latter space has the homotopy type of (an orbifold) circle bundle over $M_g(1)$: the equivalence is defined by the map which assigns to a surface with embedded disk, the same surface marked at the specified point with the nonzero tangent vector defined by the derivative of the embedding map at the origin. For the purposes of homotopy theory, we can thus regard this connected sum operation as mapping the product of a circle bundle over $M_g(1)$ and a circle bundle over $M_h(1)$, to M_{g+h}, cf. [10]. Perhaps this is where I should note that this construction admits a symmetry with respect to the circle group: if we rotate both embedded disks by the same amount, then we get an isomorphic connected sum.

3.1. However, there is reason to think that this construction can be improved. Mumford and Deligne have constructed a nice compactification $\hat{M}_g$ of the moduli space of genus g surfaces, and their techniques have been extended by Knudsen [8] to construct analogous compactifications $\hat{M}_g(1)$. [More recently, Penner [11] and Kontsevich [9,§2] have constructed versions of these spaces which look more familiar to topologists.] The resulting spaces are not smooth, but they are rational homology manifolds, and they can be regarded as objects of a category of correspondences if we are willing to tensor all the relevant cohomology groups with the rationals. To be more precise, let L_g be the (orbifold) tangent bundle along the fiber to the forgetful map $\hat{M}_g(1) \longrightarrow \hat{M}_g$, and let

$$P_{\mathbf{C}}(L_g \oplus L_h) \longrightarrow \hat{M}_g \times \hat{M}_h$$

be the 2 - sphere (orbifold) bundle obtained by projectifying the product of these two line bundles. We can define a connected sum map

$$P_{\mathbf{C}}(L_g \oplus L_h) \longrightarrow \hat{M}_{g+h};$$

or, more correctly, we can define a map from a homotopy - equivalent space defined by pulling back this bundle over a product of moduli spaces of surfaces with embedded disks, which assigns to a pair of marked surfaces and the ratio of two tangent vectors, the surface constructed by excising the embedded disks and glueing the two sufaces together along an interposed plane annulus with conformal parameter given by the tangent ratio . Putting these two maps together, we get a correspondence (of grade two) which maps $\hat{M}_g(1) \times \hat{M}_h(1)$ to

$\hat{M}_{g+h}$. This seems to be the simplest way to construct a monoid structure on the compactified moduli spaces: but the result is an object which is a monoid in the category of rational correspondences, rather than in the category of spaces. Conceivably, the rationalization might be avoided by working with equivariant correspondences constructed from moduli spaces with level structures, along lines suggested by Looijenga.

4   Singularities of maps

4.0. Finally, note that Thom's theory of singularities fits in very naturally with the machinery of correspondences. For simplicity I will deal only with the complex case, and I will write $J^r(p,q)$ for the (complex-linear) space of r-jets of holomorphic map-germs

$$(\mathbf{C}^p, 0) \longrightarrow (\mathbf{C}^q, 0),$$

while $L^r(p)$ will denote the group, under composition, of r-jets of invertible map-germs from $(\mathbf{C}^p, 0)$ to itself. Thus the group

$$L^r(p,q) := L^r(p) \times L^r(q)^{\mathrm{op}}$$

acts (on the left) on the space $J^r(p,q)$, cf. [14, Levine II §8].

In Thom's terminology, the singularity types are just the orbits of this action, but for our purposes it will be convenient to call any proper complex-oriented $L^r(p,q)$-equivariant map

$$S \to J^r(p,q)$$

of (complex) codimension s, a (generalized, complex) singularity type (of real codimension $2s$); the desingularizations Z_k of the closures of the spaces of rank k maps in $J^1(p,q)$ [14, Porteous §1] define an interesting class of examples of such things. The cobordism classes of these generalized singularity types define a graded module

$$MU^{2*}_{L^r(p,q)}(J^r(p,q))$$

about which little seems to be known.

Proposition 4.1. *An element S of degree $2s$ in this module defines a homomorphism*

$$C \mapsto C_S : \mathrm{mor}^*(P,Q) \longrightarrow \mathrm{mor}^{*+2s}(P,Q).$$

To see this, recall that the space $J^r(P,Q)$ of r-jets of complex-oriented maps from P to Q is a $J^r(p,q)$-bundle over $P \times Q$, associated to a principal bundle $L^r(P,Q)$ with structure group $L^r(p,q)$. There is thus a proper complex-oriented map

$$S(P,Q) := S \times_{L^r(p,q)} L^r(P,Q) \longrightarrow J^r(P,Q)$$

of complex codimension s; but because the space $J^r(P,Q)$ is homotopy-equiv-
alent to $P \times Q$, this map defines a class $[S(P,Q)]$ in $MU^{2s}(P \times Q)$, and the
homomorphism

$$\mathrm{mor}^*(P,Q) = MU^{*+2q}(P \times Q) \longrightarrow MU^{*+2q+2s}(P \times Q) = \mathrm{mor}^{*+2s}(P,Q)$$

is just multiplication by $[S(P,Q)]$.

To see that this construction is reasonably natural, recall that if $f : P \to Q$
is a smooth map, then its r-jet is a lift of the graph morphism

$$\mathrm{gr}(f) : P \longrightarrow P \times Q$$

to a map

$$j_r(f) : P \longrightarrow J^r(P,Q).$$

If S is a singularity type in the sense of Thom, and $j_r(f)$ is transversal to the
imbedding of $S(P,Q)$ in $J^r(P,Q)$, then the intersection $S(f)$ [i.e., the fiber
product over $J^r(P,Q)$] of these maps defines a morphism

$$S(f) \longrightarrow P \times Q \longrightarrow Q,$$

which in Thom's theory is the restriction of the map f to the subspace of P
where the map f has a singularity of type S. But the cobordism class of the
fiber product $S(f)$ can just as well be described as the fiber product over the
homotopy equivalent space $P \times Q$, and in that guise the map $S(f) \to P \times Q$ is
precisely the product of $[\mathrm{gr}(f)]$ and $[S(P,Q)]$ in the cobordism ring of $P \times Q$.
It is thus not unreasonable to describe the operation S as the assignment to
a correspondence C, the correspondence C_S defined by the 'locus' at which C
exhibits a singularity of type S.

Remarks.

i) There is a kind of suspension operation on map-germs, defined by Carte-
sian product with the (germ of the) identity map $(\mathbf{C}, 0) \to (\mathbf{C}, 0)$, which yields
an inverse system

$$MU^*_{L^r(p,q)}(J^r(p,q)) \longrightarrow MU^*_{L^r(p+1,q+1)}(J^r(p+1,q+1));$$

elements of the limit group define what might be called 'stable' singularity operations. Porteous's classes are of this sort; in fact they can be expressed in terms
of characteristic classes for the 'stable' group $U(|p-q|)$. This provides another
enrichment of the category of correspondences, in which the hom-sets become
modules over the ring generated by such geometric singularity operations.

ii) Even the ordinary equivariant cohomology of $J^1(p,q)$ (or its projectifi-
cations) seems not to have ever been considered very seriously; for all I know,
its stable limit (in the above sense) is the polynomial algebra generated by
Porteous's classes. Perhaps [3] is relevant here.

References

[1] A. Baker, J. Morava, MSp localized away from 2 and odd formal group laws, preprint, Johns Hopkins 1994

[2] R.L. Cohen, J.D.S. Jones, G.B. Segal, Floer's infinite dimensional Morse theory and homotopy theory, A. Floer memorial volume, ed. H. Hofer, C. Taubes, A. Weinstein, E. Zehnder, Birkhaüser, to appear

[3] C. de Concini, C. Procesi, Cohomology of compactifications of algebraic groups, Duke Math. J. **53** (1986) 585-594

[4] S.R. Costenoble, S. Waner, Equivariant Poincaré duality, Michigan Math. J. **39** (1992) 324-351

[5] P. Deligne, Categories Tannakiennes, in the Grothendieck Festschrift, vol II, ed. P. Cartier et al, Birkhauser (1990)

[6] M. Demazure, Motifs des variétés algebriques, Sem. Bourbaki no. **365** (1969/70)

[7] V.G. Drinfel'd, On quasitriangular quasiHopf algebras and a group closely connected with $Gal(\bar{\mathbf{Q}}/\mathbf{Q})$, Leningrad Math. J. **2** (1991)829-860

[8] F.F. Knudsen, The projectivity of the moduli space of stable curves II, Math. Scand **52**(1983)161-199

[9] E. Looijenga, Intersection theory on Deligne-Mumford compactifications [after Witten and Kontsevich], Séminaire Bourbaki no.768 (1993)

[10] J. Morava, Primitive Mumford classes, in Mapping class groups and moduli spaces of Riemann surfaces, ed. C.F. Boedigheimer, R. Hain, Contemporary Math. **150** (1993) 291-302

[11] R.C. Penner, Perturbative series and the moduli space of punctured surfaces, J. Diff. Geom. **27** (1988) 35-53

[12] D. Quillen, Elementary proof of some properties of cobordism using Steenrod operations, Adv. in Math. **7** (1971) 29-56

[13] T. tom Dieck, Transformation groups, de Gruyter Studies in Mathematics **8**, Berlin-New York (1989)

[14] C.T.C. Wall (ed.), Proceedings of Liverpool Singularities – Symposium I, Springer Lecture Notes in Mathematics **192** (1971): H.I. Levine, Singularities of differentiable mappings, p. 1-89; I.R. Porteous, Simple singularities of maps, p. 286-307

Simply Connected 6–Dimensional Manifolds with Little Symmetry and Algebras with Small Tangent Space

Volker Puppe

The following statement and questions of F. Raymond and R. Schultz appeared in a collection of problems in algebraic topology edited by W. Browder and W.C. Hsiang, "Raymond–Schultz: It is generally felt that a manifold 'chosen at random' will have very little symmetry. Can this intuitive notion be made more precise? In connection with this intuitive feeling, we have the following specific question.

Question. Does there exist a closed simply connected manifold on which no finite group acts effectively?
(A weaker question, no involution?)" (s.[3])

As Raymond and Schultz point out there exist plenty of closed manifolds without any symmetry, i.e. not admitting any non trivial action of a finite group. But all known examples have non trivial fundamental group; in fact, in many of these cases the lack of symmetry is forced upon the manifolds by the 'complexity' of its fundamental group (cf. the comments and references given by Raymond and Schultz in [3] for more detailed information in this direction.)

This note makes a few steps towards answering the above questions, more precisely it is shown that in a certain class of closed, simply connected, 6–dimensional spin–manifolds (namely those with vanishing cohomology in odd degrees) for any given prime p 'most' of these manifolds do not admit an effective, orientation preserving action of Z_p, the cyclic group of order p. Note that though for odd primes p any action of Z_p is automatically orientation preserving, this does not hold for $p = 2$ and hence our result, which exhibits the existence of many closed, simply connected manifolds that do not admit orientation preserving involutions, falls short of even answering the 'weaker question' above. On the other hand it is shown that there exist closed, simply connected manifolds on which no finite group can act effectively and orientation preserving.

I want to thank W. Browder, A. Edmonds, H. Kraft, M. Kreck, T. Petrie, T.A. Springer and A. Suciu for stimulating conversations and helpful comments. I am particularly grateful to T. Iarrobino for several enlightening discussions on the algebraic part of the matter, for providing the example at the end of this note and to T. Iarrobino and A. Suciu for carrying out the necessary computer calculations in connection with this example.

Let X be a paracompact, finite dimensional Z_p–space and $X^G := \{x \in$

X; $gx = x$ for all $g \in Z_p\}$ its fixed point set. By $H^*(-)$, resp. $H^*(-; Z_p)$ we denote Čech cohomology with coefficients Z and Z_p, respectively.

Proposition 1 *If $H^*(X)$ is free over Z and $H^{\mathrm{odd}}(X) = 0$, and if the induced action of Z_p on $H^*(X)$ is trivial, then there exists a filtration on $H^*(X^G; Z_p)$*

$$
\begin{aligned}
0 &= \mathcal{F}_{-1}(H^*(X^G; Z_p)) \subset \mathcal{F}_0(H^*(X^G; Z_p)) \subset \ldots \subset \mathcal{F}_n(H^*(X^G; Z_p)) \\
&= H^*(X^G; Z_p) \\
&\quad \text{with } \mathcal{F}_{2q}(H^*(X^G, Z_p)) = \mathcal{F}_{2q+1}(H^*(X^G, Z_p)),
\end{aligned}
$$

such that the associated graded algebra is isomorphic to $H^*(X; Z_p)$.

Proof. Already for degree reasons (and the assumption that Z_p acts trivially on $H^*(X)$) one gets that the Leray–Serre spectral sequence (with Z coefficients) of the Borel construction $X \to EG \times_G X \to BG$, $G = Z_p$, collapses and $E_\infty^{*,*} = E_2^{*,*} \cong H^*(BG, H^*(X)) \cong Z[t]/(pt) \otimes H^*(X)$, $\deg(t) = 2$. It follows that $H_G^*(X) = H^*(EG \times_G X)$ and $Z[t]/(pt) \otimes H^*(X)$ are isomorphic as $Z[t]/(pt)$-modules, and – since the restriction to the fibre $H_G^*(X) \to H^*(X)$, $t \mapsto 0$, is multiplicative – the cup product on $H_G^*(X)$ coincides with the usual (componentwise) product on $Z[t]/(pt) \otimes H^*(X)$ up to terms of lower filtration, where the filtration is defined by

$$
\mathcal{F}_q(Z[t]/(pt) \otimes H^*(X)) := Z[t]/(pt) \otimes \left(\bigoplus_{i \le q} H^i(X) \right).
$$

On the other hand the Localization Theorem (see e.g.[1]) gives that the map $H_G^*(X) \to H_G^*(X^G)$, induced by the inclusion $X^G \subset X$, becomes an isomorphism if one inverts $t \in Z[t]/(pt) = H^*(BG)$. It follows that if one evaluates at $t = 1$ one also gets an isomorphism, i.e.

$$
Z_\eta \underset{H^*(BG)}{\otimes} H_G^*(X) \xrightarrow{\cong} Z_\eta \underset{H^*(BG)}{\otimes} H_G^*(X^G),
$$

where Z_η denotes Z_p together with the $H^*(BG)$-module structure given by $\eta : H^*(BG) \to Z_p$, $\eta(t) = 1$.

By a theorem of Heller [8] and Swan [15] one has $H^{\mathrm{odd}}(X^G; Z_p) = 0$, so by the Künneth formula one sees that $H^*(X^G)$ has no p-torsion. Therefore

$$
H_G^*(X^G) = H^*(BG; H^*(X^G)) \cong H^*(BG) \otimes H^*(X^G),
$$

and

$$
Z_\eta \underset{H^*(BG)}{\otimes} H_G^*(X^G) \cong H^{(*)}(X^G; Z_p)
$$

as a $\mathbf{Z}_2$-graded algebra, where $H^{(*)}(-) := H^{ev}(-) \oplus H^{odd}(-)$. (Note that $\mathbf{Z}_\eta \otimes_{H^*(BG)} H^*_G(-)$ inherits a $\mathbf{Z}_2$-grading since $\deg(t) = 2$.) The above filtration on $H^*_G(X)$ ($\cong H^*(BG) \otimes H^*(X)$ as $H^*(BG)$-modules) induces a filtration on

$$\mathbf{Z}_\eta \underset{H^*(BG)}{\otimes} H^*_G(X) \cong \mathbf{Z}_\eta \underset{H^*(BG)}{\otimes} H^*_G(X^G) \cong H^*(X^G; \mathbf{Z}_p).$$

The associated graded algebra of this filtration on $H^*(X^G; \mathbf{Z}_p)$ is isomorphic to $H^*(M; \mathbf{Z}_p)$, since the cup product on $H^*_G(X)$ corresponds to the componentwise product on $H^*(BG) \otimes H^*(X)$ up to terms of lower filtration.   □

By a result of Chang–Skjelbred [4] and of Bredon [2] if X fullfils Poincaré duality over $\mathbf{Z}_p$ then so does each component F_ν of X^G. Now, the strategy to find a closed, simply connected manifold M, which does not admit an effective $\mathbf{Z}_p$-action is as follows:

One wants to find a graded algebra $A^* = \bigoplus_{i=0}^n A^i$ over $\mathbf{Z}$ such that A^* is free over $\mathbf{Z}$, $A^{odd} = 0$, and A^* fulfils Poincaré duality with the following additional properties:

(1) A^* does not admit an algebra automorphism of order p

(2) A^* can not be obtained as the graded associated algebra of a filtration on an algebra $B^* = \Pi_\nu B^*_\nu$, where $fd(B^*_\nu) < fd(A^*)$ for all ν ($fd(-)$ denotes the formal dimension of a Poincaré duality algebra)

(3) A^* can be realized as the cohomology algebra $H^*(M)$ of a closed, simply connected manifold.

If one succeeds in finding such an A^* then the corresponding manifold M (with $H^*(M) = A^*$) does not admit an effective $\mathbf{Z}_p$-action, for if a $\mathbf{Z}_p$-action on M is given, then because of property (1) above Proposition 1 applies. Therefore by property (2) there must be a component $F_\nu \subset M^G$ such that $fd(H^*(F_\nu; \mathbf{Z}_p)) = fd(H^*(M; \mathbf{Z}_p))$. But this implies that $F_\nu = M$ and hence the action must be trivial.

The smallest dimension n (in which these properties could possibly be achieved is $n = 6$, because for $n = 4$ it is known (s.[6]) that any simply connected 4–manifold admits a non trivial $\mathbf{Z}_p$-action at last for all $p \geq 3$ (and for $n = 2$ there is only S^2 left). Fortunately simply connected 6–dimensional spin–manifolds M (i.e. the second Stiefel–Whitney class $w_2(M) = 0$) such that $H^*(M)$ is free over $\mathbf{Z}$ have been classified by Wall (s.[16]) in the differentiable case (cf.[11] and [17] for extensions of this result).

Let $\mathcal{M}$ be the class of simply connected, closed, 6-dimensional spin-manifolds M with $H^3(M) = 0$. This implies that $H^*(M)$ is free over $\mathbf{Z}$ and concentrated in even degrees.

Theorem (s.[16]). *The diffeomorphism classes of elements of $\mathcal{M}$ correspond bijectively to isomorphism classes of invariants*

1. *H free $\mathbf{Z}$–module of finite rank (corresponding to $H^2(M)$ for $M \in \mathcal{M}$)*

2. *$\mu \colon H \times H \times H \to \mathbf{Z}$ symmetric, trilinear form (corresponding to the cup product in $H^*(M)$)*

3. *$\mathcal{P} \colon H \to \mathbf{Z}$ linear map (corresponding to the dual of the first Pontrjagin class) subject to the following conditions:*

   (a) $\mu(x, x, y) \equiv \mu(x, y, y) \pmod{2}$ for $x, y \in H$

   (b) $\mathcal{P}(x) \equiv 4\mu(x, x, x) \pmod{24}$ for $x \in H$.

(Note that the product structure of a graded algebra $A^* = \bigoplus_{i=0}^{6} A^i$ with A^* free over $\mathbf{Z}$, $A^{\mathrm{odd}} = 0$, which fulfils Poincaré duality is completely determined by the symmetric, trilinear form corresponding to the triple products of elements in A^2, since by duality $A^4 \cong \mathrm{Hom}(A^2, \mathbf{Z})$, and hence the product $A^2 \times A^2 \to A^4 \cong \mathrm{Hom}(A^2, \mathbf{Z})$ corresponds to a symmetric, trilinear form $A^2 \times A^2 \times A^2 \to \mathbf{Z}$ on A^2. On the other hand any symmetric, trilinear form on A^2 can be 'extended' in a unique way to yield a product structure on $A^* = \bigoplus_{i=0}^{6} A^i$, $A^{\mathrm{odd}} = 0$, which fulfils Poincaré duality.)

In particular, Wall's theorem settles the problem with property (3) in the above list, i.e. given a free $\mathbf{Z}$–module H of rank m and a symmetric, trilinear form $\mu \colon H \times H \times H \to \mathbf{Z}$ such that $\mu(x, x, y) \equiv \mu(x, y, y) \pmod{2}$ for $x, y \in H$, then there is an $M \in \mathcal{M}$ such that $H^2(M) \cong H$ and the cup product on $H^*(M)$ corresponds to the symmetric, trilinear form μ. For given H and μ such that (a) above holds, one can always choose a linear map $\mathcal{P} \colon H \to \mathbf{Z}$, such that the compatibility condition (b) holds too. In fact, there are even infinitely many different choices for a compatible $\mathcal{P} \colon H \to \mathbf{Z}$, i.e. there are infinitely many different diffeomorphism types in $\mathcal{M}$ which all 'realize' the same pair (H, μ) (but different Pontrjagin classes) provided μ fulfils condition (a). We hence can concentrate on properties (1) and (2) in the above list.

Let $\mathcal{F}(m; \mathbf{Z})$ denote the space of symmetric, trilinear forms on $A \cong \mathbf{Z}^m$. Note that $\mathcal{F}(m; \mathbf{Z})$ is a free $\mathbf{Z}$–module of rank $\binom{m+2}{3}$. In fact, $\mathcal{F}(m; \mathbf{Z})$ is iso-morph – in a non canonical way – to the $\mathbf{Z}$–module of homogeneous polynomials of degree 3 in m variables (with integer coefficients). Note that the canonical map which assigns to a form μ the polynomial $f(x) := \mu(x, x, x), x \in A$, only becomes an isomorphism, if one extends the coefficients in such a way that 6 becomes invertible; so this holds e.g. for coefficients in $\mathbf{Q}$, $\mathbf{R}$ or $\mathbf{C}$. There is a canonical map in the other direction, the polarization, which assigns to a homogeneous, cubic polynomial f the symmetric, trilinear form μ, defined by

$$
\begin{aligned}
\mu(x, y, z) \;:=\;\; & f(x + y + z) - \big(f(x + y) + f(y + z) + f(z + x)\big) \\
& + \big(f(x) + f(y) + f(z)\big), \quad x, y, z \in A.
\end{aligned}
$$

The compositions of these two canonical maps in both ways give just multiplication by 6. The general linear group $\mathrm{GL}(m; \mathbf{Z})$ acts on $\mathcal{F}(m; \mathbf{Z})$, induced by the standard action on $A \cong \mathbf{Z}^m$. The orbits of this action on $\mathcal{F}(m; \mathbf{Z})$ correspond to isomorphism classes of symmetric, trilinear forms on A. The canonical maps mentioned above are equivariant with respect to the $\mathrm{GL}(n; \mathbf{Z})$–action induced by linear substitution (base change). For any $\mu \in \mathcal{F}(m; \mathbf{Z})$ there is a corresponding graded algebra $A^* = \bigoplus_{i=0}^{6} A^i$ such that A^* is free over $\mathbf{Z}$, $A^{\mathrm{odd}} = 0$, $A^2 = \mathbf{Z}^m$, and A^* fulfils Poincaré duality. Assume that $\alpha \colon A^* \to A^*$ is an orientation preserving (i.e.: $\alpha|_{A^6} = id$) algebra automorphism of order p. There is a corresponding element of order p in $\mathrm{GL}(m; \mathbf{Z})$, which we denote by α, too, obtained by restricting the automorphism to $A^2 = A \cong \mathbf{Z}^m$. That $\alpha \colon A^* \to A^*$ is an orientation preserving algebra automorphism means precisely that μ is invariant under the corresponding element $\alpha \in \mathrm{GL}(m; \mathbf{Z})$. In particular, if the isotropy group of $\mu \in \mathcal{F}(m; \mathbf{Z})$ (with respect to the above action of $\mathrm{GL}(m; \mathbf{Z})$ on $\mathcal{F}(m; \mathbf{Z})$) does not contain a subgroup of order p, then the algebra corresponding to μ does not admit an orientation preserving automorphism of order p.

Our aim now is to show that 'most' algebras of the above type (i.e. corresponding to elements in $\mathcal{F}(m; \mathbf{Z})$) do not admit any non trivial orientation preserving automorphism of finite order. Note that this can be thought of as an algebraic analogue (for a certain restricted class of algebras) of the expectation formulated by Raymond and Schultz that manifolds 'chosen at random' will have little symmetry (i.e. 'few' finite subgroup in their automorphism group). To give a precise statement we need some further notation. Let $\mathcal{A}(m)$ be the subset of symmetric, trilinear forms $\mu \in \mathcal{F}(m; \mathbf{Z})$, such that the isotropy group, $\mathrm{GL}(m, \mathbf{Z})_\mu$, of μ contains non trivial finite subgroups, in other words such that the algebra corresponding to μ admits non trivial, orientation preserving automorphisms of finite order, and let $\mathcal{F}^N(m; \mathbf{Z})$ be the subset of those $\mu \in \mathcal{F}(m; \mathbf{Z})$, such that for all coefficients μ_{ijk} of μ (with respect to the canonical basis of $\mathbf{Z}^m$) one has: $-N \le \mu_{ijk} < N$, $N \in \mathbf{N}$. Note that while $\mathcal{A}(m)$ is a union of certain orbits of the $\mathrm{GL}(m; \mathbf{Z})$ action on $\mathcal{F}(m; \mathbf{Z})$, the subset $\mathcal{F}^N(m; \mathbf{Z})$ is not closed under this action (for $N > 0$). Let $\sharp B$ denote the number of elements of the finite set B.

Proposition 2 *For $m \ge 6$*

$$\lim_{N \to \infty} \frac{\sharp\{\mathcal{A}(m) \cap \mathcal{F}^N(m; \mathbf{Z})\}}{\sharp\{\mathcal{F}^N(m; \mathbf{Z})\}} = 0.$$

Proof. H. Kraft and T.A. Springer have pointed out to me how one could obtain the above result from the work of A.M. Popov (s. [12]) on principal isotropy group of representations of Lie groups. The following different and rather elementary proof relies on the well known results (s. [5]) on integral representations resp. rational representations of $\mathbf{Z}_p$. One only needs the fact that a representation of

Z_p on a finite dimensional vector space over the rationals Q decomposes into a direct sum of the following two types of representations:

    i. Q, i.e. the trivial representation, and

    ii. $Q[\lambda] := \ker(Q[Z_p] \xrightarrow{\epsilon} Q)$, where ϵ is the usual augmentation of the group ring $Q[Z_p]$

(Clearly $Q[Z_p] \cong Q[\lambda] \oplus Q$.)

    Let $\mathcal{A}^p(m)$ be the subset of those forms $\mu \in \mathcal{F}(m; Z)$ such that the corresponding algebra admits an orientation preserving automorphism of order p. Then $\mathcal{A}(m) = \bigcup_p \mathcal{A}^p(m)$, but since Z_p can act on Z^m non trivially only if $p \leq m + 1$, one only needs to consider a finite set of primes for a fixed $m \in N$, namely: $\{p; p \leq m + 1\}$. The idea of the proof is to show that $\mathcal{A}^p(m)$ has 'positive codimension' in $\mathcal{F}(m; Z)$ and hence the same should hold for $\mathcal{A}(m) = \bigcup_{p \leq m+1} \mathcal{A}^p(m)$. To make this a reasonable statement we extend the coefficients to R. Let ρ be a representation of Z_p on Z^m. We denote the extension to Q^m and R^m by the same symbol. Clearly the submodule, $\mathcal{F}(m; Z)^\rho$, of forms invariant under ρ is contained in $\mathcal{F}(m; R)^\rho \cap \mathcal{F}(m; Z)$, and hence

$$\mathcal{A}^p(m) \quad \subseteq \quad \mathcal{A}^p(m; R) \cap \mathcal{F}(m; Z), \text{ where}$$
$$\mathcal{A}^p(m; R) \quad := \quad \bigcup_\rho \mathrm{GL}\,(m; R)\mathcal{F}(m; R)^\rho,$$

and ρ varies over a set of representatives of equivalence classes of non trivial representations of Z_p on Z^m. Since we extend the coefficients to R, it suffices to consider equivalence classes over Q (or R) here.

    We want to show that $\mathcal{A}^p(m; R)$ has positive codimension in $\mathcal{F}(m; R) = \mathcal{F}(m, Z) \otimes R$.

    We first consider the case $p = 2$. Any representation ρ of Z_2 on Q^m is equivalent to one of the form $Q^r \oplus Q^s$, $r + s = m$, where Q^s is a trivial Z_2-module, and the generator of Z_2 acts on Q^r through multiplication by (-1). Extending the coefficients from Q to R gives a corresponding decomposition $R^r \oplus R^s$. The dimension of the fixed subspace of the action on $\mathcal{F}(m; R)$ induced by ρ, $\mathcal{F}(m; R)^\rho$, is given by

$$\dim \mathcal{F}(m; R)^\rho = s \binom{r + 1}{2} + \binom{s + 2}{3}.$$

(Note that if $x_1, \ldots, x_r$ and $y_1, \ldots, y_s$ are bases of R^r and R^s, respectively, the monomials of degree 3 in $x_1, \ldots, x_r, y_1, \ldots, y_s$ form a basis of $\mathcal{F}(m; R)$ and the action induced by ρ is given by multiplying such a monomials by $(+1)$ or (-1). Hence the invariant monomials under this action form a basis of $\mathcal{F}(m; R)^\rho$.)

The subspace $\mathcal{F}(m; R)^\rho \subset \mathcal{F}(m; R)$ is invariant under the action of the subgroup

$$\mathrm{GL}(r; R) \times \mathrm{GL}(s; R) \subset \mathrm{GL}(m; R).$$

Therefore

$$\dim(\mathrm{GL}(m; \boldsymbol{R}) \; \mathcal{F}(m; \boldsymbol{R})^\rho)$$
$$\leq \dim(\mathrm{GL}(m; \boldsymbol{R})/ \; \mathrm{GL}(r; \boldsymbol{R}) \times \mathrm{GL}(s; \boldsymbol{R})) \dim \mathcal{F}(m; \boldsymbol{R})^\rho.$$

One gets:

$$\dim\left(\mathrm{GL}(m; \boldsymbol{R})\mathcal{F}(m; \boldsymbol{R})^\rho\right) \;\leq\; m^2 - (r^2 + s^2) + s\binom{r+1}{2} + \binom{s+2}{3}$$

$$\leq\; 2rs + s\binom{r+1}{2} + \binom{s+2}{3}$$

Let $\Delta := \dim \mathcal{F}(m; \boldsymbol{R}) - \dim\left(\mathrm{GL}(m; \boldsymbol{R}) \; \mathcal{F}(m; \boldsymbol{R})^\rho\right)$ denote the codimension of $\mathrm{GL}(m; \boldsymbol{R})\mathcal{F}(m; \boldsymbol{R})^\rho$ in $\mathcal{F}(m; \boldsymbol{R})$, then

$$\Delta \;\geq\; \binom{r+s+2}{3} - (\binom{s+2}{3} + s\binom{r+1}{2} + 2rs)$$

$$\geq\; \binom{r+2}{3} + r\binom{s+1}{2} - 2rs = \frac{(r+2)(r+1)r}{6} + \frac{r(s+1)s}{2} - 2rs$$

$$\geq\; \frac{(r+2)(r+1)r}{6} + rs\left(\frac{s-3}{2}\right)$$

Hence $\Delta > 0$ if ($r \geq 1$ and $s \geq 3$) or ($r \geq 2$ and $s \geq 0$). In particular, for a non trivial representation $\Delta > 0$ if $m = r + s \geq 4$.

The subset $\mathcal{A}^2(m; \boldsymbol{R})$ is contained in a union of subsets of the form

$$\mathrm{GL}(m; \boldsymbol{R})\mathcal{F}(m; \boldsymbol{R})^\rho,$$

where ρ runs over a finite set of representatives of all equivalence classes of non trivial representations of Z_2 on $\boldsymbol{Q}^m$. Therefore, for $m \geq 4$, $\mathcal{A}^2(m; \boldsymbol{R})$ has positive codimension in $\mathcal{F}(m; \boldsymbol{R})$. Note that the above bound for Δ is not best possible, e.g. if $r = 1$ and $s = 0$ then obviously $\Delta = 1 > 0$.

For odd p the argument is similar, though a little bit more complicated, and there is a slight difference between $p = 3$ and $p > 3$.

Let ρ be a representation of Z_p on $\boldsymbol{Q}^m$, which is equivalent to $(\boldsymbol{Q}[Z_p])^r \oplus \boldsymbol{Q}^s$, $pr + s = m$, where Z_p acts trivially on $\boldsymbol{Q}^s$ and diagonally on $(\boldsymbol{Q}[Z_p])^r$. Let $x_1^1, \ldots, x_p^1, x_1^2 \ldots x_p^2, \ldots, x_1^r, \ldots, x_p^r, y_1, \ldots, y_s$ be a basis of $\boldsymbol{Q}^m$ corresponding to the direct sum decomposition above in such a way that the generator $g \in Z_p$ permutes the elements of the basis. Then g also permutes the elements of the basis of $\mathcal{F}(m; \boldsymbol{R})$ given by all monomials of degree 3 in the variables $x_1^1 \ldots y_s$.

Hence, for $p > 3$, one has

$$\dim \mathcal{F}(m; \boldsymbol{R})^\rho = \frac{1}{p}\left[\binom{m+2}{3} - \binom{s+2}{3}\right] + \binom{s+2}{3}$$

(in fact, $\mathcal{F}(m; R) = (Q[Z_p])^\mu \oplus Q^\nu$ as Z_p-module, with $\mu = \binom{m+2}{3} - \binom{s+2}{3}$ and $\nu = \binom{s+2}{3}$).

The subspace $\mathcal{F}(m; R)^\rho \subset \mathcal{F}(m; R)$ is invariant under the action of the subgroup $\mathrm{GL}(r; R) \times \mathrm{GL}(s; R)$, where $\mathrm{GL}(s; R)$ is considered as the subgroup of $\mathrm{GL}(m; R)$ acting on the last s coordinates, and $\mathrm{GL}(r; R)$ acts via the diagonal action on the first pr coordinates, put together in p blocks of r elements each, in other words $\mathrm{GL}(r; R) \xrightarrow{\cong} \mathrm{Aut}(R^r) \to \mathrm{Aut}((R[Z_p])^r)$, where the second map is given by extending the ground ring from R to $R[Z_p]$, tensoring each automorphisms $R^r \to R^r$ with $id_{R[Z_p]}$. We therefore get

$$\dim(\mathrm{GL}(m; R)\, \mathcal{F}(m; R)^\rho) \le m^2 - (r^2 + s^2) + \frac{1}{p}\left[\binom{m+2}{3} - \binom{s+2}{3}\right] + \binom{s+2}{3},$$

and

$$\begin{aligned}
\Delta \;:=\; & \dim \mathcal{F}(m; R) - \dim(\mathrm{GL}(m; R)\, \mathcal{F}(m; R)^\rho) \\
\ge\; & \binom{m+2}{3} - m^2 + r^2 + s^2 - \frac{1}{p}\left[\binom{m+2}{3} - \binom{s+2}{3}\right] - \binom{s+2}{3} \\
\ge\; & \frac{p-1}{p}\left[\binom{m+2}{3} - \binom{s+2}{3}\right] - m^2 + r^2 + s^2 \\
\ge\; & \frac{p-1}{p}\left[\binom{pr+2}{3} + \binom{pr+1}{2}s + pr\binom{s+1}{2}\right] - (pr+s)^2 + r^2 + s^2
\end{aligned}$$

Hence

$$\begin{aligned}
6p\Delta \;\ge\; & (p-1)(pr+2)(pr+1)pr + 3(pr+1)prs \\
& +3pr(s+1)s - 6p((p^2-1)r^2 + 2prs) \\
\ge\; & (p-1)pr(pr+2)(pr+1) - 6(p+1)r) \\
& +3prs((p-1)((pr+1) + (s+1)) - 4p) \\
\ge\; & (p-1)pr(p^2r^2 - 3pr - 6r + 2) + 3prs((p-1)(pr + s + 2) - 4p))
\end{aligned}$$

It follows that $\Delta > 0$ if $r \ge 1$ and $p \ge 5$.

We next consider a representation ρ of Z_p on Q^m, which is equivalent to $(Q[Z_p])^r \oplus (Q[\lambda])^s$. Note that because of $Q[\lambda] \oplus Q \cong Q[Z_p]$ (and the cases already discussed above) it is not necessary to consider additional summands Q with trivial Z_p-action.

We want to reduce this case to the previous one, i.e. we add s trivial summands Q and extend the coefficients from Q to R to obtain

$$V := V' \oplus V'' \cong (R[Z])^{r+s}, \qquad \text{with } V' = (R[Z_p])^r \oplus (R[\lambda])^s$$
$$\text{and } V'' = R^s \text{ (with trivial } Z_p\text{-action).}$$

The space of invariant elements of a representation V (over $\mathbf{Q}$ or $\mathbf{R}$) is just the image of the norm homomorphism $\mathcal{N} : V \to V$, $\mathcal{N}(v) = \frac{1}{p}\sum_{i=0}^{p-1} g^i v$, where g generates $\mathbf{Z}_p$. We have already calculated $\dim \mathcal{N}(V)$, and we want to use this to get $\dim \mathcal{N}(V')$. From the splitting $V = V' \oplus V''$ one gets the following splitting of $\mathcal{F}_3(V)$, the symmetric, trilinear forms on V, which is compatible with the $\mathbf{Z}_p$-action and hence also with the norm $\mathcal{N}$:

$$
\begin{array}{ccccccccc}
\mathcal{F}_3(V) & = & \mathcal{F}_3(V') & \oplus & \mathcal{F}_{2,1}(V',V'') & \oplus & \mathcal{F}_{1,2}(V',V'') & \oplus & \mathcal{F}_3(V'') \\
\mathcal{N}\downarrow & & \mathcal{N}\downarrow & & \mathcal{N}\downarrow & & \mathcal{N}\downarrow & & \mathcal{N}\downarrow \\
\mathcal{F}_3(V) & = & \mathcal{F}_3(V') & \oplus & \mathcal{F}_{2,1}(V',V'') & \oplus & \mathcal{F}_{1,2}(V',V'') & \oplus & \mathcal{F}_3(V'')
\end{array}
$$

where $\mathcal{F}_{2,1}(V',V'')$ is generated by monomials of degree 3 of the form $v_1' v_2' v''$ with $v_1' v_2' \in V'$, $v'' \in V''$, and $\mathcal{F}_{1,2}(V',V'')$ is analogously defined. Since V'' is a trivial $\mathbf{Z}_p$-module, so is $\mathcal{F}_3(V'')$ and $\mathcal{N}: \mathcal{F}_3(V'') \xrightarrow{\cong} \mathcal{F}_3(V'')$.

Also

$$
\dim \mathcal{N}(\mathcal{F}_{2,1}(V',V'')) = \dim \mathcal{N}(\mathcal{F}_2(V'))\dim(V'')
$$

and

$$
\dim \mathcal{N}(\mathcal{F}_{1,2}(V',V'')) = \dim(\mathcal{N}(V'))\dim(\mathcal{F}_2(V'')),
$$

where $\mathcal{F}_2(V')$ denotes the quadratic forms on V' etc. Using the same procedure for $\mathcal{F}_2(V')$ (resp. $\mathcal{F}_2(V' \oplus V'')$) instead of $\mathcal{F}_3(-)$ one gets

$$
\dim \mathcal{N}(\mathcal{F}_2(V')) = \dim \mathcal{N}(\mathcal{F}_2(V' \oplus V'')) - \dim \mathcal{N}(\mathcal{F}_{1,1}(V'.V'')) - \dim \mathcal{N}(\mathcal{F}_2(V'')),
$$

and

$$
\dim \mathcal{N}(\mathcal{F}_2(V'')) = \dim(\mathcal{F}_2(V'')) = \binom{s+1}{2}
$$

$$
\dim \mathcal{N}(\mathcal{F}_{1,1}(V',V'')) = \dim(\mathcal{N}(V'))\dim(V'') = rs
$$

$$
\dim \mathcal{N}(\mathcal{F}_2(V' \oplus V'')) = \frac{1}{p}\binom{m+s+1}{2}.
$$

Therefore

$$
\dim \mathcal{N}(\mathcal{F}_{2,1}(V',V'')) = \left(\frac{1}{p}\binom{m+s+1}{2}\right)s,
$$

and

$$
\dim \mathcal{N}(\mathcal{F}_{1,2}(V',V'')) = r\binom{s+1}{2}
$$

Altogether one has

$$
\dim \mathcal{F}(m; \mathbf{R})^\rho = \frac{1}{p}\left[\binom{m+s+2}{3} - \binom{m+s+1}{2}s\right]
$$

$$
+ \; rs^2 + \binom{s+1}{2}s - r\binom{s+1}{2} - \binom{s+2}{3}
$$

And by a similar calculation as above one gets:

$$\Delta := \dim \mathcal{F}(m; \mathbf{R}) - \dim(\mathrm{GL}(m; \mathcal{F}(m; \mathbf{R})^\rho) > 0 \text{ if } r \geq 0 \text{ or } s \geq 1.$$

It follows that $A^p(m; \mathbf{R})$ has positive codimension in $\mathcal{F}(m; \mathbf{R})$ for $p \geq 5$ and any m.

The calculation for $p = 3$ is again similar, but there is a slight difference in the formula for the dimension of $\mathcal{F}(m; \mathbf{R})^\rho$. If ρ is a representation of Z_3 on Q^m, which is equivalent to $(Q[Z_3])^r \oplus Q^s$, then:

$$\dim \mathcal{F}(m; \mathbf{R})^\rho = \frac{1}{3}\left[\binom{m+2}{3} - \binom{s+2}{3} - r\right] + \binom{s+2}{3} + r.$$

Using this formula one gets

$$\Delta := \dim \mathcal{F}(m; \mathbf{R}) - \dim(\mathrm{GL}\ (m; \mathbf{R})\ \mathcal{F}(m; \mathbf{R})^\rho) > 0$$

for $(r \geq 1$ and $s \geq 3)$ or $(r \geq 2$ and $s \geq 0)$.

In case ρ is equivalent to $(Q[Z_3])^r \oplus (Q[\lambda])^s$ one obtains (along the same line of argument as in case $p \geq 5$ above)

$$\Delta := \dim \mathcal{F}(m; \mathbf{R}) - \dim(\mathrm{GL}(m; \mathbf{R})\ \mathcal{F}(m; \mathbf{R})^\rho) > 0$$

$r \geq 2$ or $s \geq 2$.

In particular $A^3(m; \mathbf{R})$ has positive codimension in $\mathcal{F}(m; \mathbf{R})$ if $m \geq 6$. Putting the results for the different primes together gives, that

$$A(m; \mathbf{R}) := \bigcup_{p \leq m+1} A^p(m; \mathbf{R})$$

has positive codimension in $\mathcal{F}(m; \mathbf{R})$ if $m \geq 6$.

The claim of Proposition 2 can now be shown for example by the following argument, which was suggested to me by M.Nüsken. Given $\epsilon > 0$ one can divide $\mathcal{F}(m; \mathbf{R}) \cong \mathbf{R}^{\alpha(m)}$ into equal size (closed) subcubes Q_ν of side length $\frac{1}{N}$ with edges parallel to the coordinate axes in such a way that $0 \in \mathbf{R}^{\alpha(m)}$ is the center of one of these cubes. One can choose N large enough so that $A(m; \mathbf{R}) \cap [-1, 1]^{\alpha(m)}$ is contained in a union of subcubes $\bigcup_{\nu \in J} Q_\nu$ with volume Vol $(\bigcup_{\nu \in J} Q_\nu) < 2^{\alpha(m)}\epsilon$.

This is possible since $A(m; \mathbf{R}) \cap [-1, 1]^{\alpha(m)}$ is a closed (and hence compact) subset of $[-1, 1]^{\alpha(m)}$ of measure zero. Since $A(m; \mathbf{R})$ is a cone in $\mathcal{F}(m; \mathbf{R})$ one gets that

$$A(m) \cap [-N, N]^{\alpha(m)} \subset (A(m; \mathbf{R})) \cap [-N, N]^{\alpha(m)}$$

is contained in $N(\bigcup_{\nu \in J} Q_\nu)$, which is the union of cubes NQ_ν with Vol $(NQ_\nu) = 1$, and the center of each Q_ν has integer coordinates. Clearly

$$\sharp J = \text{Vol}\ (N(\bigcup_{\nu \in J} Q_\nu)) = N^{\alpha(m)}\ \text{Vol}\ (\bigcup_{\nu \in J} Q_\nu) < N^{\alpha(m)} 2^{\alpha(m)}\epsilon$$

and

$$\text{Vol } (N(\bigcup_{\nu \in J} Q_\nu))/ \text{ Vol } ([-N, N]^{\alpha(m)}) < \epsilon.$$

Hence

$$\#\{\mathcal{A}(m) \cap \mathcal{F}(m; N)\} < N^{\alpha(m)} 2^{\alpha(m)} \epsilon,$$

and

$$\frac{\#\{\mathcal{A}(m) \cap \mathcal{F}(m; N)\}}{\#\{\mathcal{F}(m; N)\}} < \epsilon. \qquad \square$$

Let $\mathcal{R}(m; \mathbf{Z}) := \{\mu \in \mathcal{F}(m; \mathbf{Z}); \mu(x, x, y) \equiv \mu(x, y, y) (\text{mod } 2) \text{ for all } x, y \in \mathbf{Z}^m\}$, and $\mathcal{R}(m; N) := \mathcal{R}(m) \cap \mathcal{F}(m; N)$. By the above result of Wall the algebra structures, which correspond to elements of $\mathcal{R}(m; \mathbf{Z})$, are precisely those which can be realized as cohomology algebras of manifolds M in $\mathcal{M}$ (with $rkH^2(M) = m$).

Corollary 1 *For $m \geq 6$*

$$\lim_{N \to \infty} \frac{\#\{\mathcal{A}(m) \cap \mathcal{R}(m; N)\}}{\#\mathcal{R}(m; N)} = 0.$$

Proof. If $x_1, \ldots, x_m$ is a basis of $\mathbf{Z}^m$ then $\mathcal{R}(m; \mathbf{Z})$ is determined by the $\binom{m}{2}$ linearly independent equations

$$\mu(x_i, x_i, x_j) - \mu(x_i, x_j, x_j) \equiv 0 \, (\text{mod } 2) \quad \text{for} \quad i < j.$$

Therefore

$$\frac{\#\{\mathcal{R}(m; N)\}}{\#\{\mathcal{F}(m; N)\}} \geq \frac{1}{2^{\binom{m}{2}}},$$

where equality holds for even $N \in \mathbf{N}$. Hence Corollary 1 follows from Proposition 2. $\qquad \square$

A vague interpretation of Corollary 1 is that for most $M \in \mathcal{M}$ the cohomology algebra $H^*(M)$ does not admit a non trivial automorphism of finite order (cf. (1) above). We want to show next that most of these cohomology algebras mod p, $H^*(M; \mathbf{Z}_p) = H^*(M) \otimes \mathbf{Z}_p$, cannot be obtained as graded associated algebras of a filtration on an algebra $B^* = \Pi_\nu B^*_\nu$, where the B^*_ν are Poincaré duality algebras of formal dimension 0,2 or 4 (cf. (2) above).

Let $\mathcal{G}(m; \mathbf{Z}_p) \subset \mathcal{F}(m; \mathbf{Z}_p)$ denote the subspace of those forms in m variables over $\mathbf{Z}_p$ such that the corresponding algebra structure (over $\mathbf{Z}_p$) can be obtained as graded associated algebra of a filtration on some algebra $B^* = \Pi_\nu B^*_\nu$ of the form described above.

Proposition 3

$$\frac{\#\mathcal{G}(m; \mathbf{Z}_p)}{\#\mathcal{F}(m; \mathbf{Z}_p)} \leq \frac{p^{\beta(m)}}{p^{\alpha(m)}}, \text{ where } \alpha(m) := \binom{m+2}{3} \text{ and } \beta(m) = 4m^2 + 8m + 5,$$

in particular

$$\lim_{m \to \infty} \frac{\sharp \mathcal{G}(m; \mathbf{Z}_p)}{\sharp \mathcal{F}(m; \mathbf{Z}_p)} = 0.$$

Proof. On B_ν^* with $fd(B_\nu^*) = 0, 2$ or 4 the product structure is completely determined by the orientation $B_\nu^* \xrightarrow{\mathcal{O}_\nu} \mathbf{Z}_p$ and the bilinear symmetric form $B_\nu^* \times B_\nu^* \to B_\nu^* \xrightarrow{\mathcal{O}_\nu} \mathbf{Z}_p$, where the first map is the product on B_ν^*.

Hence the product structure of $B^* = \Pi_\nu B_\nu^*$ with $\dim B^* = 2m + 2$ is completely described by the orientation $B^* = \Pi_\nu B\nu \xrightarrow{\mathcal{O} = \{\mathcal{O}_\nu\}} \mathbf{Z}_p$ and a $(2m + 2) \times (2m + 2)$–matrix M consisting of blocks along the main diagonal which correspond to the factors B_ν^* of B^*. The block corresponding to B_ν^* is

the 1×1–matrix (1), if $fd(B_\nu^*) = 1$;

the 2×2–matrix $\begin{pmatrix} 0 & 1 \\ 1 & 0 \end{pmatrix}$, if $fd(B_\nu^*) = 2$; and an $n_\nu \times n_\nu$–matrix of the form

$$\begin{pmatrix} 0 & \cdots & 01 \\ & & 0 \\ \overline{M}_\nu & \vdots & \\ & \vdots & \\ 10 & \cdots & 0 \end{pmatrix},$$

if $fd(B_\nu^*) = 4$ and $\dim B_\nu^* = n_\nu$.

In the last case $\overline{M}_\nu$ is a symmetric matrix with coefficients in $\mathbf{Z}_p$. Hence a (rather rough) upper bound for the number of isomorphism classes of algebras $B^* = \Pi_\nu B_\nu^*$ as above with $\dim B^* = 2m + 2$ is $p^{\binom{2m+3}{2}} p^{2m+2}$, where the first factor is an upper bound for the number of bilinear, symmetric forms on $\mathbf{Z}_p^{2m+2}$ (i.e. matrices M above) which can occur in this context, and the second factor bounds the number of possible orientations on $\mathbf{Z}_p^{2m+2}$. We also need a bound for the number of different filtrations

$$0 = \mathcal{F}_{-1}(B^*) \subset \langle 1 \rangle = \mathcal{F}_0(B^*) \subset \mathcal{F}_2(B^*) \subset \mathcal{F}_4(B^*) = \ker(\mathcal{O}) \subset \mathcal{F}_6(B^*) = B^*,$$

with $\mathcal{F}_{2q}(B^*) = \mathcal{F}_{2q+1}(B^*)$ and $\dim \mathcal{F}_2(B^*) = m + 1$,

more precisely: we want to know the number of different choices for a basis $x_1, \ldots, x_m$ of $\mathcal{F}_2(B^*)/\mathcal{F}_0(B^*)$. The form $\mu \in \mathcal{F}(m; \mathbf{Z}_p)$ on $\mathbf{Z}_p^m$ which corresponds to the graded associated algebra of this filtration is then given by $\mu(x_i, x_j, x_k) := \mathcal{O}(x_i x_j x_k)$ for $x_i, x_j, x_k \in (\mathcal{F}_2(B^*)/\mathcal{F}_0(B^*)) \cong \mathbf{Z}_p^m$.

Hence the number of different filtrations on B^* including a choice of a basis on $\mathcal{F}_2(B^*)/\mathcal{F}_0(B^*)$ is bounded by $p^{(2m+1)m}$. Again this is a rather rough estimate for several reasons; only multiplicative filtrations on B^* need to be considered and – depending on the product in B^* – the condition '$\mathcal{F}_2(B^*)\mathcal{F}_2(B^*) \subset$

$\mathcal{F}_4(B^*)'$ gives restrictions for the relevant filtrations. Also, clearly not every m–tuple of elements in $\mathcal{F}_2(B^*)/\mathcal{F}_0(B^*)$ forms a basis. Nevertheless the above estimates are good enough to yield Proposition 3, namely:

$$\sharp\mathcal{G}(m; \mathbf{Z}_p) \leq p^{\binom{2m+3}{2}}p^{(2m+2)}p^{(2m+1)m} = p^{\beta(m)} \quad \text{with}$$

$$\beta(m) = \frac{1}{2}((2m+3)(2m+2)) + (2m+2) + (2m+1)m$$
$$= 4m^2 + 8m + 5$$

It follows that

$$\frac{\sharp\mathcal{G}(m; \mathbf{Z}_p)}{\sharp\mathcal{F}(m; \mathbf{Z}_p)} \leq \frac{p^{\beta(m)}}{p^{\alpha(m)}} = \frac{1}{p^{\alpha(m)-\beta(m)}}, \text{ and hence}$$

$$\lim_{m \to \infty} \frac{\sharp\mathcal{G}(m; \mathbf{Z}_p)}{\sharp\mathcal{F}(m; \mathbf{Z}_p)} = 0, \text{ since } \lim_{m \to \infty} (\alpha(m) - \beta(m)) = \infty. \qquad \square$$

Let $\mathcal{G}^p(m) := \pi_p^{-1}(\mathcal{G}(m; \mathbf{Z}_p))$ denote the preimage of $\mathcal{G}(m; \mathbf{Z}_p)$ with respect to the reduction mod p, $\pi_p \colon \mathcal{F}(m; \mathbf{Z}) \to \mathcal{F}(m; \mathbf{Z}_p)$.

Corollary 2 *Given a prime p and $\epsilon < 0$, then there exists an $m_0 \in \mathbf{N}$ such that for all $m \geq m_0$ there exist arbitrarily large $N \in \mathbf{N}$ with*

$$\frac{\sharp\{(A^p(m) \cup \mathcal{G}^p(m)) \cap \mathcal{R}(m; N)\}}{\sharp\mathcal{R}(m; N)} < \epsilon.$$

Less technical and less precise: Most manifolds in $\mathcal{M}$ do not admit an effective, orientation preserving $\mathbf{Z}_p$–action.

Proof. For $N = pK$ it follows from Proposition 3 that

$$\frac{\sharp\{\mathcal{G}^p(m) \cap \mathcal{F}(m; N)\}}{\sharp\mathcal{F}(m; N)} \leq \frac{p^{\beta(m)}}{p^{\alpha(m)}}.$$

On the other hand

$$\frac{\sharp\mathcal{R}(m; N)}{\sharp\mathcal{F}(m; N)} \geq \frac{1}{2\binom{m}{2}}. \quad \text{(s. proof of Corollary 1)}.$$

Hence

$$\frac{\sharp\{\mathcal{G}^p(m) \cap \mathcal{R}(m; N)\}}{\sharp\mathcal{R}(m; N)} \leq \frac{p^{\beta(m)+\binom{m}{2}}}{p^{\alpha(m)}}.$$

Therefore there exists $m_0 \in \mathbf{N}$ (with $m_0 \geq 6$) such that

$$\frac{\sharp\{\mathcal{G}^p(m) \cap \mathcal{R}(m; N)\}}{\sharp\mathcal{R}(m; N)} < \frac{\epsilon}{2} \quad \text{if } m \geq m_0.$$

Given such an m one has, by Corollary 1 (note that clearly $\mathcal{A}^p(m) \subset \mathcal{A}(m)$), that

$$\frac{\#\{\mathcal{A}^p(m) \cap \mathcal{R}(m; N)\}}{\#\mathcal{R}(m; N)} < \frac{\epsilon}{2} \quad \text{if } N \text{ is large enough .}$$

Hence

$$\frac{\#\{(\mathcal{A}^p(m) \cup \mathcal{G}^p(m)) \cap \mathcal{R}(m; N)\}}{\#\mathcal{R}(m; N)} < \epsilon \quad \text{for } m \geq m_0,$$

if $N = pK$ is large enough. $\qquad\qquad\square$

The following remarks describe different attempts to improve Corollary 2.

Remarks 1

1. Along the same line of arguments one can prove a statement analogous to Corollary 2 where the fixed prime p is replaced by a finite set of primes.

2. The problem with getting an analogous result for the set of all primes is in finding a good bound for $\#\{\bigcup_p \mathcal{G}^p(m)) \cap \mathcal{R}(m; N)\}$. If the above inequality

$$\frac{\#\{\mathcal{G}^p(m) \cap \mathcal{F}(m; N)\}}{\#\mathcal{F}(m; N)} \leq \frac{p^{\beta(m)}}{p^{\alpha(m)}},$$

would hold simultanously for all primes if N is large enough (so it would hold, in particular, for primes much larger than N), which means that in a certain sense $\mathcal{G}^p(m)$ is uniformly distributed in $\mathcal{F}(m; \mathbf{Z})$, then one could imitate the above arguments to show that most manifolds in $\mathcal{M}$ do not admit any effective, orientation preserving action of a finite group.

3. To answer even the 'weaker question' of Raymond and Schultz mentionend above, one has to take into account also orientation reversing involutions. A certain part of the above arguments can be imitated in this case, too. One essential difference is, that the formal dimensions of components of the fixed point set can be odd.

   An orientation reversing involution on $M \in \mathcal{M}$ induces a non trivial involution, α, on $H^*(M) =: A^*$, in particular the restriction of this involution to A^6 is multiplication by (-1), and the form $\mu \in \mathcal{F}(m; \mathbf{Z})$, which corresponds to the algebra A^*, is mapped to $(-\mu)$ under the coordinate change, which corresponds to the involution on A^2. Now, any A^* (corresponding to some $\mu \in \mathcal{F}(m; \mathbf{Z})$) admits such an involution $\alpha\colon A^* \to A^*$, namely:

   $$\alpha|_{A^0} = id, \ \alpha|_{A^2} = -id, \ \alpha|_{A^4} = id, \ \alpha|_{A^6} = -id.$$

   But by arguments similar to those given for Proposition 2 'most' A^* do not admit any other non trivial involution. Mod 2 the above 'universal' involution becomes trivial. Let $M \in \mathcal{M}$ be a manifold with an involution that induces the 'universal' involution on $A^* := H^*(M)$ and hence the trivial involution on $A^* \otimes Z_2 = H^*(M; Z_2)$. Looking at the Serre spectral sequence of the Borel construction with coefficients Z_2 one gets

$$E_2^{*,*} = H^*(BZ_2; H^*(M; Z_2)) \cong H^*(BZ_2; Z_2) \otimes H^*(M; Z_2).$$

Since $H^i(BZ_2; Z_2) \neq 0$ for all $i \geq 0$ there is now room for non trivial differentials, which would correspond to non trivial derivations of the algebra $A^* \otimes Z_2 = H^*(M; Z_2)$. It is not difficult to check that for most $\mu \in \mathcal{F}(m; Z)$ the corresponding algebra $A^* \otimes Z_2$ does not admit any non trivial derivations. In that case the spectral sequence must collapse. In a similar way as above one can show that most of these algebras $A^* \otimes Z_2$ can be obtained as associated graded algebras of filtered Poincaré duality algebras B^* of formal dimension < 6 only is the following way: B^* is isomorphic to $A^* \otimes Z_2$ with a shift in degree, i.e. $B^i \cong A^{2i} \otimes Z_2$, and the filtration on B^* is given by

$$\mathcal{F}_{2q}(B^*) = \mathcal{F}_{2q+1}(B^*) = \bigoplus_{i=0}^{q} B^i.$$

Putting things together one obtains that for most manifolds $M \in \mathcal{M}$ the only non trivial involutions which might exist on M would have to be orientation reversing and such that the cohomology algebras $B^* := H^*(M^{Z_2}; Z_2)$ of the fixed point sets are isomorphic to $A^* \otimes Z_2 = H^*(M; Z_2)$ with the degree shift $B^i \cong A^{2i} \otimes Z_2$.

It would be interesting to realize such examples by 3–dimensional complex manifolds (defined by equations with real coefficients) where the involution can be described as 'complex conjugation' (compare the complex conjugation on $\mathbb{C}P^3$ with fixed point set $\mathbb{R}P^3$).

For the above results we have used the fact that $H^*(M; Z_p)$ is the graded associated algebra of a filtration on $H^*(M^{Z_p}; Z_p)$ if a Z_p-action on $M \in \mathcal{M}$ is given which is trivial on $H^*(M; Z)$. The relation between the cohomologies of M and M^{Z_p} can also be expressed by the following proposition (cf.[13],Corollary 1 and [14]).

Proposition 4 *If Z_p acts cohomologically trivial on $M \in \mathcal{M}$, then $H^*(M^{Z_p}; Z_p)$ (as a filtered algebra) is a deformation of negative weight of the graded algebra $H^*(M; Z_p)$.*

As a corollary one gets (cf.[14]):

Corollary 3 *If $H^*(M; Z)$ does not admit any orientation preserving automorphism of order p and if $H^*(M; Z_p)$ does not admit non trivial deformations of* *negative weight over $\mathbf{Z}_p[t]$ (as parameter space), then M does not admit orientation preserving, non trivial $\mathbf{Z}_p$-actions.*

(Note that the triviality of the deformation in general only implies that $H^*(M^{\mathbf{Z}_p}; \mathbf{Z}_p)$ is isomorphic to $H^*(M; \mathbf{Z}_p)$ as an ungraded - actually as a filtered - algebra (cf.[13],Prop.5). But in the case at hand (the corresponding cubic form μ must be non trivial) this means that they have to be isomorhpic already as graded algebras.)

The condition on the non existence of deformations in the above corollary follows, if $H^*(M; \mathbf{Z}_p)$ has no non trivial infinitesimal deformations of negative weight (see e.g.[7] for background material). The infinitesimal deformations of a graded algebra A^* correspond to the commutative part $H_c^{2,*}(A^*, A^*)$ of $H^{2,*}(A^*, A^*)$, the second Hochschild cohomology of A^* (with coefficients in A^*), which has an internal grading, coming from the grading of A^*. The infinitesimal deformations of negative weight correspond to the subgroup $H_c^{2,<0}(A^*, A^*)$, where the internal degree is negative. From a different point of view (see e.g.[9]) the infinitesimal deformations can be interpreted as the tangent space of A^* (in the variety of algebra structures on a vector space of dimension equal to dim A^*). That A^* has 'small tangent space' in the sense of [9] implies that it has no non trivial infinitesimal deformations of negative weight. Hence it follows from the 'small tangent space argument' in [9], that the subset $S(m; \mathbf{R}) \subset \mathcal{F}(m; \mathbf{R})$ of forms which correspond to algebras that admit non trivial infinitesimal deformations of negative weight (over $\mathbf{R}$) has positive codimension if $m \geq 6$. By an argument similar to the second part of the proof of Proposition 2 one gets the following result.

Proposition 5 *For $m \geq 6$*

$$\lim_{N \to \infty} \frac{\#\{S(m) \cap \mathcal{F}(m; N)\}}{\#\mathcal{F}(m; N)} = 0, \quad \text{where } S(m) := S(m; \mathbf{R}) \cap \mathcal{F}(m; \mathbf{Z}).$$

So by an argument analogous to the proof of Corollary 2 one gets the following theorem.

Corollary 4 *For $m \geq 6$*

$$\lim_{N \to \infty} \frac{\#\{(A(m) \cup S(m)) \cap \mathcal{R}(m; N)\}}{\#\mathcal{R}(m; N)} = 0.$$

A universal coefficient argument gives that if the $\mathbf{Z}$-algebra A^* correspond to an element in $S(m; \mathbf{R}) \cap \mathcal{F}(m; \mathbf{Z})$, i.e. $H_c^{2,<0}(A^*, A^*) \otimes \mathbf{R} = 0$, then for all but a finite number of primes p one has $H_c^{2,<0}(A^* \otimes \mathbf{Z}_p) = 0$ (cf.[14]).Hence for all but a finite number of primes $A^* \otimes \mathbf{Z}_p$ does not admit a non trivial deformations of negative weight. It follows that a manifold $M \in \mathcal{M}$ with $H^*(M) = A^*$ can admit non trivial $\mathbf{Z}_p$-actions for at most finitely many primes p.

We follow a suggestion of the referee and collect our results, which in one way or another indicate that 'most' manifolds (in $\mathcal{M}$) have little symmetry, in the following Theorem using the notion of a density function.

Let $\tilde{\mathcal{M}}_m$ denote the set of manifolds in $\mathcal{M}$, such that $rkH^2(M) = m$, together with a specified basis of $H^2(M)$. By Wall's Theorem above $\tilde{\mathcal{M}}_m$ can be identified with a certain subset $\tilde{\mathcal{R}}_m$ of $\mathcal{F}(m; \mathbf{Z}) \times \mathbf{Z}^m \cong \mathbf{Z}^{\gamma(m)}$ where $\gamma(m) = \alpha(m) + m$ (i.e. the set of pairs $(\mu, \mathcal{P})$ which fulfil Wall's conditions (a) and (b) above).

Define the density $d_m(\tilde{\mathcal{C}})$ of a subset $\tilde{\mathcal{C}}$ in $\tilde{\mathcal{R}}_m$ by

$$\limsup_{N\to\infty} \frac{\#\{\tilde{\mathcal{C}} \cap [-N, N)^{\gamma(m)}\}}{\#\{\tilde{\mathcal{R}}_m \cap [-N, N)^{\gamma(m)}\}}.$$

Cleary this density function has the following properties:

(1) $d_m(\tilde{\mathcal{C}}_1 \cup \tilde{\mathcal{C}}_2) \leq d_m(\tilde{\mathcal{C}}_1) + d_m(\tilde{\mathcal{C}}_2)$, in particular a finite union of subsets of density zero has zero density, too.

(2) $d_m(\tilde{\mathcal{C}}) < 1$ implies $\tilde{\mathcal{R}}_m \setminus \tilde{\mathcal{C}} \neq \emptyset$.

Theorem

1. *The subset of $\tilde{\mathcal{M}}_m$ given by those manifolds which admit a cohomologically non trivial, orientation preserving action of a finite group has density zero.*

2. *For a given prime p let $\tilde{\mathcal{C}}_m^p \subset \tilde{\mathcal{M}}_m$ denote the subset given by those manifolds which admit a non trivial, orientation preserving $\mathbf{Z}_p$-action. Then*

$$\lim_{m\to\infty} d_m(\tilde{\mathcal{C}}_m^p) = 0.$$

3. *The subset of $\tilde{\mathcal{M}}_m$ given by those manifolds which admit non trivial $\mathbf{Z}_p$-actions for infinitely many primes has density zero.*

Proof. The three parts of the Theorem are immediate consequences of the Corollaries 1,2 and 4, respectively. Note that the information given by the first Pontrjagin class $\mathcal{P}$ was not really used to prove the above corollaries. Already the information given by μ was enough to derive the given result. On the other hand the fact that $\mathcal{P}$ is always invariant under the action gives further restrictions on the possible automorphisms of the pair $(\mu, \mathcal{P})$ (compare the discussion of the example below). $\square$

Remarks 2

1. Note that while part 3 of the Theorem says that most $M \in \mathcal{M}$ have $\mathbf{Z}_p$-symmetries (i.e. admit non trivial $\mathbf{Z}_p$-actions) for at most finitely many

primes p, it is not clear which primes actually could occur or could be excluded. On the other hand part 1 says that for a given prime p most manifolds $M \in \mathcal{M}$ do not have orientation preserving Z_p-symmetries, but the situation for primes different from this p (or more general, different from a finite set of primes that can be fixed in advance) is not clear.

2. In [10] there is an example of an algebra A^* with small tangent space, corresponding to a form in $\mathcal{F}(6; Z)$, which can be realized as the cohomology $H^*(M)$ of a manifold $M \in \mathcal{M}$. Hence this manifold M has Z_p-symmetries for only finitely many primes p.

3. The following example of a symmetric, trilinear form in $\mathcal{F}(6; Z)$ is due to A.Iarrobino. He and A.Suciu checked on a computer, that the corresponding algebra mod p, $A^* \otimes Z_p$, has small tangent space for all primes. This example is discussed in greater detail in an algebraic context in the forthcoming note by A.Iarrobino and A.Suciu:'A Gorenstein Artin algebra with no jump deformation in any characteristic'.

Example. Let $\mu \in \mathcal{F}(6; Z)$ be defined in such a way that the corresponding cubic polynomial f (with $f(x) = \mu(x, x, x)$) is given by

$$
\begin{aligned}
f(x_1, \ldots, x_6) \;=\; & 6(x_1 x_4^2 - x_1^2 x_4 + x_2 x_4^2 + x_2^2 x_4 - x_2^2 x_5 + x_2 x_5^2 \\
& + x_3^2 x_4 - x_3 x_4^2 + x_3^2 x_6 + x_3 x_6^2 + x_5^2 x_6 + x_5 x_6^2 \\
& + x_1 x_2 x_4 + x_1 x_2 x_5 + x_1 x_3 x_6 + x_2 x_4 x_6 + x_3 x_5 x_6 + x_4 x_5 x_6 \\
& + x_4^3 + x_6^3)
\end{aligned}
$$

This example fulfils Wall's 'realization condition' (a). It is easy to check (using a computer) that the automorphism group is finite. (Clearly, if one considers automorhisms of prime order p, at most $p = 2, 3, 5$ or 7 could possibly occur.) If $M \in \mathcal{M}$ is a manifold, such that the cup product on $H^*(M; Z)$ is given by the above $\mu \in \mathcal{F}(6; Z)$ and $h\colon M \to M$ a homeomorphism (of finite order), then h^* not only keeps the form μ invariant, but also the first Pontrjagin class $\mathcal{P}\colon H^2(M; Z) \to Z$. Since, given μ, any Pontrjagin class, which fulfils the compatibility condition (b) $\mathcal{P}(x) \equiv 4\mu(x, x, x) \pmod{24}$ (s. Wall's theorem above), can be realized by a manifold in $\mathcal{M}$, it is possible to choose $\mathcal{P}$, and hence M, in such a way that no non trivial element of the isotropy subgroup $G_\mu \subset GL(G; Z)$ of μ fixes $\mathcal{P}$. (Choosing a basis $x_1, \ldots, x_6 \in H^2(M; Z)$ the possible Pontrjagin classes compatible with μ can be written as $\mathcal{P}(x_i) = 4\mu(x_i, x_i, x_i) + \tilde{\mathcal{P}}(x_i)$, $(i = 1, \ldots, 6)$, where $\tilde{\mathcal{P}}(x_i) \in 24Z$; in other words: any linear map $\tilde{\mathcal{P}}\colon H^2(M; Z) \to 24Z$ can occur this way. Over the rationals the set of linear maps from $H^2(M; Z)$ to Q, which have the property that they are fixed by some non trivial element of G_μ, is the complement of finite union (since G_μ is finite) of vector spaces of dimension < 6. So in a similar sense as above 'most' linear maps from $H^2(M; Z)$ to Q are not invariant under a non trivial element in G_μ. Multiplying by appropriate constants one obtains a linear maps

$\tilde{\mathcal{P}}: H^2(M; \mathbf{Z}) \to 24\mathbf{Z}$, which are not fixed by a $g \in G_\mu \backslash \{1\}$. In fact, again 'most' maps have this property. The same holds for the corresponding Pontrjagin classes $\mathcal{P}: H^2(M; \mathbf{Z}) \to 24\mathbf{Z}$. Hence the automorphism group of the pair $(\mu, \mathcal{P})$ is trivial for such a $\mathcal{P}$. (Added in proof: G.Nebe has shown – using reduction modulo 5 and computer calculations – that already the automorphism group of μ itself is trivial.)

It follows that any action of a finite group on the corresponding manifold $M \in \mathcal{M}$ must be cohomologically trivial. Since – by the computations of A.Iarrobino and A.Suciu – the algebra $H^*(M; \mathbf{Z}_p) = H^*(M; \mathbf{Z}) \otimes \mathbf{Z}_p$ has only trivial infinitesimal deformations of negative weight for all primes p, one has the following consequence.

Corollary 5 *There exist simply connected, closed, differentiable 6–dimensional spin-manifolds on which no finite group can act effectively and orientation preserving.*

I want to thank the referee for his suggestions, in particular, concerning the formulation of the above Theorem.

References

[1] Allday, C. and Puppe, V. 1993: Cohomological Methods in Transformation Groups. Cambridge Studies in Advanced Mathematics 32, Cambridge University Press, Cambridge.

[2] Bredon, G.E. 1973: Fixed point sets of actions on Poincaré duality spaces, Topology 12, 159-75.

[3] Browder, W. and Hsiang, W.C. 1978: Some problems on homotopy theory, manifolds and transformation groups, Proceedings of Symposia in Pure Mathematics. 32, 251-267.

[4] Chang, T. and Skjelbred, T. 1972: Group actions on Poincaré duality spaces, Bull.Amer.Math. Soc. 78, 1024-1026.

[5] Curtis, C.W. and Reiner, I. 1962: Representation Theory of Finite Groups and Associative Algebras, Wiley Intersience, New York.

[6] Edmonds, A.L. 1987: Construction of group actions on four–manifolds, Trans.Amer.Math.Soc. 299, 155-170.

[7] Gerstenhaber, M. and Schack, S.D. 1988: Algebraic cohomology and deformation theory, M. Hazewinkel and M. Gerstenhaber (Eds), Deformation Theory of Algebras and Structures and Applications, Kluwer Academie Publishers, Dordrecht–Boston–London, 11-264.

[8] Heller, A. 1954: Homological resolutions of complexes with operators, Ann. of Math. 55, 283-303.

[9] Iarrobino, A. and Emsalem, J. 1978: Some zero–dimensional generic singularities; finite algebras having small tangent space, Compostio Mathematica 36, 145-188.

[10] Iarrobino, A. 1985: Compressed algebras and components of the punctual Hilbert scheme, Algebraic Geometry, Sitges 1983, Proceedings. Springer Lecture Notes in Math. 1124, 146-166.

[11] Jupp, P.E. 1973: Classification of certain 6–manifolds, Proc. Camb. Phil. Soc. 73, 293-300.

[12] Popov, A.M. 1986: Finite isotropy subgroups in general position of simple linear Lie groups, Trans. Moscow Math. Soc., 3-63 (Translated from the Russian original).

[13] Puppe, V. 1978: Cohomology of fixed point sets and deformation of algebras, manuscripta math. 23, 343 -354.

[14] Puppe, V. 1988: Simply connected manifolds without S^1-symmetry, Algebraic Topology and Transformation Groups, Proc. Göttingen (1987), T. tom Dieck (Ed.), Lecture Notes in Math, 1361, Springer–Verlag, Berlin–Heidelberg–New York, 261-268.

[15] Swan, R.G. 1960: A new method in fixed point theory, Comment. Math. Helv. 34, 1-16.

[16] Wall, C.T.C. 1966: Classification problems in differential topology.V On certain 6–manifolds, Invent. math. 1, 355-374.

[17] Žubr, A.V. 1988: Classification of simply connected topological 6–manifolds, Topology and Geometry–Rohlin Seminar, O.Ya. Viro (Ed.), Lecture Notes in Math. 1346, Springer–Verlag, Berlin–Heidelberg–New York, 325-339.

Speculations on Gromov Convergence of Stratified Sets, and Riemannian Volume Collapse

Frank Quinn

Abstract

This paper proposes some topological background intended for the study of limiting behavior of Riemannian manifolds. The first conjecture suggests that Gromov limits of certain stratified sets are themselves stratified. The second asserts that stratified sets near such limits are mutually homeomorphic, and the third suggests structure for Riemannian manifolds with a lower curvature bound near such limits.

1 Introduction

The principal questions posed here concern Gromov convergence of stratified sets. These speculations were evoked by questions about the limiting behavior of Riemannian manifolds. We begin with the Riemannian motivation.

Gromov [G] introduced a definition of distance between abstract metric spaces, extending a definition given by Hausdorff for subsets of a space. There is a simple criterion for a set of metric spaces to have compact closure (with respect to this distance function) in the collection of all metric spaces. Finiteness theorems are an important application. If a function can be shown to be locally constant on the closure of a precompact set, then it takes only finitely many distinct values. This has been used to prove finiteness theorems for homology properties, homotopy type, homeomorphism and diffeomorphism type, in various collections of Riemannian manifolds. A discussion in the general metric space context is given in [PV].

This compactness principle has focused attention on the structure of Gromov limits, i.e. points in the closure of the set of manifolds. To use it to prove that a function of manifolds takes finitely many values one must:

(1) extend the definition of the function to limit points, and

(2) show it is locally constant.

For this one must understand something about the limit spaces. In fact many of the applications so far are in cases where nothing much happens in the limit: limit spaces are still manifolds, and are homeomorphic to nearby Riemannian manifolds. Genuine topology-changing convergence is still mysterious.

This work was partially supported by the National Science Foundation.

In the geometry community, Gromov convergence to a non-manifold limit is usually referred to as "volume collapse." Typically the problem is simplified a bit by assuming the diameters are bounded above. In this case dramatic (topological) changes can only occur if the volume goes to zero.

The current situation is roughly summarized in Figure 1. Assume for this that dimension is constant, diameter and volume are bounded above, and sectional curvature is bounded below. The remaining region in the volume-curvature plane is divided into four subregions depending on whether volume is bounded away from 0, and whether curvature is bounded above. When both these bounds are imposed then smoothly no collapse occurs: the limits are smooth manifolds diffeomorphic to sufficiently close-by Riemannian manifolds ([C]). When curvature is bounded then near a volume-0 limit the spaces look locally like manifolds with actions of compact nilpotent Lie groups, and the limit is the quotient space ([Fa]). A powerful local analysis of manifolds near volume-0 limits has been carried out in ([CFG]). When curvature is allowed to be unbounded above but volume is bounded away from 0 then topologically no collapse occurs: the limit is a topological manifold homeomorphic to sufficiently nearby Riemannian manifolds. These limits were shown to be homology manifolds, and nearby manifolds were shown to be mutually homeomorphic, in [GPW]. The limits were shown to actually be manifolds by G. Perlmann. This leaves a mostly unknown region where curvature is unbounded above at the same time as the volume goes to 0.

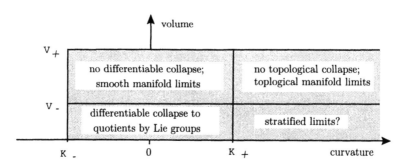

Figure 1. Gromov convergence phenomena in the volume-curvature plane

In this paper we provide some conjectural material on the topology of stratified sets which—if substantiated—may shed some light on convergence in this region. The principal statement is a non-collapse criterion for stratified sets. This would allow explanation of structure near a stratified limit space, and pinpoints what must happen for further collapse to occur. It would not, however, imply that all limit spaces are stratified in the sense used here. It may also permit extension of the Gromov-compactness technique to stratified contexts, for instance giving finiteness of homeomorphism types in collections of algebraic varieties with singularities.

2   Conjectures

The proposal describes structure of some limits of stratified spaces. It is a non-collapse condition in the sense that the limit is much like nearby spaces. The spaces used are the "homotopically stratified sets" defined in [Q1]. In these sets adjacent strata are related by homotopy conditions rather than geometric or fibration conditions as required in Whitney or Thom stratified sets. The point is that homotopy conditions pass through limits better than geometric ones.

The key ingredient in [GPW] for preventing collapse in manifolds was careful control of local contractibility. We extend this to the stratified setting. For this we define a *contractibility function* to be $\rho [0, R] \to [0, \infty)$ which satisfies:

(1) ρ is continuous and monotone increasing;

(2) $\rho(0) = 0$ and $\rho(r) \geq r$

These conditions are for convenience: more general functions can be used but the statements must be more elaborate to allow for non-monotonicity, etc. The upper endpoint R of the domain of ρ is an important part of the function. Typically things will be required to happen only for points closer together than R.

Definitions of "ρ contractible" and the "ρ uniform local fundamental group condition" are given in §2.

Conjecture 1. *Suppose a contractibility function ρ and dimension n are fixed. Suppose X is a Gromov limit of compact homotopically stratified sets with ANR homology manifold strata, all of which satisfy the ρ stratified local contractibility and uniform local fundamental group conditions, and all have dimension $\leq n$. Suppose further that X is finite dimensional. Then X is also an ANR homology manifold stratified set, and given $\epsilon > 0$ there is $\delta > 0$ depending only on ρ and n so that if Y is such a stratified set within δ of X then there is a stratum-preserving ϵ homotopy equivalence between the homotopy intrinsic stratifications of X and Y.*

Remarks.

(1) The main point in "stratum-preserving" is that the complements of skeleta are preserved, not just the skeleta themselves. In other words, under these hypotheses global convergence implies stratum-wise convergence.

(2) Note that the equivalence is between intrinsic stratifications. A stratification can always be subdivided. Since this does not change the underlying space or metric it will not affect convergence. The intrinsic stratification [Q2] is a maximal one from which all others are obtained by subdivision, so it is the only one which has a chance of surviving convergence.

(3) Following up on this point, there should be a relation between contractibility functions. Given ρ there should be τ so that if X has contractibility function ρ then the intrinsic stratification of X has contractibility function τ. In this case the limit will also have contractibility function τ.

(4) Finally we remark on the finite dimensionality requirement. In general it is possible for limits to be infinite dimensional [E]. What was found in the manifold case (see the erratum to [GPW]) is that a refinement of the estimates used to show that the set of manifolds has compact Gromov closure, also shows that the limits are finite-dimensional. Ferry [Fy] has shown that many conclusions can be recovered even if the limit space is not finite dimensional.

The second conjecture relates stratified sets near a limit.

Conjecture 2. *Suppose ρ and n are fixed, and X is a finite-dimensional Gromov limit as in conjecture 1. Then given $\epsilon > 0$ there is $\delta > 0$ such that if Y_1 and Y_2 are stratified sets with manifold strata satisfying the local ρ conditions and lying within δ of X then there is an ϵ homeomorphism $Y_1 \simeq Y_2$.*

Remarks.

(1) The new condition, which should allow a homeomorphism conclusion, is that the strata are manifolds. Homotopically stratified sets with manifold strata have a lot of topological structure, see [Q1].

(2) The homeomorphisms may not preserve the given stratifications. They will, of course, preserve the intrinsic stratifications.

(3) Current technology will require additional hypotheses on low-dimensional strata (dimensions 3 and 4).

(4) A relative version would give a uniqueness statement.

Finally we speculate on the structure of manifolds near a stratified set.

Conjecture 3. *Suppose X is a compact metric ANR homotopically stratified set with homology manifold strata, and suppose n and K_- are given. Then there is $\delta > 0$ so that if M is a Riemannian manifold with dimension n and sectional curvature $> K$, within δ of X in the Gromov metric, then there is a map $p\, M \to X$ which is a stratified system of approximate fibrations whose fibers are manifolds of almost nonnegative curvature.*

Remarks.

(1) There is some contractibility function ρ so that X satisfies the local contractibility and π_1 conditions. One would expect this function to play a large role in a proof, perhaps in essence by reducing it to questions over open strata, with growth conditions near the ends.

(2) This would imply that Gromov limits near X are obtained by partially collapsing fibers in a system of approximate fibrations. It does not imply that these limits are all stratified, but seems to come as close to that as possible in a topological context. It clarifies the geometry that must be done to refine the result in the Riemannian context: it must be shown that this partial collapse takes place along strata in a subdivision.

(3) This is supposed to be a partial replacement for the local analysis in [CFG] for the bounded curvature case. The local contractibility structure of the space X provides a subsitute for the universal constants which mark the onset of collapse in the bounded case.

(4) Stratified systems of fibrations are defined in [Q3]. They are maps whose restrictions over each stratum of X are fibrations, with a homotopy compatibility condition between strata. Approximate fibrations are weakened versions of fibrations, satisfying only an approximate version of the homotopy lifting property. Actually it may be necessary to go to something weaker yet: maps satisfying a local homological condition similar to that of fibrations. The problem is that fundamental groups of preimages may change from point to point.

(5) This statement should be refined to include a size restriction on the map p, and a relative version is needed.

(6) Manifolds with almost nonnegative curvature are ones with metrics of bounded diameter but curvature bounded below by numbers arbitrarily close to 0. (Different metrics for different bounds.) This condition on preimages should arise as follows: choose a projection so preimages are very small submanifolds. The curvature is still bounded below by K. Scale so the diameter becomes 1. Then the curvature scales to be bounded below by a number close to 0. Uniqueness should imply that there are projections with preimages which are smaller but still diffeomorphic. This gives other metrics on the manifold with lower curvature bounds closer to 0, and demonstrates that it is "almost nonnegatively curved."

3 Definitions

We describe the local contractibility and fundamental group conditions used in Conjecture 1. For this we assume $\rho\, [0, R]@ >>> [0, \infty]$ is continuous and monotone, $\rho(0) = 0$, and $\rho(r) \geq r$. Such functions are used to control slippage in various geometric constructions; data of diameter less than r gives a conclusion of larger diameter, $\rho(r)$.

We will also use a brief notation for "balls." Suppose $r > 0$ and Y is a subset of a metric space. Then $\langle Y \rangle^r$ denotes "Y enlarged by r": all points whose distance to Y is less than r. In particular the open ball of radius r about the point y is denoted $\langle y \rangle^r$.

A space is locally contractible if points have neighborhoods which contract inside larger neighborhoods. We make this more precise by specifying how much larger the second neighborhood must be, and also require that the retraction preserve the stratification.

Definition. A filtered metric space $X^n \supset X^{n-1} \supset \cdots \supset X^0$ is defined to be *stratified locally ρ-contractible* if for each $x \in X$ the following holds: let j be such that $x \in X^j - X^{j-1}$, then for $0 < r < R$ such that $\rho(r) < d(x, X^{j-1})$ the ball of radius r about x has an almost stratum-preserving deformation retraction to x inside the ball of radius $\rho(r)$.

To be explicit, "almost stratum-preserving" means there is a deformation

$$R \langle x \rangle^r \times [0, 1] \longrightarrow X$$

so that $R(y, 1) = y$, $R(y, 0) = x$, and if y is in a stratum $X^j - X^{j-1}$ then so is the half-open arc $R(y, (0, 1])$.

Note the locally ρ-contractible condition restricts the speed at which strata can complicate near lower strata. For instance consider a cusp of an arc in the plane (see figure 2). The cusp point is a point in the 0-skeleton not near any other such point, so balls about it up to radius R must contract. For a point on the arc the size of the ball depends on the distance to the cusp point. Note that a contracting ball on one branch must be disjoint from the other branch of the arc.

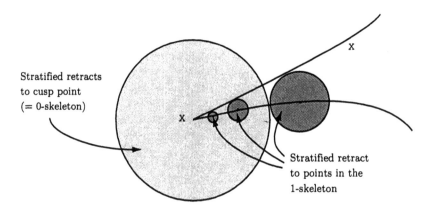

Figure 2. Stratified contractible balls in a stratification of the plane

The uniform local fundamental group condition ensures that local fundamental groups of a stratum behave in a regular way near a lower stratum. If a loop in the complement of a stratum is nullhomotopic in the complement, we can describe how close the nullhomotopy has to come to the lower stratum, in terms of how close the loop itself comes to the lower stratum.

Definition. X satisfies the ρ *uniform local fundamental group condition* provided for each point $y \in X^j - X^{j-1}$ and r such that $\rho(r) < d(y, X^{j-1})$ the following holds: if a loop is contractible in $\langle y \rangle^r \cap (X^k - X^{k-1})$ then there is a contraction in $\langle y \rangle^{\rho(r)}$ disjoint from $\langle X^j \rangle^t$ for any t such that the loop itself is disjoint from $\langle X^j \rangle^{\rho(t)}$.

We describe an example illustrating the role of the uniform local fundamental group condition. Suppose U is a homology h-cobordism; a manifold with two boundary components $\partial_0 U$ and $\partial_1 U$, fundamental group π, and the relative homology groups $H_*(U, \partial_i U; Z[\pi])$ vanish for $i = 0, 1$. Suppose also that $\pi_1 \partial_1 U \neq \pi = \pi_1 \partial_0 U$. For example U might be a homology h-cobordism from a nonsimply connected homology sphere to the genuine sphere. U deformation retracts to $\partial_0 U$. Define X to be U union a collar on $\partial_1 U$ and a cone on $\partial_0 U$. Then there are metrics d_n on X in which U union the cone has diameter less than $1/n$, and so that the metric spaces (X, d_n) converge to the standard metric on the cone on $\partial_1 U$. These satisfy all the hypotheses of the second conjecture except for the uniform local fundamental group condition. The conclusion of the conjecture is false since X is not homeomorphic to the cone (the fundamental groups of the complements of the cone point is different).

4   More detail

This section contains suggestions for proofs of conjectures 1 and 2.

Conjecture 1. In this we assume that stratified spaces Y_i have their homotopy intrinsic stratifications. Basically this means the strata are homology manifolds, and there is no coarser stratification with this property.

We work backwards, from the conclusion into the technicalities. The next-to-last step is to show that for every $\epsilon > 0$ there is $\delta > 0$ so that if Y_1 and Y_2 are stratified sets within δ of X then there is a stratum-preserving homotopy equivalence of radius $< \epsilon$ from Y_1 to Y_2. If we know this then the last step should be to choose Y_i converging to X, stratified equivalences $f_i\, Y_i \to Y_{i+1}$ so that the compositions $f_j \circ \cdots f_1$, and compositions of inverses, form Cauchy sequences of functions. The limit should give a homotopy equivalence $Y_1 \to X$ so that if we define a stratification on X to make it stratum-preserving then X is a stratified set of the required type.

Backing up a step, to get stratified equivalences it should be sufficent to prove a relative version of: any δ map $Y_1 \to Y_2$ is ϵ homotopic to a stratum-preserving map. This in turn should follow by downward induction on skeleta: suppose $g_{m+1}\, Y_1 \to Y_2$ preserves j-skeleta (and complements) for $j \geq m + 1$, then show that it is homotopic to g_m which also preserves the m-skeleton (and complement).

Changing g_{m+1} to preserve the m-skeleton is based on the fact that there is an arbitrarily small homotopy equivalence of Y with the pushout of the diagram

$$Y^m \leftarrow \text{holink}(Y, Y^m) \rightarrow Y - Y^m.$$

The unstratified version is [Q1], Lemma 2.4. But to preserve the stratification achieved so far (respecting j-skeleta, $j > m$) $Y - Y^m$ should be regarded as a stratified set, and the stratified homotopy link of [Q2] should be used. The point is that to get a stratum-preserving map it is sufficient to construct a map of holink diagrams. This would be done using two lemmas:

Lemma 1. *Given n and ρ there is a contractibility function τ $[0, T] \rightarrow [0, \infty)$ so that if Y has dimension $\leq n$ and satisfies the ρ contractibility and local π_1 conditions then for every j and $\delta < T$,*

i) *there is a stratum-preserving homotopy of $Y - Y^j$ into $Y - \langle Y^j \rangle^\delta$ of radius $< \tau(\delta)$, and*

ii) *there is an almost stratum-preserving homotopy rel Y^j of $< \langle Y^j \rangle^\delta$ into Y^j, of radius $< \tau(\delta)$.*

Statement (ii) is a globalization of the stratified local contractibility condition defined in section 3. In particular "almost stratum-preserving" has the same meaning as in that definition. Statement (i) is a controlled version of the "reverse tameness" condition of [Q1]. It should follow from tameness, duality, and the uniform local fundamental group condition.

Lemma 2. *Given n and ρ there is a contractibility function τ $[0, T] \rightarrow [0, \infty)$ with the following property: if Y_1, Y_2 are homotopically intrinsically stratified, have dimension $\leq n$ and satisfy the ρ contractibility and local π_1 conditions, and $g\ Y_1 \rightarrow Y_2$ preserves j-skeleta and complements for $j > m$, and is a $\tau(\delta)$ homotopy equivalence for $\delta < T$ then*

i) $g(Y_1^m) \subset \langle Y_2^m \rangle^\delta$, *and*

ii) $g(Y_1 - \langle Y_1^m \rangle^\delta) \subset Y_2 - Y_2^m.$

These are used as follows: Y_1 maps to the pushout of

$$Y_1^m \leftarrow \text{holink}(Y_1, Y_1^m) \rightarrow Y_1 - Y_1^m.$$

Lemma 1 gives a map from this to the pushout of

$$\langle Y_1^m \rangle^{\tau^2(\delta)} \leftarrow \langle Y_1^m \rangle^{\tau^2(\delta)} - \langle Y_1^m \rangle^{\tau(\delta)} \rightarrow Y_1 - \langle Y_1^m \rangle^{\tau(\delta)}$$

Lemma 2 implies that this maps to the pushout of a similar diagram for Y_2;

$$\langle Y_1^m \rangle^{\tau^3(\delta)} \leftarrow \langle Y_1^m \rangle^{\tau^3(\delta)} - \langle Y_1^m \rangle^\delta \rightarrow Y_1 - \langle Y_1^m \rangle^\delta$$

and finally Lemma 1 gives a map from this to the pushout

$$Y_2^m \leftarrow \text{holink}(Y_2, Y_2^m) \rightarrow Y_2 - Y_2^m,$$

which is equivalent to Y_2.

The key to the whole thing is lemma 2, which asserts that intrinsic skeleta and complements are almost preserved. For this we use the characterization: given $Y \supset Y^{n-1} \supset \cdots \supset Y^{m+1}$ then Y^m is the subset of Y^{m+1} where the stratified homotopy link $\text{holink}^s(Y, Y^{m+1})$ fails to be locally constant. For technical reasons one might add "or Y^{m+1} fails to be a homology manifold." In fact if Y has some stratification with homology manifold strata then this follows from local constancy of the homotopy link, but it may be easier to add it to the list of conditions than to prove it.

Suppose $x \in Y^{m+1}$. Then the fiber at x of the homotopy link of Y^{m+1} in Y can be approximated by a neighborhood of x minus a smaller neighborhood of Y^{m+1}. In terms of a contractibility function τ, something like $\langle x \rangle^{\tau(\delta)} - \langle Y^{m+1} \rangle^{\delta}$. The total space of the homotopy link over a subset $U \subset Y^{m+1}$ can similarly be approximated. If $x \in U$ then the inclusion of the fiber into the total space is approximated by

$$\langle x \rangle^{\tau^2(\delta)} - \langle Y^{m+1} \rangle^{\tau(\delta)} \rightarrow \langle U \rangle^{\tau^3(\delta)} - \langle Y^{m+1} \rangle^{\delta}$$

In particular the function τ can be arranged so that if U is sufficiently far from Y^m, or x is in Y^m, then the image on π_1 and homology with π_1 coefficients is the same as the image of the fiber of the holink. (Since these are all stratified, more precisely we want the π_1 and homology of each stratum to have the right image.) The local constancy condition is that these image should be the same for each $x \in U$.

The first observation is that the local constancy condition is satisfied when has U of diameter $< R$ and is far from Y^m. This is because the holink is a fibration, and therefore a product over U since U contracts inside $Y^{m+1} - Y^m$. On the other hand it fails for U near Y^m. Specifically if U contains both a point in Y^m and a point sufficiently far away, then these both have holinks whose homologies are obtained from the approximations. But the holinks are different, so cannot have the same image.

Finally we observe that this local constancy condition is preserved by δ equivalences for sufficiently small δ. This shows Y^m and its complement are both approximately preserved, as required. Or at least we conjecture that it can be shown that way.

Conjecture 2. To get to homeomorphisms from conjecture 1 one could follow the general outline of the manifold case ([GPW]):

(1) modify the homotopy intrinsic filtration of X at a few isolated points to get a filtration homotopy equivalent to the topological intrinsic filtration of Y. These points are where the stratum is a homology manifold but the local fundamental group of the complement is not abelian;

(2) observe that the strata of X are resolvable ANR homology manifolds, so in fact $X \times T^2$ is a manifold homotopically stratified set;

(3) show that if Y is close enough to X then $Y \times T^2$ is homeomorphic to $X \times T^2$. Here T^2 is the 2-torus. In contrast to Conjecture 1 "close enough" cannot be specified just in terms of ρ; it seems to depend on manifold structure features of $X \times T^2$. This should be proved by first using the fact that close enough manifolds are homeomorphic to get homeomorphisms outside a neighborhood of lower strata. Then controlled topology is used to extend this to the neighborhood.

(4) From this one concludes that if Y_1 and Y_2 are two such nearby manifold homotopically stratified sets there is a homeomorphism $Y_1 \times T^2 \simeq Y_2 \times T^2$. Then one shows such a homeomorphism can be "desuspended" to a homeomorphism $Y_1 \simeq Y_2$. This step is a glorified application of the h-cobordism theorem. It is in this step that the dimension restrictions on strata are necessary.

References

[CFG] Jeff Cheeger, Kenji Fukaya, and Mikhael Gromov, Nilpotent structures and invariant metrics on collapsed manifolds, *J. Am. Math. Society* **5** (1992), 327–372.

[C]   Jeff Cheeger, Finiteness theorems for Riemannian manifolds, *Am. J. Math.* **92** (1970), 61–74.

[E]   Teresa Engel, Gromov-Hausdorff convergence to non-manifolds, preprint, Ithaca College 1991.

[Fy]   Steven Ferry, Topological finiteness theorems for manifolds in Gromov-Hausdorff space, *Duke J. Math* **74** (1994), 95–106.

[Fa]   Kenji Fukaya, A boundary of the set of Riemannian manifolds with bounded curvatures and diameters, *J. Diff. Geom.* **28** (1988), 1–21.

[G]   Mikhael Gromov, Groups of polynomial growth and expanding maps, *Publ. Math. I.H.E.S.* **53** (1981), 183–215.

[GPW] Karsten Grove, Peter Petersen V, and Jyh-Yang Wu , Geometric finiteness theorems via controlled topology, *Invent. Math.* **99** (1990), 205–213; Erratum, *Invent. Math.* **104** (1991), 221–222.

[PV]   Peter Petersen V, A finiteness theorem for metric spaces, *J. Diff. Geometry* **31** (1990), 387–395.

[Q1]   Frank Quinn, Homotopically stratified sets, *J. Am. Math. Society* **1** (1988), 441–499.

[Q2]   Frank Quinn, Intrinsic skeleta and intersection homology of weakly stratified sets, *Geometry and Topology, Lecture Notes in Math* **105**, Marcel Dekker, New York, 1987.

[Q3]   ———, Ends of maps, II, *Invent. Math.* **68** (1982), 353–424.

Bordism of Automorphisms of Manifolds From the Algebraic L-Theory Point of View

Andrew Ranicki

Introduction

Among other things, Browder [1] initiated the application of surgery theory to the bordism of automorphisms of manifolds and the related study of fibred knots and open book decompositions. In this paper the bordism of automorphisms of high-dimensional manifolds is considered from the point of view of the localization exact sequence in algebraic L-theory.

The mapping torus of an automorphism $f: M {\longrightarrow} M$ of a closed n-dimensional manifold is a closed $(n+1)$-dimensional manifold

$$T(f) \ = \ M \times [0,1]/\{(x,0) = (f(x),1) \,|\, x \in M\} \ .$$

Given a space X let $\Delta_n(X)$ be the bordism group of pairs (M, f) with M a closed oriented n-dimensional manifold (in one of the standard categories O, PL, TOP) and $f: M {\longrightarrow} M$ an orientation-preserving automorphism, togetherwith a map $g: M {\longrightarrow} X$ and a homotopy $gf \simeq g: M {\longrightarrow} X$. The mapping torus construction of Browder [1, 2.28] defines a morphism of abelian groups

$$T \ : \ \Delta_n(X) \ {\longrightarrow} \ \Omega_{n+1}(X \times S^1) \ ; \ (M, f) \ {\longrightarrow} \ T(f)$$

to the bordism group $\Omega_{n+1}(X \times S^1)$ of closed oriented $(n+1)$-dimensional manifolds N with a map $N {\longrightarrow} X \times S^1$. The relative group $AB_n(X)$ in the exact sequence

$$\ldots \ {\longrightarrow} \ AB_n(X) \ {\longrightarrow} \ \Delta_n(X) \ \overset{T}{{\longrightarrow}} \ \Omega_{n+1}(X \times S^1) \ {\longrightarrow} \ AB_{n-1}(X) \ {\longrightarrow} \ \ldots$$

is the bordism group of oriented $(n+2)$-dimensional manifolds with boundary $(W, \partial W)$ with a map $W {\longrightarrow} X \times S^1$, such that $\partial W = T(f)$ for some representative (M, f) of an element of $\Delta_n(X)$.

López de Medrano [6] applied the Witt group of automorphisms of symmetric forms to the study of $\Delta_*(\text{pt.})$. Neumann [7] computed the automorphism Witt group over $\mathbb{Z}$, and Kreck [5] used this to compute $\Delta_*(\text{pt.})$ for $* \geq 5$. The automorphism bordism groups $\Delta_*(X)$ are closely related to open book decompositions, as considered by Winkelnkemper [17].

Quinn [8] developed a surgery theory for open book decompositions, in which the obstruction groups are defined for any ring with involution A by

$$W_n(A) = \begin{cases} \text{Witt group of nonsingular asymmetric forms over } A \text{ if } n \text{ is even} \\ 0 \text{ if } n \text{ is odd.} \end{cases}$$

The automorphism bordism groups $AB_*(X)$ are geometrically isomorphic to the open book bordism groups $BB_*(X)$ of [8]

$$AB_*(X) = BB_*(X).$$

The main result of [8] identifies

$$BB_n(X) = W_{n+2}(\mathbb{Z}[\pi_1(X)]) \quad (n \geq 5)$$

for any space X with finitely presented fundamental group $\pi_1(X)$. The main result of this paper (Theorem 3.1) obtains a different algebraic expression for the high-dimensional groups $AB_*(X) = BB_*(X)$, as the surgery obstruction groups of Wall [16]

$$AB_n(X) = L_{n+2}(\Omega^{-1}\mathbb{Z}[\pi_1(X \times S^1)]) \quad (n \geq 5)$$

of the following (noncommutative) localization $\Omega^{-1}\mathbb{Z}[\pi_1(X \times S^1)]$ of the Laurent polynomial extension of the group ring $\mathbb{Z}[\pi_1(X)]$

$$\mathbb{Z}[\pi_1(X \times S^1)] = \mathbb{Z}[\pi_1(X)][z, z^{-1}] \quad (\bar{z} = z^{-1}).$$

A $k \times k$ matrix ω in the Laurent polynomial extension $A[z, z^{-1}]$ of a ring A is *Fredholm* if the $A[z, z^{-1}]$-module morphism $\omega : A[z, z^{-1}]^k \longrightarrow A[z, z^{-1}]^k$ is injective and the cokernel is f.g. projective as an A-module. In Proposition 1.7 it will be proved that the localization $\Omega^{-1}A[z, z^{-1}]$ inverting the set Ω of Fredholm matrices in $A[z, z^{-1}]$ has the property that a finite f.g. free $A[z, z^{-1}]$-module chain complex C is A-finitely dominated (i.e. A-module chain equivalent to finite f.g. projective A-module chain complex) if and only if

$$H_*(\Omega^{-1}A[z, z^{-1}] \otimes_{A[z, z^{-1}]} C) = 0.$$

Now suppose that $A = \mathbb{Z}[\pi]$ is a group ring, so that

$$A[z, z^{-1}] = \mathbb{Z}[\pi \times \mathbb{Z}].$$

If N is a connected finite CW complex with universal cover $\widetilde{N}$ and fundamental group $\pi_1(N) = \pi \times \mathbb{Z}$ the infinite cyclic cover $\overline{N} = \widetilde{N}/\pi$ of N is a connected CW complex with $\pi_1(\overline{N}) = \pi$. The infinite CW complex $\overline{N}$ is finitely dominated if and only if the $\mathbb{Z}[\pi][z, z^{-1}]$-module chain complex $C(\widetilde{N})$ is $\mathbb{Z}[\pi]$-finitely dominated, if and only if

$$H_*(N; \Omega^{-1}\mathbb{Z}[\pi][z, z^{-1}]) = 0.$$

If N is a closed $(n+1)$-dimensional manifold with $\pi_1(N) = \pi \times \mathbb{Z}$, and $\overline{N} = \tilde{N}/\pi$ is finitely dominated, then the fibering obstruction $\Phi(N) \in Wh(\pi \times \mathbb{Z})$ of Farrell [4] and Siebenmann [14] is defined, such that $\Phi(N) = 0$ if (and for $n \geq 5$) only if $N = T(f)$ for an automorphism $f : M \longrightarrow M$ of a codimension 1 submanifold $M \subset N$ such that $f_* = 1 : \pi_1(M) = \pi \longrightarrow \pi$. The mapping torus function

$$
\begin{aligned}
T \; : \; &\{\text{closed } n\text{-dimensional manifolds } M \text{ with an automorphism } f : M \longrightarrow M \\
&\quad \text{such that } f_* = 1 : \pi_1(M) = \pi \longrightarrow \pi\} \\
&\longrightarrow \{\text{closed } (n+1)\text{-dimensional manifolds } N \text{ such that } \pi_1(N) = \pi \times \mathbb{Z}, \\
&\quad H_*(N; \Omega^{-1}\mathbb{Z}[\pi][z, z^{-1}]) = 0 \text{ and } \Phi(N) = 0 \in Wh(\pi \times \mathbb{Z})\} \; ; \\
&(M, f) \longrightarrow T(f)
\end{aligned}
$$

is thus a bijection for $n \geq 5$. The relative bordism group $AB_n(X)$ can thus be viewed as the bordism group of oriented $(n+2)$-dimensional manifolds with boundary $(W, \partial W)$ with a map $W \longrightarrow X \times S^1$, such that

$$
\begin{aligned}
\pi_1(W) &= \pi_1(\partial W) = \pi_1(X) \times \mathbb{Z} \;, \\
H_*(\partial W; \Omega^{-1}\mathbb{Z}[\pi_1(X)][z, z^{-1}]) &= 0 \;, \quad \Phi(\partial W) = 0 \in Wh(\pi_1(X) \times \mathbb{Z}) \;.
\end{aligned}
$$

In Theorem 3.1 the bijection will be used to identify

$$
AB_n(X) = L_{n+2}(\Omega^{-1}\mathbb{Z}[\pi_1(X)][z, z^{-1}]) \quad (n \geq 5) \;.
$$

The open book surgery of Quinn [8] is replaced here by the homology surgery of Cappell and Shaneson [2] and Vogel [15]. Combining the identification of 3.1 with the result of [8] gives geometric identifications

$$
L_*(\Omega^{-1}\mathbb{Z}[\pi_1(X)][z, z^{-1}]) = W_*(\mathbb{Z}[\pi_1(X)]) \;.
$$

In Ranicki [13] we shall give direct algebraic identifications

$$
L_*(\Omega^{-1}A[z, z^{-1}]) = W_*(A)
$$

for any ring with involution A.

I am grateful to the Royal Society of Edinburgh for the travel grant which enabled me to attend the conference.

1   Finite domination

Given a ring A let $A[z, z^{-1}]$ be the Laurent polynomial extension, the ring of polynomials $\sum\limits_{j=-\infty}^{\infty} a_j z^j$ with coefficients $a_j \in A$ such that $\{j \in \mathbb{Z} \,|\, a_j \neq 0\}$ is finite.

Definition 1.1. An $A[z, z^{-1}]$-module chain complex C is *A-finitely dominated* if it is A-module chain equivalent to a finite f.g. projective A-module chain complex.

The Novikov completions $A((z))$, $A((z^{-1}))$ of $A[z, z^{-1}]$ are the rings of formal power series defined by

$$A((z)) = \{ \sum_{j=-\infty}^{\infty} a_j z^j \mid \{j \leq 0 \mid a_j \neq 0 \in A\} \text{ finite}\},$$

$$A((z^{-1})) = \{ \sum_{j=-\infty}^{\infty} a_j z^j \mid \{j \geq 0 \mid a_j \neq 0 \in A\} \text{ finite}\}$$

with $A((z)) \cap A((z^{-1})) = A[z, z^{-1}]$.

Proposition 1.2 (Ranicki [12]). *A finite f.g. free $A[z, z^{-1}]$-module chain complex C is A-finitely dominated if and only if*

$$H_*(A((z)) \otimes_{A[z,z^{-1}]} C) = H_*(A((z^{-1})) \otimes_{A[z,z^{-1}]} C) = 0.$$

The two conditions of 1.2 can be united, using the diagonal ring morphism

$$A[z, z^{-1}] \longrightarrow A((z)) \times A((z^{-1})) \; ; \; x \longrightarrow (x, x).$$

A finite f.g. free $A[z, z^{-1}]$-module chain complex C is A-finitely dominated if and only if

$$H_*((A((z)) \times A((z^{-1}))) \otimes_{A[z,z^{-1}]} C) = 0.$$

Proposition 1.3. *The following conditions on a $k \times k$ matrix ω in $A[z, z^{-1}]$ are equivalent:*

(i) *the $A[z, z^{-1}]$-module morphism $\omega : A[z, z^{-1}]^k \longrightarrow A[z, z^{-1}]^k$ is injective and the cokernel is a f.g. projective A-module,*

(ii) *ω becomes invertible in $A((z)) \times A((z^{-1}))$,*

(iii) *the 1-dimensional f.g. free $A[z, z^{-1}]$-module chain complex*

$$C : C_1 = A[z, z^{-1}]^k \xrightarrow{\omega} C_0 = A[z, z^{-1}]^k$$

is A-finitely dominated.

Proof. (i) $\Longrightarrow$ (iii) C is A-module chain equivalent to the 0-dimensional f.g. projective A-module chain complex P defined by $P_0 = \mathrm{coker}(\omega)$.
(ii) $\Longleftrightarrow$ (iii) Immediate from 1.2.

(iii) $\Longrightarrow$ (i) The $A[z, z^{-1}]$-module morphism $\omega : A[z, z^{-1}]^k \longrightarrow A[z, z^{-1}]^k$ is injective (i.e. $H_1(C) = 0$) since $A[z, z^{-1}] \longrightarrow A((z)) \times A((z^{-1}))$ is injective. We have to prove that $H_0(C) = \operatorname{coker}(\omega)$ is a f.g. projective A-module. Let

$$\omega = \sum_{j=-N^+}^{N^-} \omega_j z^j$$

with ω_j a $k \times k$ matrix in A, and $-N^+ \le 0 \le N^-$. Let C^+ be the $A[z]$-module subcomplex of C defined by

$$d^+ = \omega| : C_1^+ = \sum_{i=0}^{\infty} z^i A^k \longrightarrow C_0^+ = \sum_{j=-N^+}^{\infty} z^j A^k ,$$

and let C^- be the $A[z^{-1}]$-module subcomplex of C defined by

$$d^- = \omega| : C_1^- = \sum_{i=-\infty}^{-1} z^i A^k \longrightarrow C_0^- = \sum_{j=-\infty}^{N^--1} z^j A^k .$$

The intersection $C^+ \cap C^-$ is the 0-dimensional f.g. free A-module chain complex with

$$(C^+ \cap C^-)_0 = \sum_{j=-N^+}^{N^--1} z^j A^k ,$$

and there is defined an exact sequence

$$0 \longrightarrow C^+ \cap C^- \longrightarrow C^+ \oplus C^- \longrightarrow C \longrightarrow 0$$

with

$$A[z, z^{-1}] \otimes_{A[z]} C^+ = A[z, z^{-1}] \otimes_{A[z^{-1}]} C^- = C .$$

As in the proof of 1.2 there are defined short exact sequences

$$0 \longrightarrow C^+ \longrightarrow (A[[z]] \otimes_{A[z]} C^+) \oplus C \longrightarrow A((z)) \otimes_{A[z,z^{-1}]} C \longrightarrow 0 ,$$
$$0 \longrightarrow C^+ \cap C^- \longrightarrow C^+ \oplus (A[[z^{-1}]] \otimes_{A[z^{-1}]} C^-)$$
$$\longrightarrow A((z^{-1})) \otimes_{A[z,z^{-1}]} C \longrightarrow 0 .$$

By hypothesis

$$H_*(A((z)) \otimes_{A[z,z^{-1}]} C) = H_*(A((z^{-1})) \otimes_{A[z,z^{-1}]} C) = 0 ,$$

so that there are defined A-module isomorphisms

$$H_0(C^+) \cong H_0(A[[z]] \otimes_{A[z]} C^+) \oplus H_0(C) ,$$
$$H_0(C^+ \cap C^-) \cong H_0(C^+) \oplus H_0(A[[z^{-1}]] \otimes_{A[z^{-1}]} C^-)$$
$$\cong H_0(A[[z]] \otimes_{A[z]} C^+) \oplus H_0(C) \oplus H_0(A[[z^{-1}]] \otimes_{A[z^{-1}]} C^-) .$$

Thus $H_0(C)$ is (isomorphic to) a direct summand of the f.g. free A-module $H_0(C^+ \cap C^-)$, verifying that $H_0(C)$ is a f.g. projective A-module. $\qquad \square$

Definition 1.4. A square matrix ω in $A[z, z^{-1}]$ is *Fredholm* if it satisfies any one of the equivalent conditions of 1.3.

Example 1.5. (i) If $\omega = z - h$ for an invertible $k \times k$ matrix h in A then ω is Fredholm: the $A[z, z^{-1}]$-module morphism $\omega : A[z, z^{-1}]^k \longrightarrow A[z, z^{-1}]^k$ is injective with cokernel the f.g. free A-module A^k (z acting by h).
(ii) If $\omega = 1 - zp$ for a projection $k \times k$ matrix $p = p^2$ in A then ω is Fredholm: the $A[z, z^{-1}]$-module morphism $\omega : A[z, z^{-1}]^k \longrightarrow A[z, z^{-1}]^k$ is injective with cokernel the f.g. projective A-module $\text{im}(p)$ (z acting by 1).

Cohn [3, pp. 254-255] defines the localization $\Sigma^{-1}R$ for any ring R and any set Σ of square matrices with entries in R to be the ring with generators all the elements of R and all the entries m'_{ij} in formal inverses $M' = (m'_{ij})$ of the matrices $M \in \Sigma$, subject to all the relations holding in R as well as

$$MM' = M'M = I \ (M \in \Sigma) .$$

The canonical ring morphism $i : R \longrightarrow \Sigma^{-1}R$ has the universal property that any ring morphism $f : R \longrightarrow S$ such that $f(M)$ is invertible for each $M \in \Sigma$ has a unique factorization

$$f : R \xrightarrow{\ i\ } \Sigma^{-1}R \longrightarrow S .$$

In general, $i : R \longrightarrow \Sigma^{-1}R$ may not be injective – for example, if $0 \in \Sigma$ then $\Sigma^{-1}R = 0$ is the zero ring.

Definition 1.6. Let Ω be the set of Fredholm matrices in $A[z, z^{-1}]$, and let $\Omega^{-1}A[z, z^{-1}]$ be the localization of $A[z, z^{-1}]$ inverting Ω.

The diagonal ring morphism

$$A[z, z^{-1}] \longrightarrow A((z)) \times A((z^{-1})) ; \ x \longrightarrow (x, x)$$

is injective, and has a factorization

$$A[z, z^{-1}] \xrightarrow{\ i\ } \Omega^{-1}A[z, z^{-1}] \longrightarrow A((z)) \times A((z^{-1}))$$

so that the canonical ring morphism $i : A[z, z^{-1}] \longrightarrow \Omega^{-1}A[z, z^{-1}]$ is injective.
Given an $A[z, z^{-1}]$-module chain complex C let

$$\Omega^{-1}C = \Omega^{-1}A[z, z^{-1}] \otimes_{A[z, z^{-1}]} C$$

be the induced $\Omega^{-1}A[z, z^{-1}]$-module chain complex.

Proposition 1.7. *The following conditions on a finite f.g. free $A[z, z^{-1}]$-module chain complex C are equivalent:*

(i) $H_*(\Omega^{-1}C) = 0$,

(ii) $H_*((A((z)) \times A((z^{-1}))) \otimes_{A[z,z^{-1}]} C) = 0$,

(iii) C is *A-finitely dominated*,

(iv) C is $A[z, z^{-1}]$-*module chain equivalent to the algebraic mapping cone* $C(z - h : P[z, z^{-1}] \longrightarrow P[z, z^{-1}])$ *for an automorphism* $h : P \longrightarrow P$ *of a finite f.g. projec*   *tive A-module chain complex* P.

Proof. (i) $\Longrightarrow$ (ii) Immediate from the existence of a ring morphism

$$A[z, z^{-1}] \longrightarrow A((z)) \times A((z^{-1})) .$$

(ii) $\Longrightarrow$ (i) Choose a basis for each C_r, writing

$$C_r = A[z, z^{-1}]^{k_r} \quad (r \geq 0) .$$

There exist $A[z, z^{-1}]$-module morphisms $\Gamma : C_r \longrightarrow C_{r+1}$ $(r \geq 0)$ such that the $A[z, z^{-1}]$-module endomorphisms

$$\omega = d\Gamma + \Gamma d : C_r = A[z, z^{-1}]^{k_r} \longrightarrow C_r = A[z, z^{-1}]^{k_r} \quad (r \geq 0)$$

induce automorphisms over $A((z)) \times A((z^{-1}))$. Each ω has a Fredholm matrix in $A[z, z^{-1}]$ by 1.3, so that $H_*(\Omega^{-1}C) = 0$.
(ii) $\Longleftrightarrow$ (iii) This is 1.2.
(i) $\Longrightarrow$ (iv) Assume that C is n-dimensional, and let

$$C_r = A[z, z^{-1}]^{k_r} \quad (0 \leq r \leq n) .$$

There exist $A[z, z^{-1}]$-module morphisms $\Gamma : C_r \longrightarrow C_{r+1}$ $(0 \leq r \leq n-1)$ such that the $A[z, z^{-1}]$-module endomorphisms

$$\omega = d\Gamma + \Gamma d : C_r = A[z, z^{-1}]^{k_r} \longrightarrow C_r = A[z, z^{-1}]^{k_r} \quad (0 \leq r \leq n)$$

are defined by Fredholm matrices ω in $A[z, z^{-1}]$, with $\Gamma^2 = 0$. The $A[z, z^{-1}]$-module morphism

$$d + \Gamma = \begin{pmatrix} d & 0 & 0 & \cdots \\ \Gamma & d & 0 & \cdots \\ 0 & \Gamma & d & \cdots \\ \vdots & \vdots & \vdots & \ddots \end{pmatrix} :$$

$$C_{odd} = C_1 \oplus C_3 \oplus C_5 \oplus \ldots \longrightarrow C_{even} = C_0 \oplus C_2 \oplus C_4 \oplus \ldots$$

is injective with f.g. projective A-module cokernel. The $(n-1)$-dimensional f.g. projective A-module chain complex P defined by

$$P_r = \begin{cases} \operatorname{coker}(d + \Gamma : C_{odd} \longrightarrow C_{even}) & \text{if } r = 0 \\ \operatorname{coker}(\omega \oplus \omega \oplus \ldots : C_{r+1} \oplus C_{r+3} \oplus \ldots \longrightarrow C_{r+1} \oplus C_{r+3} \oplus \ldots) \\ \qquad\qquad \text{if } r = 1, 2, \ldots, n-1 , \end{cases}$$

$$d \; : \; P_r \longrightarrow P_{r-1} \; ; \; [x] \longrightarrow \begin{cases} [0,x] & \text{if } r = 1 \\ [(d+\Gamma)(x)] & \text{if } r = 2,3,\ldots,n-1 \end{cases}$$

is equipped with an automorphism

$$h \; : \; P \longrightarrow P \; ; \; [x] \longrightarrow [zx]$$

such that C is $A[z,z^{-1}]$-module chain equivalent to the algebraic mapping cone $C(z-h:P[z,z^{-1}]\longrightarrow P[z,z^{-1}])$.
(iv) $\Longrightarrow$ (iii) The algebraic mapping cone $C(z-h)$ is A-module chain equivalent to P.
(iv) $\Longrightarrow$ (i) Each of the $A[z,z^{-1}]$-module morphisms

$$z - h \; : \; P_r[z,z^{-1}] \longrightarrow P_r[z,z^{-1}] \; (r \geq 0)$$

induces an $\Omega^{-1}A[z,z^{-1}]$-module isomorphism

$$z - h \; : \; \Omega^{-1}P_r[z,z^{-1}] \longrightarrow \Omega^{-1}P_r[z,z^{-1}] \; ,$$

so that

$$H_*(\Omega^{-1}C) \; = \; H_*(z-h:\Omega^{-1}P[z,z^{-1}]\longrightarrow \Omega^{-1}P[z,z^{-1}]) \; = \; 0 \; . \qquad \square$$

Example 1.8. If $A = K$ is a field then Ω consists of all the square matrices in $K[z,z^{-1}]$ with non-zero determinant, and $\Omega^{-1}K[z,z^{-1}] = K(z)$ is the function field of K. A finite f.g. free $K[z,z^{-1}]$-module chain complex C is K-finitely dominated if and only if $H_*(K(z) \otimes_{K[z,z^{-1}]} C) = 0$, if and only if the homology K-vector spaces $H_*(C)$ are finite-dimensional.

Definition 1.9. (i) The *automorphism category of A* is the exact category $\mathrm{Aut}(A)$ in which an object (P,h) is a f.g. projective A-module P together with an automorphism $h : P \longrightarrow P$, a morphism $f : (P,h) \longrightarrow (P',h')$ is an A-module morphism $f : P \longrightarrow P'$ such that $h'f = fh$, and a sequence $(P,h) \longrightarrow (P',h') \longrightarrow (P'',h'')$ is exact if the A-module sequence $P \longrightarrow P' \longrightarrow P''$ is exact.
(ii) The *automorphism class group* $\mathrm{Aut}_0(A)$ is the class group of the automorphism category

$$\mathrm{Aut}_0(A) \; = \; K_0(\mathrm{Aut}(A)) \; .$$

Proposition 1.10. *The torsion group of $\Omega^{-1}A[z,z^{-1}]$ is a direct sum*

$$K_1(\Omega^{-1}A[z,z^{-1}]) \; = \; K_1(A[z,z^{-1}]) \oplus \mathrm{Aut}_0(A) \; .$$

Proof. The relative term $K_1(i)$ in the localization exact sequence of algebraic K-theory

$$\cdots \longrightarrow K_1(A[z,z^{-1}]) \xrightarrow{\;i\;} K_1(\Omega^{-1}A[z,z^{-1}]) \xrightarrow{\;\partial\;} K_1(i)$$
$$\longrightarrow K_0(A[z,z^{-1}]) \longrightarrow \cdots$$

is the class group of the exact category of f.g. $\Omega^{-1}A[z, z^{-1}]$-torsion $A[z, z^{-1}]$-modules of homological dimension 1, which is isomorphic to the automorphism category $\mathrm{Aut}(A)$, with

$$\partial \; : \; K_1(\Omega^{-1}A[z, z^{-1}]) \longrightarrow K_1(i) \; = \; \mathrm{Aut}_0(A) \; ; \; \tau(\Omega^{-1}C) \longrightarrow [C, \zeta]$$

sending the torsion $\tau(\Omega^{-1}C)$ for an A-finitely dominated finite based f.g. free $A[z, z^{-1}]$-module chain complex C to the class of the A-module automorphism

$$\zeta \; : \; C \longrightarrow C \; ; \; x \longrightarrow zx \; .$$

The morphism ∂ is a surjection which is split by

$$\Delta \; : \; \mathrm{Aut}_0(A) \longrightarrow K_1(\Omega^{-1}A[z, z^{-1}]) \; ;$$
$$[P, h] \longrightarrow \tau(z - h : \Omega^{-1}P[z, z^{-1}] \longrightarrow \Omega^{-1}P[z, z^{-1}]) \; .$$

The morphism

$$i \; : \; K_1(A[z, z^{-1}]) \longrightarrow K_1(\Omega^{-1}A[z, z^{-1}]) \; ; \; \tau(C) \longrightarrow \tau(\Omega^{-1}C)$$

is an injection which is split by

$$\Phi \; : \; K_1(\Omega^{-1}A[z, z^{-1}]) \longrightarrow K_1(A[z, z^{-1}]) \; ; \; \tau(\Omega^{-1}C) \longrightarrow \Phi(C)$$

with C an A-finitely dominated ($= \Omega^{-1}A[z, z^{-1}]$-contractible) based f.g. free $A[z, z^{-1}]$-module chain complex and $\Phi(C) \in K_1(A[z, z^{-1}])$ the algebraic fibering obstruction (Ranicki [11, §20]).   □

Example 1.11. Let N be a connected finite CW complex with universal cover $\widetilde{N}$, such that $\pi_1(N) = \pi \times \mathbb{Z}$ and the infinite cyclic cover $\overline{N} = \widetilde{N}/\pi$ is finitely dominated. Let

$$\Lambda \; = \; \mathbb{Z}[\pi_1(N)] \; = \; \mathbb{Z}[\pi][z, z^{-1}]$$

and let $\zeta : \overline{N} \longrightarrow \overline{N}$ be a generating covering translation, inducing $z : C(\widetilde{N}) \longrightarrow C(\widetilde{N})$ on the Λ-module chain level. The cellular Λ-module chain complex $C(\widetilde{N})$ is $\Omega^{-1}\Lambda$-contractible, and the $\Omega^{-1}\Lambda$-coefficient Whitehead (or rather Reidemeister) torsion of N is given by

$$\begin{aligned}
\tau(N; \Omega^{-1}\Lambda) \; &= \; \tau(\Omega^{-1}C(\widetilde{N})) \\
&= \; (\Phi(N), [\overline{N}, \zeta]) \; = \; (\Phi(C(\widetilde{N})), [C(\widetilde{N}), \zeta]) \\
&\in \; K_1(\Omega^{-1}\Lambda) \; = \; K_1(\Lambda) \oplus \mathrm{Aut}_0(\mathbb{Z}[\pi]) \; .
\end{aligned}$$

2 Localization in L-theory

We refer to Ranicki [9] for an exposition of the quadratic and symmetric L-groups $L_*(A)$, $L^*(A)$ of a ring with involution A, which we take to be defined using (unbased) f.g. free A-modules. The quadratic L-group $L_n(A)$ of Wall [16] was identified in [9] with the cobordism group of n-dimensional quadratic Poincaré complexes (C, ψ) over A, with C a f.g. free A-module chain complex.

Extend the involution on A to $A[z, z^{-1}]$ by $\bar{z} = z^{-1}$. The conjugate transpose of a Fredholm matrix $\omega = (a_{ij})$ in $A[z, z^{-1}]$ is a Fredholm matrix $\omega^* = (\bar{a}_{ji})$ in $A[z, z^{-1}]$, with

$$\operatorname{coker}(\omega^*) = \operatorname{Hom}_A(\operatorname{coker}(\omega), A) ,$$

so that $\Omega^{-1} A[z, z^{-1}]$ is also a ring with involution.

We refer to Ranicki [10, §3] for the localization exact sequence in algebraic L-theory, which applies also to $\Omega^{-1} A[z, z^{-1}]$:

Proposition 2.1. *For any ring with involution A there is defined an exact sequence*

$$\cdots \longrightarrow L^{n+2}(A[z, z^{-1}]) \xrightarrow{\ i\ } L^{n+2}(\Omega^{-1} A[z, z^{-1}]) \xrightarrow{\ \partial\ } L\mathrm{Aut}^n(A)$$
$$\xrightarrow{\ T\ } L^{n+1}(A[z, z^{-1}]) \longrightarrow \cdots$$

with $L\mathrm{Aut}^n(A)$ the cobordism group of automorphisms of f.g. projective n-dimensional symmetric Poincaré complexes over A, T given by the algebraic mapping torus, and

$$\partial \ : \ L^{n+2}(\Omega^{-1} A[z, z^{-1}]) \longrightarrow L\mathrm{Aut}^n(A) \ ; \ \Omega^{-1}(C, \phi) \longrightarrow (\partial C, \partial \phi, \zeta)$$

for any $\Omega^{-1} A[z, z^{-1}]$-Poincaré f.g. free $(n+2)$-dimensional symmetric complex (C, ϕ) over A, with $\partial C = C(\phi_0 : C^{n+2-*} \longrightarrow C)_{*+1}$, $\zeta : x \longrightarrow zx$.

3 Bordism of automorphisms of manifolds

Let X be a connected space with universal cover $\tilde{X}$.

The *symmetric signature* map

$$\sigma^* \ : \ \Omega_n(X) \longrightarrow L^n(\mathbb{Z}[\pi_1(X)]) \ ; \ (M \longrightarrow X) \longrightarrow \sigma^*(M) = (C(\widetilde{M}), \phi)$$

is defined as in Ranicki [9], with ϕ the symmetric Poincaré duality structure on the cellular $\mathbb{Z}[\pi_1(X)]$-module chain complex $C(\widetilde{M})$ of the pullback cover $\widetilde{M}$ of the oriented n-dimensional manifold M, so that

$$\phi_0 = [M] \cap - \ : \ C(\widetilde{M})^{n-*} \longrightarrow C(\widetilde{M}) .$$

There are corresponding symmetric signature maps on the automorphism bordism groups $\Delta_*(X)$, $AB_*(X)$ defined in the Introduction.

The *symmetric signature* map on $\Delta_*(X)$ is defined by

$$\sigma^* \ : \ \Delta_n(X) \longrightarrow L\mathrm{Aut}^n(\mathbb{Z}[\pi_1(X)]) \ ;$$
$$(M \longrightarrow X, f : M \longrightarrow M) \longrightarrow (C(\widetilde{M}), \phi, \tilde{f} : C(\widetilde{M}) \longrightarrow C(\widetilde{M})) .$$

(For $\pi_1(X) = \{1\}$, $n = 2k$ this is the automorphism Witt invariant of López de Medrano [6]).

The *symmetric signature* map on $AB_*(X)$ is defined by

$$\sigma^* : AB_n(X) \longrightarrow L^{n+2}(\Omega^{-1}\mathbb{Z}[\pi_1(X)][z, z^{-1}]) \ ;$$
$$((W, \partial W) \longrightarrow X \times S^1, \partial W = T(f : M \longrightarrow M)) \longrightarrow \Omega^{-1}(C(\widetilde{W}, \partial\widetilde{W}), \delta\phi/\phi) \ .$$

Here, $(C(\widetilde{W}, \partial\widetilde{W}), \delta\phi/\phi)$ is the $(n+2)$-dimensional symmetric complex over $\mathbb{Z}[\pi_1(X)][z, z^{-1}]$ obtained by collapsing the boundary in the $(n+2)$-dimensional symmetric Poincaré pair $(C(\partial\widetilde{W}) \longrightarrow C(\widetilde{W}), (\delta\phi, \phi))$ over $\mathbb{Z}[\pi_1(X)][z, z^{-1}]$ associated to the $(n+2)$-dimensional manifold with boundary $(W, \partial W = T(f : M \longrightarrow M))$. The induced $(n+2)$-dimensional symmetric complex over $\Omega^{-1}\mathbb{Z}[\pi_1(X)][z, z^{-1}]$ is Poincaré since $H_*(\partial W; \Omega^{-1}\mathbb{Z}[\pi_1(X)][z, z^{-1}]) = 0$.

Theorem 3.1. *The symmetric signature maps define a natural transformation of exact sequences*

$$
\begin{array}{ccccccc}
\cdots \longrightarrow \Omega_{n+2}(X \times S^1) \longrightarrow & AB_n(X) & \longrightarrow \Delta_n(X) & \xrightarrow{T} & \Omega_{n+1}(X \times S^1) \longrightarrow \cdots \\
\Big\downarrow \sigma^* & \Big\downarrow \sigma^* & \Big\downarrow \sigma^* & & \Big\downarrow \sigma^* \\
\cdots \longrightarrow L^{n+2}(\Lambda) \xrightarrow{i} & L^{n+2}(\Omega^{-1}\Lambda) & \xrightarrow{\partial} LAut^n(A) & \xrightarrow{T} & L^{n+1}(\Lambda) \longrightarrow \cdots
\end{array}
$$

with

$$A = \mathbb{Z}[\pi_1(X)] \ , \quad \Lambda = \mathbb{Z}[\pi_1(X \times S^1)] = A[z, z^{-1}] \ .$$

If $\pi_1(X)$ is finitely presented the symmetric signature maps

$$\sigma^* : AB_n(X) \longrightarrow L^{n+2}(\Omega^{-1}\Lambda) \quad (n \geq 5)$$

are isomorphisms, and the automorphism bordism groups $\Delta_*(X)$ fit into an exact sequence

$$\cdots \longrightarrow L^{n+2}(\Lambda) \longrightarrow \Delta_n(X) \longrightarrow LAut^n(A) \oplus \Omega_{n+1}(X \times S^1)$$
$$\longrightarrow L^{n+1}(\Lambda) \longrightarrow \cdots \ .$$

Proof. The unit

$$u = (1 - z)^{-1} \in \Omega^{-1}\Lambda$$

is such that $u + \bar{u} = 1$, so there is no difference between the quadratic and symmetric L-groups of $\Omega^{-1}\Lambda$

$$L_*(\Omega^{-1}\Lambda) = L^*(\Omega^{-1}\Lambda) \ .$$

Let $(V, \partial V)$ be an $(n + 1)$-dimensional manifold with boundary with a π_1-isomorphism reference map $(V, \partial V) \longrightarrow X$ such that

$$\pi_1(V) = \pi_1(\partial V) = \pi_1(X) \ .$$

By the realization theorems of Wall [16], Cappell and Shaneson [2] and Vogel [15] every element

$$x \in L^{n+2}(\Omega^{-1}\Lambda) = \Gamma_{n+2}(\Lambda \longrightarrow \Omega^{-1}\Lambda)$$

is the $\Omega^{-1}\Lambda$-homology surgery obstruction $x = \sigma_*(F,B)$ of a normal map of $(n+2)$-dimensional manifolds with boundary

$$(F,B) : (W, \partial W) \longrightarrow (V, \partial V) \times S^1$$

with

$$\pi_1(W) = \pi_1(\partial W) = \pi_1(X) \times \mathbb{Z}$$

and such that $\partial F : \partial W \longrightarrow \partial V \times S^1$ is a $\Omega^{-1}\Lambda$-homology equivalence, i.e. such that the pullback infinite cyclic cover of ∂W

$$\overline{\partial W} = (\partial F)^*(\partial V \times \mathbb{R})$$

is finitely dominated. Use the direct sum decomposition given by 1.10

$$K_1(\Omega^{-1}\Lambda) = K_1(\Lambda) \oplus \mathrm{Aut}_0(A)$$

to express the $\Omega^{-1}\Lambda$-coefficient Whitehead torsion of ∂F as

$$
\begin{aligned}
\tau(\partial F; \Omega^{-1}\Lambda) &= (-)^n \tau(\partial F; \Omega^{-1}\Lambda)^* \\
&= (\Phi(\partial W), [\overline{\partial W}, \zeta]) - (\Phi(\partial V \times S^1), [\partial V \times \mathbb{R}, 1 \times \zeta_{\mathbb{R}}]) \\
&= (\Phi(\partial W), [\overline{\partial W}, \zeta]) - (0, [\partial V, 1]) \\
&\in K_1(\Omega^{-1}\Lambda) = K_1(\Lambda) \oplus \mathrm{Aut}_0(A) ,
\end{aligned}
$$

with $\zeta : \overline{\partial W} \longrightarrow \overline{\partial W}$ a generating covering translation. Moreover, for every $\mu \in K_1(\Omega^{-1}\Lambda)$ it is possible to vary (F,B) by an $\Omega^{-1}\Lambda$-coefficient homology cobordism with torsion μ, changing $\tau(\partial F; \Omega^{-1}\Lambda)$ by $\mu + (-)^n \mu^*$. The duality involution on $K_1(\Omega^{-1}\Lambda)$ defined by the conjugate transposition of matrices $(a_{ij}) \longrightarrow (\bar{a}_{ji})$ is of the form

$$
* = \begin{pmatrix} * & \beta * \\ 0 & * \end{pmatrix} \quad : \quad K_1(\Omega^{-1}\Lambda) = K_1(\Lambda) \oplus \mathrm{Aut}_0(A)
$$

$$\longrightarrow K_1(\Omega^{-1}\Lambda) = K_1(\Lambda) \oplus \mathrm{Aut}_0(A)$$

with

$$* : \mathrm{Aut}_0(A) \longrightarrow \mathrm{Aut}_0(A) ; [P, h] \longrightarrow [P^*, (h^*)^{-1}] ,$$
$$\beta : \mathrm{Aut}_0(A) \longrightarrow K_1(\Lambda) ; [P, h] \longrightarrow \tau(-zh : P[z, z^{-1}] \longrightarrow P[z, z^{-1}]) ,$$
$$\beta* = - * \beta : \mathrm{Aut}_0(A) \longrightarrow K_1(\Lambda) .$$

Now β maps the automorphism class group $\text{Aut}_0(A)$ onto the direct summand $K_1(A) \oplus K_0(A)$ in the Bass decomposition

$$K_1(\Lambda) = K_1(A) \oplus K_0(A) \oplus \widetilde{\text{Nil}}_0(A) \oplus \widetilde{\text{Nil}}_0(A) .$$

The duality involution on $K_1(\Lambda)$ interchanges the two $\widetilde{\text{Nil}}_0(A)$-summands, so that every self-dual element $\tau = \pm\tau^* \in K_1(\Omega^{-1}\Lambda)$ can be expressed as

$$\tau = (\mu_1, \mu_2) \pm (\mu_1, \mu_2)^* + (0, \mu_3) \in K_1(\Omega^{-1}\Lambda) = K_1(\Lambda) \oplus \text{Aut}_0(A)$$

for some $\mu_1 \in K_1(\Lambda)$, $\mu_2 \in \text{Aut}_0(A)$, $\mu_3 = \pm\mu_3^* \in \ker(\beta)$. Applying this to

$$\tau = \tau(\partial F; \Omega^{-1}\Lambda) \in K_1(\Omega^{-1}\Lambda)$$

shows that every element $x \in L_{n+2}(\Omega^{-1}\Lambda)$ is realized as the $\Omega^{-1}\Lambda$-coefficient surgery obstruction $\sigma_*(F, B)$ of a normal map $(F, B) : (W, \partial W) \longrightarrow (V, \partial V) \times S^1$ with fibering obstruction

$$\Phi(\partial W) = 0 \in Wh(\pi_1(X) \times \mathbb{Z}) ,$$

so that $\partial W = T(f)$ is the mapping torus of an automorphism $f : M \longrightarrow M$ of a closed n-dimensional manifold M. The surgery obstruction is the difference of the symmetric signatures

$$\sigma_*(F, B) = \sigma^*(W, \partial W) - \sigma^*(V \times S^1, \partial V \times S^1) \in L_{n+2}(\Omega^{-1}\Lambda) = L^{n+2}(\Omega^{-1}\Lambda) .$$

The construction defines an isomorphism

$$L^{n+2}(\Omega^{-1}\Lambda) \longrightarrow AB_n(X) ;$$
$$x = \sigma_*(F, B) \longrightarrow (W, \partial W = T(f)) - (V \times S^1, \partial V \times S^1$$
$$= T(1 : \partial V \longrightarrow \partial V))$$

inverse to the symmetric signature map $\sigma^* : AB_n(X) \longrightarrow L^{n+2}(\Omega^{-1}\Lambda)$. $\square$

References

[1] W. Browder, *Surgery and the Theory of Differentiable Transformation Groups*, Proc. Conf. on Transformation Groups (New Orleans, 1967), Springer (1969)

[2] S. Cappell and J. Shaneson, *The codimension two placement problem, and homology equivalent manifolds*, Ann. of Maths. 99, 277–348 (1974)

[3] P. M. Cohn, *Free rings and their relations*, Academic Press (1971)

[4] F. T. Farrell, *The obstruction to fibering a manifold over a circle*, Indiana Univ. J. 21, 315–346 (1971)

[5] M. Kreck, *Bordism of diffeomorphisms and related topics*, Springer Lecture Notes 1069 (1984)

[6] S. López de Medrano, *Cobordism of diffeomorphisms of $(k-1)$-connected $2k$-manifolds*, Proc. Second Conference on Compact Transformation Groups, Springer Lecture Notes 298, 217–227 (1972)

[7] W. D. Neumann, *Equivariant Witt rings*, Bonner Math. Schriften 100 (1977)

[8] F. Quinn, *Open book decompositions, and the bordism of automorphisms*, Topology 18, 55–73 (1979)

[9] A. Ranicki, *The algebraic theory of surgery*, Proc. London Math. Soc. (3) 40, I. 87–192, II. 193–287 (1980)

[10] ——, *Exact sequences in the algebraic theory of surgery*, Mathematical Notes 26, Princeton University Press (1981)

[11] ——, *Lower K- and L-theory*, London Math. Soc. Lecture Notes 178, Cambridge University Press (1992)

[12] ——, *Finite domination and Novikov rings*, Topology (to appear)

[13] ——, *The algebraic theory of bands* (to appear)

[14] L. Siebenmann, *A total Whitehead torsion obstruction to fibering over the circle*, Comm. Math. Helv. 45, 1–48 (1972)

[15] P. Vogel, *On the obstruction group in homology surgery*, Publ. Math. I. H. E. S. 55, 165–206 (1982)

[16] C. T. C. Wall, *Surgery on compact manifolds*, Academic Press (1971)

[17] H. E. Winkelnkemper, *Manifolds as open books*, Bull. A. M. S. 79, 45–51 (1973)

Exterior d, the Local Degree, and Smoothability

Dennis Sullivan

In the late forties Whitney considered cochains on R^n which were bounded linear functionals on polyhedral chains for the norm

$$|a|_{\text{Whitney}} = \inf_b \ (|a - \partial b| + |b|)$$

where $|\cdot|$ denotes mass (length, area, volume, etc.). Whitney showed 1) these cochains (on which coboundary is a bounded operator) could be identified with the integration of differential forms ω so that both ω and $d\omega$ (in the sense of generalized derivatives) have bounded measurable coefficients and 2) this class was invariant by quasiisometries, $x \to x'$ such that

$$(1/L \text{ distance } (x,y) \leq \text{distance } (x',y') \leq L \text{ distance } (x,y))$$

(see [1] [2] [11]).

Thus we have Whitney cochains or Whitney forms, exterior d, and wedge product on any topological manifold provided with charts so that overlap homeomorphisms are quasiisometries. The usual coordinate change calculations are possible because of the a.e. differentiability of quasiisometries and they are valid, [2] chapter 9.

Let us call a manifold equipped with quasiisometrically related charts and its Whitney forms a *Whitney manifold*[1].

We will study now two questions,

  i) when does a Whitney manifold have a subsystem of smoothly related charts

  ii) what characterizes the vector bundle which is the cotangent bundle for such a smoothing.

For a smooth manifold with cotangent bundle T and exterior bundle ΛT we have a natural map by integration

  i) smooth sections $(\Lambda T) \xrightarrow{\gamma}$ Whitney forms $(=$ bounded cochains for the Whitney norm$)$

  ii) for any smooth torsion free connection ∇ on ΛT a commutative diagram

---

[1] We reserve the term "Lipschitz" for functions and other objects satisfying the Lipschitz condition, e.g. vector bundles over metric spaces. In section 2 we discuss Whitney spaces.

$$\text{sections } \Lambda T \quad \xrightarrow{\nabla} \quad \text{sections } \Lambda T \otimes \text{sections } T \quad \xrightarrow{\wedge} \quad \text{sections } \Lambda T$$

$$\downarrow \gamma \qquad\qquad\qquad\qquad\qquad\qquad\qquad\qquad\qquad\qquad \downarrow \gamma$$

$$\text{Whitney forms} \quad \xrightarrow{d} \quad \text{Whitney forms .}$$

To answer questions i) and ii) we will abstract the properties i) and ii) of a smooth manifold. A new element *the local degree* appears.

For a Whitney manifold M one can consider Lipschitz vector bundles E over M, namely the overlap functions are Lipschitz mappings into $G\ell(n, R)$. We also have connections defined by Whitney 1-forms because these have restrictions a.e. to rectifiable arcs, [2] chapter 9.

Definition. A *cotangent structure* for a Whitney manifold M is a Lipschitz vector bundle E of dimension equal to dimension M and an embedding γ: Lipschitz sections of $E \xrightarrow{\gamma}$ Whitney one forms satisfying

i) γ is bounded from the Lipschitz norm to the Whitney norm, γ is a module map over the Lipschitz functions, and the induced map $\Lambda\gamma$ on sections of the top exterior powers sends a positive Lipschitz section to a bounded measurable volume form for M with an a.e. positive lower bound. (In the non orientable case we ask this last property locally on a finite covering.)

ii) near each point of M there is at least one connection ∇ on E defined by Whitney 1-forms which is *torsion free* in the sense that the following diagram commutes.

$$\text{Lipschitz sections of } E \quad \xrightarrow{\nabla} \quad \begin{array}{c}\text{bounded measurable} \\ \text{sections of } E\end{array} \otimes \text{Whitney 1-forms}$$

$$\downarrow \gamma \otimes \text{Identity}$$

$$\begin{array}{c}\text{bounded measurable} \\ \text{one forms}\end{array} \otimes \text{Whitney 1-forms}$$

$$\downarrow \gamma \qquad\qquad\qquad\qquad\qquad\qquad \downarrow \text{wedge product}$$

$$\text{Whitney one forms} \quad \xrightarrow{\text{exterior } d} \quad \text{bounded measurable 2-forms.}$$

We have made use of the fact that the embedding γ extends to all bounded measurable sections (up to a.e. equivalence) using the action of L^∞ functions on each.

We will now study the more precise

Question. When is a cotangent structure (E, γ) over a Whitney manifold M near p locally isomorphic to that of a smooth structure on M near p?

Example. Consider the map of R^2 $(r, \theta) \rightarrow (r, 2\theta)$ and pull back the standard cotangent structure of R^2. Using the module structure over Lipschitz functions on the domain one obtains a new cotangent structure over R^2. We claim this cotangent structure is not locally isomorphic to that of smooth structure near zero because the "local degree" to be defined below is 2 instead of one. Away from the origin this cotangent structure is equivalent to a smooth structure and there the "local degree" is one.

Now we discuss *the local degree of a cotangent structure*. The definition is rather easy but the possibility to make the definition depends on Reshetnyak's theory of *mappings of bounded distortion* [10].

Let us work near a point p in M and choose for each pair of points $\{x, y\}$ near p a rectifiable arc (x, y) varying in a Lipschitz way for the Whitney norm on chains. For example pull back straight line arcs in a chosen quasiisometrical chart near p. For any local Lipschitz framing $\rho = (\rho_1, \ldots, \rho_n)$ of E near p denote the corresponding Whitney 1-forms $\gamma\rho = (\gamma\rho_1, \gamma\rho_2, \ldots, \gamma\rho_n)$ and form the mapping

$$(\text{Neighbourhood of } p, p) \xrightarrow{\rho_p} (R^n, 0)$$

defined by $\rho_p(x) = \int_{(p,x)} (\gamma\rho_1, \gamma\rho_2, \ldots, \gamma\rho_n)$, where (p, x) is the arc from p to x mentioned above. We have used here the property mentioned above that Whitney 1-forms have restrictions to rectifiable arcs [2] chapter 9.

Theorem 1. *For each choice of arc systems (x, y) and local framing ρ_p of E the corresponding mapping ρ_p (Neighborhood p, p) $\xrightarrow{\rho_p}$ $(R^n, 0)$ has zero as an isolated value near p. The local degree of ρ at p is defined, belongs to $\{1, 2, 3, \ldots\}$ and, given p and the cotangent structure (E, γ), is independent of the choices.*

Proof. The proof will be given in §1. We remark here that the burden of the proof rests on a nontrivial reverse inequality (true near p)

$$\text{distance } (\rho_p(x), 0) \geq (\text{constant}) \text{ distance } (x, p)$$

which in our Lipschitz case comes from the general structure of Reshetnyak's mappings of bounded distortion [10]. The possibility to use [10] arises because we can approximate $(\rho_1, \rho_2, \ldots, \rho_n)$ in Whitney norm by closed 1-forms $(\rho'_1, \ldots, \rho'_n)$ using the connection and the Poincaré lemma in a familiar way.

The rest of the proof relies on the simple idea that if two maps of (Neighborhood of p) $- \{p\}$ into $R^n - 0$ are much closer at each point than their distance from zero they will simultaneously satisfy (or not satisfy) the reverse inequality and have the same degree.

Definition. The *local degree* $i(E, \gamma, p)$ *of a cotangent structure* (E, γ) at the point $p \in M$ is the local degree of any choice of maps ρ_p in Theorem 1. We choose orientations so that degrees are positive. In the non orientable case the local degree is the absolute value of $i(E, \gamma, x)$ for local choices of orientation.

There is a corollary to the proof of Theorem 1. Let g be a measurable Riemannian metric on M determined by γ and a Lipschitz inner product on E. For each p in M and $\varepsilon_i \to 0$ consider the sequence of metrics g_i obtained by rescaling by $1/\varepsilon_i$ the metric on ε_i balls about p. (For convenience here we define an r ball to be a ball from a coordinate system which has g-volume r^n.)

Say a measurable metric is on an open set U in R^n is *branched Euclidean* if it is obtained by pulling back the Euclidean metric by a nondegenerate branched covering – namely a Lipschitz mapping $F : U \to R^n$ so that determinant (DF) is a.e. positive with a positive lower bound.

Corollary. *The sequence of rescaled metrics* g_i *of the* ε_i *balls about any* p *in* M *has limits in the sense of Gromov (see [7]) and every such limit is branched Euclidean with branching degree at* p *equal to* $i(E, \gamma, p)$.

Proof. The ρ' of the proof of Theorem 1 have Jacobians which are $O(\varepsilon_i)$ quasiisometries (see f) g) h)). Their $1/\varepsilon_i$ rescalings are precompact by f). The rest is definition, [7], and stability of local degree. □

The approximation ideas used in Theorem 1 draw attention to a metrical structure on the set of equivalence classes of cotangent structures cf. [6]. Choose metrics on M and E.

Definition. A cotangent structure (E', γ') is ε close to (E, γ) if there is a Lipschitz bundle isomorphism $E \xrightarrow{i} E'$ so that γ and $\gamma' \cdot i$ differ by at most ε in operator norm.

Theorem 2. *The local degree of a cotangent structure* $i(E, \gamma, p)$ *as a function to the positive integers is continuous and therefore constant for* ε *close cotangent structures. It is equal to one for* p *in an open dense set whose complement has topological dimension at most* $(n - 2)$.

If $i(E, \gamma, p) = 1$ *for all* p *in* M, *there is a sequence of cotangent structures* (E_i, γ_i) *converging to* (E, γ) *which are individually smooth. (We say that* (E_i, γ_i) *is smooth if there is a smooth structure* α_i *on* M *inside the Whitney quasiisometrical charts so that the standard cotangent structure associated to* α_i *is isomorphic to* (E_i, γ_i).)

Proof. The proof is given in §1. □

Remark. We conjecture that the ε-closeness mentioned above determines an actual metric in the set of equivalence classes of cotangent structures and that $i(E, \gamma, p) \equiv 1$ implies that (E, γ) itself is smooth. The idea for the first statement is that the (lower bound for volume) part of the definition of cotangent structure should imply the image by γ of the Lipschitz sections of E is a closed subspace of Whitney forms for the Whitney norm. The idea for the second part should be that the construction of the smooth charts for (E_i, γ_i) approximating (E, γ) actually yields metrics of bounded curvature. Thus a Gromov limit can be considered as in [7].

In this metric the smooth structures would fill out certain of the uncountably many components distinguished by the different local degree functions $i(E, \gamma, p)$ and one would obtain the analogue of a Teichmüller metric on smooth structures (with equality as isomorphism), cf. [6].

Remark. The motivation for writing this note at this time was the recent activity in four dimensional smooth manifolds using the nonlinear equation of Seiberg-Witten which in turn is based on Dirac operators.

Now exterior d and thus its adjoint d^* and the associated signature operator $d + d^*$ can be constructed using algebraic topology and metrics which are locally Euclidean in the quasiisometric sense using the work of Whitney [2] for d and the work of Teleman [13] to see that d and d^* have a common dense domain in L^2 and that the signature operator $d + d^*$ is essentially self adjoint. Yang Mills theory and Donaldson invariants can be constructed as well for Whitney manifolds, [3]. In fact the known Donaldson invariants of smooth manifolds are actually invariants of the local quasiisometry or Whitney structure.

It was noticed long ago that these Lipschitz or Whitney manifolds cannot have a "Dirac package – spinors, Dirac operator, index formula,..." because there is a Whitney M^8 which is homologically like the quaternionic plane (and so the second Stiefel Whitney class is zero) but where the $\hat{A}$-genus is not an integer.

We conjecture that Whitney manifolds with a "full-Dirac package" (to be defined) are actually smooth. A corollary conjecture is then that the new gauge theory for 4-manifolds requires the underlying smooth structure.

A further speculation is that the new gauge theory produces smooth invariants which are not biLipschitz invariants in dimension 4.

The idea for the conjecture and the definition of "full-Dirac package" would be the following. The only known construction of Dirac operators (as opposed to d or the signature operator) uses a connection on an abstract vector bundle (the spinor bundle) and an action of the forms on that bundle (Clifford multiplication). These two elements of structure constitute a refinement of the pair (E, γ) in the cotangent structure above.

Namely choose an orthogonal structure on E. Then choose an orthogonal connection on E which is torsion free in the sense of part ii) of the definition of cotangent structure. (This is done by projecting torsion free connections to skew symmetric connections.) Construct the Clifford algebra of sections of ΛE and over any open set where $\omega_2(E) = 0$ a spinor module S over Clifford, the associated connection ∇_S and the associated Dirac operator, $\Gamma(S) \xrightarrow{\nabla_S} \Gamma(S) \otimes$

$\Omega^1 \underset{\text{multiplication}}{\overset{\text{Clifford}}{\longrightarrow}} \Gamma(S)$, cf. [9].

On Lipschitz sections with compact support such a Dirac operator is formally self-adjoint, namely it satisfies $(Df, g) = (f, Dg)$. However, it was shown by Chou [14] that there is a defect in the essential self adjointness for branched Euclidean metrics (in the polyhedral setting). If in the example above we use $(r, \theta) \rightarrow (r, \ell\, \theta)$ the spin structure extends correctly if ℓ is odd but the Dirac operator sees that ℓ is not 1 in its spectral properties, and these spectral properties obstruct self adjointness.

Thus we are lead to the following

Conjecture. *Over open sets where the cotangent structure (E, γ) admits spin structures the associated Dirac operators should be essentially selfadjoint rel boundary (a concept which can be formulated locally) and then the manifold is smoothable with cotangent structure (E, γ).*

Acknowledgements. I am grateful for conversations with Alain Connes, Mike Freedman, Misha Gromov, Blaine Lawson, and Stephen Semmes about the text of this paper. Also I am thankful to my thesis advisor Bill Browder for planting the seed of the question "what is a smooth manifold?" thirty years ago at Princeton and to Alain Connes for the intuition at IHES today to try to answer the question using operators, in this case exterior d and Dirac; see the philosophy of [12] chapter VI.

1   The Proofs

Preparation. 1) Divide R^n in the standard way into congruent cubes and thicken these slightly and congruently into a cover. If we multiply the picture by $\varepsilon_i \rightarrow 0$ we obtain a family U_i' of finer and finer covers of R^n with locally constant shape and geometry.

2) Choose a finite cover U_α of M by quasiisometrically related contractible charts so that for chosen compact subsets $K_\alpha \subset U_\alpha$ the interiors of K_α also cover. For i sufficiently large all the little (thickened) cubes of U_i which intersect image K_α are contained in image U_α. Thus they lift back to M and define there a system of fine covers $U_{i,\alpha}$ of M.

3) Choose a Lipschitz inner product on E and a Lipschitz trivialization of E over the original cover U_α.

4) Choose a connection in E over U_α satisfying ii) of the definition of cotangent structure. This may logically entail rechoosing the U_α at the start, but such connections may be added using a partition of unity to make them global. These connections don't need to be orthogonal.

Proof of Theorem 1

a) For an open set B_j of the cover at level i push the framing at a central point out along rays using the connection. (The connection is defined by a matrix θ_j of Whitney 1-forms for the original framing and Whitney 1-forms have well defined restrictions to rectifiable arcs a.e. for arc length measure [2] chapter 9.) This new framing will differ from the old one by a gauge transformation σ_j of the form Identity $+0(\varepsilon_i)$. The original curvature matrices $\Omega_j = d\theta_j + \theta_j \wedge \theta_j$ are Whitney bounded and are conjugated by $\sigma_j = I + 0(\varepsilon_i)$ to obtain the new curvature matrices.

Lemma. *The new connection matrices* $\theta'_j = \sigma_j^{-1} d\sigma_j + \sigma_j^{-1} \theta_j \sigma_j$ *are small in* L^∞ *norm, on the order of* ε_i.

Proof. The holonomy along a very short polygonal arc a of size a in the coordinate system can be estimated by considering the very narrow triangle obtained by coning a to the center.

We break the triangle into pieces of area at most a^2 and observe the holonomy around each is on the order $I + 0(a^2)$ since curvature is bounded, cf. [2] chapter 5. These are conjugated by bounded transformations and then multiplied to get the holonomy around the triangle. Altogether the holonomy around the triangle is estimated by $O(\text{area of triangle}) = 0(\varepsilon_i \cdot a)$. Since this equals the holonomy along a by construction of our frame which is parallel along the long sides of the triangle we deduce our estimate of the L^∞ norm of θ'_j on a polygonal arc is $O(\varepsilon_i)$. This completes the proof of the lemma. $\qquad \square$

Note we are not claiming the Whitney norm of θ'_j is $O(\varepsilon_i)$. In fact this is not true since $\Omega'_j = d\theta'_j + \theta'_j \wedge \theta'_j = d\theta'_j + 0(\varepsilon_i^2)$ and Ω'_j could well be of order 1 in L^∞ norm so $d\theta'_j$ is expected to be of order 1 in L^∞ norm. The conjugation step in a discussion like the above is what foils such an $O(\varepsilon_i)$ estimate for the Whitney norm.

b) For this radially parallel frame on B_j consider the corresponding 1-forms $(\rho_1, \rho_2, \ldots, \rho_n) = \rho$ using the embedding γ of property i) of cotangent tructure. By property ii) $(d\rho_1, d\rho_2, \ldots, d\rho_n) = d\rho = \theta_j \wedge \rho$. Now $d\rho$ is closed and by a) has L^∞ norm $O(\varepsilon_i)$ when restricted to any planar triangle thus the Whitney norm of $d\rho$ is $O(\varepsilon_i)$. (Whitney norm and L^∞ norm on planar triangles are equivalent for closed 2-forms by definition [2] chapter 5.)

c) Now apply the cone Poincaré lemma operator of [2] chapter 7 Lemma 10b to construct a Whitney form $\eta = (\eta_1, \eta_2, \ldots, \eta_n)$ of Whitney norm $O(\varepsilon_i)$ so that $d\eta = d\rho$. This is possible because the Whitney norm of $d\rho$ is $O(\varepsilon_i)$ by b). Now we consider the closed form $\rho' = (\rho'_1, \rho'_2, \ldots, \rho'_n) = \rho - \eta$.

d) Now we consider the mapping $\rho'(x) = \int_{(p,x)} (\rho'_1, \rho'_2, \ldots, \rho'_n)$. This is a Lipschitz mapping because we have Whitney forms and the triangle estimate of a) applies (in the more trivial abelian form). Its differential is determined by the closed 1-forms $(\rho'_1, \rho'_2, \ldots, \rho'_n)$. By the third property of part i) of the definition of cotangent structure the wedge product $(\rho'_1 \wedge \rho'_2 \wedge \ldots \wedge \rho'_n)$ has a definite positive lower bound. Since ρ' is a $O(\varepsilon_i)$ perturbation of ρ and since ρ is bounded from above (by the first part of property i)) it follows that $\rho'_1 \wedge \rho'_2 \wedge \ldots \wedge \rho'_n$ has essentially the same positive lower bound for ε_i sufficiently small. It follows that the mapping $x \to \rho'(x)$ is a *mapping of bounded distortion* in sense of [10] chapter 1 §4.

e) By developing some results for certain nonlinear elliptic PDE with bounded measurable coefficients and the ideas of conformal capacity geometric information about mappings f of bounded distortion is deduced in [10]. For example

i) the value $f(q)$ at an interior point q of the domain is taken on uniquely near q, Theorem 6.3 [10].

ii) on a small ball centered at q the image of small concentric spheres about q is contained in spherical shells centered about $f(q)$ with ratio of radii controlled by the derivative data which in our case is L^n/d. Here L is the Lipschitz constant, n is the dimension, and d is the lower bound on the Jacobian determinant. Theorem 7.2 [10].

iii) the local degree at q, well defined by i), is a positive integer and equals 1 precisely when f is a local homeomorphism on a neighborhood of q, Theorem 6.1 and Theorem 6.6 of [10]. (We have arranged orientations so Jacobian determinants are positive.)

f) Let us return to slightly simpler case of ρ', which is our Lipschitz mapping of bounded distortion. Applying e) we see that ρ' maps a small ball of radius r about p with positive degree so that the boundary stays in a controlled shape spherical shell of radius r'. By applying the coarea formula (cf. Theorem 2.2 of [10]) we obtain $r^n \sim$ (local degree at p) $(r')^n$ where the constants are controlled by L^n and d of e). By Lemma 4.8 of [10] and Arzela Ascoli the set of mappings with derivative data controlled by L and d as defined in e) and with $\rho'(p)$ bounded is compact. Thus if we had a sequence of such mappings $\rho^1, \rho^2, \ldots$ with L, d control and with the local degree tending to infinity we could rescale domain and range by the same amount to keep r constant. This doesn't change L or d. But then r' would have to tend to zero and the limit map would be constant. This contradicts the fact that d stays positive in the limit by Lemma 4.8 of [10] alluded to above. We conclude the local degree at p is between 1 and a positive integer controlled by L and d.

(Note this upper bound on degree doesn't hold in the K-quasiconformal context for $n = 2$ (consider $z \to z^N$) and may not be known for $n > 2$, to the best of my knowledge.)

g) A corollary to the (L^n, d) upper bound on local degree in f) is that r and r' are of the same size with constants depending only on L^n and d. This shows that the mapping ρ' of f) satisfies a reverse inequality

$$|\rho'(x) - \rho'(p)| \geq \text{constant}(L^n, d) \, |x - p|$$

on a small ball about p (whose size cannot be estimated – see the example of introduction).

h) For ε_i small compared to the constant in g) the mapping ρ of Theorem 1 also satisfies the reverse inequality for essentially the same constant because $|\rho(x) - \rho'(x)|$ is $O(\varepsilon_i$ distance $(x, p))$. (Recall ρ' depends on ε_i as $\varepsilon_i \to 0$ while ρ is defined once and for all by integrating along the rays from p.) Thus the local degrees are the same. One last remark is that a constant (positive determinant) linear change of the framing of E doesn't alter the local degree. This completes the proof of Theorem 1.   □

Proof of Theorem 2

a) The continuity of the local degree under approximation follows immediately from the principle used in h) of the proof of Theorem 1 that two maps which are closer than their distance from the zero have the same local degree.

b) The second statement is due to Chernavskii see Theorem 6.7 chapter 2 of [10] together with the observation that since the forms defining ρ' are closed we can use ρ' to compute the index at a point near p as well.

c) Here is the proof of the third statement. By Theorem 6.6 [10] if the local degree is 1 then the ρ' of d) e) f) ... of the proof of Theorem 1 is a local homeomorphism. In fact this will be true for ρ' which are perturbations of ρ coming from the orthogonal trivializations of the bundle E from the preparation.

Such ρ' will provide immersions of the cover at level i which will be $(I + 0(\varepsilon_i))$ quasiisometries for the measurable metric on M associated to the orthogonal structure on E – i.e. the derivatives will be $O(\varepsilon_i)$ perturbations of the canonical isomorphism between the orthogonal trivialization of E and the canonical basis in R^n. Thus the overlap homeomorphisms for these charts will have $O(\varepsilon_i)$ almost constant, $O(\varepsilon_i)$ almost isometric Jacobians between open sets in Euclidean space. By proposition 1 p.76 of [6] smooth mappings which are regularizations of these will be local diffeomorphisms. Thus one can choose fine handle decompositions and work inductively and relatively with standard averaging procedures (see [6] part II for details in the averaging procedure) to smooth these overlap homeomorphisms.

This constructs a smooth structure α_i whose cotangent structure is $O(\varepsilon_i)$ close to (E, γ), and completes the proof of Theorem 2.   □

2   Whitney spaces

The existence of forms as above with d and wedge was also produced by Whitney in [2] on metric spaces which are locally quasiisometric to polyhedra. The construction is so elegant one can begin it *for any metric space*. The steps are the following

   i)   define the mass $|\ |$ of one-chains, two-chains, etc.

  ii)   introduce the Whitney norm $|\ |_\omega$ as above as $|a|_\omega = \inf_b \ (|a - \partial b| + |b|)$

 iii)   the "Whitney forms" are the continuous linear functionals on this space

 iv)   prove or assume a Whitney bounded Poincaré lemma (see Chapter VI of [2]). Gromov has suggested that this should be true for algebraic varieties with the induced metric using semialgebraic triangulations.

  v)   prove or assume the cup product formulae converge under subdivision to a well defined product (by definition the wedge).

Definition. *A* **Whitney space** *is a metric space where the above steps can be carried out* (compare [11]).

Remark. It is known that all manifolds outside dimension 4 have Whitney structures which are related by isotopies close to the identity, [4]. In dimension 4 this fails because of the Donaldson-Freedman theory and its extension in [3].

   It is also known that Whitney structures exist on all open 4-manifolds [5] because there are always smooth structures. For example compact 4-manifolds less one point have (non-unique) Whitney structures. The ambiguity can be made to be countable [8].

Remark. Suppose a topological 4-manifold has a metric making it into a *Whitney space*. We can consider vector bundles with connections over such a space. To develop Donaldson invariants one basic further ingredient is required. We need to be able to pick out from the 2-forms ω a class of positive forms so that the pointwise norm of ω is estimated by $(\omega \wedge \omega)^{1/2}$.

   Given a Freedman topological 4-manifold which by Donaldson theory admits no smooth (and therefore no Whitney structure as in the introduction) it is interesting to wonder which axiom in the above chain (§2 only) fails.

References

[1]   H. Whitney : "r-dimensional integration in space", *Proceedings of the ICM*, Cambridge, 1950.

[2]   H. Whitney : "Geometric Integration", Princeton University Press, 1956.

[3]   S. Donaldson and D. Sullivan : "Quasiconformal 4-manifolds", *Acta Math.* **163** (1989), 181-252.

[4]   D. Sullivan : "Hyperbolic geometry and Homeomorphisms", *Georgia Topology Conference Proceedings*, 1978. Editor J. Cantrell.

[5]   M. Freedman and F. Quinn : "Topology of 4-Manifolds", Princeton University Press, 1990.

[6]   S. Shikata : "A distance fonction on the set of differentiable structures", *Osaka Journal of Math.*, 1966, p. 65-79.

[7]   R. Greene and H. Wu : "Lipschitz convergence of Riemannian manifolds", *Pacific J. Math.* **131** (1988), 119-141.

[8]   M. Freedman and L. Taylor: "A universal smoothing of four-space", *J. Diff. Geom.* **24** (1986), 69-78.

[9]   B. Lawson and M.L. Michelson: "Spin Geometry", book, Princeton University Press, 1989.

[10] Yu. G. Reshetnyak: "Space mappings with bounded distortion", Translations of the AMS, vol. 73, 1982.

[11] H. Whitney: "Algebraic Topology and Integration", Proc. of Nat. Acad. of Sciences 1950's, cf. Whitney's collected works, Vol. II, p. 432.

[12] A. Connes: "Noncommutative Geometry", Academic Press, 1994.

[13] N. Teleman: "The index of signature operators on Lipschitz manifolds", IHES Publications Mathématiques n⁰ 58 (1983), 39-78, MR 85 f:58112.

[14] S. Chou: "Dirac operators on spaces with conical singularities", *Trans. AMS* (1985), "Dirac operators on pseudo-manifolds", preprint I.A.S. 1984.

Contributors

Alejandro Adem
University of Wisconsin

Amir H. Assadi
International Centre for Theoretical Physics, Trieste
University of Wisconsin

M. Bökstedt
Aarhus University

Sylvain E. Cappell
Courant Institute of the Mathematical Sciences
Institute for Advanced Study

Ruth Charney
Boston College

Michael W. Davis
Ohio State University

Peter J. Eccles
University of Manchester

Michael H. Freedman
University of California at San Diego

Ian Hambleton
McMaster University

Jean-Claude Hausmann
Mathématiques-Université, Geneva

Sören Illman
University of Helsinki

Gabriel Katz
Massachusetts Institute of Technology

Matthias Kreck
Johannes Gutenberg-Universität

Wolfgang Lück
Johannes Gutenberg-Universität

I. Madsen
Aarhus University

R. James Milgram
University of New Mexico

Jack Morava
John Hopkins University

Erik K. Pedersen
State University of New York at Binghamton

Volker Puppe
University of Konstanz

Frank Quinn
Virginia Tech

Andrew Ranicki
The University of Edinburgh

Julius L. Shaneson
University of Pennsylvania

Dennis Sullivan
Graduate School
City University of New York
Inst. des Hautes Etudes Sci.

Peter Teichner
University of California

Zhenghan Wang
University of Michigan

Shmuel Weinberger
University of Chicago
University of Pennsylvania